Endorsements For
To Caress the Air
By C. David Gierke

"A tale of everlasting importance, *To Caress the Air*, tells the story of little known American aviation pioneer Augustus Herring, who flew a manned, powered aeroplane five years before the Wright Brothers."

ROBERT BROWN
President, Academy of Model Aeronautics

* * *

To Caress the Air is historical fiction that contains factual information about Augustus Herring, the aviation pioneer who challenged the early aeroplane establishment with new ideas about powered flight and the design of airplanes. While secrecy between inventors, concerns about patent rights and court cases delayed Herring's recognition, the story's seamless transition between fact and fiction keeps the reader engaged in the pioneering days of powered flight and the men who challenged the air.

RICHARD BROX
Director Emeritus, Aero Club of Buffalo

* * *

To Caress the Air is a remarkable account of how little known engineer, Augustus Herring, overcame long odds to build and fly one of the world's first successful, powered aeroplanes. The story, enhanced by the author's keen portrayal of America in the late 19th and early 20th centuries, is told with meticulous detail and a gripping prose.

WALTER BOYNE
Colonel, USAF (Ret), Author, Historian

* * *

In *To Caress the Air,* Augustus Herring rises from obscurity as a powerful multi-dimensional character who just happened to be the first to fly. A more engaging, history of pioneer flight does not exist.

Debra Cleghorn
Executive editor
Air Age Media

* * *

A fascinating illumination of an extremely progressive and dynamic period of early American aviation, much of which took place in Western New York.

Hugh M. Neeson
Vice President: Bell Aerospace Textron
Trustee: Niagara Aerospace Museum ('55 – present)

* * *

Based on facts that will rock established perceptions of pioneer aviation, *To Caress the Air* tells the story of Augustus Moore Herring, the first man to achieve heavier-than-air, powered flight. Fictionally dramatized, Gierke's narrative is rich with technical and biographical detail and is extensively documented with footnotes.

Tom Atwood:

Former editor-in-Chief of *Model Airplane News* magazine; co-founder of *Flight Journal* magazine; founder of *Fly RC, RC Heli Pilot* and *Robot* magazines, Atwood is currently the Executive Director of the *National Robotics Education Foundation* (NREF).

* * *

A provocative work that challenges the conventional wisdom on the first flight of the Wright Brothers. The saga of Augustus Herring is one that deserves to be told.

MAURY KLEIN
Professor Emeritus of History, University of Rhode Island
Author of: *The Power Makers*

* * *

The brave and foresighted men who, more than a century ago, believed that man could fly had to fight for credibility, for funding, and, occasionally, for their legal rights in court. C. David Gierke's novel *To Caress the Air* recounts the fascinating, and largely forgotten, story of one such real-life pioneer: Augustus Herring, who may have beaten the Wright brothers into the air by five years.

DAVID M. FRIEDMAN
Author of *The Immortalists: Charles Lindbergh, Dr. Alexis Carrel and their Daring Quest to Live Forever.*

* * *

When law and engineering intersect, I pay attention. As they evolve together, can a legal system resolve conflicts of inventive minds? Gierke's richly sourced account of Herring's intense courtroom struggle reverberates in American courts today.

JOSEPH J. TERRANOVA
Trial lawyer

TO CARESS THE AIR

AUGUSTUS HERRING AND THE DAWN OF FLIGHT

Book 2 of 2

A biographical novel by: C. David Gierke

Published by Write Associates, LLC
1276 Ransom Rd., Lancaster, N.Y. 14086
Copyright © 2018 by Write Associates, LLC

All Rights Reserved
No part of this book may be reproduced or transmitted in any form or by any means, electronic or mechanical, including photocopying, or by an information storage and retrieval system without permission in writing from the publisher.

ISBN: 978-0-9990457-6-3 (paperback)
ISBN: 978-0-9990457-7-0 (hardcover)

www.davegierkebooks.com

To Caress the Air, required seven years to write, rewrite and edit. Throughout, one person stood by my side – my wife Carolyn (Gwitt) Gierke. Because of Carolyn's suggestions, encouragement and assistance (she possesses a Masters in Library Science degree, and is certified as a School Library Media Specialist), I was able to recount the invention of the aeroplane from the perspective of Augustus Herring... I couldn't have done it without her.

AUTHOR BIOGRAPHY

C. David Gierke is an acknowledged internal combustion engine authority, having written three well-received non-fiction books and numerous magazine articles on the subject. A retired high school teacher and dedicated pioneer aviation historian, Gierke has received extensive recognition for his work, including the *1978 New York State Teacher of the Year* award and a citation from the Buffalo, New York chapter of the American Society of Mechanical Engineers. Dave was inducted into the national *Model Aviation Hall of Fame* in 2003. He and his wife, Carolyn, divide their time between Lancaster, New York and Sarasota, Florida.

Also By C. David Gierke

- "Langley's Steam-Powered Flying Machines", *Aviation History*, July 1998, Volume 8, Number 6, pp. 50-56+
- *2-Stroke Glow Engines*, Volume 1
- *2-Stroke Glow Engines (Beyond the Basics)*, Volume 2
- *Airplane Engine Guide*
- *Nitro Engine Guide*
- *R/C Pilot's Handbook* (Chapter 11, "Maximizing Engine Performance")
- *Ultimate RC Flight Guide* (Chapter 6 "Getting the Most From Your Engine")

CONTENTS

Chapter 56: **Rochester, New York**. 1
Chapter 57: **Rochester, New York**. 18
    Rochester, New York. 18
    Chicago, Illinois 23
    Chicago, Illinois 32
Chapter 58: **Rochester, New York**. 36
    St. Joseph, Michigan 37
    Rochester, New York. 42
    Dune Park, Indiana 49
Chapter 59: **Miller, Indiana** 62
Chapter 60: **Rochester, New York**. 78
    Rochester, New York. 79
    St. Joseph, Michigan 83
    Rochester, New York. 92
Chapter 61: **St. Joseph, Michigan** 99
    Rochester, New York. 108
    St. Joseph, Michigan 111
    St. Joseph, Michigan 115
Chapter 62: **Rochester, New York**. 118
    St. Joseph, Michigan 122
    St. Joseph, Michigan 127
Chapter 63: **Rochester, New York**. 141
    Chicago, Illinois 142
    Rochester, New York. 148
    Rochester, New York. 153
    Rochester, New York. 157
Chapter 64: **St. Joseph, Michigan** 160
    Rochester, New York. 165
    St. Joseph, Michigan 168
    St. Joseph, Michigan 170
    St. Joseph, Michigan 173
Chapter 65: **Rochester, New York**. 179
    St. Joseph, Michigan 183
    St. Joseph, Michigan 189
    St. Joseph, Michigan 192

Chapter 66: Chicago, Illinois 197
    Rochester, New York 199
    St. Joseph, Michigan 201
    Rochester, New York 204
    St. Joseph, Michigan 207
Chapter 67: St. Joseph, Michigan 209
    Rochester, New York 212
Chapter 68: Rochester, New York 227
Chapter 69: St. Joseph, Michigan 243
    Rochester, New York 254
Chapter 70: Rochester, New York 257
    Rochester, New York 264
    Rochester, New York 267
Chapter 71: Rochester, New York 268
    Rochester, New York 269
    Rochester, New York 273
    Rochester, New York 283
    Kill Devil Hills, North Carolina 284
    Kill Devil Hills, North Carolina 291
    Rochester, New York 297
Chapter 72: Rochester, New York 303
Chapter 73: Rochester, New York 322
Chapter 74: Rochester, New York 340
    Rochester, New York 342
    Kill Devil Hills, North Carolina 350
Chapter 75: Kill Devil Hills, North Carolina 359
Chapter 76: Rochester, New York 374
    Kill Devil Hills, North Carolina 376
    Rochester, New York 383
    Rochester, New York 389
Chapter 77: St. Joseph, Michigan 395
    Rochester, New York 398
    St. Joseph, Michigan 400
    St. Joseph, Michigan 406
Chapter 78: Rochester, New York 413
    St. Joseph, Michigan 416
    Rochester, New York 426
    Rochester, New York 426
Chapter 79: St. Joseph, Michigan 431
    Rochester, New York 446
    Stevensville, Michigan 447
Chapter 80: Stevensville, Michigan 455
Chapter 81: Rochester, New York 459
    New York, New York 463
    Rochester, New York 466
    New York, New York 474

Chapter 82: Rochester, New York.......................... 476
    New York, New York................................. 479
    Rochester, New York................................ 481
    Freeport, Long Island, New York 484
    New York, New York................................. 486
Chapter 83: Somewhere south of Chicago, Illinois............. 501
    St. Louis, Missouri 513
Chapter 84: Rochester, New York.......................... 524
    St. Louis World's Fair 536
    Rochester, New York................................ 545
Chapter 85: Rochester, New York.......................... 550
Chapter 86: Rochester, New York.......................... 561
Chapter 87: Manhattan, New York, New York................ 584
    Freeport, Long Island, New York 587
    Manhattan, New York, New York.................... 590
    Freeport, Long Island, New York 594
Chapter 88: Mineola, Long Island, New York 597
Chapter 89: Mineola, Long Island, New York 607
    Mineola, Long Island, New York..................... 611
    Mineola, Long Island, New York..................... 615
Chapter 90: Mineola, Long Island, New York 618
    Rochester, New York................................ 620
Chapter 91: Rochester, New York.......................... 633
    Rochester, New York................................ 654
Chapter 92: Freeport, Long Island, New York................ 656
    Freeport, Long Island, New York 656
    Freeport, Long Island, New York 657
    Freeport, Long Island, New York 659
Epilogue 661
Footnotes 685
Significant Characters: Book 2............................ 752
Appendix 1 754
Appendix 2... 757
Acknowledgements 760
Truth or Speculation 764
Request to Reader...................................... 765

CHAPTER 56

Rochester, New York

Tuesday, October 25, 1921

As Maloney prepared to offer his apologies to the judge, Martin took a seat next to O'Grady at the plaintiff's table.

"Begging the court's pardon, Your Honor, Mr. Martin has been traveling since late last night. Unfortunately, his train arrived in Rochester a few minutes late."

Peering over the top of his spectacles, Judge Sawyer nodded. "Under the circumstances, Counselor, that's understandable. I was about to address some unfinished business from yesterday's session, so there's time for Mr. Martin to catch his breath before beginning his testimony... if, indeed, he's called upon to do so."

Turning his attention to the papers in front of him, the judge focused on a list of points related to the previous day's objection. Standing, he plucked the reading glasses from his face and tossed them onto his desk before speaking. "I have considered the defense counsel's objection to hearing additional testimony from Mr. Martin concerning Mr. Herring's patent application. In his opening statement, Mr. O'Grady accused the United States Patent Office of contributing to the demise of the Herring-Curtiss Company by rejecting Mr. Herring's 1896 powered-aeroplane application without proper justification... as well as stonewalling his subsequent attempts to have it revived. The plaintiff contends that Patent Office examiners exhibited inconsistent and confused analysis of Herring's legitimate claims, thus depriving his company the ability to defend against infringement lawsuits.

"The plaintiff further contends that the company was subsequently forced into an unfounded and unnecessary bankruptcy by Curtiss, et al., resulting in the current lawsuit, a decade later. Because of these and other mitigating factors, the court has decided to hear further testimony within this area of contention. To this end, the defense's objection and its request for a mistrial are summarily *denied*.

"Mr. O'Grady, call your witness to the stand for swearing-in."

*

Thirty-six-year-old James V. Martin looked much the same as he had a dozen years earlier when he served as Augustus Herring's assistant. Except for a liberal smattering of gray, the six-foot, 215-pound engineer retained a full head of wavy red hair and a well-groomed full mustache and goatee.

James V. Martin (1921); Stephane Sebile Collection.

At first glance, Martin's tweed sport jacket appeared to have been borrowed from one of his old Harvard professors... but his hand-tooled leather cowboy boots suggested otherwise. Pleated, dark brown trousers complemented his shirt, which was embroidered with images of Arizona cacti in mint green. A dark brown string tie was cinched loosely at the neck by a flashy silver and turquoise clip. Martin's one concession to the court was not to wear his suede cowboy hat. When wearing the hat, Martin reminded Bill Maloney of the legendary Indian fighter Buffalo

Bill Cody, whom the attorney had met as a child almost a half century earlier.

"Mr. Martin," O'Grady began, "please begin by telling the court about your military and educational background."

Martin leaned back and tented his fingers.

"Back in 1900, at the age of 16, I joined the Merchant Marines. Ten years later, I received my Marines Certificate, achieving the rank of Captain. While still in the military, I attended the University of Virginia, where I received my undergraduate degree in engineering. In 1908, I was accepted into Harvard's graduate program, from which I earned a master's degree in 1912.

"Early in 1910, I organized the Harvard Aeronautical Society, and served as its first director. Later that year, acting through the Society, I organized and directed the first *international air meet* to be held in the United States. In early 1917, I became part of the American aeroplane movement by forming the Martin Aeroplane Company. My goal was to organize, promote, and license two of my nine patented inventions to select aeroplane manufacturers—"

"Excuse me," O'Grady broke in, "what were these inventions?"

"The automatic stabilizer and the retractable landing gear."

"Thank you. Please continue."

"Unfortunately for my company, the *Patent Cross-License Agreement of 1917* and the Manufacturers Aeronautical Association put an end to that dream. As a result of the ongoing conspiracy to monopolize the aeroplane industry by the big corporations, my company's current activities are limited to the manufacture of efficient aeroplane designs and related components. For example, one of my designs currently holds the world record for *aeroplane efficiency*."

"What does that mean?" O'Grady asked.

"Efficiency relates to an aeroplane's fuel economy at various air speeds. Its determination depends upon many factors, including the machine's aerodynamic drag and the performance of its engine."

O'Grady slipped from behind the podium and approached Martin. "Sir, how did you become involved with manned flying machines?"

Martin settled back into his seat. "While studying at the University of Virginia, I read a newspaper article that featured Augustus Herring and his adventures with flying machines. Having successfully completed my degree in the spring of '08, and with graduate school not scheduled to begin until September, I wrote to Mr. Herring and asked for summertime work. He responded with a note suggesting that I travel to his New York City shop, where we could discuss the matter further.

"When it came to aeroplanes, I was a relative novice, but Mr. Herring liked my enthusiasm and willingness to learn... so he hired me as his assistant!"

"How long did you work for Mr. Herring?"

"For almost two years; I worked weekends, recesses, vacations, you name it. I even managed to wrangle a leave of absence from Harvard to help him with the Plum Island flight trials of 1910."

"What were your duties?"

"At first, I helped to build his twin-engine, two-surface machine that he intended to use for the U.S. Army Signal Corps competition. Other than the Wright brothers' entry, ours was the only other bid accepted for the Fort Myer, Virginia, showdown."

"How did Herring's government machine turn out?"

Martin leaned forward. "Due to the formation of the Herring-Curtiss Company, during the winter of 1909, work on the Signal

Corps machine slowed to a crawl as Mr. Herring designed his company's first aeroplane... the *Golden Flyer*. As a member of the Aero Club of America and the Aeronautic Society of New York, Herring had encouraged the latter's membership to purchase the yet-to-be-built machine. When Glenn Curtiss began fabricating the *Golden Flyer* in his Hammondsport shops, Herring's attention shifted to designing the *Rheims Racer*, the aeroplane that won the Gordon Bennett Trophy at—"

"Objection!" Robbins shouted. "These recollections are flawed, Your Honor! Herring had nothing to do with the design of those two Curtiss machines!"

Sawyer turned to Martin. "Mr. Martin, did you actually *observe* Mr. Herring working on the design for these two aeroplanes?"

"Yes, Your Honor! For weeks on end, I watched as Mr. Herring poured himself into the task, while Mr. Brock and I worked on the government machine. During breaks, I would peek over the boss's shoulder to see how things were progressing.[1]

"On one occasion, Mr. Curtiss made an appearance at the Broadway shop to pick up some blueprints and converse with the boss. The two men spent the entire afternoon discussing the new aeroplanes before Curtiss rushed off to catch a train back to Hammondsport.

"I recall a humorous incident that occurred during that meeting. It was a hot summer day, so when Curtiss walked into the shop carrying a bottle of Coca-Cola, Augustus chided him by saying that the drink was nothing more than a dopey stimulant—"

"Objection!" Robbins interrupted. "Nothing more than a transparent attempt to slander Mr. Curtiss—"

Banging his gavel, Sawyer had heard enough. "Mr. Robbins, the court will decide if the defendant has been slandered! Save your indignation for cross-examination! Your objection is duly noted and overruled. The witness may continue."

Martin shot Robbins a cynical stare. "Thank you, Your Honor. Due mainly to bad luck, the government plane crashed during its initial test hop at the Hempstead Plains test site—"

A torrent of laughter erupted from the defense table, accompanied by an unsolicited remark from Wheeler, "A typical Herring flight!"

Frowning, the judge banged his gavel. "Mr. Robbins... you are flirting with a contempt charge! Control your people, sir!"

Leaning back in his leather chair, Sawyer pointed at Judge Wheeler. "Defendant Wheeler... of all people! You are a long-time judge within the State's legal system, so the court expects exemplary conduct from you. By adding to the defense's disruptive outburst, you have disappointed this court!"

Pivoting his chair in O'Grady's direction, Sawyer abruptly announced a 30-minute recess.

*

When proceedings resumed, Martin was allowed to continue with his testimony.

"What action did Mr. Herring take after the crash?" O'Grady asked.

"He asked for and received a nine-month delay, and we began working on a more conventional design... one that Herring simply called the *Shop Machine*. Although we tried our best to finish this aeroplane, Gus was eventually forced to abandon the venture and forfeit his $2,000 deposit."[2]

"In your opinion, what influence did Mr. Herring have on those first two Herring-Curtiss Company aeroplanes – the *Golden Flyer* and the *Rheims Racer*?"

His eyes bright, Martin seemed pleased with the question. "Herring's influence was substantial and obvious! Even an aeronautical novice would have recognized his contributions. In this regard, Glenn Curtiss' only offering to the *Rheims Racer* design was to *saw off* the outer portion of each wing panel, at the expense of—"

"Objection!" Robbins screamed, this time stomping his foot for emphasis. "Your Honor, the record is clear, Mr. Curtiss was wholly responsible for the design of his Gordon Bennett Cup-winning racer!"

"Theatrics aside, Mr. Robbins," the judge said, "Mr. Martin provides the court with a unique perspective and insight concerning these early flying machines. Objection overruled. Continue, Mr. O'Grady."

"What were Herring's contributions to the *Golden Flyer* and *Rheims Racer*?"

"Aero*curve* lifting sections, superposed wings, rectangular wing planform, squared-off wingtips, upper surface wing covering, low-drag struts with guy-wire wing trussing, a true cruciform tail... and more.

"Mr. Herring's other designs, such as the 1892 rubber-powered model, the *Double Decker* glider of 1896, the 1896 U.S. Patent Application plane, and the powered, man-carrying flyer of 1898... all of them displayed similar features.

"If you compare Curtiss' 1907 *June Bug* to both the *Golden Flyer* or *Rheims Racer*, there are damn few similarities! The *June Bug* had triangular wingtips and used both dihedral and anhedral wings; the bottom wing bent upward and the upper wing bent downward. Everything else had been anticipated by Herring, except for Curtiss' penchant for infringing on the Wright Brothers' lateral control patent... a choice that Gus opposed."[3]

"Did you observe Mr. Herring working on other projects for the Herring-Curtiss Company?"

Martin closed his eyes in recollection. "While designing the *Rheims Racer*, Gus occasionally stopped to work on a new cylinder-head design for Curtiss' four-cylinder aero-engine. I remember prying open a wooden crate containing a set of new cylinder heads from the Hammondsport factory. Gus scrutinized those heads for hours. He measured, he made sketches, and he transferred his ideas to drafting-board illustrations. After weeks of work, he produced a set of drawings that described how these heads could be modified to incorporate water cooling, a hemispherical combustion chamber, and canted overhead intake and exhaust valves.[4]

"During Christmas week of 1909, Gus received a recently manufactured Herring-Curtiss V-8 engine from Hammondsport for him to test and tune. I remember the joy on his face after seeing that his new cylinder-head design had been used. Later, after purchasing the engine from the company, Herring used it on one of the first Herring-Burgess machines of early 1910."

"Objection! More lies, Your Honor!"

"Overruled. Continue, Mr. O'Grady."

"What became of Herring's second government aeroplane – the *Shop Machine*?"

"In November of '09, when it became clear that Curtiss and Herring couldn't get along, Gus teamed up with yacht builder Starling Burgess of Marblehead, Massachusetts. Herring moved the almost finished *Shop Machine* and most of his tools to Burgess' shop, where it was completed and re-designated the *H-B #1*, or *Flying Fish*. Two more Herring-Burgess aeroplanes of similar design would also be constructed there."

"Tell the court about your participation in the Herring-Burgess flight trials," O'Grady prodded.

"I was present on February 28, 1910, when Mr. Herring flew the *H-B #1* from a frozen lake near Essex, Massachusetts. That flight is now recognized as the first public flight of a manned, heavier-than-air machine in New England.[5]

"Two months later, at Burgess' remote Plum Island, Massachusetts test site, I assisted Herring in successfully flying a replica of his 1896 patent application aeroplane, a machine that was also fabricated in Burgess' shops."

Standing, defense attorney Robbins waited to address the judge. "Your Honor, the defense respectfully requests that Counselor O'Grady be directed to address Mr. Herring's failed 1896 patent application. Thus far, the morning court session has been frittered away reviewing Mr. Martin's dubious association with Herring... a waste of valuable court—"

Herring-Burgess Flying Fish at takeoff (1910); Library of Congress

Cutting Robbins off in mid-sentence, Judge Sawyer had other thoughts. "Since we are approaching the noon hour, I'm going to suspend further testimony until the afternoon session. When we return, I am confident that Mr. O'Grady will address Mr. Herring's patent application with his witness.

"This court is hereby in recess until 1 o'clock."

*

After lunch, O'Grady addressed the bench. "Your Honor, the plaintiff has two additional questions regarding the witness's background. May I proceed?"

"Make it brief, Counselor."

"Mr. Martin, why does the *Manufacturers Aircraft Association* consider you to be an outlaw?"

Martin shrugged. "I have sued both the Manufacturers Aircraft Association and the United States government for conspiring to monopolize the aviation industry. So, yes; they consider me to be an outlaw."

"Explain the circumstances."

Martin wagged his head mockingly. "Glenn H. Curtiss, with the help of his attorneys, helped to influence the congressional subcommittee of the National Advisory Committee for Aeronautics, also known as NACA, to concoct the *Patent Cross-License Agreement of 1917*. This agreement was openly condemned by the United States Senate for extracting large profits from the government during a time of war... but it passed anyway. To this day, the Manufacturers Aircraft Association controls Army and Navy Air Service policy and contracts, while it crushes any attempt by independents, such as myself, to market our own aeroplane-related designs and inventions."

Leaping to his feet, Robbins said with a sneer, "Objection! What, pray tell, does this have to do with Herring's pathetic, failed patent application? Your Honor, this witness's testimony has wandered all over the countryside! Is the plaintiff ever going to get to the point?"

"Mr. O'Grady," the judge pleaded, "where is this line of questioning headed?"

"Your Honor, having exposed the present-day inequities of the Cross License Agreement, we will now step back 25 years to the genesis of the *great aviation patent conspiracy*... beginning with the rejection of Augustus Herring's 1896 patent application."

Leaning back in his chair, Judge Sawyer removed his spectacles. "In terms of direction, I have allowed the plaintiff's side a good deal of leeway in the questioning of this witness. However, there are limits to my patience! Get to the point, Counselor! Objection overruled."

O'Grady continued with a degree of haste. "Mr. Martin, you have analyzed all phases of Mr. Herring's patent application from

1896. What are your impressions of the U.S. Patent Office's handling of this landmark case?"

Martin cocked his head. "As indicated by the Patent Office's official record, serial number 615,353 was diligently prosecuted through *five* successive actions, after which the examiner stated that 'the apparatus as a whole was *incapable of practical use.*'

"In response to this decision, Herring addressed an inquiry to the Patent Office in September 1897, asking specifically what demonstration would be considered satisfactory for a power-driven aeroplane with a man aboard. In response, he received the *final rejection*, dated January 4, 1898."

Pausing for O'Grady to place the Patent Office's letter into evidence, Martin tried to focus on the core of his testimony. Try as he might, thoughts of past injustices interfered with his concentration. Mercifully, O'Grady thrust the letter into his hand to read aloud.

" 'A.M. Herring

Care of: George P. Whittlesey, Esq.

Chicago, Illinois

'Applicant's information has been considered.

The entire invention rests upon a theory that has never been satisfactorily demonstrated. So far as the examiner is aware, no power driven aeroplane has yet been raised into the air with the aeronaut, or kept its course wholly detached from the earth for such considerable length of time as to constitute proof of practical usefulness.

'Up to the present time the applicant's invention rests in mere theory.

'The application is rejected as a whole upon grounds of inoperativeness. It is obvious that no claims whatever can be allowed.

W.W. Townsend, Exr.' "

"Mr. Martin, after reading this letter, what are your thoughts concerning Mr. Herring's treatment by the U.S. Patent Office?"

Martin spoke haltingly. "As an American citizen, Mr. Herring was denied his constitutional right to secure a patent for a new and useful device. This inexplicable denial of his privileges has carried over to virtually *all* of Herring's subsequent aeroplane-related patent applications."

"Objection, Your Honor!" Robbins shouted. "Setting aside the witness's flowery language, the key term here is 'useful device.' The defense agrees with the Patent Office... Herring's patent application aeroplane was not useful. In fact, it was *useless*!"

After a moment of dramatic tension, Martin responded. "Counselor, of what use is a *newborn baby*?"

Judge Sawyer leaned forward and arched his eyebrows, "An introspective observation, Mr. Martin! Continue your questioning, Mr. O'Grady."

"Mr. Martin, as a student of aviation history, what are your thoughts concerning the American system of securing patents and the effect it has had on the aeroplane manufacturing industry in this country?"

"Objection!" Robbins shouted. "Calls for a conclusion!"

"Overruled. The court wishes to hear the witness's expert opinion. You may answer, Mr. Martin."

"The United States Patent Office has been an abject failure in protecting American inventors – *especially* in the field of aeronautics. To be specific, in 1898, the agency made a grievous error by refusing to grant Mr. Herring protection for his powered aeroplane. A dozen years later, in a decision dated August 22, 1910, the Commissioner of Patents missed an opportunity to correct the mistake when he failed to revive the Herring application.

"By issuing the Wright brothers their patent in 1906, the Patent Office reluctantly acknowledged the reality of manned, powered, heavier-than-air flight. A few years later, the agency insisted it *would* have revived Herring's application if he had petitioned them in '... a timely fashion'! The commissioner explained the grounds for this arbitrary deadline by saying, '... the delay in prosecuting this case has *not* been shown to be *unavoidable*.' Double negatives aside, after waiting a decade, Mr. Herring's request for revival was deemed to be ineligible by a matter of *months*!"

O'Grady pressed on, "Mr. Martin, explain how Augustus Herring's rejected patent of 1896 allowed the Wright brothers to stifle the American aviation industry."

Martin did not hesitate to offer his response. "In retrospect, the Wright brothers' landmark patent for lateral control *eliminated* all of their American competition in the manufacture and sale of aeroplanes. Those who chose to fight the Wright patent, such as the Herring-Curtiss Company, were slapped with restraining orders and infringement lawsuits that prevented them from doing business. Therefore, in the years that preceded the Great War, America's stillborn aviation industry decayed, while European entrepreneurs flourished in an atmosphere of innovation and competition.

"Herring's rejected patent application alone left more than *20* primary claims exposed to aeronautical pirates. For example... several of his claims originated from his use of multiple propellers acting in conjunction with superposed lifting surfaces and the aero*curve*. There were additional claims for trussing and guy-wire combinations, aerodynamically shaped wing struts, and component-fastening techniques that others took for granted within a few years. Had Mr. Herring's claims been rightfully secured by a U.S. patent, entrepreneurs who came along later – such as the brothers Wright – would have been obliged to negotiate with—"

"Objection! Argumentative!" Robbins interrupted. "Your Honor, the Wrights controlled the industry because they were granted a patent for *lateral control* – something other entrepreneurs *did not* have. Herring's application didn't even bother to address that particular aspect of heavier-than-air flight!"

Turning toward the witness, Judge Sawyer raised an eyebrow. "What do you have to say about that, Mr. Martin?"

"Apparently, Mr. Robbins doesn't understand the issues!" Martin sniffed. "Back in 1896, Mr. Herring *didn't* concern himself with lateral control in his application – there were 20 other claims of greater priority to be dealt with! Allow me to clarify this point: the record clearly indicates that *all* of the Wright aeroplanes – gliders and powered machines alike – incorporated features that infringed upon Mr. Herring's pioneering claims! In this regard, the Wrights' patent drawings showed *flat planes* being used for the wing's lifting section. Did this mean that the brothers were ignorant about the advantages of the aero*curve*? Of course not! Then again, the Wrights were patenting their method of lateral control and not the aero*curve* used for the lifting surfaces.

"Had Herring rightfully obtained his patent in 1898, the Wrights would have been forced to *negotiate* with Herring to hammer out a mutual *fair usage* agreement. As a result, the American aviation industry would have profited during those pre-war years."

Nodding, Judge Sawyer seemed to agree. "Objection overruled. You may continue, Counselor."

O'Grady read from his notes, "In 1906, the Wrights were awarded a *pioneer patent* for lateral control. This enabled the brothers to charge American aeroplane manufacturers a *fee* for using wing warping and *all* of its variations. The high cost of licensing discouraged many potential manufacturers from becoming involved in the fledgling aeroplane industry. Mr.

Martin, what was the cost to our government from the Wrights' intentional suppression of the American aeroplane industry?"

"Objection! Calls for speculation, Your Honor!" Robbins shouted.

"Overruled," the judge countered. "The court is interested in hearing Mr. Martin's response."

Martin cleared his throat. "If Herring's 1896 patent application had been granted, the stagnant conditions that subsequently existed within the domestic aeroplane industry would never have occurred! Competition among manufacturers would have flourished, and *better* aeroplanes would have been produced. With America leading the way in aeroplane design, technology, and manufacturing capacity, the world conflict might have ended *sooner*! Sad as it is to admit, America's primary contribution to the air war was Curtiss sending 'flaming coffins' of foreign design, to our European—"

"Objection!" Robbins interrupted. "This witness continues to—"

"Sustained," Sawyer cut in. "Strike the witness's last statement."

*

Anxious to move to other aspects of James Martin's expertise, O'Grady flipped the pages of his legal pad to a new section, from which he read: "It's a matter of record that Mr. Herring and Mr. Curtiss had a falling-out as partners in the Herring-Curtiss Company. The conflict came to a head in late December of 1909, when company President Monroe Wheeler forwarded affidavits from various company officials to a New York State judge, declaring that Herring had *no* valuable inventions. The affidavits resulted in a questionable injunction action against Herring that deprived him of any voice in the company, from voting or selling his stock, or retrieving four of his patent applications.[6]

"That being said, Mr. Martin, what were the ramifications of this injunction in terms of Mr. Herring's patent applications?"

Martin surveyed the room with an air of polite detachment. "With the injunction against Herring in place, Curtiss and his cohorts then forced the company into bankruptcy, which they finally achieved toward the end of 1910—"

"Objection, Your Honor!" Robbins shouted. "A conclusion not based on fact."

"Sustained; strike from the record. Continue, sir."

"As you recall," Martin said, "it was during this tumultuous period in Herring's life that the Patent Commissioner refused to revive his 1896 application—"

"Why didn't Herring appeal the Patent Commissioner's decision?" O'Grady interrupted.

"Because the bankruptcy proceedings deprived Augustus of the funds necessary to take the case to the United States Supreme Court – his *only* remaining recourse.

"By 1911, Curtiss had succeeded in acquiring the assets of the bankrupt company for a pittance, including the four Herring patent applications.[7] As a result, Herring was stripped of his patents until 1917 – eight years after the injunction was issued. That's when the statute of limitations finally ran out on the improperly obtained document.

"Objection!" Robbins interrupted. "Another conclusion not based on fact."

"Sustained; strike from the record. Mr. Martin, any references *contrary* to the findings of a previous court ruling cannot be admitted into the record of this trial. The words 'improperly obtained' meet that criterion. You may continue."

"Your Honor," Martin said, "the point of my testimony is to highlight the actions that were taken against Mr. Herring's patent interests. For instance, there were the numerous unsubstantiated rejections by the Patent Office, and then there was Curtiss' injunction that prevented Herring from prosecuting his patent applications. To those of us who have investigated the injustices heaped upon this true aviation pioneer, there can be but one conclusion: there was an ongoing *conspiracy* directed against Gus Herring!"

Robbins bellowed, "Yet another conclusion, Your Honor!"

"Sustained," the judge said. "Strike the witness's last statement. The witness may continue."

Distressed over his inability to present what he considered to be important facts, Martin glared at the judge. "During the period leading up to the war's end, Curtiss and his wife are reputed to have received 13 million dollars directly and indirectly from the company that Augustus Herring started. To date, Mr. Herring has not received a single cent in return for his years of struggle and accomplishment—"

"Mr. Martin," O'Grady interrupted, "please summarize Augustus Herring's U.S. Patent Office experience."

"Gladly," Martin said through clenched teeth. "The People's laws were used to impede and discredit Augustus Herring. The allegation that his inventions were of no value and that a short delay on his part was the sole excuse for not reviving his original patent application defies all logic and fairness. Furthermore, for eight years, an injunction from a *US federal court of equity* restrained him from taking action against the Patent Office to establish the true value of these applications."

CHAPTER 57

Rochester, New York

Wednesday, October 26, 1921

Court resumed with Judge Sawyer's appraisal of the trial. "Since yesterday's recess, I have devoted much thought to the progress, or lack thereof, of this civil suit since its inception three weeks ago.

"Standard procedure dictates that I call the attorneys into chambers for my remarks, but this time, I have elected to break with tradition. Instead, I will communicate directly with all of the participants, including the defendants, the witnesses, and even the courtroom reporters. I'll begin by evaluating the plaintiff's side of the suit.

"The attorney for the plaintiff appears to favor the roundabout method of questioning witnesses. As a result, Mr. O'Grady has managed to frustrate both the defense attorney and the Court. I have often found myself wondering, *What direction is he headed in?* This strategy also encourages witnesses to state their opinions and make conclusions that cannot be allowed to stand in the official record. This technique may prove effective during a jury trial, but since the Court is the sole determiner of this suit's outcome... what is the point?"

After a momentary lull, the judge continued. "The incessant objections, bickering, and blatant lack of respect for this court, as displayed by the defense attorney and the defendants, have led me to conclude that their side of the aisle is attempting to push this suit toward a *mistrial.*

"Listen carefully... there will be *no* mistrial on my watch! My trials have *always* produced a decision – no matter how long it takes. If either party is unhappy with the results, the New York State Court of Appeals is your remedy, and it will be ready and willing to hear your complaints."

In an effort to lift the threatening cloud that had descended over the courtroom, the judge then softened his remarks. "Gentlemen, the plaintiff set the wheels of justice into motion by filing court papers in 1917... more than four years ago! If we hope to have a verdict rendered within our lifetimes, I implore that there be explicit questioning and fewer objections."

Swiveling his chair toward the podium, Sawyer nodded to O'Grady. "Counselor, you may begin your questioning."

"Your Honor, with the court's permission, I will now ask Mr. Martin a series of questions concerning Mr. Herring's activities during the 1897 calendar year – a period when a U.S. Patent Office examiner deliberated the fate of the inventor's original 1896 powered aeroplane application—"

"Objection! This is hearsay testimony! The plaintiff's 'witness' was not present during this period of—"

Rather than wait for the judge to rule, O'Grady cut his adversary's argument short. "Your Honor, I would have preferred that Mr. Herring be present to answer questions about his activities during this crucial time period, but circumstances have confined him to a hospital bed. Keeping the court's schedule in mind, the plaintiff requests that Mr. Martin be allowed to summarize this period of Mr. Herring's life. Upon his return, the plaintiff agrees to allow the defense to question Mr. Herring at length, to either confirm or disprove the accuracy of Mr. Martin's testimony."

"Your Honor," Robbins said, "this arrangement is unacceptable!"

"Objection overruled."

Disturbed that his opening remarks had apparently fallen on deaf ears, Sawyer nonetheless explained his ruling to the defense attorney. "Although somewhat unusual, I am going to allow the witness to provide his recollection of the time period in question. However, as suggested by the plaintiff's attorney, the defense will have the opportunity to interrogate Mr. Herring. You may continue, Mr. O'Grady."

"Mr. Martin, having submitted his powered-aeroplane application to the U.S. Patent Office on December 11, 1896, what did Mr. Herring do during the subsequent weeks?"

"He and his family celebrated Christmas and the New Year, while Gus continued to work at his part-time engineering job with the Chicago boat company. During his free time, he began fabrication work on his twin-cylinder engine. Toward the end of January, the Patent Office sent a letter of rejection. Although disturbed by their abrupt denial, Mr. Whittlesey assured Herring that this was indeed standard procedure.[1]

"Throughout this period, Mr. Herring pursued a multitude of potential sponsors for the construction of his proposed aeroplane and engine. He wrote letters, sent telegrams, and traveled to meet with officials from various organizations, such as the Hearst Newspapers – all to no avail.[2]

"At the end of May, Herring was laid off by the boat works; the job had lasted only nine months. Seeing an opportunity, Chanute proposed that Gus conduct center-of-pressure experiments on the carcasses of soaring birds that the old man had borrowed from various museums, including the Smithsonian. Chanute hoped that information gleaned from Herring's research could be used to upgrade his *Katydid* glider. After two weeks of toiling alone in Chanute's home workshop, Herring, bored stiff, quit the job. He then returned home to continue work on his engine project.

"Throughout the spring, Herring, Chanute, and their patent attorney worked on the British and French applications for the powered machine." Martin paused to check his notebook before continuing. "Because of inevitable delays, the filings were deferred until June 25. On the Friday before, Herring told me he had received an envelope with an Elmira, N.Y. return address. It was the first of many letters to be exchanged between Herring and his future benefactor, Matthias C. Arnot."

Pulling Arnot's letter from a folder, Martin handed it to O'Grady, who passed it to the judge. After the usual inspection, Sawyer instructed the court clerk to enter the document into evidence. Having been assigned an exhibit number, the letter was returned to Martin to be read aloud.

"'Dear Mr. Herring,

Having read of your wonderful flying success at the Indiana Dunes this past summer and fall, I am writing to inquire as to the possibility of purchasing a copy of your two-surface glider for my personal use.

'As for my background, I was an 1891 graduate of Yale, with degrees in business and engineering. In my workshop, I have constructed a boat with an inboard engine that has run successfully. I have set up my own telegraph line to learn its operation. I built and operated the first amateur telephone in Elmira. I also constructed an acetylene engine to propel a bicycle. As you can see, I am interested in everything new.

'Although I am not an aeronautical expert on a level with you, I was Elmira's first gliding enthusiast. In the early 1890s I constructed and flew a homemade contraption consisting mostly of wings and a frame. I hung by my hands from this frame and leapt from high hills so I might glide back to the ground. I also had the opportunity to take the glider to Italy where I found better cliffs for my purposes.

'If you are so inclined to build a replica of your "double-decker," write to me at the address at the top of this letter. If possible, please include a list of specifications, an estimate of the time required, and the price.³

'Hoping to hear from you soon.

Yours in aviation,

Matthias C. Arnot

Vice President

Chemung County Trust Co.' "

O'Grady wandered out from behind the podium. "What did Herring do with Mr. Arnot's request?"

"He immediately wrote back, providing the glider's specifications, along with a quote of $400 for its construction and delivery. In the meantime, another rejection letter arrived from the U.S. Patent Office. The examiner had disallowed several of Herring's claims, including one for multiple propellers driven from a single power source. Attorney Whittlesey subsequently changed the claim to read... wait a second, let me check the wording... 'tandem propellers turning in opposite directions' to avoid the perceived conflict."

"Did Mr. Arnot respond to Herring's offer?"

Martin closed his eyes, as if searching his memory. "A second letter arrived from Elmira in early July. I remember the date because of its proximity to Independence Day."

"What effect did the letter have on Mr. Herring's work?"

"Well, first of all, there was a very welcome cashier's check for $400, and a request to meet with Mr. Herring. Subsequently, Gus sub-contracted the glider's construction to Bill Avery's carpentry shop and made arrangements for Arnot's trip to Chicago. The *New Two-Surface Machine*, as Gus called it, would be similar to his '96 glider with several subtle, but important differences.⁴

"While awaiting Arnot's arrival, Gus tended to another pressing need – securing a full-time job. Responding to an ad in the Chicago *Tribune*, he scheduled an interview with the Truscott Boat Works, of St. Joseph, Michigan. On Tuesday, July 6, he crossed the toe of Lake Michigan aboard a lake steamer, arriving barely in time for his 11 o'clock appointment. Truscott officials were impressed with his educational and engineering background, and they hired Herring on the spot as a mechanical engineer – to begin work on August first. Gus quickly found a suitable home for his family; a two-story house at 413 Church Street, within walking distance to the Boat Works."

Chicago, Illinois

Thursday, July 8, 1897

Herring stood on the platform as the New York Central passenger train pulled into the South Chicago terminal. As travelers began to disembark, he held up a pasteboard sign with Arnot's name in large, hand-painted black letters. Within seconds, a thin young man in his late 20s approached with his hand outstretched. His firm handshake was accompanied by the tipping of his black bowler, revealing a neatly groomed head of dark brown hair, parted down the middle.

"Mr. Herring," Arnot said with a broad smile, revealing very white teeth beneath his carefully trimmed, full mustache. "It is a distinct pleasure to meet you, sir!"

Dressed in a three-piece charcoal-gray business suit with a starched, high-collar white shirt, and a dark green bow tie, Arnot fit Herring's image of an Eastern banker. At 5 feet, 11 inches tall, Arnot looked gaunt. His drawn appearance was accentuated by his protruding cheekbones and closely set brown eyes. Dark circles beneath those eyes suggested a life spent squinting at ledger sheets, with little if any time devoted to outdoor activities.

Carrying one of Arnot's two Gladstone bags, Herring directed his guest to a waiting hansom cab. Fifteen minutes later, they arrived at Herring's front sidewalk, where young William hopped down from the porch to greet his father. After playing an abbreviated game of catch with his son, Herring escorted his guest into the house.

"Do you have children, Matthias?" Herring asked.

"No, we didn't. Before Elizabeth's untimely death, she had been informed by her doctors that she was incapable of bearing children."

Matthias C. Arnot (1898); Library of Congress

An uncomfortable few seconds passed before Lillian entered the parlor. As Herring would learn, although polite to a fault, Arnot most always said exactly what he was thinking. With introductions complete, the two men retired to the far end of the parlor where Herring maintained his modest study and drawing table. "Tell me, Matthias, what prompted your interest in one of my aero*curve* gliders?"

Making himself comfortable in the overstuffed chair closest to the drawing table, Arnot arched an eyebrow. "Your success with the two-surface glider convinced me that I needed such a machine for my own experiments. I'm afraid that the 'flying bug' bit me during my brief forays into the air back in Elmira and later in Italy.

"Two years ago, I was in Berlin to see Herr Lilienthal fly, but the weather was unfavorable. Needless to say, I left for home disappointed."

"Did you meet The Master?"

"Briefly, at his factory... but he was quite busy."

"Do you speak German?"

"Yes, I handle all of the German transactions for our bank."

"Aha, then we have something else in common! By the way, where did you learn of my activities?" Gus asked.

"Late last year, there was an article in the New York *Times*. It was authored by a Mr. Dienstbach – he's quite an admirer of yours."

"Ah yes, Carl is certainly an accomplished aviation journalist. He attended all of our flight trials at the dunes last year. Do you have any long-term goals in aviation, Matthias?"

"Besides learning to fly," Arnot said, "I would like to become involved in the flying-machine movement... perhaps as an aeroplane manufacturer, or the producer of lightweight aero engines."

Herring nodded his approval while offering words of caution. "I believe that the manufacture of manned, heavier-than-air flying machines will eventually come to pass in this country. Working toward that end, I am planning to add an engine to an apparatus similar in appearance to my *Double Decker* glider. However, as with all great ideas that have yet to be perfected; there are problems—"

"What specifically are the problems?" Arnot interrupted.

Herring settled into the stuffed chair across from his guest. "The answer depends upon whom you are questioning! Before his unfortunate death, Herr Lilienthal was in the process of outfitting his standard glider with thrust-producing wingtip pinions. A small engine of his own design was intended to deliver the necessary flapping action. Professor Langley maintains that a larger engine and a simple *scaling up* of his steam-powered aerodrome model will provide the answer to powered, manned flight. The expatriate, Sir Hiram Maxim, contends that a more powerful engine will solve the problem all by itself, with virtually little else considered! Octave Chanute believes that attaining automatic stability should be every experimenter's first priority. If he were in charge, there would be no manned, dynamic flight attempts made until the equilibrium issue had been *completely* resolved. There are other opinions... but are you prepared to listen for the rest of the afternoon and well into the evening?"

Before Arnot could answer, Lillian popped her head around the corner and said, "Gentlemen, lunch is ready. Augustus, show Mr. Arnot where he can refresh himself."

After enjoying sandwiches of homegrown tomatoes, potato salad, and tall glasses of freshly squeezed lemonade, the men again retreated to the parlor to resume their conversation. Striding to his drawing table, Herring removed the protective oilskin covering, revealing a multiple-view drawing of Arnot's updated two-surface glider. "Matthias, come look at your glider! I've made some design changes that I believe will make it better than last summer's machine."

As the men studied the illustration, Arnot asked questions at a furious pace. "What changes are you referring to? Why will the new machine be better? What is the wingspan? Why are you using the arc-type curve? How does the tail regulator work? What material will you use to cover the surfaces? How much will the completed machine weigh?"

For more than an hour, the men were lost in the technology of gliding flight. As Arnot's enthusiasm grew, so did Herring's. Before long, the two were literally dancing around the drawing table, mixing laughter with technical discussion. As Lillian listened to parts of the conversation from her kitchen, she couldn't help but think,...*These men are two peas in a pod. I think that Augustus has found a believer!*

Around 3 o'clock, Arnot meandered back to his original question. "What problems remain to be solved before a manned, powered, heavier-than-air machine can be successfully demonstrated?"

Settling into his chair, Herring looked intently into Arnot's eyes. "Matthias, as a lender of money, you realize how important adequate funding is to any new venture. In that regard, experimenting with flying machines is no different. For example, there are costs for tooling, costs for materials, costs for obtaining patents, costs associated with transportation, living expenses, and maybe an assistant or two. There is no getting around it, *financing* is the greatest obstacle to overcome."

Arnot seemed unconcerned about the financial aspect of the problem. "What is your next biggest problem?"

"Protecting ideas. Costs are one thing, but I am becoming disillusioned with the U.S. Patent Office – the bureau's examiners don't follow their own rules! They insist upon rejecting my claims for no other reason than that they are unaware of any successful manned, powered flights having been made!"

For the next hour and a half, the men discussed the ramifications of a failed patent application. Arnot, as it turned out, had also experienced disappointing results with two of his applications.

"To be honest," Herring said, "I'm thinking about withdrawing my application and going underground with the power-plane experiments. By nature, I'm opposed to secrecy, but I'm being forced to keep my claims to myself while developing the next aspect of dynamic flight. My hope is… after I demonstrate my powered machine to the world, that the Patent Office will have no choice but to award me a *pioneering* patent.

"But on the negative side, by withdrawing my application, I will leave more than 20 claims unprotected. If someone steals them, I'll be out of luck… and for me, good luck has always been in very short supply!"

"If you don't mind me asking, Augustus, how have you managed to pay for the prosecution of your patents up until now?"

Slouching, Herring thought, *How much should I tell him? I've just met the man for Christ's sake.* Throwing caution to the wind, Herring decided to divulge his agreement with Octave Chanute. "This is a bit complicated, Matthias... but I'll try to explain. Mr. Chanute tried to claim more responsibility for the design of the two-surface machine than I was willing to concede. On the other hand, I had engineered and fabricated the swing-wing mechanism for his *Katydid* glider.

"After some squabbling, a trade-off was negotiated. He *gave* me the opportunity to patent my manned, power aeroplane in the United States, and I agreed to allow *him* to prosecute all aspects of his swing-wing system in America and Europe. As a third component of the deal, we applied jointly in Britain and France for the power-plane patent. In return for these concessions, Chanute is paying for the majority of our attorney and patent application fees."

Arnot looked confused. "If Chanute was against powered flight, why does he want to join you in the pursuit of European patent applications?"

Herring shrugged. "Your guess is as good as mine! I can only surmise that the old man is trying to cover all of the bases. He could be thinking, 'If anything comes of this cockamamie idea, I can claim partial responsibility for its success.' "

"So... money hasn't been a problem for you in the prosecution of the U.S. patent application?"

"Thus far, I've spent almost $200 of my own money... it's been a burden."

Rising to his feet, Herring shuffled back to the drawing table. Thumbing through several more drawings, he located what he was looking for and pulled it from the stack. "Look here, Matthias, this is what I envision the power-plane to look like."

As Arnot's eyes panned over the multiple-view drawing and two-point perspective, he became lost in its details. Soon, he was mumbling to himself in a barely audible whisper. "Wheeled undercarriage, twin cylinder engine, two inline propellers, three superposed surfaces, truss bracing, regulator-controlled cruciform tail.

"Gus, in order to handle the additional weight of the engine and propellers, why didn't you simply extend two superposed surfaces, rather than adding a third wing?"

"An excellent question," Herring said, grinning. "When the machine's wingspan exceeds 18 or 19 feet, the weight-shifting method of lateral control becomes increasingly unmanageable. To prevent that from occurring, I simply added a third, relatively short-span superposed surface. Of course, the third lifting surface raises the machine's center of gravity and center of resistance, which requires other design changes."

Herring flipped to a new set of drawings that featured individual engine components along with construction notes and calculations. An accompanying plate featured an artistic-looking exploded view of the opposed, twin-cylinder gasoline engine. Overwhelmed by the magnitude of Herring's new prime mover, Arnot had many questions pertaining to its operational cycle, design, and construction. "Your engine has an overhead poppet valve and a row of what appear to be exhaust ports low on the cylinder. Does it operate on the four-stroke Otto cycle or the two-stroke Clerk cycle?"

"Another astute observation," Herring replied, smiling. "Actually, she works on a slightly modified Otto cycle. The single overhead poppet valve performs the same operations as the intake and exhaust valves in Nicholas Otto's old cast-iron block design... but my adaptation is much lighter and less complicated. The circumferential ring of drilled ports is borrowed from Joseph Day's two-stroke design. They allow the cylinder to

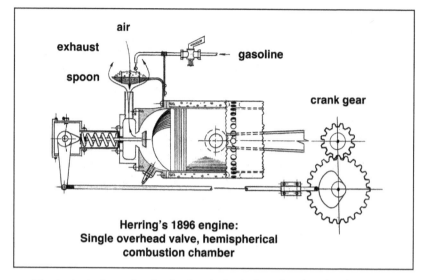

Herring's single overhead valve, hemispherical combustion chamber engine (1897); drawing by author

begin scavenging exhaust gasses before the piston begins moving toward the hemispherical combustion chamber head on the exhaust stroke."

Arnot pointed to a feature on the plan. "That bowl-shaped combustion chamber is unusual; when did you begin using that geometry?"

The question set Herring's mind wandering. "I began experimenting with the hemispherical chamber back at Stevens Institute in the late '80s—"

Their attention was diverted by the sound of the door opening. Lillian appeared and called the men to supper. Afterwards, Herring ushered Arnot out to his workshop behind the house. Barely noticing the confined quarters of the shed, Arnot hovered over the partially completed aero engine. "When do think you'll have her running?"

"Depends how long my money holds out. I'm waiting on a quote for a Ruhmkorff coil and six Hellesen dry cells. I've decided against trying to use a small dynamo or magneto ignition system at this time; they're too complicated, too heavy and too expensive! I'll stick with a battery ignition system for the time being. Once I get back to work in St. Joseph, I'll be able to order all the ignition pieces that I need."

Squinting, Arnot seemed puzzled. "St. Joseph? Are you moving?"

After accounting for his move to Michigan, Herring emphasized the positive aspects of the change. "I've already identified two machine shops that appear to do excellent work. The house that I'm renting has a large storage shed behind it that I can use for a shop. And St. Joseph has a long, wide beachfront where I anticipate testing the new power machine."

Herring hesitated before continuing. "Matthias, in terms of testing and getting used to your new glider, I suggest we take her to the old dunes site – it's closer to St. Joseph than it is to us here in Chicago!"

"When do you expect the machine to be ready for trials?"

"According to Bill Avery... about the middle of August."

"When will you be moving to St. Joseph?"

"Looks like the last week in July."

"How will you get the glider there?"

"I'll have Bill ship it to me by rail, then I'll assemble her at my home. We should plan flight trials for at least a week in September... when the weather is good. Can you drag yourself away from work long enough to attend?"

Pacing the dirt floor of Herring's shed, Arnot considered the proposal. "I'll have to look at my schedule. I'll let you know as soon as I get back to Elmira. Uncle Matthias is quite strict about me tending to business, but you can count on me, Gus. When it's time to fly the new machine... I'll be there!"

"Good! We're going to need the help of a third person. I'll write Henry Clarke, an old friend from Philadelphia. He helped with some of my experiments back East. If he's not available, I'll contact Andrew Vassallo, a mechanical engineer friend and former assistant who lives in New York City. Andy helped me fly the Lilienthal-type gliders back in '94."

Herring hesitated before continuing. "I'm getting ahead of myself! All our plans hinge on Bill Avery getting the glider built and delivered to me before the end of August! We'll stroll over to his shop in the morning."

Chicago, Illinois

Friday, July 9, 1897

Their breakfasts finished, Herring and Arnot walked the few blocks to Avery's East Superior Street shop. Entering at the east end of the long, single-story woodworking facility, the men were met by Bill himself, who had been working since daybreak. "Gus! It's not every day you honor us with your presence! Who's your friend?"

After introductions, Avery escorted Arnot to the opposite end of the building, where he had set up to work on the new glider. "As you can see, your glider is well underway. I made most of the ribs from Sitka spruce that was left over from last year's two-surface machine, but I had to chase down the rest of the wood last Monday. By the way, Gus, for some unknown reason, Sitka spruce is impossible to obtain right now, and I had to settle for black spruce. Other than a few dark streaks through its heartwood, I can't tell the difference between the two."

After inspecting lengths of the new wood, Herring shrugged. "I've used black spruce before. It has nice grain structure – long and straight... it'll work."

For the next hour, Gus and Bill explained the glider's construction to Arnot, including how the ribs, spars, and vertical struts would be assembled using piano-wire as trussing. At precisely 10 o'clock, a tubby figure waddled into the shop. Peering over his ever-present reading glasses, Octave Chanute squinted to see who was standing at the other end of the 50-foot structure. As he shuffled over to join the conversation, the old man paused briefly to acknowledge J.S. Avery, Bill's father.

After Herring introduced Arnot, Chanute got to the point of his visit. "Bill informed me that you would be paying him a visit this morning, Mr. Arnot. I simply wanted to make your acquaintance before you returned to New York. From what I've been told, your enthusiasm for glider flying is commendable for such a young fellow. When and where will you test the new machine?"

Before Herring could deflect Chanute's question, Arnot spoke up. "Mr. Herring and I hope to fly the glider at your old Indiana dunes site sometime in September. It's the Dune Park Station site, right, Gus?"

Herring nodded meekly. "I'm excited!" Arnot continued. "Can't wait to get started!"

Chanute held up an index finger. "Young man, allow me to offer my assistance. Telegraph me on the morning that you depart from Elmira, and I'll meet you at the dunes to serve as your official photographer. God willing, you'll have some excellent glass-plate images to document your efforts."

"Wonderful! All the more reason to be excited about our trials!"

Later, as the men strolled back to Herring's flat for lunch, Matthias commented about his newfound friend. "What a fine fellow, this Chanute!"

Biting his lip, Herring gathered his thoughts. "Matthias, your enthusiasm for aviation is a good thing. I know meeting you has been invigorating for me. But based upon my years of experience in the field, I must caution you to be careful about sharing too freely. Octave can be charming, but I must be candid: due to our collaboration in the pursuit of patents, we are still cordial to one another. However, make no mistake, there is much animosity between the old man and myself. Remember this, Matthias... when Chanute pats you on the back, he's really looking for a place to stick the knife!

"Because I was once his paid assistant, he thinks that my knowledge and technical achievements are his personal property. The two-surface machine is the primary bone of contention. He claims most of the credit, while quite the opposite is true...."

Catching himself wandering off topic, Herring returned to the issue of photographs. "I've seen prints from the glass plate photographs that Chanute exposed at last summer's trials at the Indiana dunes. Most are out of focus and underexposed – especially those of the two-surface machine in flight. Therefore, I suspect that his offer to act as your personal photographer may contain a hidden motive."

"What possible motive could there be?"

"I'm surmising," Herring said, "but knowing the old man as I do, he needs good photographs to illustrate his lectures and magazine articles."

"How can our new glider help him illustrate the 1896 experiments?"

"He'd simply substitute photographs of your machine for those of his own."

Arnot hesitated before questioning that explanation. "The new machine will have changes. How will Chanute explain the differences?"

Herring stopped walking and seized Arnot's arm. "There's something you should know, Matthias. Chanute is 65 years old and he's becoming senile. He has difficulty understanding the simplest of concepts. For instance, he still doesn't understand how my tail regulator works... even though I've explained it to him over and over again. I think that he will just ignore the differences between the '96 and '97 machines, and most people won't know the difference. If Chanute is going to take photographs, *you must be sure to purchase all of the supplies* – otherwise he will claim ownership of the glass plates, and then you'll have a difficult time obtaining prints for yourself down the road."

Arnot slowly shook his head. "Sounds as if I made a mistake by inviting Chanute to our trials. I admit being somewhat overwhelmed by the man's reputation, but..."

"That's water over the dam, Matthias. He would have learned about our plans sooner or later."

CHAPTER 58

Rochester, New York

Thursday, October 27, 1921

As James V. Martin's testimony entered its third day, his bloodshot eyes told a story of stress and tension. Martin would later admit that much of the pressure was related to his bedside visit with Herring the previous evening. Recuperating from an attack of apoplexy... Gus described it as "a troubling but minor episode."

Although Herring was anxious to hear about Martin's day in court, he was more concerned with providing an array of complex details to weave into his friend's forthcoming testimony. By the end of visiting hours, Martin, overloaded with new information, was mentally fatigued.

The next morning, as Martin lethargically sprinkled baking soda onto his toothbrush, he wondered, *How can I be expected to remember everything?*

*

Sitting at the plaintiff's table, attempting to organize his thoughts, Martin was jolted by the bailiff's booming voice. "All rise! The New York Supreme Court of Monroe County is hereby in session! The honorable S. Nelson Sawyer, presiding!"

Settling himself behind the bench, Judge Sawyer got right down to business. "Mr. Martin, please take the witness stand. Be advised, sir... you are still under oath."

Taking his cue from the judge, Attorney O'Grady began his questioning. "Mr. Martin, yesterday you ended your testimony by describing how Octave Chanute invited himself to the Herring-Arnot flight trials. Please describe the circumstances leading up to those trials."

St. Joseph, Michigan

Sunday, August 29, 1897

Herring idly brushed pyroxiline dope onto the glider's vertical rudder while half-listening to the fire and brimstone sermon rumbling through the open windows of the First German Baptist Church next door.

"Now I've seen it all!" a man's voice grumbled. "Listenin' to a preacher's sermon while working on a fool flyin' machine!" As the burly man sauntered into his workshop, Gus squinted into a backdrop of brilliant sunshine and struggled to identify the stranger from his shadowy silhouette.

"Henry Clarke – I was just thinking about you! But, the suit... did you sneak away from the homily next door?"

At 5 feet, 9 inches tall, and weighing 210 pounds, Henry Clarke looked more like an enforcer at a longshoremen's convention than he did a churchgoer. Dressed in his finest single-breasted, light-gray summer suit with all the trimmings, Clarke appeared authoritative as he tipped his steel-gray bowler. Standing, Herring shook hands with his old friend before dusting off his best shop stool.

"Have a seat, Henry. You just missed some wonderful pipe organ music! In the short time that we've lived here, I've grown fond of that grand old wind instrument and the elderly gentleman who plays it. In fact, he stopped by this morning to say he could smell the dope fumes all the way down the street; claimed it reminded him of ripe bananas. When he asked what I was doing, I told him to come back after church and I'd show him, if he promised to play one of my favorite hymns."

"You have a favorite hymn?" Clarke asked, grinning.

"*Rock of Ages*!" Gus replied. "As a youngster back in Georgia, I was *required* to attend church every Sunday. I liked singin' the hymns best of all. Nowadays, this is as close as I get to going to church!"

"Well, did the old man play it?"

"Not yet... I'm still waiting!"

A week earlier, Herring had telephoned his old classmate and assistant at his Philadelphia home, hoping to lure the 30-year-old to the Indiana dunes for a week of fun and adventure. Informed of the dates and other particulars, Clarke had seemed a bit reluctant until Herring mentioned his need for a paid assistant to help with launching and retrieving the machine. Temporarily out of engineering work, Clarke readily agreed to $2 per day, plus meals, lodging, and a round-trip railroad ticket.

While attending Stevens Technical Institute, Clarke had watched Herring experiment with a propeller-testing machine and what he called aero*curve* flying models. Later, intrigued by the potential of heavier-than-air flight, he assisted Herring with many of his early projects, including the man-carrying glider that collapsed during its first foray into the air.

"After we spoke last week," Herring said, "I telephoned our old friend Andy Vassallo. He'll meet us at the dunes sometime during the week."

Subsequent to Clarke's post-graduation move back to Philadelphia, Vassallo had taken his place at Herring's Broadway Avenue shop.

After cleaning his pyroxiline dope brush in a container of solvent, Gus helped Clarke haul his luggage to the house. As they reached the back door, the church's pipe organ, accompanied by the choir, thundered a refrain that reverberated throughout the neighborhood.

Rock of ages, cleft for me, let me hide myself in thee;

When I soar to worlds unknown, see thee on thy judgment throne;

While I draw this fleeting breath, when mine eyes shall close in death.

The men sat on the stoop and enjoyed all four verses. When the hymn was finished, Herring cocked his head at Clarke and winked. "Maybe we should start going to church, Henry. Did you hear the part about 'soaring to worlds unknown'?"

Inside, Gus introduced Henry to Lillian, who informed the men that lunch would be ready in 10 minutes. Leading Clarke up the creaking wooden stairs, Gus commented about his rented house. "We've been here for almost a month, but we're still getting accustomed to all the extra space. Besides the large parlor and kitchen, we have three big bedrooms, one of which I'm going to set up as a drafting room and study. But for now, that's going to be your room. While I'm washing up, why don't you change into your work clothes? After lunch, you can help brush pyroxiline onto the cotton covering."

After they finished their sandwiches and coffee, the men shuffled back to the shop, where Herring removed a second camelhair brush from his oak tool chest. As they brushed the dope onto the cruciform tail assembly, their talk wandered far and wide.

"Did the move from Chicago go smoothly?"

Herring sighed. "About what I had anticipated. We've moved so often during the past couple of years that I've come to expect a certain amount of commotion. But I must admit, the steamboat ride across Lake Michigan was awful. We moved on the last Monday in July. Fortunately, I had had the forethought to send Lillian and the children on to St. Joseph by train. All of our household belongings, along with my crated tools and equipment, were strapped to the deck of the steamer, *Northern Eagle*.

"The steamer passage was very rough, and it took longer than expected – almost four hours. Bill Avery and I were both seasick. It got to the point where I didn't care if the whole works washed overboard! Fortunately for me, the deckhands did a good job of securing everything, and we arrived safely at the St. Joseph pier... only a stone's throw from my new employer's factory complex. Before long we had the rented horse-drawn wagon loaded, and we proceeded to navigate south on Broad Street. A quarter mile up the hill, there it was... Church Street. When Bill and I finally got to the house, we found Lillian and the kids sitting on the front stoop – I had forgotten to give her a key!

"With everything moved, I paid Bill for his help and took him to the depot for his return to Chicago. By the middle of the week, everything was pretty much arranged in the house, but my shop equipment had to sit outside while I cleaned out this barn. You wouldn't believe all the fossilized manure and garbage that had accumulated in here over the years. It took me five working days and 11 trips to the dump before—"

"Well," Clarke interrupted, "looks like you've got plenty of room now. And I like all of the windows – lots of light for daytime work. What you gonna do for heat in the winter... and what about lights for—"

"One thing at a time, Henry!" Herring broke in. "I'm planning to install a coal-burning stove and gas lamps all along the walls – one between each of those windows."

"Gas lamps? What about electric lights?"

"The electric service hasn't come this far yet, and I don't want to wait. Besides, obsolete gas-lamp fixtures are selling dirt-cheap in places like Chicago. I can outfit this shop for a fraction of what it would cost to bring in electricity, and what's more, the gas pipe is already here. Have to admit though... I miss electric motor-driven machinery. I got spoiled working in Bill Avery's shop last year. For the time being, I've been using kerosene lanterns for the nighttime work. It's far from ideal, but I have little choice; I'm tryin' to finish this glider by this coming Friday."

As Clarke continued brushing dope onto his section of the horizontal rudder, other questions came to mind. "When did you get the bare-bones machine from Mr. Avery?"

Shuffling over to his makeshift desk, Herring rummaged through some papers before finding the railroad invoice. "Here it is... I signed for the machine on August 18, a Wednesday. I've been working on her steadily for the last 10 days – mostly nights."

"Now that I'm here," Clarke said, "the job will get done faster. By the way, what exactly do you do at the boat factory?"

Herring made a face. "For now, I'm reviewing their entire boat-building operation. I've been charged with finding new ways to cut costs, while improving the quality of their product. So far, I've done nothing more than observe, take notes, and talk to supervisors. The workers think I'm a spy for management... I get lots of dirty looks.

"Tomorrow morning I'm meeting with Mr. Truscott and his sons to make long-range recommendations and suggest what short-term changes can make the most impact on their business. When I get back from taking some of my vacation time, I'll meet with them again to see what they've decided to do, and how they want me to—"

Clarke held up a finger as a signal to wait. "Let me get this straight, you've been on the job for less than a month and you're taking vacation time to experiment with this *Double Decker* flying machine? If ya don't mind me askin', how did you manage that?"

Herring continued brushing. "I made the September vacation time part of my contract, and Mr. Truscott granted my request... simple as that."

"I'm curious, Gus, what will you recommend to them?"

"Long-term, I'll suggest that they switch to gasoline-fueled internal combustion engines."

"What are they using now?"

"Steam engines. I'll also propose that they switch to a *team approach* for building their boats. The old idea that one man is responsible for building one boat has to change. Short-term recommendations are generally confined to their tooling, machinery, and the use of certain materials. Down the road, I hope to design a new type of marine engine... one with revolutionary features."

When the men finished the doping, they switched to the task of assembling the covered, but yet-to-be-doped, superposed wing panels. For the next five days and evenings, with time taken out for Herring's company presentation on Monday, the men worked at a feverish pace to complete and trial-assemble Arnot's new glider.[1]

Rochester, New York

Thursday, October 27, 1921

Refreshed by his lunch break, James Martin returned to the witness stand, anxious to describe events from the '97 Indiana dunes flight trials.

"Mr. Martin," O'Grady began, "when did Mr. Herring and his assistant actually finish the new *Double Decker*?"

Martin stared at his notes for several moments before raising his head. "Late in the evening of September 2... according to

Mr. Herring, it was a Thursday. By 10 o'clock the next morning, they had finished dismantling the glider and began building two wooden shipping crates. Shortly before noon, Matthias Arnot arrived to help complete the job."

"Did Herring notify the press about the impending trials?"

"He said that he telephoned the Chicago *Tribune,* Chicago *Record,* and Chicago *Herald,* as well as Herr Dienstbach at the New York *Times.* He informed them that he and a Mr. Matthias Arnot, of Elmira, New York, would be flying a new, two-surface glider at the Indiana dunes test site during the week of September 5."

"Contrary to Chanute's anti-newspaper campaign from the previous year," O'Grady said, "why did Herring welcome the press into their camp?"

Martin cast a knowing glance at the attorney. "Gus was pleased to have his experiments described by the reporters, along with his aeronautical theories and predictions... without interference from Chanute."

Satisfied with Martin's answer, O'Grady referred to his notes. "When did their party of three leave for the Indiana dunes?"

"Early Saturday morning. The men started loading crates, tents and provisions aboard a horse-drawn wagon borrowed from Truscott's at the beginning of the Labor Day weekend. Two round trips were required to haul everything to the St. Joseph pier. By mid-morning all of the cargo had been loaded and lashed to the deck of *Ares,* a side-wheel steamer known as a 'flat bottom.' With light winds and a low chop, the coast-hugging voyage across Lake Michigan to the Dune Park site proved to be uneventful."

"Then what happened?"

"Early afternoon was spent moving the crates to shore, and hauling them to a sheltered location among the dunes, and setting up camp. After gathering driftwood for a fire, Gus and Matthias pitched in to help Henry prepare the evening meal. After cleanup, talk around the evening campfire focused on reassembling the new glider."

"When did that work take place?"

"Gus said that he was up at daybreak. Before breakfast, he had already pried the lids off the two crates and removed pieces of the machine for inspection."

"Was there any damage?" O'Grady asked.

Martin considered this for a moment. "Apparently, there was a small tear in the nainsook covering on the vertical rudder. Gus called it 'an easy mend.' However, the next step proved to be time consuming and tedious: the establishment of a level working surface from which to position and join the superposed lifting surfaces. After achieving a level platform with the aid of adjustable-height wooden sawhorses, the upper lifting surface was fastened above its lower counterpart by their vertical struts. Temporary wooden braces were plunged deep into the sand, which helped keep the wing panels both square and plumb to each other. The laborious task of installing the piano-wire bracing took most of the remaining daylight. After supper, the cruciform tail assembly was attached to the machine's central framework by its two hinged bamboo booms.[2]

"With dusk approaching, Herring installed the cords and elastic springs that controlled the tail's automatic regulator. They carried the completed machine into the big tent; Herring still had recurring nightmares of his '96 machine being mangled in a violent storm a year earlier."

"Objection, Your Honor," Robbins said, rising from his seat at the defense table. "How can Mr. Martin possibly know all of these minute details? Must I remind the court that this so-called witness was *not present* during these trials?"

Arching his eyebrows, the judge cocked his head in Martin's direction. "I was wondering about that myself, Mr. Martin. Explain your apparently encyclopedic knowledge of such matters."

Martin shifted in his chair. "After court adjourned yesterday, I took the streetcar over to Rochester General Hospital and visited with Mr. Herring. A good portion of our time together was spent reviewing the 1897 expedition to Dune Park, and especially the difficulties associated with the assembly of the new two-surface machine. Augustus was very patient... explaining the machine's setup and testing. I only use notes to get the dates straight—"

"Your Honor, Mr. Martin's relating secondhand knowledge of Herring's alleged experiences is very disturbing to the defense! Again, I must protest the inclusion of this witness's hearsay testimony into the official record—"

"Yes, yes, Mr. Robbins, once again your objection has been duly noted and overruled! Carry on, Mr. O'Grady."

"When was the new *Double Decker* glider first flown?"

Martin checked his notes. "The following day... it was Labor Day Monday. Herring said the sky broke sunny and clear, with an 18-mile-an-hour wind out of the north – perfect flying conditions. Gus made the initial flight from a dune only 20 feet above the beach, and it turned out to be a beauty! After his usual running start, the glider coasted majestically for a distance of 199 feet in 17 seconds. Afterwards, he calculated that the first flight's one-foot drop for each 10 feet of forward progress exceeded his theoretical projections for the design... a phenomenon that Herring attributed to 'a favorable incline in the wind's direction.'

"It was a great beginning! After each of the next half-dozen flights, Herring fine-tuned the machine by adjusting the tail regulator's cords and the elastic springs. For the seventh flight, he moved to a launch point 35 feet above the beach. From there, he made a straight-ahead glide of 340 feet, landing at the edge of the lake.

Herring-Arnot Double-Decker glider. Note Herring's dogs, Rags and Tatters (1897); 1900 McClure's Magazine

"But he said that by lunchtime, the wind had shifted and was gusting out of the west – an unfavorable direction, and that ended flying for the remainder of the day."

"When did they resume the experiments?" O'Grady asked.

"On Tuesday, reporters from two Chicago newspapers, the *Tribune* and the *Record*, arrived in camp just in time to witness Herring make his first flight of the day – a 200-foot effort from a low takeoff point. With ideal flying conditions the order of the day, Mr. Arnot made his first foray into the air, flying a very commendable 175 feet, which left him both short of breath and ecstatic in his praise of the machine's docile and self-correcting nature. When interviewed by Mr. Macbeth of the *Times-Herald*, he was quoted as saying that Herring was 'a genius in the emerging era of heavier-than-air flying machines.' "

"Objection! *Double* hearsay!"

Waving a yellowed newspaper article above his head, O'Grady protested. "Your Honor, I have Mr. Macbeth's article right here... it's from Sunday, September 12—"

"Your Honor," Robbins chirped, "how can you allow this contrived drivel to be entered into evidence?"

"Overruled. Continue, Mr. O'Grady."

"Mr. Martin, tell the court about Mr. Herring's most notable flights."

Martin's eyebrows shot up. "The first took place just before noon in what appeared to be treacherous conditions. Herring launched from a spot 50 feet above the beach into a 35-mile-an-hour wind from the north. He had to quickly shift his weight hard to port, which caused the machine to yaw slowly in that direction. Then he pitched his body to starboard and forward, forcing the craft to 'crab' to the northwest. As the glider traversed to the west in what today would be called a 'quartering' position, he was able to take advantage of the wind's upward trend along the face of the dune. The machine flew on and on for almost 40 seconds, covering nearly 300 feet. Finally, Gus threw his weight hard to starboard and forward, compelling the apparatus to yaw back into the prevailing wind and allowing it to eventually alight softly on the beach.

"Later, when asked if the machine could have flown farther, Herring said: 'I might have flown to infinity had the fickle wind cooperated. I must add a lightweight *motor* – it will allow me to fly great distances without interruption.' "

"Objection! Hearsay... calls for speculation!"

"Your Honor," O'Grady sputtered, "Hearsay? This is a direct quote from—"

Standing, the judge pointed directly at Robbins. "Since you've objected Counselor, I will allow the plaintiff to enter its document into evidence! Is this what the defense considers to be a timesaving tactic? You may wish to reconsider, Mr. Robbins!"

After a brief conference with Wheeler and Curtiss, the defense withdrew its objection.

"Let's try again, shall we?" the judge said. "Continue, Counselor."

"Mr. Martin, were any other significant flights made on Tuesday?"

"Mr. Arnot made several short flights after lunch, along with one by Mr. Manley of the *Record*, who described his experience in the paper's Sunday edition. Late in the afternoon, with the wind approaching 40 miles an hour, Herring made another long, quartering flight that unexpectedly splashed into the lake. Fortunately, Herring kept the machine dry by hoisting it above his head as he waded to shore. Later, he proclaimed that the 56-second effort represented the longest time-aloft flight ever recorded for a manned, heavier-than-air machine."

Raising his hand as if to demur, Robbins had a change of heart and dropped his clenched fist to the table with a thud.

Ignoring Robbins' not-so-subtle protest, O'Grady continued. "Did Mr. Arnot stay for the entire week?"

"No. Mr. Herring indicated that Arnot left for Dune Park Station to catch an eastbound train. As he was leaving camp, Gus gave Matthias a compass for the challenging walk to the station and some advice on how to avoid snakes and wild boars. He said, 'Stay out of the swamps and walk like an Indian... otherwise, the pigs will hear you coming a mile away.' "

"How did the experiments proceed after Mr. Arnot's departure?"

"Wednesday dawned windy and wet, which kept all parties inside the big tent with little to do. While Herring kept busy by making minor repairs and adjustments to the glider, the two Chicago reporters urged him to reveal his plans for extending the machine's performance. Shrugging off the inquiry, Herring said that 'an engine of sufficient power and lightness does not yet exist. Until this shortcoming has been resolved, any such discussion would be premature.' "

Dune Park, Indiana

Thursday, September 9, 1897

Thursday dawned overcast but calm. By 10 o'clock, Herring thought the wind had increased enough to consider moving the machine from the protective confines of the big tent. To confirm his suspicions, he grabbed his anemometer case and trudged to the big dune east of camp.

Returning 15 minutes later, Gus recognized the backside of the portly Octave Chanute, who was standing just inside the tent's entrance. Having arrived by chauffeured horse and buggy a few minutes earlier, Chanute wasted little time in establishing what he considered to be the origins of the new flying machine, to anyone within earshot.

Stopping a few feet short of entering the tent where the others had gathered, Herring listened as the old man appraised his latest effort. "Tsk, tsk," Chanute clucked. "Just as I thought! The new apparatus is nothing more than a faithful copy of my *Double Decker!*"

Sensing a conflict brewing, the reporters waited until Chanute had retreated to pitch his pup tent before seeking Herring's opinion about the old man's curt appraisal of his work. Eagerly taking the bait, Herring explained the differences between the new glider and last year's machine.[3] "The new effort is a refinement of my design series that has been fermenting for the past five years – long before my acquaintance with Mr. Chanute. Although the differences in the present machine are subtle, they have resulted in significant performance gains. As to his claim to have invented the two-surface machine, I have but this to say... there are those who *take* credit and those who *deserve* it."

As the reporters scribbled in their notepads, Herring continued. "Gentlemen, I must remind you that these trials are being sponsored by my friend and colleague, Mr. Matthias Arnot of

Elmira, New York. That said, please be aware that Mr. Arnot has *invited* Mr. Chanute here to *observe* and to share our facilities... free of charge. In return, Chanute has promised to act as Mr. Arnot's *personal* photographer."[4,5,6]

Soon afterward, sensing Herring's annoyance, Chanute began a campaign of mild praise for his former assistant... a tactic he often used to bolster his own stature as the kindly old Father of Aviation. "Because my former assistant's stabilization device has proven to be somewhat reliable," Chanute said, "his machine tends to remain aloft longer, especially in strong winds... thus allowing me time to obtain satisfactory photographs."

As Chanute shuffled toward the beach with his Thornton Pickard folding half-plate camera and tripod, Herring handed him a package of dry glass-plate negatives and other photographic supplies that were paper-wrapped and cushioned within a black rucksack. Still simmering over Chanute's assertions, Herring marched back to camp thinking... *he has cheek! How dare he spread his self-promotional rubbish to the reporters! When he returns, Matthias shall be informed of this latest absurdity.*

*

Friday morning dawned sunny, but breezy and cool. As the men sat shivering around the campfire, they ravenously consumed their hot breakfasts of Canadian bacon, eggs, and steaming coffee. Shortly, the unmistakable clatter of an approaching buggy drew everyone's attention. Seconds later, Matthias Arnot pulled up, hitched the stallion to a lonely scrub bush, and walked into camp. He spotted Gus and acknowledged his compatriot's verbal welcome with a wave of his hand. Strolling over to him, Arnot reached into his front trouser pocket and pulled out a small object. "Augustus! Here, take your compass! That's the last time I'll set foot in these damn swamps! I got chased up a tree by... a huge black hog! Thought I was a goner! My obituary flashed before my eyes, 'New York banker eaten by wild pig!' "

His complaint was met by a cascade of laughter.

"How did you get away?" Herring asked, wiping a tear from his cheek.

"It was no laughing matter, Gus! I distracted the beast by throwing him my Limburger sandwich. While he was snuffling it down, I slid down the back side of the tree and ran like hell!"

Another round of raucous laughter pierced the morning stillness.

"I'm glad you survived your ordeal, my friend," Herring said, "because I have good news! The machine has proven to be a marvelous performer. I've had several flights of 500 feet or more since the hop you witnessed several days ago. Although the wind has been blustery, today promises to be excellent for putting in long flights. Sit down and have breakfast; there is much to discuss before we head out."

At exactly 9:30, Herring launched himself into a 38-mile-an-hour wind from the northwest. Using all of his acquired skill, he stayed aloft for 61 seconds, while traveling a distance estimated to be about 600 feet. With such long flights, the turnaround time lengthened to almost an hour. By the time Herring was ready for a second try, the anemometer showed that the wind had dropped off appreciably... to less than 20 miles an hour. After a brief discussion, it was decided to move the glider down to the 30-foot level, and Matthias replaced Gus as the operator. For the remainder of the morning, Arnot practiced making relatively short flights lasting between eight and 12 seconds.

Chanute was delighted with the shorter flights because they allowed him to snap additional pictures. From Herring's perspective, he was glad to have the old man isolated from the reporters.

Back in camp for lunch, Chanute tapped his fingers on the covering of the glider's lower lifting surface. "Augustus, how have you managed to make the cotton covering as tight as a drum? You weren't able to accomplish that with last year's *Double Decker*."

Herring rubbed his chin and thought, *Here we go again! The old man won't give credit where credit is due. He'll only report the details as something that he developed... I'm going to make up a whopper of a story.*

"It's a new process that I've developed," Herring said. "The first step is to boil the nainsook covering in water for five minutes. Then it's wrung out and stretched loosely over the wing's framework and nailed into place with shoemaker's tacks.

"As the cotton dries, it shrinks and eventually becomes taut... as you see here. While the shrinkage is taking place, I apply a penetrating coat of shellac to wherever the covering touches the frame; this glues the nainsook to the wood. The next step is very important: when the fabric becomes wrinkle-free, a coat of specially prepared pyroxiline is quickly applied to the covering to stop the shrinking process—"

Chanute stopped scribbling long enough to interrupt. "What do you add to the pyroxiline, and how much?"

"I was getting to that. I add a thimbleful of *linseed oil* to each pint – but only for the first coat. After allowing at least two hours for the sealer to dry, two or three additional coats of full-strength pyroxiline are applied to complete the job."

"Do you cover the wing when it's flat or after it has been bowed to the aerocurve shape?"

"We cover and tack the wing when it's flat, then we bow the ribs to their final configuration after the coat of linseed oil-modified pyroxiline has been brushed on."

While Chanute continued to scribble, Herring glanced over at a very confused Henry Clarke... and winked.

*

After the suppertime dishes had been washed and stowed away, Arnot summoned Herring to walk with him along the lakeshore. When Matthias was certain he could not be overheard, he began to speak. "Augustus, I'm convinced that your theories about manned, powered flight will soon come to pass! I have been pondering how I might expedite this momentous event. Consider the following..."

Arnot offered to pay Herring $300 a month, over the next 10 months, to build and prepare a manned, powered aeroplane for flight trials. He had but one stipulation: after Gus, he would be next in line to fly the machine. For security purposes, the Elmira banker promised to take out a $35,000 life insurance policy, with Herring named as beneficiary. The proceeds were to be used for the further development of the powered aeroplane should anything happen to Arnot. After a short discussion, the two shook hands.

Matthias had an afterthought, "Gus, you must keep these arrangements secret... tell only Lillian! There are certain members of my immediate family who would cause great trouble if they learned about my 'crazy' ideas."

With the Arnot deal sealed, Herring allowed himself a moment to dream that manned, powered flight might still be his honor to achieve. As the two strolled back to camp, Herring brought Arnot up to date on Chanute's self-serving tomfoolery.

"We should throw him out of camp," Arnot said, his face reddening.

"Hold on, Matthias. Since he has been taking photographs with supplies that you purchased... the negatives are yours. We can use good images of the machine."

Gus paused before continuing. "One more thing... under no circumstance must we divulge our partnership to Chanute. He will only use us as gossip fodder, and speculate as to our intentions... it's his *modus operandi*. For now, the only thing he needs to know is that you purchased a gliding machine for your personal experimentation, and I am assisting you in learning to fly it."

As the two men walked into camp, they were met by Carl Dienstbach and Andrew Vassallo, who had just arrived. "Augustus," Dienstbach said, "sorry for getting here so late! I was on assignment and couldn't get away until yesterday morning."

"No matter, Carl, there's plenty of flying to be done until we break camp on Sunday! Andrew, I'm glad that you could come also! Have you been introduced to Carl?"

"We rode in together on the same train! We've been swapping 'A.M. Herring stories' ever since I boarded in Buffalo. Hope the flying weather is good tomorrow... Henry Clarke has been telling us tall tales about some very long flights!"

After introducing Carl and Andy to Matthias, the foursome joined the others around the campfire for a lively, if not entirely candid, conversation.

*

It happened at 2 o'clock in the morning. When the first gusts were accompanied by the rhythmic tapping of raindrops, Herring pulled his trousers on and darted to the large marquee tent. As he ran, he shouted to the others for help. By the time Gus had unlaced the tent's entrance flap and penetrated the flailing beast, Arnot, Clarke, Vassallo, and Dienstbach were hot on his heels. Once inside, Herring directed the others to man the corner posts while he struggled to restrain the wildly oscillating center pole. Twenty minutes later, the blustery wind, estimated to have blown in excess of 60 miles an hour, had subsided. The threat ended as suddenly as it had begun.

At breakfast, Chanute and the Chicago reporters claimed to have slept soundly through the storm. When rain and gusty winds continued throughout Saturday morning, it appeared that no further flying would take place before their scheduled departure on Sunday morning.

In the meantime, anxious to confirm the quality of his photography, Chanute used his rented buggy to trot back to Miller, where he would "... have a few of my glass plate negatives developed." By 3 o'clock, the old man had returned with a wagon laden with supplies... and some bad news. "The chemically prepared dry glass plates that Mr. Arnot supplied to me were apparently defective; the photographic images were all severely underexposed and, therefore, rendered quite useless."

Outwardly aggrieved by the turn of events, Chanute explained why he had purchased the provisions. "Gentlemen, I promised you good photographs, and if I have anything to say, we shall have them! I am prepared to *bankroll* another full week of flight trials, including the provisions you see here, plus any salaries, rentals, or other expenditures that may crop up!"

Although Chanute's proposal seemed fishy to Gus and Matthias, their small assemblage of enthusiasts, especially Vassallo and Dienstbach, tentatively agreed to accept the offer... subject to Herring's ability to extend his time off from the boat works.

Matthias, along with Gus and the two Chicago reporters commandeered Chanute's wagon and returned to Miller. While Arnot waited for his train back to Elmira, Herring dispatched a telegram to Truscott, and the newspapermen relayed their stories to their editors by telephone.

As the tension mounted, Matthias suddenly began to experience one of his chronic headaches. With the next eastbound train not due for more than an hour, and Herring waiting for his reply, Matthias persuaded Gus to accompany him to Miller's lone pharmacy. As they approached the old Miller Avenue storefront, Herring noticed a small sign that announced: Photographic Supplies & Developing.

Leaving Arnot to consult with the pharmacist, Herring walked toward the rear of the store where an employee stood casually behind a plate-glass display cabinet that contained a few inexpensive box cameras and assorted photographic supplies.

"The sign out front says that you do photographic developing. Does that include dry-plate glass negatives and silver chloride paper prints?"

The clerk, doubling as the store's photographic expert, seemed startled by the stranger's more than cursory knowledge of development techniques. "Yes sir, we perform all types of developing and printing right here in the store. As a matter of fact... I'm printing from glass-plate negatives as we speak."

Turning, the clerk – a thin balding man in his early 40s – stepped behind an opaque green curtain that blocked visibility to a back room. When he returned several seconds later, he carried a half-plate-size photographic print that he held by one corner with tweezers.

"Here's an example of my work, mister," the man said with a dash of pride.

Still dripping from its water rinse, the image caused Herring to look more closely. Noting his customer's curiosity, the clerk offered an unsolicited opinion. "Yes... there are people who say it's a flying machine. For the past two summers, men who claim to be working on those things have flocked here... like barn swallows to bugs! Matter of fact, some of them are down at the dunes right now!"

Recovering from his initial shock at seeing an image of his glider, Herring was about to speak when Arnot joined him at the display case.

"What's going on, Augustus?"

Pointing at the still-dripping print, Herring spoke with degree of skepticism in his voice. "This gentleman claims this strange-lookin' contraption is a flyin' machine—"

"It's true," the clerk blurted out. "If ya don't believe me, come on back and I'll show ya some more pictures."

The pharmacy's back room was typical for a photo lab of the 1890s. Most noticeable was the absence of ambient light… all the cracks had been sealed with black "sugar paper," tacked in place. *A meticulous, time-consuming job,* thought Herring, as he glanced around the cave-like space. One wooden bench contained trays filled with liquid chemicals; another was crowded with delicate-looking white-pine frames of various sizes. These held the specially prepared printing paper and the glass plate negatives before they were exposed to a source of light, a process known as the *contact print* method. Across the room's low ceiling, rows of lightweight cotton cord spanned the open spaces. It was from those cords that photographs of the flying machine hung from miniature spring clips.[7]

Careful not to touch the drying prints, Herring and Arnot made the best of the lab's low-intensity red-light gas lantern to squint at the more than 50 images of their two-surface machine. "I notice that you don't have electric lights yet," Herring observed. "How do you expose the negatives and prints – by lantern?"

"I use natural light to do my exposures, mister. That's what these wooden frames are for – they hold the glass-plate negative against the photographic paper. I cover the frame with a black cloth, step outside, point the frame at the sun, and flip the cloth back. I use a stopwatch to get an accurate exposure time. The only problem I have is gettin' the proper exposure on cloudy days."

"Or at night," Arnot wisecracked.

Thanking the man for the informative tour of his lab, the men left the pharmacy and headed back to the Western Union terminal. Neither spoke of Chanute's deception.

"Nothing yet," Herring said as he marched out of the telegraph office.

Arnot was waiting on the oak and iron bench in front of the office. "My train won't get here for almost an hour," he said. "That's good... we have things to discuss! Augustus... what the hell is going on?"

Sliding onto the opposite end of the bench, Herring offered nothing more than a reserved smile before speaking. "The old man needed an *excuse* to get us to stay here another week. A botched set of glass plate negatives was the best he could come up with. As a bonus, he thinks he'll be getting a pristine set of glass plate negatives—"

"He could have had all the prints he wanted," Arnot interrupted. "Why is it important for us to stay here an additional week?"

Herring threw his hands up in a "who knows" gesture. "What have I been telling you for the better part of a month? The only thing that matters to the old man is his legacy! He's *buying* his way into *our* trials and *your* two-surface machine."

Arnot clenched his fists. "Let me get this straight. Because Chanute has offered to bankroll a second week at the dunes, he'll claim that the expedition was all his doing?"

"Why else would he *lie* about the negatives?"

Simmering, Arnot searched for another motive. "Perhaps he is trying to learn why this *Double Decker* design is so much better than last year's. Perhaps he feels the need for more time to study the machine."

Herring grimaced. "I doubt it... he's had days to study it. No, there's a simpler explanation. Chanute needs to be *officially* tied to these trials, and he's doing it the only way he knows how – with deception, and his money."

Arnot had heard enough. "Let's go back to camp and have it out with the old bastard! Then we'll drive him back to the pharmacy to retrieve our negatives and prints. I don't need his handouts, the lying son-of-a—"

Herring grasped his friend's shoulder. "Matthias, I'm afraid it's too late... he's already won."

"How in hell's name can that be?"

"The Chicago newspapers will report that the city's kindly old pioneer of aerial navigation has once again shown generous support for another man's flying-machine trials. They'll say that Chanute's former assistant... *me*, along with an enthusiastic newcomer... *you*, were rescued from certain failure by the much-celebrated 'Father of Aviation.'

"Furthermore, the Chicago reporters will portray Chanute as a person who selflessly used his own money for additional provisions and even offered to pay the salaries of assistants, etc., etc. That poppycock will all find its way into print... if it hasn't already."

Accepting Herring's logic, Arnot buried his face in his hands. "Augustus, how can one person be so calculating?"

For a few moments the men sat in silence, each absorbed in his own thoughts.

"It takes practice, Matthias; he's had much practice. We may not stop Chanute from claiming responsibility for our expedition, but we might be able to sway the journalists' opinions, and at least stop the old fart from latching onto our negatives and photographs!

"I have an idea," Gus continued. "While the reporters are still here, why don't we *announce* our goal for achieving manned, powered flight within the next year? Chanute's sputtering and stammering will be entertaining; the old man will be beside himself with anger and resentment!"

Arnot's eyes began to sparkle. "Good, let's do it! That should make for an interesting second week on the dunes! But, aren't you worried about what Chanute will say and write about you?"

Herring's reply came immediately. "Hell, *no*! He'll just write negative crap about me anyway! My only concern is for my domestic patent application from last year. Since Chanute is paying for the majority of its prosecution, I wouldn't be surprised if he instructed Whittlesey, his attorney, to abandon the effort... an effort the Patent Office and I have been struggling with for the past nine months! The patent situation, both domestic and foreign, is the only leverage that the old man holds over me—"

Arnot looked Herring straight in the eye. "Say no more! When Chanute breaks the agreement ... I'll assume the cost of obtaining your patents!"

Herring stood speechless. "Matthias, you don't have to—"

"Gus," Arnot interrupted, "if we're going to be successful, we can't have manipulative old men interrupting our work! Keep me informed about the patent problem."

Herring clapped his hands together, signaling a change of subject. "Make no mistake, Matthias; if our trials had ended today, the old man would still manipulate the facts. We may as well take advantage of the additional time and have some fun at his expense. Here's a ploy I've been thinking about ever since Chanute arrived in camp: since he doesn't speak German, that's the language we'll use when he's around. We'll let Carl in on the ruse... the three of us will drive him mad."

"One thing's for sure," Arnot said. "I'll have no further interaction with the old man... I'll treat him like he doesn't exist... I'll stare right through the son-of-a—"

"Listen," Herring interrupted, "if my request for additional time off is turned down, we'll reject Chanute's offer out of hand, and he can feed all of those provisions to Lake Michigan's fish.

In the meantime, I'll wait here for Truscott's response. When I have their answer, I'll telegraph you."

"Be sure to wire me right away – even if I'm on the train."

Reaching across the bench, Arnot grasped Herring's arm. "Gus, this is important! If you must leave for St. Joseph tomorrow morning, be sure to take possession of our glass plate negatives and prints! Chanute can't be allowed to leave town with them! If there's going to be a second week, I'll be back as soon as possible – probably no later than Wednesday or Thursday. On a happier note, I can't wait to get some more flying time! As for Chanute... he's got a surprise coming!"

CHAPTER 59

Miller, Indiana

Saturday, September 11, 1897

Arnot waved goodbye from the window seat of his Pullman coach as the eastbound train lumbered past the Western Union terminal. Responding with a short salute, Herring leaned back and thought, *Thanks to Matthias' generosity, I no longer need Chanute's money to pursue a domestic patent. Truth is... now that I'm out from under the old man's thumb, there's nothing to keep me from pursuing dynamic flight!*

Deciding to wait in the comfort of the Miller Hotel, Herring handed a note to the telegrapher, informing him where he could be found if and when his telegram finally arrived. Five minutes later, he stood on the veranda of the hotel's restaurant, slapping road dust from his trousers with his captain's hat. Satisfied that he looked presentable, Gus pushed open the swinging door, revealing a small lobby and reception counter to the left, and the entrance to the restaurant to the right. Saturday night was always busy in Miller, and this evening proved no exception. As he glanced around the large room for an unoccupied table, there came a shout.

"Gus! Gus Herring!"

Standing, Macbeth, the reporter from the *Herald*, waved to gain his attention. As Herring navigated his way through the maze of tables, he noticed that Manley of the *Record* was with him.

"Over here, old man! We're just about ready to order!"

*

Following Gus and Matthias's departure from camp, those who had remained enjoyed a leisurely afternoon. Chanute spent his time sketching various components of the *Double Decker* glider while conducting a running interrogation of Henry Clarke about the machine's construction. The ever-curious Andy Vassallo crawled over and under the glider's framework, inspecting fastening methods and joinery techniques that would only interest an engineer. Herr Dienstbach, the sole remaining reporter onsite, used his notes to flesh-out an upcoming story about the flight trials for the New York *Times*.

When Octave Chanute first rolled into camp on Thursday morning, he quickly discovered that the expedition had relied on dry cereal, canned tomato soup, and coffee for their first two meals of the day. An occasional catch of smallmouth bass and perch from the lake provided most of their protein during the first week of their stay.[1]

Chanute hauled in more provisions for their unscheduled second week at the site. Included was a battered hardwood box, filled with 100 pounds of block ice. Cut from the lake the previous winter, the ice nestled within a galvanized steel container, separated from the outer box by two inches of ground cork insulation. The tightly sealed and insulated door helped to keep the small food locker – filled with slabs of meat, a crock of butter, and bottles of milk – cold for up to four days, provided the box was kept in the shade.

As late afternoon passed into evening, the men prepared a grand supper of corned beef and cabbage, boiled potatoes, pumpernickel bread, and hot coffee. "Augustus will kick himself when he learns what he missed for supper," Chanute said as he slathered butter on another slice of the dark bread. "You know how much he loves his milk!"

The sun had already begun to set before the men finished their chores and settled around the campfire. Soon, Herring's request for an additional week of vacation time became the dominant topic of conversation.

"Have no fear," a confidant Chanute said, "Augustus will be here all of next week."

"I wouldn't be so sure of that," Clarke replied. "Gus needs that engineering job. Since his father's inheritance money ran out earlier this year, he and his family have been living mostly on their savings. If the boat works says 'Git back here,' he'll be gone tomorrow morning... that's my bet."

"No offense, Mr. Clarke," Chanute said with a chuckle, "but money has never stopped Augustus from making ill-advised decisions. Remember, he elected to quit as Professor Langley's assistant. That was an excellent-paying position, and he left in a huff, just before the program achieved its goal! Come to think of it, I have a letter from the professor in my files... accusing Herring of being 'headstrong, impulsive and impossible to manage.'"

Hearing no immediate rebuttal, Chanute continued. "Based upon my own experience, I might add that Mr. Herring also exaggerates, claims more credit than he deserves, and has a decidedly jealous disposition."

A tense silence settled over the other three men, interrupted only by the crackling of the campfire. Staring intently into the flames, Vassallo poked his stick at a smoldering, half-burnt log. Sighing, he leaned back and stared into the old man's faded blue eyes. "I'm not a member of the flying-machine movement, but I've known Gus Herring for more than a decade, and I've worked for him off and on for much of that time. He may be headstrong, impulsive, and impossible to manage, but I've never known him to exaggerate, claim more credit than he has a right to, or exhibit jealousy toward anyone. No offense intended, Mr. Chanute, but I don't believe *any* of that crap."

Chanute shrugged and threw up his hands. "It's unfortunate, Mr. Vassallo... but the truth hurts! You weren't there for the trials of '96; otherwise, you could have witnessed for your—"

"I was there," Dienstbach interrupted, "and I have no idea what you're talking about! In fact, I defy you to give one example of Augustus taking more credit than he deserved."

Nodding, Chanute seemed to welcome the challenge of identifying his former assistant's shortcomings. "The current patent applications, both foreign and domestic are good examples. Herring claims that he invented the two-surface wing configuration... an idea clearly anticipated by Francis Wenham of Great Britain. He also claims to be the first to have used his so-called aero*curve* – *that* was foreseen by Pénaud of France, among others—"

"Hold on!" Dienstbach broke in. "I read both of Herring's patent applications. He claimed the *combination* of multiple lifting surfaces and the aero*curve*... Wenham used flat planes on his two-surface model. Pénaud used his aerocurve on a single-surface model. William Henson of Great Britain also used flat planes on his multi-surface model. As I'm sure you are aware... these are big differences when it comes to *claims* on a patent application!"

At this point, Henry Clarke had lost his patience. "Since you're in a condemning mood, Mr. Chanute, why don't you provide us with an example of Herring's jealous disposition? In all my time spent with Gus, I can't recall a single instance—"

"It's clear by your tone of voice, sir, that you have already made up your mind regarding my comments. Nevertheless, I will be happy to accommodate your request. Mr. Herring was so *intensely* jealous of William Paul's *Albatross* soaring machine at last year's trials that he steadfastly refused to acknowledge its existence! You can say what you want, Mr. Clarke, but his jealous disposition was also noticed by individuals other than myself – such as the two Chicago reporters that are also covering these trials—"

Snapping the thread of Chanute's monolog, Dienstbach's voice began to quaver as he spoke. "As you are well aware, I observed first hand the August and September trials of '96. In my opinion, Mr. Herring displayed absolutely no jealousy. At worst, you might say that he was concerned for Butusov's well being. When you failed to heed Gus's warning about the *Albatross's* airworthiness, he decided to leave camp rather than be associated with a serious, if not fatal, mishap.

"Mr. Chanute, as a journalist, I question *your* motives. Why are you supporting Butusov, a confirmed *murderer and deserter* from the Russian Navy... even after the facts were made known to you by the U.S. State Department?"

Having waited patiently while Dienstbach rebutted his analysis of Herring's character and his support of Butusov, which he conveniently ignored, Chanute took the opportunity to attack the German's journalistic integrity.

"Herr Dienstbach, as a professional international journalist, you of all people should recognize *deception* when you see it! I believe you have been duped, my good man! As proof of my contention, a friend sent me an article that you wrote for the *Times* last year... just after my second excursion to the dunes had concluded. I must say, Carl, your writing demonstrates quite an infatuation for our friend Herring! If memory serves me, you referred to him as 'the second coming of Lilienthal.' Certainly, you must see the utter exaggeration in that proclamation, can't you, sir?"

Dienstbach's eyebrows arched as he considered Chanute's challenge. "Exaggeration? I think not! For example, Mr. Herring has accumulated more flying time in heavier-than-air machines than any other practitioner in the field – with the *possible* exception of the late Herr Lilienthal. Otto told me shortly before his death that he considered Herring to be his greatest student, someone who would someday exceed all of his personal achievements."

"Why are you so intent on demeaning Augustus?" Clarke added. "Do you actually believe that we will change our opinions? If so, you're pissing into the wind!"

"Demeaning? Oh no, that is not my intent, Mr. Clarke! I am merely identifying the deficiencies that have prevented Mr. Herring from becoming a *leader* in the field of aeronautics... God knows, he has a talent for chauffeuring flying machines about the sky for a few odd seconds."

Chanute pointed a stubby index finger at Clarke. "As we speak, another of Augustus' failings comes to mind. The record shows that Herring has already made enemies of several *important* people within the field of aerial navigation – and he's still a relatively young man! Of all his faults, perhaps the most egregious is his disrespect for the pioneers... a conspicuous reluctance to recognize the accomplishments of others – myself included."

A lengthy period of silence followed, the men retreating into their innermost thoughts. Finally, Dienstbach spoke up. "Your motives interest me, Mr. Chanute. Photography blunders and a smoldering hatred for Mr. Herring aside... why have you offered to sponsor a second week of his trials?"

"Motives? I have no motives, sir! I am simply making good on a promise to provide *usable* photographs. It's unfortunate that Mr. Arnot provided me with defective glass plates; nonetheless, I am committed to holding up my end of the bargain! Although a week's sponsorship represents a degree of financial burden, I would have done the same for any aeronautical upstart. As my beloved wife is fond of saying, 'Octave, your love of flying machines always gets the better of you.' In addition I might add, '... no matter how disagreeable the personalities.' "

*

Macbeth and Manley were still teasing Herring about his menu selection, as he finished his second glass of milk.

"Gus, I can't believe that you actually ordered *Surströmming*," Macbeth said, holding his nose.

"Why is that?" Herring mused, wiping his upper lip with his cotton napkin. "As I mentioned earlier, I spent several years in Scandinavia during my youth. That's where I acquired a taste for fermented fish... herring in particular. Besides, the red cabbage really does help to mask its pungent aroma."

"The Danish meatballs were wonderful," Manley said, nodding, "but I had to avoid breathing through my nose the entire meal!"

As the three men shared a laugh, the high-pitched voice of the Western Union boy echoed throughout the room. "Telegram for Mr. Herring!"

Tossing the boy a penny tip, Herring ripped open the envelope and read the message to himself. Shoving the transcript into his trouser pocket, he made his much-anticipated announcement. "Well boys, looks like I'll be staying another week!"

After paying the bill – a total of $4, including the tip... the men made their plans. They would stay overnight at the hotel, meet for breakfast in the morning, and share the cost of a chauffeured buggy ride back to camp. Before retiring for the evening, they all walked over to the Western Union building to send additional telegrams: the reporters to their respective newspapers, Gus to Lillian, Arnot, and Captain Werwage, owner of the *Ares*.

*

Early Sunday morning, Herring and the two Chicago reporters arrived back in camp to find the others dutifully performing their post-breakfast chores. "Good news, gentlemen!" Herring shouted. "Everything is set for an additional five days – I'll be able to stay through next Friday. Saturday morning we'll break camp and head for home."[2]

Since Chanute was paying for the second week of flight trials, he reestablished the "no work or flying on Sunday" decree from his earlier trials. With improving weather, Herring whiled away the day making minor repairs and adjustments to the glider's tail regulator – a direct violation of the old man's wishes.

Chanute, who knew better than to challenge his former assistant, spent the day fiddling with his camera and exploring the surrounding waterfront territory. With the old man away chasing butterflies, Dienstbach and Herring carried on a spirited discussion in German that touched upon everything from European developments in heavier-than-air research to Chanute's scathing condemnation of Gus the previous evening.

After supper, troubled by what his German reporter friend had told him, Herring invited Dienstbach, Clarke, and Vassallo to walk with him along the beach. When they were far enough away from camp not to be overheard, Gus revealed what he and Matthias had discovered concerning Chanute's "defective glass plates" and what they thought was the old man's ruse. After a long conversation, the men agreed not to let on what they knew about Chanute's chicanery. Instead, the group would bide its time until Arnot's return later in the week.

Later, when Chanute and the Chicago reporters had gathered around the campfire, Herring's men attended but avoided any discussion of flying machines – a respectful but decidedly anti-Chanute policy that they would employ for the remainder of their stay at the dunes. The lack of "aeroplane talk" soon became obvious to Macbeth and Manley, who probed for reasons behind the information lockout – all to no avail.

After supper each evening, Herring's group either took long walks or gathered privately in the big tent to discuss the day's activities. If Chanute didn't get the message presently, he surely would by week's end.

*

Favored by mild Midwest fall weather, Herring made more than 100 flights over the next three days. With winds between 15 and 30 miles an hour from directly off the lake, he averaged distances in excess of 350 feet, with the best approaching 500. All during this period, Chanute took his photographs, allegedly trying to duplicate the images of the previous week. Herring's time-aloft numbers averaged 20 seconds, with a 37-second effort being the longest.

Because the wind's velocity was too low to produce distance or duration records, Herring was content with making preplanned controlled flights, where he practiced transitioning from gliding directly into the wind to a quartering slide along the dune's face... and back again. In several instances, toward the end of particularly satisfying glides, he succeeded in swinging the machine completely around... to alight safely while roaming downwind.

On Wednesday, Herring was again able to coax Clarke into trying his hand at operating the two-surface machine. After a bit of additional tutoring, Henry took his running start from a point 20 feet above the beach. Once airborne, Dienstbach shouted, "*Brechen Sie den Hals und ein Bein,* Henry!" (Break your neck and a leg – a traditional German good luck blessing).

To everyone's surprise, Clarke flew well... exceptionally well! Inexplicably, he produced a beautiful floating glide that lasted 20 seconds and stretched for almost 200 feet – representing an advance of 10 feet for every foot of drop.

"That was invigorating!" Clarke exclaimed, as he and Andrew hauled the glider back up the dune. "I'm gonna build me one of these when I get back to Philadelphia!"

"I hate to burst your bubble, Henry," Herring said, "but there was a persistent upward trend in the wind during your glide. Under normal circumstances you would have flown for about 140 feet. Regardless, that had to be the longest second-time flight in the history of aerial navigation!"

After Clarke had accepted congratulations from all present, Herring had a question. "Henry, do you want to roll the dice again… or are you going to rest on your laurels?"

"I'll pass," Clarke said, laughing.

*

On Matthias Arnot's return to the dunes, his westbound Pullman from Elmira carried a man from Boston, whose destination was also Miller, Indiana. Early Thursday morning, Arnot and James Means, publisher of the prestigious *Aeronautical Annual*, stepped off the train together. Since they were the only two trudging east on Miller Avenue, the men struck up a casual conversation.

"I'm here to visit with an old friend," Means said. "He invited me here to *watch* a flying machine—"

"What a coincidence!" Arnot interrupted. "I'm heading to the dunes to *fly* that machine!"

After mutual introductions, the twosome agreed to share the cost of renting a chauffeured horse and buggy. Once underway, Arnot's curiosity got the better of him. "Tell me, Mr. Means, how did you learn of our extended stay here at the dunes?"

Means preceded his answer with a dismissive wave of his hand. "Octave telephoned me at my home. He said he was returning to camp with provisions and photographic supplies for the coming week. He promised that I would be able to see the flight trials… so here I am!"

While Arnot silently seethed, his thoughts turned to Chanute's promise. *What if Gus couldn't stay the additional week?*

Means would have found himself out in the wilderness... alone. Augustus is right... the old man is one arrogant son of a bitch!

It was almost noon when the buggy at last rolled into camp. Retrieving their luggage, the pair quickly slogged through ankle-deep sand to their destination: the big tan tent that sat between a group of low-lying dunes. The travelers' timing turned out to be impeccable; lunch was about to be served.

Means had arrived at an opportune time for another reason: he was about to witness Herring's record-setting 900-foot quartering glide into a 38-mile-an-hour wind... his longest foray into the air thus far. As Gus guided the *Double Decker* into a quartering maneuver, causing the glider to fly parallel to the line of dunes facing the lake, Arnot shouted, "*Fliegen güt* (Fly well) Gus!"

Not understanding Arnot's communiqué, Means was nonetheless beside himself with excitement. "My God... I can't believe my eyes! How does the machine maintain its equilibrium in this unholy wind?"

Leaving Arnot behind, the 44-year-old Means trudged through the sand in pursuit of the machine and its capable operator, wanting nothing more than to help carry the apparatus back to its starting point. Later, when alone with Chanute, Means encouraged his old friend to take matters into his own hands.

"Octave, it's time to act! You must add *power* to one of your own machines and win this race to fame and glory!"

Nodding, Chanute was thinking of ways to stall for time. "James, I will consider your proposal once the flight trials have concluded and I have had time to think the matter through. I will write you before the end of this month."

*

With sunny skies and temperatures in the 60s, Friday proved to be another great day for flying. With Herring's assistance, Arnot made 15 straight-ahead flights into winds gusting between 15 and 20 miles-an-hour. Gus pronounced his progress to be "most satisfactory" as he helped retrieve the glider after its last flight of the day. Standing alone with Arnot out on the beach, Herring related the story of Chanute's recent campfire tirade.

"Tomorrow," Matthias vowed, "the old man will be exposed!"

*

Saturday morning found Herring, Clarke, and Vassallo up before dawn, breaking down the big tent and dismantling the two-surface machine. By 8 o'clock, they had finished their breakfasts and were filling the crates for the return trip to St. Joseph.[3]

Meanwhile, Chanute had carefully packed his camera equipment and the new set of exposed glass plate negatives into protective boxes, preparing to load them into the two-horse wagon for the return trip to Miller. Walking over to where Arnot was packing tools, Chanute clarified his intentions. "Matthias, I will take these exposed glass plate negatives back to Chicago with me, where I'll have them developed and prints made. When they're ready, I'll mail everything to you. However, I must admit that I am reluctant to send these fragile plates through the mail. Perhaps we could arrange for them to be retrieved by you or your agent sometime in the near future?"

"Perhaps," said Arnot in a matter-of-fact manner. "For the time being, send me two sets of prints."

*

At precisely 10 o'clock, Captain Bill Werwage had beached the *Ares* to within 50 feet of the crates, tents, and other equipment. Herring and Clarke would load the scow and accompany the cargo back to St. Joseph. As if choreographed, a two-horse wagon appeared in the distance. Chanute had rented the wagon the previous Saturday to haul Means and himself, along with their luggage, back to Miller. As it turned out, there was also room for Arnot, Vassallo, and Dienstbach. The two Chicago reporters, Macbeth and Manley, had departed earlier by horse and buggy. By 10:30 a.m., the campsite was back to its natural, desolate state. The little group of aerial navigation enthusiasts had scattered like leaves to the wind.

*

When the wagon arrived at the Miller train station and the luggage had been unloaded, the men hurriedly checked the train schedules. Headed west to his Chicago home, Chanute would have to wait but an hour. The Easterners would have to wait an additional hour before their train arrived.

With little time to spare, Chanute began his trek to the pharmacy, a quarter mile to the east. To the old man's dismay, Means trotted up alongside to further promote his cause: to challenge the upstart Herring and solve the problem of manned, dynamic flight. "From my perspective, Octave, all that you need do is to borrow one of Langley's little steam engines – I can assist you with that – and mount it to your refurbished '96 two-surface machine. Your boy, Avery, will be able to handle that job lickety-split."

For reasons known only to him, Means failed to acknowledge Chanute's *opposition* to the pursuit of dynamic flight prior to solving the overriding dilemma of automatic equilibrium – the reason the old man became interested in flight experiments to begin with. Chanute resisted the urge to reiterate his long-standing position, but instead thought... *either Means didn't read my articles published in* his *Annual, or he must have a very short memory!*[4]

Ever the politician, Chanute recognized that Means was an influential player in the field of aeronautics, and his opinions had to be considered. Most important, his publisher friend represented an ally in the preservation and promotion of his status as the "Father of Aviation." Later, Octave would write to his friend after he had had time to cool his heels; besides, there was a more immediate problem... how to get rid of Means before he reached the pharmacy?

"James, do me a favor," Chanute said, handing Means four 25-cent pieces. "Since I'm running short on time, I won't be able to tip the livery stable supervisor for his excellent service. While I'm at the pharmacy, will you take him this token of my appreciation? I'll meet you back at the train station."

As Means trotted off, the old man continued the remaining 50 yards in welcome silence. When he passed through the pharmacy's front door, Chanute was relieved to see that his photographic technician was working this day. "Edward, my friend, it's good to see you again! I'll be heading back to Chicago within the hour. Do you have my goods ready? By the way, I will be taking another set of undeveloped glass-plate negatives back to Chicago for developing and printing. Do you have something I can wrap them in, to protect against breakage?"

Just then the bell on the front door jangled as someone entered. Glancing over his shoulder, Chanute was startled to see Arnot striding toward him.

"Matthias," Chanute said with a nod and a broad smile, "what brings you here?"

"I'm here to pick-up my photographic plates and prints... but don't rush, Mr. Chanute, I'll wait until after you have paid for their processing."

Flabbergasted by the sudden turn of events, and realizing that he had been caught in a lie, Chanute quickly turned and paid the $6 bill. "Thank you, Edward, I hope that the *flower and fauna* images turned out well. Madame Chanute will be gratified to see—"

"Oh no, Mr. Chanute, there aren't any photographs of flowers. These are all pictures of flyin' machines. I hope that you're happy with the quality – I think they're exceptional!"

Realizing that the jig was up, Chanute picked up the package and hurried out into the street, with Arnot in hot pursuit. Seizing the old man by the shoulder, Arnot spun him around for a face-to-face confrontation. "Unhand me, sir!" Chanute sputtered. "How dare you accost me in this manner!"

"Save your breath, old man! You've got some explaining to do! Why did you lie about these photographs not turning out?"

After some unintelligible stammering, Chanute waved his free arm above his head. "There had to be a reason for another week of trials. Otherwise you might not have accepted my offer!"

"Why was it so important to have the second week here at the dunes? We accomplished everything that we set out to do during the first week."

Hesitating, Chanute handed the bundle of glass plate negatives and prints to Arnot. "I have my reasons. Let's just say that I've always been an avid supporter of aerial navigation."

As Chanute turned to walk away, Arnot stepped in front of him, blocking the way. "Not so fast!" he growled. "You're not getting off that easily! Why don't you admit it? You thought that you were *buying* your way into our trials! You thought this would make it easier to claim credit for Mr. Herring's new machine! You'll use our trials as another way to feather your nest, and add to your precious legacy!

"I've got news for you Octave, I'll be watching all of the domestic aviation publications... the newspapers, the magazines, the professional journals, everything! If I so much as see one mention of the *Chanute trials of 1897*, I'll drag your fat ass into court, and you *know* that I have the means to back up my promise. How will your precious reputation stand up then?"[5]

Staggered by the indignity of being dressed down by a much younger man of less experience or reputation, Chanute's face turned the color of a harvest-ripe apple. "I've been threatened by better men than you, Arnot! My only response to your insolent behavior is to wish you luck!"

As Chanute tried to nudge Arnot aside, he added a parting salvo. "Before I forget... I found your group's conduct to be despicable this past week. I would compare it to biting the hand that feeds you – the childish snubbing, the use of a foreign language to exclude me from your conversations—"

Arnot grinned broadly. "Perhaps you should take a correspondence course in German... it *is* the language of science and engineering!"

Chanute muttered an obscenity under his breath as he pushed past his adversary.

"One other thing, Octave... not only did you lie about this first set of plates and prints, remember that the second set *also* belongs to me! That was the deal, sir – you were to provide me with the glass plate negatives and prints from the second week!"[6]

As Chanute wobbled unsteadily down Miller Avenue toward the train station, he shouted over his shoulder.

"Tell it to my lawyer!"

CHAPTER 60

Rochester, New York

Friday, October 28, 1921

After James Martin had finished giving his Indiana dunes testimony, Counselor O'Grady requested permission to approach the bench. A moment later, he and Mr. Robbins were staring up at Judge Sawyer.

"Your Honor, I am confident that Mr. Herring will be back in court and ready to testify on Monday morning. Therefore, with the court's permission, the plaintiff wishes to release Mr. Martin today, rather than first thing Monday. Mr. Martin has pressing personal business to attend to and can ill afford waiting until —"

"Mr. Robbins," the judge interrupted, "do you wish to cross-examine the witness? If not, is there any objection to the court releasing this witness today?"

"No objection, Your Honor, provided Mr. Herring *actually* materializes on Monday. My clients have been more than cooperative during the plaintiff's illness and want nothing more than a responsible continuation of these proceedings. We have no questions for this witness."

"Very well then, it's settled."

Rochester, New York

Monday, October 31, 1921

As Judge Sawyer called court to order on the final day of October, all eyes were focused on the man sitting next to Attorney Maloney at the plaintiff's table. Dressed in a conservative dark gray suit, white shirt, and dark green bow tie, Augustus Herring looked pale and a bit thinner than before his emergency hospitalization 10 days earlier.

"Welcome back, Mr. Herring!" the judge said. "The Supreme Court of New York wishes you a healthy and productive return to these chambers, sir! Counsel for the plaintiff has informed me that you are prepared to resume your testimony; is that correct?"

Noticing that his hands were trembling, Herring clenched his fists. "Yes, Your Honor."

"Excellent! Mr. O'Grady, call your next witness."

Shuffling to the witness stand, Herring carefully lowered himself into the chair. Straining to see as he stared into the dimly lit visitors' section, Augustus located his family: the ever-faithful Chloe, his wife Lillian, and son William. To celebrate the occasion of his discharge from the hospital and William's 28th birthday, Chloe had arranged a quiet evening dinner at the Hotel Rochester. Afterward, they would drive the Ford back to the house for coffee and cake.

"As you know, Mr. Herring," O'Grady said, "in your absence, Mr. Martin has been presenting the details of your patent applications and the 1897 excursion to the Indiana dunes. His testimony concluded last Friday with an account of the confrontation between Mr. Arnot and Mr. Chanute. This is where I would like you to pick up the story."

"Very well," Herring said in a thin, strained voice.

"As you remember, you and Mr. Clarke returned to St. Joseph from the Indiana dunes on Saturday, September 18, 1897. What are your recollections of that day?"

Leaning back in his chair, Herring rubbed his chin with a forefinger. "I recall that Mr. Clarke and I stowed the crated two-surface machine in my shop, along with the tools and other equipment used on the excursion. We then employed the horse and wagon to return the tents to the rental depot before heading to the railway station. I paid my friend the $65 I had promised for his assistance and bade him farewell. I returned the horse and wagon to the livery stable, and walked home. After supper I turned in for the evening, tired but grateful that the flight trials had gone well."

"What was next on your agenda?"

"For the next two weeks, I was consumed by my job at the boat works. On Wednesday, September 29, I traveled by train to Chicago on company business. Having completed my tasks well before the 5 o'clock departure back home, I hailed a hansom cab and headed for Avery's shop on East Superior Street – I wanted to discuss Bill's role in fabricating parts for the manned, engine-assisted aeroplane. Halfway there, I changed my mind and directed the driver to head for East Huron Street – Chanute's place.

"Out of curiosity, I wanted to hear the old man's version of his run-in with Arnot. I had heard all about the unpleasant conflict from Matthias, who telephoned me shortly after he arrived back in Elmira."

"Did Chanute offer a rebuttal to Arnot's story?"

"That was the strange part; when I asked him what had happened with the photographs, he said there had been a misunderstanding and that he didn't want to discuss it further... so I dropped the issue altogether. He seemed much more interested in discussing the status of my American patent application—"

"Had you heard from the Patent Office?" O'Grady interrupted.

"I was getting to that. Earlier that day, as I was leaving home for the train station and my trip to Chicago, the postman walked up and handed me a letter... it turned out to be another Patent Office rejection! That was on my mind when I told Chanute that I was going to withdraw the application and would try to safeguard my interests by working in secrecy. Later, I changed my mind and decided to write a letter of inquiry."[1]

"What was your question?"

"I asked Mr. Townsend, the Patent Office examiner, what *demonstration* he would consider to be satisfactory for a power-driven aeroplane with a man aboard?"

"Did you receive a response?"

"Three months later, on Monday, January 4, 1898, I received a *final* rejection notice."

"What was their explanation?"

"I'll have to paraphrase: '... the entire invention rests upon a theory that has never been satisfactorily demonstrated. The application is rejected as a whole on the grounds that it is inoperable... no claims can be allowed.' "[2]

"Objection! Your Honor, this information has already been entered into evidence!"

"Yes it has, Mr. Robbins!" the judge snapped. "We in the profession call it 'corroborating evidence.' Overruled!"

"What did you do next, Mr. Herring?" O'Grady asked.

"I telephoned Mr. Whittlesey, my patent attorney. He stated in no uncertain terms that I had reached the end of the road with this application and there could be no further appeal."

"Did you accept Mr. Whittlesey's word on this?"

"I was skeptical. Whittlesey was Chanute's lawyer and that raised a red flag, especially since the old man had agreed to finance my domestic application. I suspected that Chanute had decided not to waste any more of his money on an appeal."

"Did you discuss the patent application problem with your silent partner, Mr. Arnot?"

"With the lawyers insisting that there was no legal recourse for an appeal, Matthias and I decided to temporarily give up on the patent application. Instead, we vowed to work toward perfecting the powered aeroplane in secrecy. We were certain that pioneer patents would follow our success."

Herring pressed a hand to his forehead, as if in pain. "It's been almost a quarter century since our misguided decision, and there isn't a day that passes that I don't regret it."

O'Grady shuffled some papers to allow the court time to linger on the relevance of Herring's last statement. "What happened next?"

"My abbreviated and somewhat strained visit with Chanute prompted an early departure... allowing me to meet with Bill Avery."

"Tell the court about this meeting."

"I discussed Avery's role in fabricating the wooden components for the new machine. I explained how the apparatus would differ from the '97 glider and the changes that would be required. These included the three marginally larger superposed wings, the sturdier central framework made from steel tubing, and the addition of a lightweight, wheeled structure for rolling the heavier machine from one place to another... and possibly to aid in alighting.

"Finally, we discussed price and a projected delivery date: $300, with the crated components delivered to my St. Joseph shop no later than April 1, 1898."

"Therefore," O'Grady said, "Mr. Avery would have six months to provide you with the milled-to-size components... not the finished machine."

"That's correct. I would have to cut, fit, and assemble the components myself."

"It was the fall of '97; how was work on the engine progressing?"

Herring scowled. "Little had been accomplished since I moved to St. Joseph. I still had to identify machine shops and other companies that would help me to complete the engine."

St. Joseph, Michigan

Saturday, October 2, 1897

Herring was up and out of the house early. Although he had walked the same half-mile route to Truscott's every day, this morning was special; he wasn't going to work, he was headed to the old Engberg Machine Shop, a quarter-mile east of the boatworks property.

As he approached the three-building complex on Lake Street, he was greeted by the pungent aroma of rotting fish wafting through the industrial site. *Another inexplicable fish kill*, Herring thought. *It shouldn't be that hard to figure out when you count all of the pipes and ditches that dump God-only-knows-what into the lake.*

Engberg's concrete-block shop was capped with a shallow wooden gabled roof covered with light brown and olive green asphalt shingles. One hundred feet long and 40 feet wide, the structure contained a row of nine, soot-coated, crank-out windows running down its west side. A slightly cockeyed, three-step concrete entranceway offered a means of reaching the weather-beaten wooden door. Over the door hung a peeling wooden sign decked out with old English-style lettering that announced

Engberg's. The 80-year-old building had served as a buggy works in the early 19th century before being converted to a machine shop by William Engberg's father immediately after the Civil War.

Herring ignored the front entrance. The shop was closed for the weekend, but Engberg and his foreman were inside, working to refurbish some balky machine. In an earlier telephone conversation, Herring had been instructed to enter through the employees' door at the rear of the building. Once there, Herring peered through the coal-dust-coated window of the facility's powerhouse, where a horizontal twin-cylinder steam engine rested on its oak-beam foundation. Based on its size, Herring guessed that the mid-century behemoth was probably a non-condensing relic operating on borrowed time. Ten feet to the east of the powerhouse stood a 10-foot pile of bituminous coal for feeding the engine's burner – enough chemical energy for a month's worth of mechanical work. Ever inquisitive, Herring also noticed the single 4-inch diameter, overhead steel shaft that transferred the steam engine's power into the shop.

In a final inspection of the property, Herring's eyes scanned across the gravel driveway to the west, and focused on the dilapidated barn where Engberg kept horses and a variety of buggies, wagons, and at least one fancy carriage. *The property is a dump,* Herring thought. *Can this place produce high-quality work?*

Herring banged a fist three times on the old oak door. A few seconds later, he heard the metallic clank of a bolt-lock being opened to reveal an elderly man outfitted in an oil-darkened, full-length leather apron.

"Mr. Herring, I presume?" Bill Engberg said, grinning. "I won't shake your hand, sir... as you can see, I'm awash with grease!"

With hair the color of a two-week-old New York City snowbank, 60-year-old William Engberg looked a decade older. At 5 feet 8 inches tall and 190 pounds, he seemed at ease surrounded by the sights, sounds, and smells of his machine shop. "Come on in, Herring... maybe you can tell us what the hell is wrong with my 14-inch lathe."

As Herring entered the lair of oily black machinery, the stench of overheated lubricating oil and scorched leather dominated the shop's atmosphere. He felt right at home. Looking around, his eyes followed the overhead power shaft down the middle space to where leather belts distributed their vital motion to the lathes, milling machines, shapers, and grinders. Without fanfare, the gaunt, lanky, 6-feet 4 inch tall figure of George Housam uncoiled from behind the disassembled lathe, where he had been working. After wiping his hand on a soiled shop rag, Housam extended it to Herring.

"I can always tell a man who doesn't mind a bit of grease," the 175-pound Housam said.

"I see you don't have electric power, Mr. Engberg."

Waving his hand in a "who cares" fashion, Engberg made short work of Herring's observation. "Too damned expensive! I'd be wasting my money on these old machines—"

"He won't admit it, Mr. Herring," Housam interrupted, "but the men think these here machines came over on the *Mayflower!*"

Slowly shaking his head, Engberg reacted as if he had heard all of this before. "These machines may have come over on the *Mayflower*, Mr. Herring, but they're still accurate and well maintained – that's why we're here this mornin'. My father started the business right after the war, and here I am, still at it more than 30 years later."

"How many employees do you have, Mr. Engberg?"

Scratching his head, Engberg used his fingers to count. "Besides Housam here, there are nine other machinists plus the powerhouse engineer, and my wife... who keeps the books."

"The barn next door, is that yours, too?"

"Yes sir. The wagons are used for shop deliveries and to haul raw materials, includin' a couple of trips a week for coal. My powerhouse man also tends to the animals when he isn't shoveling coal, lugging cinders or oilin' the engine."

Engberg narrowed his eyes. "Mike Berg over at the boat works tells me that you're buildin' a flyin' machine... is that true?"

Herring shrugged, sensing what was coming next. "Nothing new there, Mr. Engberg. I've been flying these machines for the past three years – first in New York, then in Indiana, and now—"

"And now," Housam interrupted, "you're gonna fly one of them thangs right here in St. Joseph!"

"That's right... and that's why I'm here this morning. I need the services of a precision machine shop with expert operators. Your shop came highly recommended."

"You know what people say about men who try to build flyin' machines, don't you Herring?" Engberg continued.

"Gentlemen, like the *Mayflower* story, I've heard it all before; everything from *Darius Green*, to crackpot, to crazy. I consider those who hold such opinions to be ill informed... and I pay them no heed."

"That's good enough for me, Mr. Herring. As long as you pay ahead of time, we'll do our best to satisfy your needs."

With that implicit agreement, Herring showed the men drawings for two, five-inch-diameter spur gears that were intended to transfer power from the engine's crankshaft to the forward propeller shaft. They discussed the material – tool steel, along with tolerances for the mesh and shaft bore.

When Engberg retreated to his office to work out a price for the job, Herring and the 40-year-old Housam engaged in casual conversation. "You live over on Church Street, next to the German Baptist church, don't ya Mr. Herring?"

"That's right, but how did you—"

"My family goes to that church," Housam continued. "One Sunday, 'bout a month ago, the congregation smelled them bananas of yours! Ol' George, the organist, he was tellin' folks that the *professor* next door was smashin' 'em up to paint his flyin' machine... is that what you were doin', Mr. Herring?"

Gus smirked. "My homemade varnish does smell a bit like bananas, so I'm not surprised by the confusion, but where did he get the idea that I was a professor?"

Housam shrugged. "When people can't figure somebody out, Mr. Herring, they give 'em titles like that. Did you know that kids in the neighborhood call yer shop the *mystery barn*? Folks say that you keep to yourself, don't say much and chase kids away when they come snoopin'—"

"George, it sounds like you have an ear for gossip!"

"That's right!" Housam said, thrusting out his chest. "I live on Broad Street, right around the corner from you. Most days, I see you walkin' to and from Truscott's."

Herring grinned. "Well, George, you'll have to drop by sometime. There's not much going on right now, but I'll show you around – show you some models and drawings. I've also got a crated flying machine in the back—"

Returning from the office, Engberg grumbled as he sidestepped between two milling machines, requiring him to hold the scribbled estimate above his head. "Read it and weep," he said, in a voice that left no doubt about his skepticism for Herring and his flying machines.

The estimate for two complete, ready-to-install gears came to $31. Two-thirds of that amount would be required before starting the work, with the remainder due upon completion... in about two weeks. Digging into his black leather wallet, which he kept chained to his belt, Herring seemed unfazed by Engberg's request. "I'll pay *all* of it now, Mr. Engberg. With satisfactory results, this will be the first of many transactions to follow."

*

Saturday, October 16, dawned cold and overcast. Like an army intent on wiping out a weakened enemy, the north wind relentlessly separated the last leaves from the town's maple and oak trees, forcing residents to accept the inevitable: St. Joseph would soon be smothered by another winter.

The temperature inside Herring's uninsulated shop was a cool but livable 60 degrees – 25 degrees warmer than outdoors, due to the reconditioned coal-burning stove and chimney that he had installed the previous weekend.

Pleased with the progress he had made in preparing the *basswood* patterns for the engine's cast-iron cylinders and cylinder heads, Herring retreated to the house, where Lillian had prepared his lunch – a toasted cheddar-cheese sandwich, hot tomato soup, and a large glass of whole milk – his favorite meal on a cool, blustery day.

Within half an hour he was back at his workbench when the shop's door first rattled and then banged open. At first Herring thought that some brat from the neighborhood needed shagging, but as he turned to address the distraction, he was surprised to see the Ichabod Crane-like figure of Housam.

"George, you're a sight for sore eyes! Come on in and close the door. Do you have something for me?"

Bundled in a gray canvas coat, black woolen scarf and fisherman's cap, Housam made a beeline for the potbelly stove. "I got somethin' fer ya alright, Mr. Herring. It's wrapped up in this here oiled paper. You unwrap it, while I get warm!"

Hastening to the task, Herring reminded George of a child tearing the wrapping from a Christmas gift. "Aha! Look at these beautiful gears!"

Hustling over to the engine's cast iron crankcase, Herring slid one gear onto the snout of the crankshaft, checking its alignment.

"Looks perfect, George! Now let's see how the propeller shaft gear fits... then I'll check the mesh."

Five minutes later, Herring was still smiling as he turned to shake Housam's hand. "Good job! Is this your handiwork, George, or did someone else do the machining?"

Housam joined Herring at the assembly bench. "This was my project for the past two weeks, but I think the shop lost money on the job. Tool steel is a damnation to machine... took forever to mill those tooth profiles."

Running his hand over the finished crankcase and installed crankshaft, Housam was intrigued by the engine's design and fine machine work. "Where did ya have these pieces made? They're well done... well done indeed!"

"The iron crankcase was cast at McCarthy's shop on Chicago's east side. The tool steel crankshaft and crankcase casting were machined at Illinois Precision, also on Chicago's east side. McCarthy's also cast a beautiful set of babbitt – that's an alloy of tin and lead – half-shell main bearing inserts for the crankshaft and journal end of the connecting rods."

As Housam marveled at the state-of-the-art casting and machine work, Herring divulged some of his thinking that had led to the engine's lightweight design. "As you can see, the crankshaft has only one journal *offset*. This cuts down on weight while offering an alternate firing arrangement for the two opposed cylinders. The two connecting rods run adjacent to one another on the journal and are lubricated by a 'dipper' that splashes them with castor oil in the sealed crankcase."

Soon, the conversation turned to the basswood pattern on the shop's foot treadle-actuated lathe – a machine designed to make precision round objects from wood. Surrounded by two wall-mounted gas lamps, the work area was well illuminated, especially during the day when natural, diffused light flooded in from an adjacent whitewashed window. Lying on the dirt floor beneath the machine was a pile of curlicue wood shavings, a byproduct from the flat-chisel tool that Herring employed while roughing out a cylinder. Propelled by the rocking motion of the treadle, a force that he applied with his left foot, Herring had mastered the art of doing two things at once – no small feat when the accuracy of the final product hung in the balance.

"What's the little cylinder for, Gus?"

Removing the two-inch-diameter, foot-long rolling-pin-like piece from the lathe, Herring proceeded to measure its diameter with an outside caliper. "This is the core for the power cylinder pattern. As you know, it will save a lot of time for the machinist if he doesn't have to hog out all of that unnecessary material inside the casting."

Nodding, Housam walked back to the workbench in the center of the shop where Herring had been gluing quarter-round leather fillets between the power cylinder pattern and its two flat-end flanges.

"Why is there a flange on either end of the cylinders?"

Unrolling the engine's assembly drawing, Herring placed a small block of wood at each corner to keep the blueprint copy flat on the bench. "Look here," he said, pointing to the profile view of the power cylinder. "The flange at the bottom end gets fastened to the crankcase by four good-size machine screws. The top flange uses four more of these screws to sandwich an asbestos gasket between itself and the cylinder head.

"Between these evenly spaced machine screws, four long tool steel screws extend all the way from the cylinder head, through the two cylinder flanges and into the threaded holes of the crankcase casting. Once adjusted, these screws will provide rigidity to the thin-wall iron power cylinders, while eliminating the weight of a heavy casting. Now you know the secret to makin' a lightweight engine! You won't tell anybody, will you?"

George ignored Herring's weak attempt at humor. Instead, he lifted the completed cylinder-head pattern, and turned it over in his hands to inspect its many features. After comparing its dimensions to those on the assembly drawing, he turned toward Herring. "This pattern is well done, Gus, but there's somethin' I don't get. Why did you make a bowl-shape on the bottom side of the head... the end that fits into the top of this here power cylinder?"

"It's called a *hemispherical* combustion chamber. For a number of reasons, it allows the engine to develop more power than other chamber shapes. I ran a series of dynamometer tests on various combustion chamber geometries while attending Stevens Institute of Technology, and the hemisphere turned out to be the best. By placing my overhead poppet valve in the middle of the bowl – as you call it – the opening of the valve won't interfere with the piston crown when it's at the top of its stroke. Then, by swinging the sparking plug over to the side of the bowl, it avoids the reciprocating actions of both the valve and piston, while effectively igniting the gasoline and air mixture near the end of the engine's compression event.

"I also believe that the hemispherical chamber allows the gasoline-air charge to mix better before ignition takes place – something that the early Otto Cycle designs failed to take into account."[3]

Far out of his element, Housam was glad to change the subject. "Are you headin' back to Chicago to have the cylinders and heads sand cast?"

"I haven't decided yet," Herring said, scratching his head. "Is there somewhere closer that does fine work?"

George cracked his knuckles. "Yup, the *Iron Works* does first-class sand casting."

"Where are they located?'

"About half-mile down the road from Engberg's, on Lake Street. When you get there, ask for ol' Albert Mahaffey – tell him I sent ya. Be careful, though, Uncle Al ain't too keen about talkin' to cranks!"

Changing to a less contentious topic, Housam continued. "I see you've whitewashed the inside of yer windows so the kids can't see in. How's that been workin' fer ya?"

Scowling, Herring shook his head. "There's not as much snooping, but every once in a while, some snot-nose brat will try to peek through the crack in the shop door, and then I'll chase him away. It's become quite the game around here."[4]

Rochester, New York

Monday, October 31, 1921

Returning promptly from lunch, Attorney O'Grady wasted little time in asking his first question.

"Mr. Herring, in January 1898, you met with Mr. Arnot at your St. Joseph home. What took place at this winter meeting?"

Gus peeked at his notes before answering. "Three things were discussed: My progress with the new gasoline engine; how to proceed in the aftermath of the rejected U.S. patent application; and Matthias's increased support of my flying machine program."

"What was proposed?"

"In addition to our previous arrangements, Matthias proposed buying into 10 percent of the future profits of the business. For this, he offered an additional $4,000... which I accepted."

O'Grady turned to a new page in his notes. "Mr. Herring, you were able to produce an operating engine of your own design in only 14 weeks. How was that possible?"

Taking a deep breath, Herring crossed his legs before speaking. "With money on hand, I was able to utilize several of St. Joseph's manufacturing and production resources to get the work done. The Iron Works did all of the castings, while Saranac's shop split the machining with Engberg's. I was also fortunate to have use of Truscott's machine shop on most Saturdays. Although their machinists usually had weekends off, production personnel still needed steam for forming various boat parts – this allowed me to run the steam engine that powered the lathes and other shop equipment. The use of this equipment came in handy, especially when I had to make minor modifications."

"What were some of the fabrication challenges that needed to be overcome?"

Rubbing his forehead, Herring struggled to recall the major engine components that had been fabricated more than two decades earlier. "The iron power cylinders were a formidable challenge because the walls were very susceptible to breakage during machining; they were only the thickness of a ten-cent piece!

"I charged Saranac's shop with performing the delicate boring and honing operations on those cylinders. Their direct current, electric-motor-powered machines featured smooth, variable speed operation – a definite advantage over the jerky, belt-shifting speed changes experienced with steam engines. I also gave Saranac contracts to produce the overhead camshafts, poppet valves, and a timer mount for the battery ignition system.

"Engberg's shop did the machine work on the cylinder heads, half-speed cam-drive wheel, and valve train sprocket gear as well as the push-pull rods, rear propeller driveshaft and its universal-joint coupling. Toward the end of the project, they also made the lightweight aluminum flywheel.

"The sparking plugs proved to be a thorny issue until Carl Dienstbach came to the rescue. He suggested that the German company, Benz, might be willing to provide a few units from their experimental stationary engines."

"Did they send them?" O'Grady asked.

"Three weeks later, a package of six 10-millimeter sparking plugs arrived at my home from Berlin – courtesy of Herr Dienstbach!"

"Were they satisfactory?"

Herring lifted an eyebrow. "Strange things happen when you build something new and unusual. Since the German plugs used *metric* threads, none of the machine shops in St. Joseph had that size *tap* to cut threads into the two cylinder heads – normally a simple procedure. It took me three weeks and a trip to Chicago to locate that tool!"

Returning to his podium, O'Grady flipped through several pages of notes. "Mr. Herring, allow me to summarize your power-aeroplane testimony up to this juncture: You arranged to have Mr. Avery's shop in Chicago fabricate the wooden parts, along with Engberg's and two other shops making parts for a lightweight gasoline engine. In addition, you were working as a full-time engineer at the Truscott Boat Manufacturing Company... besides taking care of your family! Were there any other tasks that you were working on to further the project?"

Perspiring, Herring mopped his brow with his handkerchief. Next, he loosened his tie and pulled open the collar on his shirt. Noticing his apparent discomfort, Judge Sawyer decided to intervene. "Mr. Herring, are you all right? Do you need a brief recess?"

"Thank you, Your Honor. I'll be fine. Let's continue... I believe that I was asked about additional work on the new flying machine?"

"That is correct," O'Grady reiterated.

"I did all of the work on the cooling jackets myself. The space between the sheet-metal jackets and the power cylinders was filled with castor oil, through which waste heat was transferred to air-cooled wire *solenoids* that were wrapped around the outside of the jackets and then soldered.[5] Besides cooling, the oil acted to lubricate a vital part of the engine. As the temperature of the oil increased, it expanded, forcing its way through tiny passageways in the cylinder, where it lubricated the piston skirts.

Herring's twin-cylinder opposed gasoline engine (1897); Meiller/Herring Collection

Then there were the propellers! Two 5-foot-diameter air-screws with opposite pitches that were made from American black walnut."

"What do you mean by the term 'pitch'?" the judge asked.

"Pitch refers to a propeller's *twist* and is measured in inches. The propellers for that machine had a pitch of 36 – the theoretical distance the blades would advance through the air in one revolution. Of course, there is slippage and other variables to contend with... propeller design has proven to be very complicated and is well beyond the scope of our discussion—"

"Objection!" Robbins interrupted. "Conclusion."

Swiveling to face the defense attorney's table, Sawyer winced. "Mr. Robbins, are you suggesting that the witness, an engineer by trade, should burden the court with a drawn-out technical explanation concerning propeller design?"

"As a matter of fact, *yes*! It is the position of the defense that the witness is bluffing. We contend that he knew little or nothing about propeller design back in 1897 or '98!"

Turning his attention to O'Grady, the judge could see indignation etched across the attorney's face. "Does counsel for the plaintiff wish to rebut?"

"Your Honor, for weeks Mr. Robbins has tried to make a mockery of this trial by leveling insults and innuendoes at Mr. Herring and his work in aerial navigation. In all instances, I believe that these accusations have been thwarted and duly dismissed by Your Honor. The court has observed these tactics and has correctly labeled them as stalling maneuvers designed to drag out the testimony and exhaust the witness's will to continue. As the plaintiff's attorney, I believe that these continued actions by the defense demonstrate further contempt for the Supreme Court of New York—"

"Perhaps," the judge sighed. "Mr. Herring, what do you have to say about your competency in the matter of propellers?"

Herring shrugged. "I suggest that the court compare the airscrews used by other experimenters before and after my propeller design of 1898."

"How might that be accomplished?" Sawyer asked.

"Photographs. By comparing a series of propellers arranged chronologically over an extended period of time, I believe that even the casual observer will be able to determine who was influenced by whom!"

"Ob...Objection!" stuttered Robbins. "This so-called comparison will only be subjective in nature – where is the objective analysis?"

"This is a court of law, Mr. Robbins, not the selection committee for the Nobel Prize! Overruled! Very well, Mr. Herring, how long will it take for you to assemble your photographs?"

"Fifteen minutes, Your Honor. I have my collection of pictures here with me today."

Cracking his gavel on the oak block, Judge Sawyer stood and stretched, looking strangely like a grounded vampire bat.

"There will be a 15-minute recess. Court will resume promptly at 2:30."

*

When court reconvened, five photographs, labeled A-E, rested on O'Grady's podium.

"Well, Mr. O'Grady," Sawyer inquired, "did your witness find his photographs?"

Holding up the yellowed black and white prints, O'Grady addressed the bench. "Your Honor, the plaintiff wishes to enter these photographs into evidence."

After the usual procedural delays, including haggling with the defense over the relevance of the evidence, Judge Sawyer directed O'Grady to continue with his questioning.

O'Grady handed Herring the first exhibit. "Mr. Herring, describe the image depicted in exhibit A."

"This is a propeller from Sir Hiram Maxim's enormous steam engine-powered flying machine of 1894."

Herring handed the exhibit to the bailiff, who passed it to the judge. "As you can see, Your Honor, the propeller resembles a conglomeration of masts and sails from a sailing ship."

Herring moved to the next exhibit. "Exhibit B shows a propeller from Professor Langley's steam-powered aerodrome model of 1896. Notice that it resembles a child's pinwheel toy; it has a wooden framework covered with varnished linen.

"Exhibit C shows a propeller from my manned, powered aeroplane of 1898. It was shaped from laminated walnut and closely resembles the propulsion units in exhibits D and E."

After scrutinizing the last three photographs, Judge Sawyer stared down at Herring.

"What is the significance of the propellers in exhibits D and E?"

"Exhibit D is an image of the propeller from the *Wright Flyer* of 1903! Exhibit E shows the propeller from Mr. Curtis' *June Bug* of 1908!"

Shifting his gaze back to the photographs, the judge studied them for several more seconds. "Mr. Herring, are you contending that the Wright and Curtiss propellers are essentially the same as your earlier design?"

Herring's jaw tightened. "Look at the evidence, Your Honor. What do you think?"

"Objection! Never trust appearances, Your Honor! Closer inspection will show that the pitch and area distribution are different—"

"We're back to the question of *pitch*, are we, Mr. Robbins? I believe that Mr. Herring has made his point... objection overruled."

CHAPTER 61

St. Joseph, Michigan

Saturday, February 12, 1898

It was Abraham Lincoln's birthday... but Herring, who ignored most holidays, had important engine matters to contend with on this overcast winter morning. Tramping through a dusting of new snow, he traversed the 50-odd feet to his so-called "mystery barn" and unlocked the structure's planked, eight-foot-tall door. Before he slipped into the shadowy confines of the shop, Gus turned and waved to Lillian, who was watching from the kitchen window. Mindful of the sub-freezing temperature, Herring quickly tended to the potbelly coal stove, while mentally scrolling down a list of tasks that needed immediate attention. *Gotta see if the little carburetor is gonna work*, he thought, *but I can't check it here in the shop... could cause a fire.*

Today was *the* day. Would the little engine run... or not?

Minutes later, the access door squeaked open again – just enough to allow George Housam to squeeze through. Pulling the reluctant door closed with a thump, George turned on his heel and made a beeline for the sputtering stove. "Not much heat comin' out of this thang, Gus!"

Still looking down at his work, Herring smiled and replied, "You gotta let her warm up, George! I just fired her up a bit ago!"

George registered his disappointment by stomping his feet. "Wood-burnin' stoves heat up a lot faster! This old thang will take half a day to git goin'!"

Keeping his winter coat and woolen hat on, Housam moseyed over to where Herring was fiddling at the front of his engine. "What ya doin', professor? Why is that steel protractor fastened to the end of the crankshaft?"

Trying to concentrate on the task at hand, Gus mumbled, "I'm settin' the ignition timing... need the 'breaker points' to open at 20 degrees before top-center on the compression stroke. Once the engine gets running, I'll either 'advance' or 'retard' the spark timing until she develops the most power for the propeller I'm using."

Housam leaned closer. "How do ya know when the engine is puttin' out the most power?"

Herring hesitated for a moment before looking up. "For me, it's when the engine is running the *fastest*... you can tell by listening to the *exhaust note*, or you can use a crankshaft revolution counter and stopwatch."

Removing his hat, Housam scratched his head. "Ya know, Gus, if I'm gonna help ya with this here engine project, ya gotta explain how some of these thangs work. Most of the time, I don't understand what the hell you're doin', or why you're doin' it."

Standing, Herring peered at his assistant. "Well, where do you want me to begin?"

"You kin start anywhere – I'll stop ya if I already know what's goin' on. I bought this book called *The Gas and Oil Engine* by a fella named Dugald Clerk – the 1896 edition. I've been readin' it most every night for the past month... but I'll be damned if any of the engines in that book look anything like this thang."

Hands on hips, Herring nodded. "I'm impressed. Clerk wrote that book for engineers! All right, we've got some time...

but first let me lock down this ignition advance lever before I forget it."

A few seconds later, Gus tossed the open-end wrench and protractor onto the bench and motioned for Housam to move closer to the little engine. "First of all," Herring said, as he turned the propeller, "my inspiration for this engine came from the Benz Company in Germany. Last year they began producing a new Otto-Cycle engine for their lightweight 'Velo' automobile; they call it the 'Kontra.' It's an opposed, twin-cylinder design that has a total displacement of 100 cubic inches, and it develops five horsepower at 900 revolutions per minute. Unfortunately, the Kontra weighs almost 200 pounds! That's 40 pounds per horsepower – too heavy for a flying machine with a man aboard.

"I designed my engine to have a total displacement of 42 cubic inches – less than half the size of the Kontra... but it also needs to produce five horsepower! My goal is to reach five pounds per horsepower—"[1]

George thought for a moment before interrupting. "Then it'll only weigh 25 pounds! Is that possible?"

"We'll find out soon enough. Here are some technical things you should know, George: a modern engine's operating cycle is made up of four strokes of the engine's piston—"

"Yep! I know that."

"Good. What do you know about the term 'stroke'?"

George hesitated, buying time by removing his heavy coat. "The engine's stroke is how far the piston can move in the cylinder... the book says it's from top-center to bottom-center."

"I'll accept that. In Otto's four-stroke-cycle engine, how many revolutions of the crankshaft are needed to complete a cycle?"

"Two."

Herring nodded. "What are the four events or operations of the Otto Cycle?"[2]

George scratched his head again. "The inlet, compression, expansion, and exhaust... and each of these operations uses one stroke of the piston."

"I'm impressed! What is it that you don't understand about my engine?"

George hesitated before speaking. "The engines in Clerk's book look a lot like the steam engine back at Engberg's shop... they're big, they're heavy, and they all have big, heavy flywheels. Your engine doesn't look like any of 'em, plus it only has one valve in each of the cylinder heads.

"Another thang... what about them holes you drilled around the bottom of the cylinders? Professor Clerk calls them 'ports,' but I couldn't find any engine that uses them like you do."

Herring paused as he decided how to respond. "We should talk about poppet valves and ports. As you pointed out, my engine uses both—"

"That's what I'm talkin' about," George interrupted. "Why is one poppet valve and a bunch of ports better than two poppet valves?"

These are probing questions, Herring thought. *George is smarter than he looks.* Aloud, he said, "By doing away with the second poppet valve, the engine can be made simpler, lighter, and possibly more efficient."

For the next 15 minutes, Herring discussed several of the key issues surrounding his choice of valves and ports.[3]

Housam motioned in the general direction of the engine. "So, you're mixin' two kinds of engines – the four-stroke and the two-stroke—"

"Yes," Herring broke in, "but it's *still* a four-stroke-cycle engine, so it requires two revolutions of the crankshaft to complete all of the operations. Like I said, I've taken the best ideas from each engine type and cobbled together a design that best suits the needs of a flying machine. In the process, I've had to innovate to make some of the systems work to my liking."

"Like that single poppet valve in the cylinder head?"

"That's one. I call it an *overhead* valve. I believe that I'm the first to use a single overhead poppet valve that works in unison with cylinder exhaust ports."

"How do you know when to open and close the poppet valve?"

"The *overhead* cam holds the key to that, George."[4]

For the next several minutes, Herring pointed out the various components that made up the "valve train" of his engine. Pulling his watch from his pocket, Gus noted the time. "Time to git to work, George! Now that I know you're interested in engines, we'll continue this conversation later."

*

By 10 o'clock, the men had pushed the four-wheeled cart onto the vacant lot adjacent to the mystery barn. An additional 10 minutes were spent leveling the test stand on the frozen ground by using various thicknesses of scrap lumber from the shop. Satisfied, Herring slipped the six-inch bubble level back into his coat pocket. As the men stood back to scrutinize their work, the little machine appeared small and fragile, its overall width spanning a mere 20 inches.

"All right," Herring said, "let's anchor the test stand with these four cinder blocks. We don't want everything 'walking' around once the propeller starts to spin—"

A shout resounded from the house. "Augustus... they're ready!"

Turning, Herring observed Lillian standing inside the rear door, holding a square-bottom paper bag at arm's length. "Wait for me back in the shop, George," Herring said, as he trotted toward his spouse.

Needing no further incentive, Housam jogged straight to the stove.

Having rejoined his assistant seconds later, Herring reached into the bag with his gloved hand.

"Whatcha got," George asked, "somethin' ta eat?"

Herring ignored the query. "George, you've lived here all your life, but I'm originally from Georgia. Under the circumstances, if anyone should be cold... it should be me! Still, today's weather isn't too bad. The thermometer says it's already 25 degrees... but Lillian took pity on us!"

Pulling a snowball-sized object out of the bag, Herring quickly handed the newspaper-wrapped sphere to Housam.

"Wow! That thang is hot!" George cried, juggling the heavy ball. "What the hell is it?"

"It's a trick I learned as a child, while living in Germany... it's a baked potato!"

Dashing to the nearest workbench, George tossed the hot potato onto the tool-laden surface. Herring followed his assistant and dumped three more tubers onto the bench.

"If they're not for eatin'... what are they for?"

Leaning close to Housam's good ear, Herring whispered. "Put one in each of your coat pockets... they'll keep your hands warm!"

A serious expression crossed George's face. "For how long?"

"A couple hours."

After stuffing two potatoes into his own pockets, Herring began collecting the tools and equipment he would need outside. Pausing, he glanced over at Housam. "Come on, George, stop playing with those hand warmers and bring the gallon tin of gasoline and a funnel! I think I've got everything else."

Although Herring had only *one* of the engine's two cylinders ready to run, his anxiety over the project wouldn't allow him to delay initial testing any longer. The recently purchased magneto had yet to arrive, so he was forced to use his Ruhmkorff coil and Hellesen dry cells in a "buzzer-type" battery ignition system. "George, hold the funnel still! You're shaking so much, I can't tell where to pour the gas!"

A one-pint, tin-plated steel container served as the engine's gravity-fed fuel tank. It was mounted six inches above and slightly behind the power cylinder's overhead poppet valve. A small petcock was soldered to the tank's bottom. In turn, a length of 1/8-inch-diameter copper tubing was attached to the petcock's outlet by a short piece of black rubber tubing. The fine copper tubing terminated at what appeared to be an old *tablespoon*, which had been mounted near the poppet valve's runner port.

"As you can see," Herring said, "the spoon's handle has been bent at 90 degrees – just below the head. Two holes have been drilled through the handle, allowing the spoon to be machine-screwed to the side of the aluminum cylinder head... a temporary setup, but it should work okay for this first test." After replacing the tank's screw-on cap, Herring opened the petcock one-quarter turn and blew into the tank's short vent tube, causing gasoline to flow through the copper feed tube. Within seconds, the first dribbles of fuel began to accumulate in the spoon's bowl.

"How you gonna keep that gas from splashin' out of the spoon after the engine starts?"

Herring thought... *time to have a little fun.* "That's a trade secret, George... only flying machine men are allowed to know how that's done—"

"What?" Housam shouted.

Grinning broadly, Herring pulled a piece of sponge from his pants pocket, holding it up high for George to see. "Don't get all up in arms, I'm just joshin' with you. Besides, you know what they say, don't you? There are only two kinds of secrets... one that's *too good* to keep and one that's *not worth* keeping!"

Herring proceeded to secure the little sponge to the bowl of the spoon with a few turns of copper magnet wire. "See how the sponge soaks up the gasoline? That should keep it from splashing out of the spoon. Engine vibrations aren't all bad. They help to break up the liquid gasoline into small particles... that's called *atomizing*. Then, as the hot exhaust gasses strike the bottom side of the spoon, the atomized gas turns into a *vapor*. If all goes well, the gasoline vapors mix with the surrounding air and are sucked into the engine on the inlet stroke. I'll control the *mixture strength* with the petcock."

"What do you mean by 'mixture strength,' Gus?"

"It's a British term used to describe the ratio of air to fuel by weight. If you *strengthen* the mixture, it has more gasoline mixed with the available air... if you *weaken* the mixture, it contains less gasoline. When you get the mixture strength just right, the engine will run the fastest, with the most power."

"What do you call that spoon and sponge thang?"

"I call it a simple *carburetor*. Everyone who experiments with these new gasoline-burning engines is trying to figure out the best way to mix the gas with the air before the mixture gets sucked into the cylinders—"

"Are you gonna patent the spoon and the sponge?" George asked with a wry smile.

"Too expensive to bother with," Herring growled.

"What other tricks ya got for me, Gus?"

"No more tricks, but I need you to hold this wire onto the end of the sparking plug, while I check the ignition timing."

Housam did as he was told, and Herring flipped on the ignition toggle switch at the side of the Ruhmkorff coil box, setting off a barely audible buzzing. Moving to the front of the engine, Gus slowly rotated the propeller in a counterclockwise direction. Suddenly, Housam yelped and jumped back from the engine. George had just experienced his *first* electric shock.

"What the hell!" Housam shrieked as he tried to shake the cramp from his hand.

Laughing uncontrollably, Herring flipped the switch back to the off position. "Sorry!" Gus cried, wiping a tear from his cheek. "I just couldn't help myself!"

After explaining how the evil-but-harmless event occurred, Herring held onto the sparking plug and allowed Housam to return the favor. After Gus yelped and hopped about, the two shared a laugh before getting back to business. Within 15 minutes, everything appeared ready for the first attempt at starting the engine.

"George, I want you to stand *behind* the engine, where you won't get hurt if something breaks or flies off. Now listen, this is important! If anything bad happens, pull this string. It's connected to the ignition switch and will shut her down."

Checking that the petcock valve was closed at the bottom of the tank, Herring reached behind the ignition box to confirm that the ignition switch was in the "off" position. After visually checking that the ignition advance-retard lever was in the predetermined position for startup, Herring moved to the front of the engine. He grasped the wooden propeller and rotated it in a counterclockwise direction until he felt the unmistakable "bump" that signaled the engine's compression event. Positioning both hands on the propeller's trailing edge, about halfway out from the crankshaft's hub, Gus practiced the rapid pulling action past the compression bump that he hoped would initiate

Herring's twin cylinder opposed gasoline engine on test stand (1897); Meiller/Herring Collection

a start. At first, the act of cranking the propeller seemed foreign and unwieldy. But he felt confident that with further practice the procedure would be successful.

Reaching behind the propeller, Herring opened the tank's petcock to its predetermined position and confirmed with his forefinger that the sponge was indeed saturated with gasoline. Last, he flipped the ignition switch to the on position.

All was ready.

Rochester, New York

Tuesday, November 1, 1921

As the *Herring-Curtiss Company v. Glenn Curtiss, et al.* civil lawsuit settled into another day of testimony, the promise of an uneventful morning was shattered by an unexpected challenge from the defense. "Your Honor, after reviewing Mr. Herring's testimony concerning his engine of 1898, the defense wishes to contest the accuracy and truthfulness of his recollections by submitting a list of reservations and complaints to be entered into the official court record."

Peering over his reading glasses, Judge Sawyer appeared puzzled. "Why is the defense requesting such an unusual action?" he asked.

"When we present our case, this list will serve as the basis for expediting the testimony of our expert witnesses."

"Expediting testimony?" the judge repeated with a cynical smile. "Since when has the learned attorney for the defense *ever* concerned himself with conserving the court's time? Your petition represents a break in courtroom precedent, Mr. Robbins – one that I am not prepared to honor! Request denied!"

"Very well, Judge, in light of the court's denial, the defense reserves the right to submit its objections in regard to the witness' testimony from yesterday... Monday, October 31."

"You may proceed, Mr. Robbins."

Robbins strolled to the front of the defense table to address the judge. "Your Honor, the defense objects to the perception that Mr. Herring invented the twin-cylinder, opposed type internal combustion engine."

Sawyer's head swiveled toward the witness. "Mr. Herring, do you have a rebuttal?"

Herring coughed awkwardly. "I make no claim to the invention of the opposed-cylinder arrangement, Your Honor; I merely gained inspiration from the Benz design."

"Continue, Mr. Robbins."

"The defense also objects to the perception that Mr. Herring invented the overhead valve configuration for the internal combustion engine."

"Mr. Herring?"

"Nowhere in my testimony did I claim to have invented the overhead valve. The overhead valve in combination with the overhead cam is another matter—"

"That was not the question!" Robbins interrupted. "Move to strike, Your Honor."

"Granted. Confine your answer to the question at hand, Mr. Herring. Continue."

"The defense further objects to the perception that Mr. Herring was the first to use the combination of an overhead valve and cylinder ports. The defense contends that the two-stroke-cycle *diesel* engine used this combination well before this witness stumbled onto the combination."

"Mr. Herring? What say you, sir?"

Herring's gaze was almost smoldering. "The two-stroke diesel is an entirely different engine in both its operating cycle, and its use of cylinder ports, in addition to its use of the overhead poppet valve. In the two-stroke diesel, as the ports are uncovered by the descending piston, fresh air is blown *into* the cylinder from a crankshaft-driven compressor, forcing the exhaust gasses out of the engine via the overhead poppet valve. As I described in yesterday's testimony, my engine operated quite differently – it was unique for its day."

"That's another of the defense's objections, Your Honor. If this engine were as unique and successful as the witness suggests, surely Mr. Herring would have applied for a patent. However, our search of the Patent Office's archives found nothing. Apparently, Mr. Herring had not submitted an application during the years in question."

"Response, Mr. Herring?"

"It's true, Your Honor. I lacked the funds necessary to pursue a patent for my engine. I had recently lost my bid to patent the powered aeroplane, and funds were simply not available."

"You may continue, Mr. Robbins."

"Finally, the defense objects to the perception that Mr. Herring developed all of the systems for his 1898 gasoline engine, when in fact he *copied* and outright *stole* the ideas of others—"

"Your Honor!" O'Grady erupted. "The defense has gone too far! There is no proof of these outrageous allegations! Robbins is cross-examining my witness... he's out of order!"

Banging his gavel, the judge demanded order. "Mr. Robbins, qualify your contentions."

"Here's but one example of many, Your Honor. Throughout his testimony, Mr. Herring implies that he was the *first* to use gasoline as a fuel in both steam burners and internal-combustion engines. Our research has shown that many other experimenters preceded Mr. Herring in using this fuel."

"Mr. Herring, what say you, sir?"

Sitting rigidly upright, Herring struggled to maintain his composure. "I have *never* claimed to be the first to propose, or use, gasoline as a fuel in any form of combustion engine. I believe that distinction belongs to one of my professors of Mechanical Engineering at the Stevens Institute of Technology. To the best of my knowledge, Dr. C.F. Yates experimented with the first gasoline-fueled Otto Cycle engine back in the mid-1880s. I *will* however, take credit for designing and fabricating the high-pressure gasoline burner used in Professor Langley's aerodrome of 1896. It was a primary reason for the machine's success."

St. Joseph, Michigan

Saturday, February 12, 1898

Having flipped the ignition switch to the *on* position, Herring gave the propeller a forceful counterclockwise pull.

Nothing.

Another pull – the same result. Reaching behind the engine to its protruding crankshaft, he nudged the spark-advance lever a few degrees in a clockwise direction, effectively allowing the spark to occur earlier on the compression stroke.

"Let's try her again," Herring said, scowling.

After three more unsuccessful pulls, a weak puff of blue smoke whiffed lazily from the exhaust ports. Herring advanced the timing again, and this time, there was a loud "pop" accompanied by a plume of blue-gray smoke. Squeezing the carburetor sponge with his left index finger, Herring decided that more gasoline was needed.

"Weak mixture," he muttered, opening the petcock another quarter turn. On the next pull, the engine began what might best be called a stumbling, hit-or-miss operation, with random explosions that reminded Gus of the reports emitted from his father's old Civil War Colt revolver. Stepping back from the whirling propeller, Herring hesitated long enough to admire the long-awaited spectacle. Giving the apparatus a wide berth to avoid the almost-invisible spinning propeller, Herring slipped behind the engine, where he began the process of adjusting the spark timing and fuel mixture. Within half a minute, he had the little prime mover operating somewhat smoothly. Herring glanced over one shoulder, then the other, searching for George; but he was nowhere to be found. *No matter*, thought Gus, *George will get used to the noise and commotion soon enough*. After about two minutes of running time, Herring shut the engine down by snapping the ignition switch to the off position and then closing the petcock.

As if by some sleight of hand, George reappeared from his hiding place behind the barn. However, before he could be warned, he touched the engine's hot cylinder head with his bare finger. Yanking his hand away in pain, George let out a yelp. "Ouch! Is there any other way this 'devil' can sting me?"

Herring took the opportunity to emphasize his greatest fear. "There's one other way, George... always be careful of the spinning propeller! Never stand in front of or beside the propeller while the engine is running! Like I warned you earlier, always stand to the rear of the machine. If the propeller flies off the shaft or breaks apart, you'll be safe—"

Herring abruptly stopped speaking. He wanted to add something more, but describing the penalties for a propeller mishap made him cringe.

*

Back inside the warm shop, Gus talked about improving the engine's performance.

"Although the engine ran okay using the spoon carburetor, I have to try Truscott's *vapor tank* system."

"Why... is it a better idea?"

"I should be able to control the mixture strength better... something that is very difficult to do with my spoon carburetor."

"How does Truscott's system work?"

"By heating the gasoline to form a vapor—"

"Damn!" Housam butted in. "Won't those vapors explode? Ya remember what happened when folks tried usin' gasoline in their lanterns, don't ya?... They blew up!"

"No, George... just the opposite is true! Over at Truscott's, they're using this method of carburetion on their *Vapor Motor* that is designed to *reduce* the risk of gasoline fires or explosions. You see, gasoline vapor *can't* burn without air, and since there's no air in the tank, hose, or carburetor manifold, the vapors can only come into contact with air at the engine's poppet-valve port.

"Let me explain how the system works. There are only a few parts... a quart-size tin-plated steel tank, a length of 1-inch-inside-diameter black rubber hose, and a few soldered-copper tube fittings that make up the carburetor. When assembled, the rubber hose connects the tank to the carburetor, which is screwed to the cylinder head at the poppet valve's induction port. Other than a rudimentary air damper and tank heater, nothing else is required."

"Sounds good to me," said Housam, not wanting to appear uninformed.

"Excellent," Herring said. "Here's how the system works... when the tank is warmed by the heater, gasoline *vapors* are generated. These vapors travel through the connecting rubber hose to the carburetor. At this juncture, the gasoline vapors mix with outside air before being sucked into the engine on its inlet stroke. The amount of air allowed to mix with the vapors is controlled by an air damper – similar to the device used on my coal stove. The air damper accurately controls the engine's mixture strength and is much more precise than the hit-or-miss spoon-type carburetor ever could be!"

George thought about this for a few seconds. "How are you gonna warm up the gasoline?"

Herring waved his hand in a dismissive manner. "That's the thing... Truscott powers their tank heater with a *Henri Tudor* rechargeable lead-acid battery from the United Kingdom. Although the battery is far too heavy for flying machine use, Mr. Truscott insisted on loaning me one to try. If the vapor system works well, I may be able to use the engine's exhaust heat and do away with the battery altogether.

"One other thing... we must connect the mechanical counter to the crankcase housing. The crankshaft's Woodruff key extends just far enough to trigger the counter on each of its revolutions. That way, while I'm adjusting the engine's spark advance and mixture strength, you can operate my stopwatch to determine the engine speed in revolutions per minute.

"What do you say, George... can I depend on you not to act like a nervous chicken after the engine starts?"

Housam ignored the goading. "You've been around these devils before, Mr. Herring. That was the first time for me! It's not only me... did you see Pastor Miller's horse rear up out there on the street? I thought his buggy was gonna tip over backwards!"

Herring shrugged as he began removing his crude "spoon" carburetor. "Times are changing, George. It won't be long before automobiles will outnumber horses."

St. Joseph, Michigan

Saturday, February 19, 1898

Spring-like conditions had prevailed in St. Joseph for most of the previous week. Saturday morning dawned to more of the same: sunny skies, 50-degree temperatures, and variable winds out of the south.

Taking advantage of the unusually fine conditions, Gus breathed a sigh of relief as his hybrid twin-cylinder engine barked to life, its whirling propeller sending a blur of wintertime dust billowing across the vacant lot beside his shop. From his position behind the engine, Herring concentrated on making incremental adjustments to the ignition timing and the carburetor's air damper control. With the little prime mover still operating on only one cylinder, it nonetheless responded by increasing its speed... with no hint of a misfire.

"Start the watch, George!" Gus shouted, pointing at the revolution counter. Thirty seconds later, Herring snapped the ignition switch to the off position, thus ending the engine's second two-minute run. Before the smoke had cleared, he was busy taking the cylinder head and cooling jacket temperatures with his pocket thermometer.

"Three hundred degrees for the head... 235 for the oil jacket. What time did you get?"

Housam bent down and pointed to the stopwatch. "Let me see... 8.1 seconds for 50 revolutions. According to the chart, that's 370 revolutions per minute."

Herring frowned as he sat on the edge of the test stand. "Not good enough. I need twice that number of revolutions to fly the machine at an air speed of 24 miles per hour."

"What about when we hook up the other cylinder? That should give us more power!"

Nodding, Herring acknowledged Housam's observation.

"It will help, but I don't think it will be enough to push the revolutions up to 720... because the second cylinder also has to turn the *rear* propeller."

"How about makin' the propellers a little smaller? Won't that let the engine speed up?"

"Yes, but then the propellers won't deliver enough thrust to fly the machine at 24 miles per hour—"

George rubbed his chin. "Gus, what's so special 'bout 24 miles per hour anyway?"

Hesitating, Herring struggled to find a way to explain the horsepower, thrust, and airspeed relationship. "I designed the aeroplane to develop enough lift to fly in a 20-mile-per-hour headwind. With power being the major concern, I calculated the propeller pitch and diameter to provide enough thrust to fly only 24 miles per hour. Of course, that leaves only 4-miles-per-hour ground speed for forward progress."

"You told me earlier that if you increase the *area* of the lifting surfaces, the machine would fly in less of a headwind. Why don't you do that?"

Herring looked at him in astonishment. "George, that's a very insightful question! Here's the problem... when the lifting surface's span gets upward of 20 feet with my two-surface machine, it becomes very difficult to control *laterally* by shifting the operator's weight from side to side. The machine is already at its limit – so I can't go any further.

"No, my friend, the answer lies with the engine. It must be made to deliver more power to the propellers, and they must be able to spin at a minimum of 720 revolutions per minute."

As Herring contemplated the problem, another issue came to mind. "George, did you notice how everything was shaking, including the test stand? Those vibrations are sapping a lot of the engine's power... but I'm afraid the fix is counterproductive! I could add a heavy flywheel, but it's not practical when you're trying to build a lightweight engine.

"Let's push the engine back into the shop. Our next step will be to get the second cylinder working."

CHAPTER 62

Rochester, New York

Tuesday, November 1, 1921

Returning early from his lunch break, O'Grady was studying his notes when the penetrating voice of the court bailiff disrupted his concentration. "All rise!"

With his usual haste, Judge Sawyer hurried to his seat behind the bench. Peering over the rim of his new *pince-nez* reading glasses, he surveyed the orderly nature of his domain. Satisfied, he directed the litigants to be seated. "Mr. Herring, you may proceed to the witness box. Mr. O'Grady... it's your witness."

Displaying a fervent grin, O'Grady resumed his questioning. "Mr. Herring, according to your previous testimony, the new engine first ran on February 12, 1898. Is that correct?"

"Yes, sir. I remember the date because it was President Lincoln's birthday."

"For how long did you experiment with your new gasoline engine?"

"About two months – until the middle of April."

Stepping from behind the podium, O'Grady approached the witness. "Early in your testing, with only one propeller and one cylinder operating, the mechanism managed but 370 revolutions per minute. Mr. Herring, according to your testimony, that number was only half the speed necessary to fly the new machine. Were you able to improve upon that disappointing performance?"

Running his fingers through his hair, Herring hesitated as he pondered where to begin. "Somewhat. For example, after refining the carburetor, the shaft speed increased to 410 rpm – a 12-percent gain... but there was a more pressing problem: due to severe vibrations, the engine and the propeller drive mechanisms were in danger of shaking themselves to destruction!"

"Did you find a solution?" O'Grady asked.

Shaking his head, Herring slumped back in his seat. "By making the engine's second cylinder operational, I hoped that the shaking would diminish."

"Were there any delays in adding the second cylinder?"

"Just one. As the delivery date for the airframe parts drew near, I realized the need for additional shop space. After conferring with Mr. Arnot, I rented an area at the rear of Truscott's machine shop for six dollars a week. This was an ideal arrangement, since I was allowed to test-run the engine on company property – something I could no longer do at my Church Street home because the engine's unmuffled exhaust noise aggravated a few of the neighbors and prompted one of them to lodge a complaint with my landlord."

"When did you make the move to Truscott's?"

"On the first Saturday in March. Then I worked evenings and Saturdays for the next two weeks, preparing the engine for further testing."

"According to my notes," O'Grady said, "you tested the completed engine on the third Saturday in March. Is that correct?"

"Yes, sir."

"Tell the court about these new experiments."

Herring nodded as his mind drifted back to that hectic period. "With both cylinders operating, the shaking decreased some, but soon the engine mount and rear propeller-shaft coupling broke under the continued stress."

"Were you able to fix this problem?"

"Yes, but it required adding a 14 pound flywheel to the engine's crankshaft; this raised the engine's weight to 26-pounds. When the accessories and propellers were added in, the total came to 38 pounds – too heavy."[1]

"How much crankshaft speed did the engine produce after you added the second cylinder?"

"Well, that was it. After tuning the engine to run reliably at maximum power, it turned the tandem propellers at only 575 rpm, 80 percent of the minimum required to fly the machine. When I ran the engine on my Prony brake dynamometer without its propellers, it delivered 2-1/2 horsepower when loaded to 575 rpm, confirming my suspicions that there still wasn't enough power."

"Mr. Herring," the judge interrupted, "explain the function of a Prony brake dynamometer... in layman's terms, if you will."

"Certainly, Your Honor. A dynamometer is a device that places a load on the engine's output shaft to determine its torque, or twisting force, at any predetermined speed—"[2]

"What does torque have to do with an engine's horsepower?" the judge asked.

"Horsepower is a *derived* commodity. Essentially, it's calculated by multiplying the engine's torque by the speed of its output shaft. Because an engine's torque changes throughout its operational speed range, so does its horsepower."[3]

Apparently satisfied, Sawyer waved his hand toward O'Grady. "Continue, Counselor."

"Mr. Herring, what did you learn from your dynamometer testing?"

"I learned that by *reducing* the load on the engine, its horsepower increased to a maximum of 4-1/2 at 1,450 rpm! What was needed – and I couldn't provide – was a 2-to-1, *step-down* gearing to the propellers. In other words, the engine needed to turn two revolutions for every revolution of the propellers. With the engine turning at 1,450 rpm, the propellers would have no difficulty spinning at the required 700 rpm."

"Objection, Your Honor!" Robbins shouted from his seat behind the defense table. "This type of problem is mere child's play to a competent engineer! Herring's design error would be painfully obvious to any professional with—"

With a crack of his gavel, Judge Sawyer interrupted the attorney's tirade. "Mr. Herring, do you wish to rebut?"

Nodding, Herring leaned forward. "Many people know the answer the minute the other fellow solves the problem."

Raising an eyebrow, the judge jotted something into his notes. "The court agrees, Mr. Herring. Objection overruled. Continue, Counselor."

"Mr. Herring, since you understood the problem... why couldn't you fix it?"

"After investigating most of my options, I concluded that there was no *elegant* way to achieve the desired result... sadly, my gasoline engine, although promising, would have to be set aside."

"Was there enough time to start the process over again?"

Placing his elbows on his knees and clasping his hands, Herring shook his head. "No... however, there was one other possibility. I could convert the gasoline engine into a *compressed-air motor*."[4]

"How does the compressed-air motor differ from the gasoline engine?"

"Instead of burning a mixture of gasoline and air in the engine's cylinders to generate pressure, the new motor used the expansion of high-pressure air to accomplish the objective—"

"Could compressed air produce enough power to fly the new machine?"

Nodding vigorously, Herring continued. "To spin the tandem propellers at 720 rpm, my calculations indicated that I would require 600 psi – that's pounds per square inch – of compressed air at the motor's two overhead poppet valves."

"Was there much work required to convert the gasoline engine into a compressed air motor?"

"New cylinder heads were needed, as well as a system for timing the compressed air's entry and cutoff inside the cylinders. By removing the heavy flywheel and all of the accessories needed to operate the gasoline engine, the compressed air motor was expected to weigh considerably less, even with its air tank, valves, and connecting pipes."

"Were you able to get started right away on the necessary changes?"

Herring shook his head. "No. The shipment from Mr. Avery's Chicago shop had arrived."

St. Joseph, Michigan

Saturday, April 16, 1898

Gus had been working in his shop for almost two hours by the time George Housam squeezed through the door, making his usual beeline to the potbelly stove. "Dang! What stanks? Smells like a dead rat! Phew! Izzit this stuff you're heatin' on the stove?"

"Hello to you too, George!" Herring said. "It's *hide glue*, have you ever heard of it?"

"Ya mean *animal glue*, don't ya? If I'm not mistaken, they make it by boilin' down horse parts. How can you stand being cooped up in here with all of these doors and windows closed?"

"I don't have a choice, George. I'll put up with the stink if the glue holds my flying machine together. Old Man Truscott says the stuff is the best money can buy – it comes all the way from England."

Holding his nose, Housam managed a weak smile. "You better not have this gluepot goin' during church services tomorrow! Smellin' bananas is one thing... but this? Most of the flock thinks you're off your trolley to begin with – this animal glue stuff might prove 'em right!"

"You're a regular churchgoer, George," Herring said with a scowl. "Make sure the windows are closed!"

Housam dipped a wooden matchstick into the dark brown goo. "Why do you have to keep heatin' this stuff?"

"The wooden parts have to be clamped together before the glue cools... so it has to be applied *hot*. Besides being strong, hide glue has another big advantage for aeroplane construction – it's repairable. If a joint breaks, all I have to do is glue it together again; unlike other adhesives, hide glue sticks to itself. Aren't you glad you came by this morning, George? Look at all you've learned!"

"If I'm gonna stay, you've gotta get me one of Lillian's clothespins. If I breathe through my nose, I'm gonna get sick, sure as—"

"Here you go," said Herring, tossing a spring-type clothespin to his assistant, "I'm way ahead of you."

Breathing through his mouth, Housam turned his attention to the new flying machine. Herring had just started assembling a wing. "Is this the wood that goes into buildin' the new aeroplane? What the hell... it doesn't amount to a hill of beans!"

Housam was referring to the stack of sticks and beams that occupied one of the benches near the shop's east wall.

"Think it's worth $300?" Gus asked, smiling.

"Three hundred?" Housam gasped. "To me, it looks like a bunch of pine sticks and edging from the lumber mill's scrap pile!"

Pulling a 20-foot length of Sitka spruce from the stack, Herring carried it over to the window, where the light was better. "Look here, George, do you see the *bevel* on the narrow edge of this wing spar? It's the exact angle needed to match up with the wing ribs when they're bowed to the correct curve. If the bevel isn't perfect, the vertical struts won't match up, making it impossible to bolt the lifting surfaces together properly. I could hand-plane that bevel myself, but it would take me half a day to do it... or I could hire Mr. Avery to run the piece through his electric-motor-powered *jointer* machine. Multiply this job times the four spars that are needed, and you will begin to see the value in these 'sticks and edgings.'

"Look at the wood grain in this piece of spruce, George. It's so straight that you can follow it down the entire length, and it doesn't walk off the edge, either. Sitka spruce is by far the best wood for aeroplane construction."

"It looks to me like you got yourself an aeroplane *kit*," Housam said, "just like the rowboat kits that you can buy through the *Montgomery Ward* catalog."

"I never thought of it that way, but I guess you're right... it's a kit that *I* designed, but a kit nonetheless."

For the next few minutes, Herring described how the wing would be constructed over three workbenches that he had meticulously leveled and converted into a holding fixture for the spar, the front and rear edges and multiple aero*curve* ribs. "Here's where the hide glue comes into play. The individual aero*curves* get glued and clamped to these three 20-footers."

As the men worked at gluing and clamping the pieces together, their conversation turned to completion dates. "When do you think you'll have her ready to fly, Gus?"

"It's hard to say, but I'm aiming for the middle of June. My plan is to work on the airframe during the week and the motor on Saturdays. After we finish the airframe, I'm going to fly it on Silver Beach – as a manned *kite*."

At the time, St. Joseph was accessed from Lake Michigan by the Morrison Channel, which was fitted with docking facilities for the Great Lakes steamships that frequented the town. Directly adjacent to the channel was Silver Beach, a sandy, lakefront recreational area that extended far to the southwest in the direction of Chicago. Adjoining the beach and stretching to the northeast along the Morrison Channel stood the Truscott Boat Manufacturing Company, where Herring was employed.

"Why do we have to fly the new aeroplane like a kite?"

Herring pondered George's question for a moment. "I need to take some measurements before I fit her with the motor and propellers."

Housam couldn't contain his curiosity. "What kind of measurements are ya talkin' about?"

Nodding, Herring knew that sooner or later he would have to explain everything to his curious assistant. "The pull on the towrope, angles, the wind speed – you know... engineering measurements."

"I seen people fly kites before, but I never seen anybody try to crawl aboard the damned thang! Ya gotta admit... that's kinda crazy!"

"I'm not going to chance an unmanned crash this late in the game, so I've decided to tow the machine into the air—"

"How ya gonna do that?" George interrupted.

"Horses should do the trick. After taking the necessary measurements, I'll release the towrope and fly the machine back to the beach as a glider."

Housam thought about this for several seconds. "Horses can be pretty fickle, ya know... you got someone to drive the team?"

Finished clamping a rib to the spar, Herring checked that it was square before turning to his assistant. "I've heard through the grapevine that you're the best horseman around these parts, George. Would you consider doing the job?"

Housam scratched his chin and grinned. "Mark Twain once said he could live for two months on one good compliment – you could say I'm the same way. How much pull do you reckon the horses will have to make on your rope?"

As he thought, Herring smiled to himself. "The greatest pull will be when we start out. I'd say, no more than 300 pounds, quite a bit less, once the machine gets airborne—"

"Hell, that ain't nothin'! One old mare kin do the job!"

"Well, George, you've got two months to find the best animal available. One more thing... there has to be a small cart behind the horse to carry the man who will read the spring scale and—"

"Spring scale? Why do you need a spring scale?"

"To measure the pull on the kite's towrope."

"Why do you need to know that?"

"Everything in its own time, my friend! I'll tell you all about it before we do the experiment. For the time being, let's get back to work!"

St. Joseph, Michigan

Saturday, April 23, 1898

The sun shone brightly as Gus walked toward Truscott's, anticipating another productive Saturday work session with his compressed air motor. The temperature was already in the mid-40s, and the robins were chirping for rain as he turned north onto Broad Street for the pleasant downhill trek to Lake Street, a quarter mile away. As was his practice, Herring peered over the houses and treetops that lined the way, watching for a telltale sign of smoke... the rolling black cloud of effluent that normally issued from the company's 70-foot-tall, red-brick stack – a town landmark that also served to pinpoint the location of Truscott's central powerhouse. If smoke was indeed billowing, Herring knew that the coal-fed burner was supplying heat to the boiler and steam to the big, twin-cylinder engine... confirmation that power would be available to run the shop's machinery.

There's a land breeze blowing, Herring thought. *Lillian will be in high spirits... she can hang out her laundry today, without having to worry about the soot*!

Gus carried his bag lunch, a rolled-up set of detail drawings, and a wooden pattern for the motor's new cylinder head. Later, Housam would pick them up for delivery to the foundry. Matthias Arnot was also scheduled to arrive by train that morning; Herring was anxious to show off his latest work and discuss plans for flying the new apparatus as a kite.

Walking past a half-dozen storage buildings at the southwest end of Truscott's property, Herring was reminded of his tour of the Columbian Exposition, back in '93. He remembered being impressed that Truscott had received the "Award of Quality" designation for its scores of gondolas and small steam launches that exhibited flawless craftsmanship. Three years later, after severing his business ties with Octave Chanute, Gus marveled at his good fortune to be employed as an engineer at this same company.

The Truscott Boat Manufacturing Company was established five years before Herring's arrival in St. Joseph. English immigrant Thomas Truscott, along with his three sons, moved into the lakeshore complex originally occupied by the Morrison Tub and Pail factory. Initially, the company built small rowing craft, sailboats, and steam launches before expanding their product line around the time they hired Herring.

Continuing his walk down Lake Street, Herring passed a modest, single-story structure that served as the Truscott Company office. Determined to appear uninterested, he stared straight ahead, knowing that Old Man Truscott, recognizable by his long white beard, would be working at his desk and peering out the bay window – a practice he performed routinely every day of the week except Sunday. On the Lord's Day, the entire Truscott clan attended services at St. Joseph's Third Presbyterian Church, where Thomas served as an elder and Sunday school teacher – positions the octogenarian took very seriously.

Picking up his pace, Herring passed the massive boat-fabrication shop and at last, his destination – the three-story, red-brick machine shop – came into view. Other than its size, the building's outstanding feature was its windows: hundreds of them dominated all sides of the structure. Skylights also adorned the long, gabled roof, and stained-glass panels decorated the arched front entranceway.

Reaching the machine shop's entrance, Herring stopped to gawk at the lake freighter, *Northern Challenge*, which was docked at the Morrison Channel pier, a hundred feet to the north. As the steam-crane operator offloaded 20-foot-long bundles of cedar and cypress onto the steel-reinforced concrete dock, a crew of Truscott workers used four-wheeled carts to shuttle the lumber into two long, cinder-block storage buildings.

Across Lake Street to the east, Herring noticed a newly built, 30-foot steam-powered launch being loaded aboard a flatbed railway car for transport to parts unknown. Beyond the machine shop, a dozen cottage industries housed in ramshackle outbuildings produced the parts needed for Truscott's boats. An upholstery shop made hand-stitched seat cushions, awnings, and sails; a small foundry cast brass fittings; and a finish shop filed and polished the castings.

Another single-story shop, 75 feet long by 40 feet wide, was dedicated to assembling "vapor engines" for the current line of company launches. Within this facility, an experimental engine division specialized in the design and development of prototype engines. This is where Herring had spent most of his working days.

Entering, Gus was greeted by the familiar sights, sounds, and smells of an operational machine shop. Over the rattle and squeal of the overhead driveshaft and pulley system, three vertical milling machines were producing mounds of brass shavings, their expert operators attentive to the tasks at hand. Amid the smells of scorched leather and hot mineral oil, Herring relaxed; he was once again in his element.

Staring up at the 35-foot-high cathedral ceiling, he marveled at the bright, outdoor-like quality of the shop's interior. Sunlight beamed through skylights and the east wall, and was diffused by a multitude of metallic and whitewashed surfaces, illuminating every nook and cranny within. Where a second floor would normally have been, an 8-foot-wide, semi-cantilevered catwalk encircled the structure, gaining support and stability from its sturdy post-and-beam underlayment. Atop this platform, dozens of light- to medium-duty machines produced metal components, such as flagpole ferrules and rope grommets for decks and transoms.

Fully utilized, the machine shop could accommodate more than 100 workers with room to spare. However, America's ongoing economic depression had forced Truscott to reduce the workforce, thereby enabling Herring to rent space for his aeronautical pursuits. Striding down aisles of dormant, belt-driven machinery, Gus arrived at "his" area near the rear of the structure.

Setting his rolled plans and basswood casting-pattern onto the shop's solitary workbench, Gus then removed his canvas jacket and captain's cap, tossing them onto the shop's old wooden coat tree. His thoughts were soon interrupted by the thin, insipid voice of Edward Truscott. "Mr. Herring... I thought that was you. Working on the flying machine's engine this morning?"

Nodding, Herring reached out to shake his boss's hand. "If I don't get an early start, the day will pass before I know it!"

Edward, Thomas Truscott's youngest son, was charged with the design, development, and manufacture of engines for the company's launches. An 1888 graduate of the University of Chicago's College of Mechanical Engineering, Edward, who was the same age as Herring, also had a strong interest in exploring the promise of the internal combustion engine.

Edward stood two inches over six feet and weighed 150 pounds. When viewed from the side, he had an uncanny resemblance to a factory-fresh lead pencil... complete with a distinctive Roman nose. A younger version of his father, Edward also had a narrow face, elongated ears, and weak chin. In an attempt to establish a degree of individuality, he sported an impressive waxed handlebar mustache, which matched the color of his close-cropped, chestnut-brown hair. In his padded-shoulder tweed sport jacket and brown bow tie and trousers, Truscott seemed better attired for an afternoon at the country club rather than working in the grimy environment of his father's machine shop.

"Mr. Herring, I was looking at your little engine this morning, and I'm impressed by some of your innovations, especially the cylinder head and arrangement of the valves. I wanted to give the crank a turn, but decided not to risk getting splashed with castor oil. Father has a meeting scheduled this morning, so I decided just to look!"

Walking over to the test stand where the stripped-down engine sat, Herring beckoned Truscott to come closer. "Here it is... but I'm about to convert her into a high-pressure, compressed air motor."

Laughing, Truscott interrupted, "My God, Herring, why? This machine must have cost a pretty penny to produce! Why are you abandoning the effort?"

"It's the old story... if you're on the wrong path, what's the sense of running? With the vibration-damper flywheel in place, the engine is too heavy. Then, there's the issue of power; in its present configuration it doesn't produce enough to fly my *aerocurve* apparatus."

"Hold your horses!" Truscott exclaimed. "How do you know how much horsepower is required?"

While explaining the basis for his calculations, Herring noticed that Truscott had begun to smile as he removed his jacket and donned a shop apron. "There are people who say you should have your head examined for delving into the world of the impossible! Mind you, I'm not one of them!" Truscott exclaimed. "Actually, I'm rather impressed by your dedication and enthusiasm! Now, tell me more about this cylinder head and single overhead valve—"

For the next hour, Truscott pored over the engine. Herring removed one of the cylinder heads and explained the virtues of the hemispherical combustion chamber, recounting his experiences with the overhead valve and camshaft configuration.

Herring's gas engine of 1897-98; (1897) Aeronautical Annual

Wiping castor oil off his hands, Edward snatched his watch from a vest pocket and noted the time with a soft whistle. "Time flies!" he said, grabbing his jacket. "Father will be 'hell-bent for leather' if I miss the start of his meeting... Gus, we'll speak of this again!" With his parting proclamation ringing in Herring's ears, Edward Truscott trotted away.

*

"There you are, Augustus!" shouted Arnot, as he strolled toward the 10-inch metal lathe his friend was operating. "I thought I might be in the wrong building!"

"Matthias! I was wondering when you would arrive. What time is it?"

"Eleven o'clock. I'm a little late... I had a bit of difficulty finding this place. Fortunately, an elderly gentleman over at the business office directed me here—"

"That would be Thomas Truscott," Herring interrupted. "He owns the company."

Releasing the drive-belt tension on his machine, Herring turned to escort Arnot back to his part of the building.

"Unfortunately, Gus, I must return to Elmira this evening. There's much to discuss, so why don't you fill me in on what you've been doing?"

Herring approached the partially disassembled engine. "I wrote to you concerning the problems I was having with the gasoline engine. After conducting further tests on the dynamometer, I have concluded that the 1-to-1 speed ratio between the engine's crankshaft and the tandem propellers... is wrong. Unfortunately, I haven't been able to find an acceptable way to change this, so I have focused my attention on another type of prime mover. I'm currently modifying the engine into a compressed-air motor – an intermediate step in our quest for prolonged manned flight."

For the next several minutes, Herring explained the changes that he had made to the mechanism, while detailing the advantages of the air motor. "You see, Matthias, the twin-cylinder air motor will provide *two* power impulses for *each* revolution of the crankshaft. Combine that with the ability to vary its power by regulating the cylinder pressure, and we will have a very flexible prime mover for flying our machine."

Arnot thought about this carefully before responding. "To me, it's a question of knowing how much power is required to fly the machine. How do we know for certain, since a representative machine hasn't been flown?"

Surprised by Arnot's burst of insight, Herring quickly recovered. "It's the old chicken-and-egg argument. You can't fly the machine until you know how much power is needed... and you can't tell how much power is needed until you fly the machine!

"To make matters more complicated, there are two distinct power levels to consider. There's the power required to achieve takeoff and the power needed to keep the aeroplane flying at a steady-state condition. A much higher power level is required for takeoff, because the motor must overcome the inertia of the machine standing at rest."

"There's another problem," Matthias said, placing a hand on Gus's shoulder. "Earlier, you mentioned that you wanted to fly the powered machine as a kite before committing to any engine or motor... but, you will only be measuring the horsepower required for steady-state flight. Will that be adequate?"

Herring frowned. "It would be much harder to measure the variables. For instance, the towrope force at takeoff will vary depending on how fast the tow-horse starts off and the lifting surface's angle of attack; it's difficult to measure that.

"The experiment's not perfect, but I'm still planning to kite the machine in about six weeks. I'll add ballast to simulate the weight of the motor and propulsion system, and give her a try. I'll need you and Bill Avery to help take the measurements. After we complete the kite experiment, all that will remain is to finish work on the motor and test it for horsepower."

"What if this new motor doesn't meet your expectations?" Arnot kibitzed. "What then?"

Sinking down onto one of the shop's old wooden chairs, Herring thought, *Why is Matthias being so contentious? Something must be afoot.*

"Takeoff has always posed a problem for those contemplating powered, heavier-than-air flight. Langley used a spring catapult to *fling* his aerodrome models into a steady-state speed. Sir Hiram Maxim launched his steam-powered monstrosity downhill on a captive rail system. His idea was to overcome the machine's starting inertia by enlisting the help of gravity. Lilienthal planned to use a similar technique to overcome his lack of power... by launching the powered machine from his manmade hill."

"Why don't we do something similar? Why does the motor have to provide excess power just to overcome the startup inertia?"

Herring replied without hesitation. "Since the compressed-air motor is easier to handle than the gasoline engine, I considered launching from one of our sand dunes, but there are other concerns. For instance, say we launch the machine into an offshore breeze, switching the motor on after gaining flying speed; within seconds, I'd be flying over Lake Michigan! After witnessing firsthand the fate of Langley's aerodrome models after they alighted on – or should I say splashed *into* – the Potomac River, I would rather pursue another method. Besides, I don't swim that well!"

"What about towing the machine up to speed and altitude, as you propose to do with the kite experiment? Once you attain cruising speed, release the tow rope, switch on the motor, and fly away with adequate power."

"We'll keep that as an option. But first, let's see how the manned-kite experiment works out... then we'll decide what our next step should be. In the meantime, I've got my hands full completing the necessary modifications to the motor. I also have to find a gasoline or oil engine large enough to power a two-stage air compressor capable of producing upwards of 600 pounds of pressure.

"In that regard, I'll be heading back to Chicago to visit my former employer, the Great Lakes Boat Company. They had an old stationary engine/compressor unit that was no longer operable. I'll offer to make the necessary repairs if they'll loan me the equipment for our experiments. The need to store high-pressure air in an adequate vessel could be our most troublesome task... we'll have to wait and see.

"Speaking of vessels, one of the boilers that Truscott used in their old steam launches is the right size for our air tank. It's seven inches in diameter, two feet long, weighs seven pounds... and is rated at 1,000 psi."

At Gus's suggestion, they headed back to his Church Street shop so Arnot could view the progress being made on the new aeroplane. On the way, Herring shared his liverwurst and horseradish mustard sandwich with his friend and benefactor.

"You'll come back when I kite the machine, won't you, Matthias?"

Arnot spoke haltingly. "I'm not sure. My uncle has been unusually watchful of late. It was all I could do just to slip away for the day! He and mother have this mutual concern that my interests do not reflect those of my deceased father; as you know, banking was his life! Now that my older brother John has been diagnosed with consumption and is convalescing in an Arizona sanitarium, all of their attention has been directed at having me continue the Arnot banking tradition. To make matters worse, mother is hounding me to remarry and produce an heir!"[5]

"I understand completely," Herring said. "I have my own mother problems."

At that moment, George Housam shouted a greeting to Herring from across the street. Hopping over the piles of horse manure that littered the brick pavers, he trotted over to where the men had stopped. "Gus! I was on my way to Truscott's to pick up the new pattern and drawings... thought you said you were gonna be there all day?"

"Sorry, George, with all of the commotion, I forgot about our meeting!"

After introducing Arnot, Herring suggested that George join them back at the Church Street shop. Fifteen minutes later, as Housam stoked the fire, Arnot addressed one of Herring's modifications. "According to your latest letter, you have incorporated a system that allows the lifting surfaces to be disassembled without cutting apart the struts or piano-wire trussing. How is that possible?"

"It's one of my best improvements yet, Matthias! Come over to the bench and look at our first dry run of the lifting surfaces assembly."

As the men gathered in front of the bare, two-surface framework, Herring pointed to the aero*curve* ribs.

"The framework of each lifting surface consists of multiple flat ribs glued at regular intervals to a 20-foot-long spar, then to the wing's forward and rearward edges. Notice that the ribs are carefully aligned to be square with the spar and edges. After allowing time for the glue joints to cure, each rib is arched to its correct curvature, using cotton kite string – a procedure similar to stringing an archery bow."

Pointing to the top of a steel tubing strut, Herring continued. "Notice that the vertical strut is fastened to a hardwood tab with a bolt and nut; this is the *key* to assembling and disassembling the surfaces. In fact, if you look at the top and bottom of all 16 struts, you'll see the same attachment method—"

"Are the hardwood tabs glued to the face of the spar and the forward edge?" Arnot interrupted.

"That's correct. In case of a crash, I'd rather replace a hardwood tab or two, rather than a slew of wooden ribs and steel tubing struts!"

Sensing that Arnot was following his logic, Herring continued. "Next, the frame for the top lifting surface is temporarily braced in position with lengths of scrap pine, making sure that it's plumb and square with the lower lifting surface... then the vertical struts are bolted into place.

"Here's the best part," Gus said, pointing to one of the strut bolts. "There's a cross-drilled hole and a circumferential groove cut into the head of each bolt. The hole allows the piano-wire to be threaded through—"

"From the top of one strut to the bottom of the next," Arnot continued, "resulting in a zigzag pattern from one wingtip to the other!"

"You have it, Matthias! Not only does this make short work of the spanwise trussing operation, but it also allows us to take them apart without destroying the wood and steel components! Just remove the nuts and push out the bolts that carry the piano-wire."

"What about the 'x' trussing between the spar and forward edge?" Arnot asked.

Moving to the wingtip where the piano-wire trussing between the rear strut and the forward edge could be seen, Gus pointed to another clever assembly feature. "If you look closely, Matthias, you'll see there's a special washer between the nut and the hardwood tab at each strut end. The washer is oblong, allowing it to protrude from behind the nut. By drilling a small hole through one end of the oblong, an anchor point is established for the piano-wire."

"Ingenious!" Arnot hooted. "I only have two questions: how do you tighten the spanwise piano-wire, and how do you intend to cover the lifting surfaces?"

"As for your first question... allow me to demonstrate." Retrieving two open-end wrenches from the bench, Herring shuffled over to a vertical strut and loosened a nut. With the second wrench, he slowly turned the head of the bolt 90 degrees clockwise, effectively winding a bit of the piano-wire into the shallow groove around the fastener's head, thereby tensioning the wire; he then retightened the nut, locking everything into place. "By working your way from one wingtip to the other, piano-wire tension and surface alignment can be tuned like a stringed musical instrument!"

"Bravo, professor!" Arnot shouted, applauding. "Now, what about the covering?"

"First, we'll disassemble everything, including the bow in the aero*curve* ribs; then we'll *silk* cover the top of the wing framework while it's lying flat on the work surface; it's much easier to work that way. Then, we'll brush three coats of pyroxiline dope onto the surfaces and let it shrink overnight. Afterward, we'll re-arch the ribs, using the cotton bowstrings, and reassemble the entire structure using the original piano-wire and strut bolts. Finally, we'll brush-on the last two or three coats of dope.

"After the dope has dried for a day or two, we'll pop the cotton bowstrings off the bottom of the aerocurve ribs. As you know, the piano-wire trussing between the fore and aft vertical struts retains the aero*curve's* arch.

"One last thing," Arnot said. "When are you going to build the *third* lifting surface?"

Shaking his head, Herring exhaled with a short laugh. "There won't be a third lifting surface. It adds too much weight and generates too much drift. I estimate that the apparatus, including motor, propellers, and accessories, will weigh about 85 pounds. Adding my weight into the mix, she'll top out at 250 pounds, ready to fly. With 162 square feet of lifting surface, the wing loading will be reasonable... only 1.54 pounds per square foot. The machine will have to fly a bit faster than last year's glider to generate enough lift, but that's where the motor's thrust will come into play. Look on the bright side... we have spare parts in case of a mishap!"

Other than an occasional pop or snap from the potbelly stove, a relative silence enveloped the shop as Arnot and Housam digested Gus's reasoning.

"Here's a question for you Gus," George said. "I've heard the words *engine* and *motor* kicked around a lot lately. Is there a difference between 'em?"

"Most people think that engines and motors are the same; they're not. An engine burns fuel internally to produce mechanical power, while a motor does the same *without* the internal burning. Some examples of motors include steam, compressed air, electric, carbonic acid, and even twisted rubber strips."

"Are you sayin' that a steam engine is really a steam *motor,* and that there's no such thang as an electric *engine?*"

"You've got it, George. Most people would say 'who cares,' but mechanical engineers know the difference and usually use the correct terminology – but, like everything else, some do and some don't. Octave Chanute doesn't know the difference – but then, he spent his working life designing animal stockyards and bridges.

"Matthias," Gus continued, "speaking of your Chicago friend... in a moment of weakness, I sent him three photographs of the gasoline engine mounted on the test stand, along with some performance figures. That was back in February, and I still haven't heard back from him."

Refusing to be drawn into a discussion about Chanute, a person he considered lacking in character, Arnot drew Herring aside. Placing a trembling hand on his partner's shoulder, he confessed, "Augustus, there's bad news... I can no longer support the flying-machine project."

CHAPTER 63

Rochester, New York

Wednesday, November 2, 1921

Back on the witness stand, Herring was prepared to answer an onslaught of questions related to Matthias Arnot's abrupt withdrawal from the flying machine program.

"Mr. Herring," Attorney O'Grady asked, "why did Mr. Arnot discontinue his support of the powered flying machine?"

Leaning forward in his chair, Herring took little time in considering his response. "Although he didn't elaborate, I surmised that his older brother's illness played a significant role in his decision to—"

"Objection!" Robbins cried, leaping to his feet. "Calls for speculation! Move to strike."

"Sustained. The witness is reminded not to answer questions with assumption, hearsay, gossip, or any other form of speculation."

"Sorry, Your Honor,"

"Mr. Herring," O'Grady continued, "with Mr. Arnot's financial support gone, what were your thoughts concerning the flying machine program?"

"Initially, I was devastated! However, although the costs for continuing the project figured to be overwhelming, I decided to carry on."

"Earlier, you mentioned traveling to Chicago in pursuit of an air compressor and engine. How were you able to accomplish that task without taking time off from your day job?"

"A week after Matthias dropped his bombshell, Ed Truscott sent me to the Windy City on company business. After completing my assignment, I found time to visit my old boss at The Great Lakes Boat Company."

Chicago, Illinois

Tuesday, April 26, 1898

Art Pesch didn't disappoint. Herring knew he would find his former boss working diligently on some facet of the company's latest experimental engine. It had been almost 10 months since Herring's departure, but Pesch's letters had kept Herring up to date on the status of their brainchild.

"Still at it, I see!" Herring hollered, as he approached Pesch's workbench in the heart of the machine shop.

"Augustus! I was just thinking about you. What brings you back to Chicago?"

At 6-foot-2 and 220 pounds, the 62-year-old Pesch's short, wavy brown hair was turning gray at the temples. After 35 years of work in the field of marine transportation, he was losing both his eyesight and his hearing. He was wearing a traditional black vest and bowtie combination, protected by a denim apron, and his black-rimmed *pince-nez* reading glasses hung from a gold chain around his neck. A thorough designer and meticulous machinist, Pesch had gained a well-deserved reputation for solving difficult problems in the arenas of marine power and propulsion; it was he who had recommended the Rühmkorff induction coil to Herring. Universally credited with producing the first porcelain, two-electrode sparking plugs in North America, Pesch

– humble to a fault – routinely redirected the recognition to others. Holding firm to his Austrian roots, he reflected the dignity of a mid-19th century, university-educated engineer.

"Actually, I'm here on Truscott business," Herring said, "but I had to see how the new boat engine was coming along."

Herring was referring to a four-cylinder, Otto-cycle engine that he and Pesch had designed from scratch. A disagreement with company officials over the placement of the poppet valves had led to Herring's dismissal – although the official reason given was a staff cutback due to the country's deplorable economic condition.

Pesch said, "The engine runs fine until I try to raise the compression ratio above the minimum... then it 'rattles,' overheats, and sometimes sticks a piston. I was wondering if you have experienced similar problems over at Truscott's?"

Had Great Lakes' management not interfered with Herring's decision to place the poppet valves *overhead*, the problems now occurring might have been avoided. But Gus was not there to gloat; he was hoping to borrow a broken-down engine and compressor. Therefore, a diplomatic response to Pesch's inquiry was in order.

"Everyone seems to be having similar problems. Recently, I read that the Germans were able to raise the compression by adding a small percentage of benzene to the gasoline. This leads me to believe that the rattling sound is actually some sort of combustion flaw that's limiting the compression and power of our engines."

"What have they got you working on over at Truscott's?"

Herring hesitated, uncertain how much information he should divulge.

"Day's three-port, two-stroke-cycle design. Truscott is calling their version the 'Vapor Motor.' They've finally decided to move away from their high-pressure steam engines... but as usual, there are problems. The main stumbling block concerns safety. We've experienced some serious fires and a minor explosion during the testing of various *surface-type* carburetors, to the point where old man Truscott has ruled them out completely.[1] However, we have developed a method for delivering gasoline vapor from a remote storage tank to the engine's crankcase, without mixing it with inlet air... which pretty much eliminates the possibility of fire or explosion."

Past differences set aside, for the next hour, the two engineers talked about many of the developments in the emerging field of internal combustion engines.[2]

"Here's something that I thought you would be interested in reading," Pesch said. "I was going to include this with my next letter."

Pulling a paper from the stack on the corner of his desk, Pesch gazed at Herring and said, "This notice is from the U.S. Patent Office. Back in 1833 – 65 years ago – the office's superintendent offered his opinion about Brown and Morey's early internal combustion engine applications." Herring silently read the document.

> *There is not in nature any better power than that of steam. We are fully convinced that a better power than steam will not result from the firing of gunpowder, or of any other explosive mixture by which sudden and intermitting effects are produced.*

Shaking his head, Herring forced a weak smile. "Self-anointed experts run in the family at our Patent Office."

Thinking that the time was right, he then revealed the real reason for his visit. "Art, I have a favor to ask. I'm hoping that you will loan me the old Daimler-Maybach four cylinder, in-line engine."

Pesch squinted as he tried to recall the engine in question. "Are you talking about the 5-horsepower stationary engine – the one with a 5-inch stroke?"

"That's the one. Last time I saw it, the rings and bearings were worn out."

Pesch waved his hand dismissively. "It's back there in the junk pile, but I don't think it's easily repairable... if I'm not mistaken, there's crankshaft journal damage."

"What about the two-cylinder compressor? Do you still have it?"

"Yes, but it's in bad shape, too! The valves leak, the packing is blown out, and there's a connecting rod knock in both cylinders. What the hell do you need them for?"

After explaining his dilemma and offering to repair both the engine and the compressor for the opportunity to use them in his flying experiments, Herring anticipated further questions from his friend.

"I can appreciate your enthusiasm, Gus, but that engine's in the junk heap for a good reason. Give me a minute to light this lantern, and I'll show you what I'm talking about."

Walking to the rear of the shop, Pesch stopped at a pair of sturdy, eight-foot-high steel-clad doors, secured by a hefty brass padlock. Handing the lantern to Herring, he pulled a three-inch steel key ring from his belt. Within seconds of picking through the hodgepodge of keys, he snatched the one he wanted, inserted it into lock and snapped it open. Pulling the great doors open exposed a dark and damp place known to employees as "the boneyard". The boneyard was 40 feet deep and ran the full 70-foot width of the building. It featured a compacted-dirt floor, no windows and no provision for artificial lighting. Filled with derelict machinery, decrepit boat hulls, abandoned construction fixtures, and used-up engines, the name of the place had always seemed appropriate to Herring.

"Lucky for me nothing ever gets thrown out around here," Herring said.

"You may not feel so lucky after seeing this engine!"

Pushing aside empty pasteboard boxes, the men snaked around pallets stacked high with construction materials and supplies from a bygone age. As Pesch hoisted the lantern above his head, the shadowy image of a once-proud German industrial powerplant loomed straight ahead, lying forlornly on its patch of oil-saturated soil.

Pesch set the lantern down next to the Daimler-Maybach and pointed to two obvious deficiencies. "Look here, Gus; both Brayton-type surface carburetors are missing, and the hot-tube igniters are broken off flush with the cylinder-head casting."

"Doesn't matter, Art; I don't have any choice but to attempt a major repair. If I had funds at my disposal, I'd purchase a new engine and compressor tomorrow."

Shaking his head, Pesch thought of something else. "What pressure are you shooting for?"

"Six hundred psi, at low volume... about half a cubic foot."

"Six hundred pounds! Gus, that compressor's *only* a two-stage unit. Besides the valve problem, you gotta make new cylinder heads with a higher compression ratio... and then you will need an *intercooler* to keep the temperature in check."

Herring had already worked out the numbers.

"Here's what I think will work: I'll have to get the engine's performance back to at least four horsepower at 600 rpm. If I drive the two-stage compressor at a 3-to-1 speed reduction, I should have plenty of torque to turn both stages with compression ratios of 7-to-1. It should do the job, figuring a 10-percent loss due to leakage, heat and friction."

Herring hesitated, as he continued to think. "You're right about the intercooler, Art... I'll have to run the first stage's output through a few coils of copper tubing, while using air blown off the engine's flywheel fan—"

"That's something else that you will have to jury-rig," Pesch interrupted. "We're using the Daimler flywheel on our prototype engine."

"That's okay, it will give me a chance to design a better fan! Anyway, the intercooler should hold the temperature down before passing into the final stage of compression. If everything goes as planned, it'll only take about 20 minutes to fill my tank."

"What about the ignition system and carburetor?"

"If it's all right with you, I'll switch to spark ignition. At first, I'll use my Rühmkorff coil and carbon batteries in conjunction with make-and-break low-voltage contact points; I'll also need a mechanical distributor. If I can afford it, I'll add a *Bosch* magneto later... for prolonged running."

"You might want to give this some more thought, Gus. Spark ignition is wonderful when it's working properly, but it's expensive and adds complications. For your needs, the old hot-tube ignition system is hard to beat."

"You're right, of course, but I'm also looking at the safety issue. My friend, Carl Dienstbach, reported that Maybach was injured last month in a gasoline explosion; he was experimenting with a hot-tube ignition engine! Working around open flames with gasoline has always scared the hell out of me. To make matters worse, I'll probably run the engine and compressor indoors! I already have the Benz sparking plugs, so replacing the old hot-tube system should be a straightforward job.

"As far as the carburetor is concerned, I'm prepared to substitute Maybach's *Spritzdüsen Vergaser* – a venturi-type spray carburetor in place of the dangerous surface units."

Rochester, New York

Wednesday, November 2, 1921

Herring's under-oath description of his meeting with Art Pesch at the Great Lakes Boat Company was rudely interrupted when a photographer's flashbulb exploded, jarring everyone in the courtroom back to the present. After taking a moment to reprimand the hapless newspaperman, Judge Sawyer next turned to O'Grady.

"You may continue, Counselor."

"Did the Great Lakes Boat Company give you the opportunity to repair and use the engine and compressor in question?"

Stretching his arms over his head to relieve the tension, Herring summarized the events that followed. "Sadly, I left Chicago empty-handed."

"Then what happened?"

"A week later, I was notified that a big wooden crate awaited me at the train depot. With the help of my assistant, George, we wheeled the crate in question to my area at the rear of Truscott's machine shop. As you might imagine, the return address on the crate caused quite a stir! Years later, Art laughed when I asked him if his company had decided to send the engine and compressor because they knew it would annoy old man Truscott."

At this point in his questioning, O'Grady requested a five-minute recess to confer with his law partner. Back at the plaintiff's table, O'Grady and William Maloney reviewed their strategy. "There's a lot at stake, Bill," O'Grady whispered. "Are you sure that you want to take this phase of the—"

"Jim... it's my turn," Maloney interrupted. "Have a seat. We've talked about this long enough."

Slapping Maloney on the back, O'Grady collapsed into his chair behind the plaintiff's table. "Okay, Bill, give 'em hell!"

As the judge called the court back into session, Maloney stepped behind the podium and announced to the bench that he would be handling the next phase of questioning for the plaintiff.

"Very well, Mr. Maloney, you may proceed."

"Mr. Herring, does the name Wilhelm Maybach ring a bell?"

"Yes. I corresponded with Herr Maybach for several years in the 1890s and early 1900s."

"What was Maybach's claim to fame?"

"As a German national, he was considered to be the world's foremost engine and early automobile designer."

"How did you become acquainted with Herr Maybach?"

Glancing down at the floor, Herring collected his thoughts. "I received a letter from him requesting information about my overhead valve engine of 1897: a mutual friend gave him my address."

"Who was that mutual friend?"

"The journalist, Carl Dienstbach."

"Back in '97, you were 31 years of age. How old was Herr Maybach?"

"Objection! Irrelevant!" Robbins chirped.

"Overruled. You may answer the question, Mr. Herring."

"Let me see... Wilhelm was 21 years older than me, so that would have made him 52 in 1897."

"Did you provide Herr Maybach with the technical information about your engine?"

"Eventually... yes."

"Weren't you worried about him stealing your ideas?"

"At the time it didn't concern me, since I was completely absorbed in pursuing my flying-machine patent applications.

There simply wasn't enough time or money left over to protect the unique features of the aero engine."

"Did Maybach use any of your ideas in his engines?"

"In 1900, Wilhelm was the technical director at *Daimler-Motoren-Gesellschaft,* where he created the Mercedes racecar – an unbeatable design for its day—"

"And let me guess," Maloney interrupted, "the Mercedes engine incorporated your overhead valve, hemispherical combustion chamber idea."

"Objection! Counsel is leading the witness!"

After administering a mild scolding, Sawyer permitted Maloney to reconfigure his statement into a question.

"Mr. Herring, what did you learn about the technology behind Herr Maybach's Mercedes racecar engine?"

"It contained my overhead valve and hemispherical combustion chamber innovation."

"Objection! Your Honor, there's no proof that Herring contributed anything to this engine."

"Your Honor," Maloney shot back, waving several sheets of paper above his head, "in this letter dated January 10, 1901, Maybach acknowledges Mr. Herring's contributions!"

As the letter reached the defense table, prior to being admitted into evidence, Robbins objected again. "Judge, Maybach's letter is written in German! Who was responsible for the English translation?"

"Mr. Maloney, who translated the letter?" Sawyer asked.

"Mr. Herring, Your Honor."

"Objection! For obvious reasons, the defense does not wish to accept the witness's translation! Therefore, we petition the court to disqualify the correspondence from being entered as evidence."

Studying his notes, Judge Sawyer took more than a minute before responding. "I have decided to take the defense counsel's objections under advisement, pending an official translation of the Maybach letter. The court clerk will arrange to have the translation performed immediately – today, if possible. In the meantime, the attorney for the plaintiff may continue with his questioning."

Maloney ambled over to his witness. "Mr. Herring, do you *know* who patented the overhead valve, hemispherical combustion chamber engine?"

Herring knitted his eyebrows. "Mr. A.R. Welch was granted a U. S. patent for the design in 1907."

"Did you know Mr. Welch?"

"Edward Truscott introduced Allie Welch to me at my rented shop space, back in April of '98 – he was anxious to see my twin-cylinder aero engine. For the next three years, Welch collaborated with Truscott to produce the first overhead valve, hemispherical combustion chamber engine to be fitted into one of the old man's boats."

"Objection! Argumentative! This witness has offered no proof as to the authenticity of these outrageous—"

"Your Honor," Herring interrupted, "the proof is right here in the Truscott catalog for 1901. May I read Edward Truscott's quote regarding the hemispherical combustion chamber engine?"

Judge Sawyer tapped his gavel. "No, Mr. Herring! Not until the catalog in question has been entered into evidence!"

Holding up a manila folder full of drawings, photographs and letters, Maloney reflected on the judge's ruling. "Your Honor, at this time, the plaintiff wishes to introduce evidence concerning Mr. Herring's development of the overhead valve, hemispherical combustion chamber engine!"

Since the noon hour was at hand, Sawyer decided to break for lunch. He recessed the morning session with a tap of his gavel.

*

Entering Herring's engine evidence required over an hour of court time. Four sets of drawings, a dozen glass-plate photographs, and six letters and catalogs had to be processed. Mildly fatigued by the tedious process that included multiple failed objections from the defense, the judge ordered Herring's return to the witness stand and the resumption of his testimony.

"Mr. Herring," Maloney said, "I am handing to you the Truscott Boat Manufacturing Company's catalog for 1901. Please read Edward Truscott's quote concerning a new engine that his company had recently tested."

Returning to his place behind the podium, Maloney waited while Herring found the underlined quote within the eight-page, black-and-white catalog. With a nod from the judge, Gus began to read:

> " 'Although the overhead valve, hemispherical combustion chamber engine performed better than my other engines, the open valves threw oil onto the passengers; for that reason we decided not to pursue its production at that time.' "

Maloney followed with a question. "What did you think about your employer using *your* invention?"

"What could I do? By 1900, I was no longer employed by Truscott, and I didn't possess the necessary engine-related patents or patent applications to help protect my ideas!"

"Did you complain to Edward Truscott?"

"I did. He informed me that since I developed the engine during my term of employment, their company legally retained all of its patent rights. When I advised him that I had been developing the hemi-engine idea since my undergraduate days at the Stevens Technical Institute, Truscott remained steadfast. He reminded me that I had agreed to the arrangement when I signed my contract of employment. Since I didn't have the resources or backing to argue the point further – I had to let it go."

Maloney used the newly entered evidence, including dated drawings, photographs, and letters, to make the plaintiff's point: Herring had once again been cheated of the opportunity to reap the benefits of his own creative labors.

With the five o'clock hour approaching, Judge Sawyer adjourned court proceedings for the day.

Rochester, New York

Thursday, November 3, 1921

Having completed his regimen of morning tasks, Judge Sawyer began the day's proceedings by waving several sheets of court stationery in the general direction of the litigants. "The court has just received the English language translation of the Maybach letter from the appointed translator. At this time, I ask that all parties exercise their right to review the document, prior to making motions."

This time there were no objections, and the letter was summarily entered into evidence. Handing the letter to Herring, Maloney wasted no time in referring to its contents.

"Mr. Herring, what was the gist of Herr Maybach's letter of January 10, 1901?"

Shrugging, Herring slouched in his chair. "Other than friendly conversation, Wilhelm expressed his appreciation for the engineering information that I provided to him over the previous three years."

"What engineering information did Herr Maybach refer to? If you will, Mr. Herring... please quote from the letter."

Slipping on his reading glasses, Herring scanned the two-page letter before finding what he was looking for.

"Ah yes, here we are. Maybach wrote,

> 'Dynamometer tests on an experimental 4-stroke cycle engine back in '99 confirmed the superiority of your hemispherical combustion chamber innovation when used in cooperation with overhead inlet and exhaust valves...'"

"What else does Mr. Maybach disclose in his letter?"

"Based on this early test, Maybach designed, built, and tested the prototype 35-horsepower, four-cylinder engine for the Mercedes racing automobile. This trendsetting machine was produced shortly after the untimely death of Gottlieb Daimler, Maybach's longtime mentor and technical associate... in 1900."

Finally, Maloney probed the uncanny similarities between Maybach and Herring concerning relationships with their former employers. According to Maybach, many of Daimler's early patents were based on his assistant's inventions, but Maybach was also not allowed to share in the fruits of his ideas. In this regard, Maybach was sympathetic toward Herring's situation. Gus continued,

> " 'Since you are now an independent, there is a golden opportunity to shine in your own right, because all of your ideas will now be attributed to you!' "[3]

"Objection! Irrelevant!" Robbins hollered, "Your Honor, what in God's name does this testimony have to do with the case of the Herring-Curtiss Company versus Glenn Curtiss?"

"I'll answer that, Your Honor!" Maloney huffed. "The plaintiff has consistently demonstrated that this lawsuit is about the *people* who have prevented Augustus Herring from realizing the fruits of his study and labor in the field of aerial navigation! This testimony has proven that Mr. Herring's internal-combustion engine ideas were pirated... only the latest indignity to be forced upon this man in his quest to develop the world's first manned, powered, flying machine!

"The defense counsel questions the relevancy of this testimony? By now it should be clear to all who have been following these proceedings that Glenn Hammond Curtiss is the foremost pirate of them all! Not only have he and his band of thieves pillaged Mr. Herring's ideas, investments, and rewards for years of labor, but Curtiss has also fleeced the citizens of this great country for millions of dollars that—"

"This is an outrage!" Robbins cried, over the thunderclaps of Judge Sawyer's gavel. "This nincompoop is making a mockery of the court! The defense demands that his argumentative and childish remarks be stricken from the record!"

Peering over at Maloney, who was standing rigidly at his podium, Robbins pointed a bony finger at the hulk of a man.

"How dare you – you bloody Neanderthal!"

With the calmness of a veteran warrior set to act, Bill Maloney strode purposefully toward the diminutive defense attorney, his face and neck reddening to the occasion, his sledge-like fists clenched at the ready.

"I'll give you Neanderthal, you sawed-off son of a bitch!" Maloney raged as he booted an obstructing wastepaper basket across the courtroom's varnished floor.

With the judge howling for the bailiff to maintain order, Curtiss, Wheeler, and Baldwin beat a path to the courtroom's rear door, leaving their chairs scattered about like roof shingles in the wake of a tornado. Robbins tried to flee, but wasn't fast enough. As O'Grady and the bailiff rushed over, trying to slow Maloney's advance, one of the big Irishman's hands grabbed a corner of the defense table and flipped it over. As it crashed to the floor, Maloney seized Robbins by the neck and lifted him off the floor, his limbs flailing wildly, like those of a doomed chicken. It took two convincing whacks of the bailiff's Billy club to subdue the enraged attorney.

*

When he awoke half an hour later, Maloney found himself behind bars with a throbbing head. "You've really done it this time, William," O'Grady said, trying and failing to restore the torn pocket on his gray tweed sport jacket. "Furthermore... this jacket is only a month old!"

"To hell with the jacket! I'll buy you another," Maloney said, rubbing the crown of his head. "What happened to Robbins? Did he soil himself?"

"Probably. The last I saw him, he was running like a rat to the rear exit. After the four bailiffs took you away, the judge called court back to order, just long enough to adjourn for the day!"

"It took four coppers to take me away, eh? Even though I was out cold?"

"Yeah, I guess they heard the commotion and came running from the other courtrooms. Someone said it would take four of them if you woke up, or somethin' to that effect."

"What was Herring doing during the hullabaloo?"

"He just sat there grinnin' – like a Cheshire cat!"

"What did Old Man Sawyer have to say?"

"Like I said, you really did it this time. He was so pissed off he could hardly speak, and for him, that's saying something! He did manage to spit out something about dealing with you tomorrow."

Rochester, New York

Friday, November 4, 1921

With the exception of Irish Bill Maloney, all the litigants and attorneys were present and accounted for. The courtroom's spectator section was packed with newspaper reporters, photographers, and curious citizens, all hoping to witness more fireworks. Chloe Herring, a regular at the proceedings, didn't hesitate to voice her opinions concerning the previous day's brouhaha, which again made her the darling of the press corps.

As always, Judge Sawyer nonchalantly breezed into his courtroom. As he settled into his big armchair behind the bench, an air of anticipation hung over the place like an early morning haze. Addressing the assemblage, Sawyer minced no words.

"The chaotic breakdown of our judicial system, as exemplified by yesterday's thuggery, should be an embarrassment to all civilized Americans. As a result of his acute contempt of court, Attorney Maloney will be confined to the county lockup for the next seven days. I have also filed a report with the New York State Bar Association in regard to his deplorable conduct. Furthermore, for the duration of this civil action suit, Mr. Maloney is barred from actively participating in the presentation of the plaintiff's case. *If* and *when* I decide to allow him to return to these chambers, he may only serve as a legal consultant to the plaintiff.

"Be advised, Mr. Robbins, your remarks from yesterday have been duly noted by the court stenographer, and have also been forwarded to the Bar Association, along with my opinion that you must share part of the responsibility for having instigated Mr. Maloney's outrage.

"However, since you are the sole litigator for the defense, I am prepared to move forward with this civil action, providing I have your assurance that there will be no further unseemly outbursts. What say you, sir?"

"You have my word, Your Honor," Robbins pledged, his left arm cradled in a white linen sling.

"If you break your promise, Mr. Robbins, I will consider declaring a mistrial and charge the defense with *all* of the court costs, including the salaries for the plaintiff's attorneys. Have I made myself clear?"

"Yes, Your Honor. Very clear."

"Very well. Mr. Herring, please assume the witness stand. Mr. O'Grady, you may proceed with the questioning."

"Your Honor, before yesterday's unfortunate interruption, counsel for the plaintiff was establishing Mr. Herring's credibility as an expert in the field of combustion engines. I will continue along similar lines.

"Mr. Herring, in the year 1898, what was your primary area of expertise?"

"Because of my job at Truscott's, most people considered me to be to be a power engineer – a professional who worked with motors and engines. My time spent with flying machines was considered to be an eccentric hobby."

"If you had patented your overhead valve, hemispherical combustion chamber engine, is it possible that you would have become a wealthy man?"

"Objection," Robbins whimpered. "Counsel is asking the witness to speculate."

"Sustained."

Having made his point, O'Grady moved on. "In 1898, did you consider the development of an adequate *engine* to be more important than the further development of the flying machine's *control system*?"

"Absolutely! My two-surface glider designs of 1896 and '97 proved to be controllable by the combination of weight shifting and the automatic regulation of the tail surfaces to upheavals in the wind. It was my contention then, as it is today, that the next significant advance in the pursuit of manned flight would be the addition of a lightweight engine and propulsion system."

"Your decision to temporarily abandon your gasoline engine in favor of the compressed-air motor represented a step backward in this pursuit... did it not?"

Nodding, Herring sat straight in his chair. "Yes... it probably did. However, the compressed-air motor still offered the potential to demonstrate the merits of adding power to a flying machine of proven gliding ability. It would allow an operator to take off from level ground and fly into a wind of significant speed, while advancing relative to the ground, under complete control – something that no other manned, experimental apparatus had been able to accomplish up until that time."

O'Grady signaled a change in his questioning by moving out from behind the podium. "Describe your subsequent progress with the new flying machine."

Herring perked up. "Like I mentioned earlier, working on the aeroplane was more like my hobby, whereas the engine and motor work was my job. After Arnot's demoralizing news, I buried myself in the construction details of the apparatus. One thing was for certain... she was a real beaut! For the first time, one of my aero*curve* machines resembled a piece of expertly crafted furniture.

"By the end of May, the machine was nearing completion, and I was contemplating contacting Bill Avery and Henry Clarke to come help with the kite experiments. That's when I received an unexpected letter from Matthias, accompanied by a check for $300!"

"What did Mr. Arnot have to say?"

" 'When do you need me?' "

CHAPTER 64

St. Joseph, Michigan

Friday, May 27, 1898

One by one the flying-machine men gathered at Herring's Church Street shop. First to arrive was Henry Clarke of Philadelphia, followed by Bill Avery of Chicago and Matthias Arnot of Elmira. Already present and accounted for were George Housam and Gus Herring, who were busy putting the finishing touches on the machine's towrope-release mechanism.

As they surrounded the completed apparatus, the group's collective attention was focused on some features not previously seen.

"Jiminy! Look at the baby-buggy wheels!" Avery said with a chuckle. Walking over to the strut-mounted spoke wheels with their vulcanized, solid-rubber tires, Avery bent down and inspected the piano-wire bracing.

"Since she's going to fly faster than the '97 glider," Herring observed, "I added the wheels to help reduce the shock of a hard alighting... besides, they'll make the machine easier to move around."

"What else is new?" Arnot inquired.

"Here's somethin'," Housam said, as he crawled into the operator's position. "Now there's a place for Gus to sit! See the little wooden seat danglin' from four wires?"

Lowering himself onto the seat, Housam demonstrated its range of motion. "It swings back and forth, but you can also sway side to side when you have to."

Following a brief discussion about the seat, Herring slid underneath the machine's lower lifting surface, taking care not to become entangled in the alighting gear's support wires. A moment later, he emerged behind the central framework, where he pointed out one of the new machine's more subtle features. "Just above the lower wing, there's a detachable steel tubing frame for the compressed air motor and rear propeller shaft; it's fastened to, but independent from, the central framework. When this sub-frame is removed, the apparatus can be transformed into a glider or kite, whatever the need may be."

Herring then urged the men to join him behind the machine's lifting surfaces. "This next feature improves the automatic regulator's ability to control the cruciform tail. The rear end of the four supporting steel tubes is now fitted with an *aluminum* universal joint, making the action smoother and more precise than it was with the old leather part."

Lifting the tail to shoulder height, Herring pointed to the bottom of the vertical rudder. "The bottom of the rudder has been fitted with an aluminum runner to help protect it from damage when it drags in the sand. If you will recall, Matthias, that was our biggest repair problem on your glider."

Drumming his fingers on the gossamer-like contraption's lower lifting surface, Avery shook his head. "Egad, Gus... this covering job turned out great! How did you get it so taut?"

"A combination of things: first we moved two of the three spars to the front and rear edges of the lifting surface... this gave the silk covering something to grab onto at the perimeter. Then we used a new formula of pyroxiline dope to keep the silk both taut and flexible."

"How is this formula different?" Arnot asked.

Grinning, Herring strode over to a nearby workbench, where he lifted an eight-ounce medicine bottle from amid the clutter. "I added half a percent of castor oil into the mix. The dope still shrinks, but now it seems to stay pliable – much the same as *goldbeaters skin* – but it's better able to ward off stone strikes and other minor bumps."[1]

*

After moving their luggage into Herring's spare bedrooms, the men gathered in the kitchen, where Lillian had set up a buffet-style lunch of soup, sandwiches, and hot coffee. While they dined, talk gravitated to news of the month-old Spanish-American War.

"Why did it take the president almost two months to declare war?" Bill Avery asked. "Damn! They sank the *Maine*... and what's worse, more than 200 of our sailors died."

"Apparently, the sinking of our battleship in Havana Harbor wasn't enough to convince McKinley," Arnot lamented. "The Democrats had to push him into it."

Even George had an opinion. "Somebody's gonna pay... I hear that Teddy Roosevelt and his Rough Riders are headed for Puerto Rico."

Between bites of his bologna sandwich, Herring relayed his latest gossip. "I had a letter from Huffaker last week. There's a rumor circulating that the War Department has asked Professor Langley to build a man-carrying aerodrome... our generals want a powered flying machine for *reconnaissance*."

"Poppycock!" Clarke snorted. "From what you've told me about Langley, the war will be over before he gets all of the brass fittings polished!"

After a good laugh, the group of five returned to the shop, where the talk turned to testing the new flying machine as a kite.

"Gentlemen," Herring said, "we only have three days to get this machine tested. Since we may experience some bad weather, the apparatus must be ready to fly at a moment's notice. If we're fortunate, data collection could be completed in a matter of minutes. On the other hand, if things go wrong... all of our work could go for naught."

"Where are we going to test, Gus?" Avery asked.

"On Silver Beach... right here in St. Joseph. There's a long stretch of sand running along the southeastern shore of the lake. I'm hoping for a southwest wind, but that's not the prevailing direction for this time of year."

Henry Clarke had a question. "Where we gonna keep the machine while we're waitin' for good conditions?"

Herring waved his hand in a dismissive manner. "Problem solved. I've rented the beach pavilion for the next three days. The building isn't ideal, but it's located less than half a mile from here; right behind the two water slides."

"What's wrong with the place?" Clarke asked.

Herring shook his head. "The building only has a 10-foot roll-up door; it's a little too narrow... so we'll have to take off the tail to get the machine in and out."

Herring had had a similar problem with the shop door behind the house. He and George had often joked about having to tear the barn down in order to move the flying machine outside.

Herring continued, "I rented a horse and wagon from Mr. Engberg, George's boss. Early tomorrow, we'll load everything and head for the pavilion to set up shop. When it comes time to fly, the mare will act as our tow engine. Speaking of towing, I want to explain how the towrope system works in getting the apparatus airborne."

At first glance, the method seemed simple. The horse, attached to a 100-foot rope, would tow the flying machine into the air, whereupon four critical measurements would be recorded. But there were concerns. At the conclusion of the experiment, how would the towrope be disengaged, allowing the apparatus to descend to the beach as a simple glider? Then, if an errant tow occurred, could the rope be disengaged quickly enough to avoid a crash? And, there was the thorny issue of *where* to attach the towrope to the flying machine. Based on his experience with smaller kites, Herring knew that the proper attachment point would be critical. It would either allow the apparatus to soar smoothly into the air or cause a nose-high, zigzag race for the clouds – the result of a location that was too far aft.

According to Herring's most recent calculations, the center of gravity needed to be approximately six inches below and 12 inches aft of the lower lifting surface's forward edge. For safety's sake, Gus decided that the initial tow should be performed from an attachment point farther forward. If the tow produced a flat, straight ahead launch, he would disengage the rope and land. By reattaching the lanyard several inches aft, the team would be able to simply try again. The process would be repeated until the desired soaring climb to equilibrium occurred.

In keeping with the long tradition of kite flying, Herring had chosen to use a rope lanyard to divide the towing load between the port and starboard sides of the machine. The free ends of the lanyard would be attached to the two horizontal wooden bars that were bolted into place between the machine's fore and aft vertical struts – the same forward struts that supported the alighting wheels. In the lanyard's final configuration, the two attachment ends were fitted with a wide loop of half-inch-thick India rubber, which would act as shock absorbers during the initial phase of the launch.

The loop of 3/8-inch lanyard rope, when fully extended to the front of the machine, formed a "v" that was attached to a hook-and-latch release mechanism. A small loop at the end of the 100-foot towrope provided a connection to the hook. A

1/4-inch cotton cord ran from the mechanism's latch to a position that was within the operator's reach.

Rochester, New York

Friday, November 4, 1921

Lunch with Chloe was relaxing, but all too brief, Herring thought, moments before being jostled back to reality.

"Augustus Herring, to the witness stand!" the judge directed. "You are still under oath, sir!"

Bypassing the preliminaries, O'Grady steered Gus's recollections to the events of May 1898. "Mr. Herring, when was the flying machine moved from your Church Street shop to the Silver Beach pavilion?"

"Saturday morning, just after daybreak. There was no traffic on Church or Broad Streets because the neighbors were still asleep. We had the wagon unloaded before anyone realized we had left the shop."

"Were you able to fly the machine right away?"

"No. There was no wind, so we worked on getting the equipment ready, and installed a tow hitch on the wagon. By noon, the wind was still under 10 miles an hour and blowing from exactly the wrong direction. After George returned from grazing the mare at the beach's edge, we ate the bag lunches that Lillian had prepared for us. Immediately afterward, I gathered the men together and clarified the purpose of the experiment."

"What did you say?" O'Grady asked.

"I tried to keep it simple. I said that we needed to determine *five* things: the mare's walking speed during the tow, the anemometer reading aboard the apparatus – which represented the sum of the wind speed and the mare's walking speed, the amount of pull on the towrope, the angle that the towrope formed with the ground, and the angle formed between the machine's aero*curve* and the horizon. I explained that with this data, simple calculations would reveal the thrust required from the machine's propellers, as well as the horsepower needed from the compressed air motor."

"Did you assign tasks for the experiment, or did you ask for volunteers?"

"I assigned some of the tasks based on experience. George and Henry were responsible for getting the machine airborne. George Housam, an expert horseman, was charged with controlling the mare and estimating her towing speed; there could be no lurching, rearing, or bucking – a smooth start was vital to avoid damage to the apparatus.

"Henry Clarke was to run alongside the lower port-side lifting surface, helping to guide the machine during the first critical seconds after launch. If it veered from its intended flight path, he would be required to correct the deficiency by pushing or pulling the surface in the desired direction. Based on the scores of launches that Henry had presided over during the glider flights of '97, I had great faith in his abilities.

"Although collecting data had nothing to do with the success of the flight, it had everything to do with the success of the experiment. Matthias Arnot volunteered to measure the pull force on the towrope. He would crouch in the rear of the wagon, spring scale in hand, waiting for Mr. Avery's signal.

"After helping to ensure a successful launch, Henry would then sprint to the wagon where he would also await Mr. Avery's signal before measuring the angle formed between the ground and the towrope; a protractor and bob-weight had been hung from the rope for that very purpose. As the flying machine's operator, I was required to obtain its air speed from the machine's onboard anemometer.

"Bill Avery's measurement would determine the success or failure of each flight. While observing from atop a 60-foot dune, 200 feet to the left of the launch site, he was entrusted to measure the aero*curve's angle of attack*, with the same instrument he had used at the Indiana dunes. That instrument," Herring continued, "consisted of a tripod-mounted and -leveled wooden beam, a stationary protractor, and a pivoting metal rule. Avery would manipulate the rule's angle until its edge ran precisely through the *forward* and *rearward* edge of the flying machine's aero*curve*, thus obtaining its angle of attack to the wind.

"Any aero*curve* angle greater than *four* degrees would generate too much aerodynamic drift, thus signaling a failed attempt. This determination required Avery to alert the others by blowing a brass army whistle – twice in rapid succession. A single whistle would indicate a successful attempt."

O'Grady referred to his notes. "Did the wind pick up enough to attempt a flight on Saturday?"

"No… but it was blowing lightly out of the southwest, a good sign for Sunday."

"With flying ruled out for the rest of the day, what on earth did you do next?"

Gus could sense O'Grady's growing impatience. "We decided to make a dry run! With the mare harnessed to the wagon, George coaxed the animal out to our likely starting point on a compacted section of the beach. Next, we strung out the towrope, hooking one end to the wagon's tow hitch, and the other to the lanyard's release mechanism. Assuming the role of the flying machine, Henry and I each grabbed an end of the lanyard, he on the starboard, me at the port. With Matthias manning the wagon and George driving, I whistled for him to start the process of moving down the shoreline. Four to five miles an hour was the objective."

"Did anything of significance happen?"

Herring's jaw tightened. "The damned horse *lurched*, causing Matthias to take a tumble, while the lanyard was yanked out of my hands. After three more botched tries at a smooth start, George had an idea."

"Let me guess!" O'Grady shouted. "Shoot the damn horse!"

A ripple of laughter rolled through the courtroom.

"Not quite that drastic," Herring said, grinning. "George pulled a carrot from his jacket and teased the mare, allowing it but a single bite."

"What effect did that have?"

"It must have done something, because she started up nice and smooth the next time. Oddly enough, this happened every time he used the *carrot trick!*"

St. Joseph, Michigan

Saturday, May 28, 1898

After devouring Lillian's meal of roast turkey, dressing, mashed potatoes, cranberry sauce, and hard-crusted Italian bread served with pots of hot coffee, the men retired to the comfort of the parlor to relax and play cards. Herring and Arnot had business to discuss and decided to take a leisurely stroll around the block before dark.

"Matthias, I can only imagine how difficult it must have been for you to get away—"

"Yes," he interrupted through clenched teeth, "but those problems are a thing of the past. In that regard, I am happy to announce that I am no longer a central figure in the Arnot banking dynasty!"

Stunned by Arnot's announcement, Herring pressed for details. "Stop me if I'm prying, Matthias, but what happened?"

Arnot's face lost its color, as he began to perspire. "After having to tell you that I would no longer be involved with the flying-machine project, I spent the next month moping about. My unhappiness finally got the attention of the bank president, my 64-year-old uncle, who took the matter to Mother. During the ensuing confrontation, I decided to call their bluff."

"Bluff? What are you talking about?"

"There are a few things that I should clarify. My older brother, who is childless, is dying of consumption in an Arizona sanitarium... I may have mentioned this. Several years ago, my beloved wife Alice died suddenly. Gus, she was only 21... I believe that you knew this, as well. However, what you didn't know concerned my mother. She is frantic to have a male heir – someone to carry on the Arnot name!

"After some serious thought, I decided to go for broke. Leaving out the grisly details, here are the results of our negotiations, in their order of importance to me: first, I've been released from the day-to-day operations of the family's banking business. Next, I will retain the symbolic title of vice president and realize an annual salary commensurate with the position. In return for my freedom and financial security, I have agreed to an *arranged* marriage. Furthermore, I have contractually agreed to produce a minimum of four offspring with my new wife, while allowing Mother a free hand in their upbringing. If female descendants are conceived, the number of children must increase until at least two healthy males become part of the mix!... How is that for selling your soul to the devil, Augustus?"

Shaking his head, Herring continued to walk with his hands buried deep in his pockets. "What about your uncle, the bank president? Doesn't he have children?"

"Uncle Matthias never married... he's a poof."

St. Joseph, Michigan

Sunday, May 29, 1898

The men stood by silently as the padlocked pavilion door rattled in the gusting wind. While Herring fumbled to find the key, a thought ran through his mind. *The weather had better cooperate... I need a southwesterly wind!*

Moments after opening the building, Gus and Bill Avery trotted to the shoreline where the wind speed and direction could be checked without obstruction.

"What's the speed, Gus?"

"Fifteen, gusting to 18; a moderate breeze."[2]

Grimacing, Herring pointed to the northwest. "Look how dark it is over the lake; a front is moving in. If we're gonna fly, it'll have to be soon!"

"The direction is perfect!" Avery said. "It's blowin' right down the shoreline. I'll bet we see a fresh breeze by 11 o'clock. Let's take the machine outside and attach the tail – what do you say?"

"First, I must check the balance. Let's get to it!"

While Gus and Bill were checking the wind conditions, Housam and Clarke lashed three small sandbags to the kite-glider's central frame. The sandbags would stand in for the combined weight of the air motor, storage tank, propellers, and accessories. Returning to the pavilion, Gus dangled from the operator's position as he checked the machine's balance point. Afterward, he moved one of the bags an inch or two forward before declaring the apparatus ready to fly. Working together, the crew carefully maneuvered the kite-glider through the pavilion door. Waiting patiently to finalize adjustments to the regulator, Gus watched as Henry and George reattached the cruciform tail.

As George soothed and tempted Engberg's mare with tasty carrots, the rest of the team carried the machine to the shoreline.

From there, they wheeled it to the designated starting point, 300 feet to the southwest. While Avery and Clark tended to the machine, Herring uncoiled the towrope and lanyard. With the horse and wagon in position, Gus slipped the looped end of the towrope over the hitch before hustling back to the flying machine to check the attachment of the lanyard to the tow bars.

"Gentlemen," he shouted, "take your positions!"

An air of excitement permeated the cool, overcast morning air, as the men scrambled about. A last check showed Herring that the wind speed had indeed increased to almost 20 miles an hour, with gusts to 22.

Conditions were perfect.

Scrambling to the operator's position, Herring secured the anemometer to a forward frame with elastic bands before lifting the 88-pound apparatus off the sand. As the machine teetered to the rear, he shuffled his hands an inch or two aft on the horizontal bars until it balanced comfortably. Glancing around, he noted that everything seemed ready, including the towrope that stretched out like a newly strung clothesline.

With adrenaline flowing through his bloodstream and butterflies fluttering in his stomach, Herring whistled for Housam to begin the launch... meaning George would begin to feed the mare! While that was taking place, Herring had a quick thought: *I must be crazy to trust my life to a horse!*

Housam shrugged before jumping back onto the wagon. Twisting in his seat, he caught Herring's eye and nodded, then turned his attention back to the mare. With a flick of the reins and a barely audible, "Hee-ah," the manned kite's first flight was under way!

An instant later, Herring's head jerked to the rear as he and the contraption soared into the air. With Clarke sprinting at the port wingtip, the apparatus rose to 15 feet above the damp sand, before abruptly leveling off and dropping its nose. Instinctively, Herring thrust his body aft to counteract the shallow dive. Within a few seconds, the machine had begun to overtake the tow horse and wagon. As the towrope slackened, Herring sensed impending danger, and he yanked the release cord. At once, the towrope disengaged from the lanyard and fell away, thus preventing a potentially catastrophic whipping of the machine to the sand below.

Seconds later, the machine darted over and beside the mare. Startled, she reared, tipping the wagon and dumping Arnot onto the sand. He tumbled awkwardly before scrambling to his feet. Unfazed, Housam reined in the horse, bringing her and the skidding wagon to a grinding halt.

Meanwhile, Herring continued gliding for another 150 feet, eventually alighting on virgin compacted sand near the water's edge. Jumping from the confines of the steel tubing and wire structure, he quickly tipped the tail upward, protecting the apparatus from the unpredictable wind.

"Damn! That was an adventure!" Avery shouted, as he ran down from the dune. "From where I was watching, the machine's glide ratio was very impressive!"

"There must have been an incline to the wind," Herring said. "Otherwise, I would have landed closer to the launch point. Starting the glide from a greater altitude will tell the story... for now, let's get her back to the pavilion. I must decide how much farther back to move the lanyard attachment points."

While the others drank coffee and listened to Herring's muttering as he agonized over the machine's tow point, Housam trotted the mare back to Engberg's barn to collect a set of *blinders*.

"She won't rear if she doesn't see the flyin' machine whizzin' by her ear!"

Arnot, the trial's only casualty, suffered a mild ankle sprain and spent the remainder of the day warming one of the pavilion's dilapidated wooden chairs. "I may reconsider banking," he teased. "It's safer!"

As the men sat down to their bag lunches, Herring ventured outside to check the weather. A few minutes later, he returned with bad news. "The wind has picked up. There's a strong breeze blowing off the lake... a driving rain should be here in a matter of minutes."

St. Joseph, Michigan

Memorial Day: Monday, May 30, 1898

Herring was wide awake as the rain buffeted his bedroom window. *At least it's blowing from the right direction*, he thought. *Can't sleep – I'm goin' down to the pavilion.*

After dressing in the dark, Herring tiptoed into the kitchen for a glass of milk, which he sipped while scribbling a note to Lillian. Donning his black canvas jacket and captain's cap, he grabbed an umbrella and slipped out the front door. Working in the dark, Gus lifted the kerosene lantern from its hook and gently removed the glass chimney. Advancing the wick a quarter inch above the tarnished brass holder, he struck a wooden match and ignited the fuel-soaked, braided fibers. Replacing the chimney, Herring deftly adjusted the wick height, changing the smoky orange flame into a vibrant yellow-white intensity that illuminated his front yard.

As dawn broke beyond Truscott's powerhouse smokestack, Gus found himself standing inside the deserted pavilion. Working by lantern light, he double-checked his earlier calculations. There could be no mistakes; today was his final chance to gather critical data.

By 6 a.m., the steady rain had eased to an intermittent drizzle. As he stood at the shoreline, staring out at the great expanse of gray water, Gus was buoyed by the direction and strength of the wind. *Twenty-five to 30, out of the southwest*, he thought. *If it drops a bit, there might be a chance later this morning.*

A little after 8 o'clock, a horse-drawn wagon rattled to a halt outside the door. As team members hustled inside, George, the last to enter, snickered. "Gus, when you get back to your shop tonight, you better watch out where you step... I didn't have time to clean up after Sandy this mornin'!"

While the men indulged themselves with hot coffee and buttered pumpernickel bread, Herring's fouled shop provided fodder for some spirited ribbing of their leader. Before long, though, the mood turned serious as Gus discussed his plans for the day. "Gentlemen, this is our last chance for success this weekend. If the weather doesn't cooperate, we may not have the opportunity for multiple flights. Therefore, risky as it may be, I have decided to move the lanyard's attachment back to the calculated fore and aft tow point... a 12-inch change from yesterday. If I have calculated correctly, the aero*curve's* angle of attack will prove to be adequate—"

"What if the attachment points prove to be too far aft?" Avery interrupted, with a tinge of anxiety in his voice.

Herring acknowledged his friend's apprehension. "Then it'll be one hell of a ride! I'll have my hand on the release cord. If she gets wild during the tow, I'll cut her loose—"

"That could be too late!" Matthias cut in. "If the machine stands on its tail at low altitude, there won't be time to get it straightened out before there's a mishap. I'm against such a substantial change... it's too risky!"

Arnot's fear for his partner's well-being opened a floodgate of opinion. During the next half hour, everyone contributed to the heated discussion. In the end, Herring threw in the towel, agreeing to move the attachment point only nine inches aft. It was a small concession from someone whom many considered to be stubborn, but a compromise nonetheless.

Herring's compressed air-powered aeroplane rigged to fly as a kite (1898); Richard Thompson illustration

By 9 o'clock, all was in place for another trial. On Herring's command, Housam urged the mare forward and the machine leapt into the air, weaving unsteadily into the overcast sky. Seven seconds after takeoff, the flying machine/kite had reached its equilibrium altitude.

From his observation point high atop the dune, Avery struggled with his tripod-mounted instrument as the kite porpoised and swayed at the end of its tether. After almost a minute of trying, he finally decided on a representative angle – 8 degrees – too great to provide characteristic towrope force readings. Seizing the whistle that hung from a cotton string around his neck, Avery took a deep breath and blew *twice* in rapid succession.

*

Everyone heard the whistle and stopped what they were doing. Craning their necks upward, Clarke and Arnot wondered why Herring had not yet disconnected from the towrope. What was stopping him? It was only after Arnot's prolonged arm waving that Herring realized the test was over. Afterward, he would explain that the shrieking from the machine's truss wires and framework – similar to the howl produced from the *shrouds* of a ship in a storm – prevented him from hearing the signal.[3]

When Gus yanked the release cord, the elastic shock absorbers at the two attachment points provided another surprise: they propelled the machine – like a marble out of a slingshot – to yet a higher altitude. Reacting to his unexpected sendoff, Herring instinctively thrust his body weight forward, forcing the nose of the apparatus down. At an estimated altitude of 80 feet above the lake, the liberated flying machine advanced down the Silver Beach shoreline, in the general direction of Indiana.

Holding his trusty stopwatch high in the air, Matthias watched from the tow wagon as the aeroplane's image began to fade from view. Moments later, it appeared to alight smoothly, without a bounce. Later, after helping return the apparatus to the pavilion, Housam returned to the beach, where he stepped off a distance of 790 feet for the gliding portion of the flight. When combined with Arnot's time-aloft figure of 89 seconds, Herring calculated the machine's ground speed to be 8.9 feet per second, or about six miles an hour.

"The machine showed a glide ratio of 10 to 1!" Avery shouted, slapping his thigh.

"I would expect a 10 to 1 glide ratio from one of my parabolic aero*curves*," Herring said, "but not from the arc-type—"

"I wouldn't be too dismissive, Augustus," Arnot interrupted. "Our machines have exceeded the theoretical limits before. I'll take the performance, ascending wind or no!"

"Maybe you're right, Matthias. Let's move the lanyard attachment forward a couple of inches."

"What about those rubber shock absorbers?" Clarke queried. "They made the flying machine act like a jackrabbit jumpin' out of a foxhole!" Clarke's dramatic observation gave rise to a serious discussion concerning the heavy-duty rubber bands. Before long, a consensus emerged: eliminate the shock absorbers altogether, and trust the mare not to lurch during the tow.

"Then there's the matter of Augustus not being able to hear Bill's whistle!" Matthias reminded his comrades. "I recommend that you keep an eye on me, Gus. When I *wave*, the test is over... for better or worse!"

By noon, the flying machine was back on the beach. Adjusting his position between the horizontal bars, Herring took one last look at the anemometer before turning his attention to the tow wagon. Whistling, he thought... *be good, Sandy!*

An instant later, the machine was airborne and climbing steadily toward the clouds. This time, Herring was prepared when his mount suddenly leveled out, in equilibrium with the forces of nature. In less than 30 seconds, he observed Matthias leaning back, straining against the resistance of the spring scale. Trotting six feet behind the wagon, Clarke checked where the bob weight string crossed the protractor scale, which indicated the towrope's angle with the ground. While all this was happening, Bill Avery was seen scrambling down the face of a distant dune, tripod in hand.

As the sun broke through the scattered clouds, Gus was momentarily distracted by the kite's flickering shadow on the beach below. Waving frantically, Arnot finally caught his partner's eye. As Herring gently released the kite from the restriction of its towrope, he settled in for what he hoped would be another long glide. Feeling confident, he swung his weight first to one side and then to the other, causing gentle turns in the process. *No doubt about it,* he thought, *this is my best machine yet.*

Thirty seconds after another smooth alighting, team members leapt from the rolling wagon toward the stilled machine. There were back slaps and handshakes all around as the men congratulated Herring and one another. Anxious to obtain the data, Herring called out to each man:

"Bill, what was the angle on the aerocurve?"

"Three degrees and steady!"

"Matthias, what was the pull on the towrope?"

"She pulled 56.6 pounds. I checked it twice!"

"Henry, what was the towrope angle?"

"Forty-five degrees. Right on the button!"

"George, how fast do you estimate Sandy was walking?"

"Near as I could tell... 'bout four miles an hour."

"Okay," said Herring, "the anemometer indicated the air speed at about 24 miles per hour... so that tells me the wind was blowing about 20 miles per hour."

"We didn't fly quite as far down the beach this time," Arnot observed. "Without the slingshot effect, you didn't have the altitude, and the turns eliminated some of the machine's forward progress—"

"No matter," Herring interrupted. "We have a successful test!"

*

By the time the team had returned the flying machine to the pavilion, it was almost 2 o'clock, and everyone was starving. As they gulped hot coffee and plunged into their lunches, Herring, with assistance from his slide rule, worked on his calculations. Before anyone even had time to belch, he had the results. "The apparatus produced 40 pounds of drift when flying in what amounts to a 24-mile-an-hour wind, the sum of the wind and the horse's towing speed. This means that we'll need 40 pounds of thrust from the propellers to realize a ground speed of four miles an hour. Generating that much thrust will require the compressed air motor to produce 2.72 brake horsepower."[4]

CHAPTER 65

Rochester, New York

Monday, November 7, 1921

Herring and his daughter had spent the weekend in their Rochester home trying to relax. While Chloe busied herself preparing homemade bread and chicken soup, her father reviewed his wastebook notes from the months of June through September 1898 – a particularly frustrating period, filled with setbacks and complications regarding his flying machine's air motor and the borrowed stationary engine and air compressor.

As Herring took the witness stand that morning, feelings of sorrow and remorse washed over him; this was testimony he would rather not furnish.

"Mr. Herring," O'Grady began, "Friday you described the successful manned kite experiments that your team conducted over the 1898 Memorial Day weekend. When, sir, did you finally attempt a manned, powered flight?"

That was abrupt! Herring thought. *Why didn't he try to set the stage?*

"The short answer to your question is: four-and-a-half months later! My first attempt at powered flight didn't occur until the second week in October of '98."

Slipping from behind the podium, O'Grady paused for dramatic effect. "What took you so long?"

Startled, Herring straightened in his seat.

"Setbacks!" he snapped, with obvious annoyance. "There were problems with all of the machinery, but the stationary engine was the worst—"

"Why, might I ask, was that?" O'Grady interrupted.

"I should have listened to Arthur Pesch when he said that engine was scrap! I still get hot under the collar thinking about the mess I got myself into."

"What mess was that?"

Herring leaned back in his chair. "Cleaning off the grime took two days. Disassembly took another three. Once I could see what I had, the situation went from bad to worse. In addition to the galled crankshaft journals, worn-out valve train and reciprocating parts, I discovered a *crack* in the engine's cast-iron cylinder head!

"After considering my options, I was obliged to try a partial restoration. Engberg's shop tried to repair the crankshaft, and the Iron Works agreed to braze the cracked cylinder head – with no guarantees. The other parts would either have to be remanufactured or refurbished."

"How did this plan work out?"

"Not too well. Although the engine was ready to run before Labor Day, we couldn't move it to the bathing pavilion until after the season-ending holiday... then, other issues popped up, including difficulties with the spark ignition system and carburetor."

"Were you working on anything else at that time?" O'Grady asked.

"The flying machine's motor provided its own collection of problems, but for the most part, it was up and running on the test stand before the stationary engine was ready."

"If there was no compressed air... how did you test the air motor?"

"I used Truscott's low-pressure shop air for the preliminary testing, making sure that everything worked as planned."

"How did the motor perform?"

"After stripping away all of the unnecessary parts, such as the breaker-points, battery, ignition coil, sparking plugs, and carburetor, the motor weighed only 12 pounds. The new air induction valve along with the piston-oiling system, a carryover from the gasoline engine, *seemed* to work as designed."

"What about the borrowed air compressor?"

Herring grimaced. "The air compressor was *almost* the last straw. With autumn's superior flying weather rapidly approaching, it became evident that the equipment wouldn't be ready in time… so, I hired George Housam to help me on evenings and weekends."[1]

"Did you receive any *good* news during that period?"

Rubbing his chin, Herring looked solemn – then he brightened. "Toward the end of August, while I was preparing the Maybach-Daimler engine for its initial run, I got a telegram from George Whittlesey, Chanute's attorney. Our British patent had been approved!"

"Which one of your applications had been granted?"

"The joint application for the powered machine—"

"Mr. Herring," O'Grady interrupted, "although Octave Chanute was passionately opposed to powered flight, he nonetheless joined you in filing for this foreign patent—"

"Objection!" Robbins intruded. "Besides being argumentative, counsel for the plaintiff is leading the witness! There is no evidence that Mr. Chanute was opposed to obtaining patents for powered flight!"

"That's not the point, Your Honor!" O'Grady said, raising his voice. "Mr. Herring's testimony has shown time and again that although Octave Chanute despised dynamic flight, he didn't miss an opportunity to capitalize on its potential – both from an economic and personal perspective. By riding his former assistant's powered-flight coattails, Chanute once again demonstrated his hypocritical, legacy-driven nature!"

Judge Sawyer took a moment to consider O'Grady's argument.

"The defense's objection is sustained. You may continue, Counselor... but with fewer dramatic interludes!"

Rummaging through his notes, O'Grady thought *Dramatic interludes, my ass*!

"Mr. Herring, what did the British powered-flight patent mean to you at the time?"

Herring seemed preoccupied. "The occasion was bittersweet. We had an English patent, but now, every astute experimenter had the opportunity to study my ideas. Then it occurred to me: why continue to work in secret? Why not call the newspapers and let the journalists in on a good thing? In exchange for my generosity and cooperation, perhaps they would help me establish precedent for my work, something I couldn't seem to accomplish with our Patent Office."

"Did you follow through with your plan?"

"I did, but not for the machine's first attempt to fly under power."

"Why not?"

"If the machine failed to perform, the newspapers would have had a field day at my expense. If I were to expect positive press, the flying would have to be successful.

"However, I'm getting ahead of myself. There were still many problems to overcome with the air compressor and stationary engine before any powered flight could be attempted."

St. Joseph, Michigan

Saturday, September 10, 1898

With George's help, Herring moved the stationary engine and compressor from his rented space at Truscott's to the Silver Beach bathing pavilion, now closed for the season. Two hours later, their horse-drawn wagon arrived at the Church Street shop, where the tailless flying machine and compressed-air motor were loaded in preparation for the return trip.

Floor space inside the pavilion was too limited for both the flying machine and its support equipment to be stored simultaneously. To ease the problem, Herring devised a rope-and-pulley system to hoist the aeroplane high into the building's rafters for temporary storage while they arranged the space below.

"George, take the horse and wagon back to Engberg's. While you're gone, I'll get to work on the stationary engine." By 11 o'clock, Housam had returned from his half-mile hike back to the pavilion, just in time to assist in lowering the flying machine to the shop floor.

"George, give me a hand mounting this air motor," Herring said, as he struggled to maneuver the powerplant into position within the aeroplane's framework. "I'll hold, you slip the four mounting bolts through the holes in the frame, then add a flat washer and hex nut on each one… I'll do the rest." Minutes later, Housam was busy working outdoors, coupling the Maybach-Daimler engine to the two-stage compressor. Using sheaves and double v-belts, a lever-actuated idler pulley would serve to engage and disengage the air compressor from the engine's output shaft.

"Gus," he hollered through the pavilion door, "do ya think a 4-to-1 speed reduction will get the compressor runnin' fast enough?"

Turning from the air motor, Herring tramped over to where Housam was working. "With a 7-to-1 compression ratio built into both stages of the compressor, it'll take a lot of engine torque to turn its input shaft. I'm keeping my fingers crossed that the Maybach engine will only load down to about 480 rpm. If that's the case, the compressor should turn about 120. That should be fast enough to fill the half-cubic-foot tank in about 15 minutes... if we can coax them to run that long!"

Returning to the task of mounting the high-pressure tank within the flying machine's sub-frame, Herring was reminded that he would now be operating the apparatus while seated behind a potential *bomb*. Housam was having similar thoughts. Struggling to his feet, the big man stretched before shuffling over to where his boss was working.

"I gotta ask ya about these welds," George said, pointing at the tank's end caps and fittings. "Don't they scare ya just a bit?"

Herring shrugged his shoulders. "After the valve and pressure-gauge flanges were welded into place, I had the tank pressure tested to 1,000 psi. Everything seemed good."

"What if there's an accident during takeoff. Let's say that big ol' propeller whacks that pressure gauge—"

"Yeah!" Herring cut in. "There could be consequences! All kinds of things could go wrong. For instance, what if a quick stop on alighting throws me into the spinning propeller? What if I get hit by a piece of broken propeller blade?

"If reasonable precautions have been taken... then it's up to fate. The way I see it, if it's your time to go, it's your time to go! Hell, George, you could slip on a patch of horse manure and get run over by a trolley!"

Nodding, Housam mumbled as he retreated to the stationary engine, "Just askin'..."

*

At 1 o'clock, the men stopped for lunch. After a few minutes of relaxing silence, Herring began thinking out loud. "The compressed-air motor has features that are superior to the Otto cycle engine. For instance, our two-cylinder motor gives two power impulses for every revolution of the crankshaft – the twin-cylinder gasoline engine delivers only one!

"Then there's the question of horsepower. It's difficult to increase the horsepower of a gasoline engine without increasing its displacement – making it bigger. I can alter the horsepower of the compressed-air motor by adjusting the delivered air pressure with an onboard reducing valve."

Listening attentively, Housam had a question. "Why not just use the full 600 pounds of tank pressure?"

Gus shook his head. "There would be too much power at the beginning of the flight and not enough toward the end. After we get the compressor to fill the tank to 600 psi, I want to test the air motor and propellers on the *pendulum machine*. Then I'll adjust the reducing valve to a pressure where the motor will spin the propellers fast enough to produce a minimum static thrust of 40 pounds."

*

By 2 o'clock, George declared the air compressor's installation to be complete. "She's ready to go, Gus! Why don't you stop what you're doin' and we'll see if the old Maybach will run?"

Herring was ready for a break. He had just finished installing the pusher propeller's rear driveshaft. Everything else was in place and plumbed, including the air tank and cylinder jackets. Shuffling over to the refurbished engine and compressor, Gus frowned as he calculated the time and money required to work toward such an uncertain result. With a specific weight of 66 pounds per horsepower, the four-cylinder Maybach-Daimler stationary workhorse would never be mistaken for a flying machine engine.

"The engine has some good points," Herring said. "There's a power event every half-revolution of its two-throw crankshaft. That makes for a smooth flow of torque to the air compressor."

Fitted with a six-volt battery made up of four Hellesen dry cells, a Hightenson induction coil, and Benz sparking plugs, the jump-gap ignition system was state of the art for 1898. On the negative side, the Maybach spray-type carburetor was a complete unknown to Herring, and adjusting it would be yet another learning experience. The engine's liquid-cooled block and cylinder head offered the possibility of prolonged running without overheating, although the question of water leaks added to the list of potential problems.

"I've spent so much time with this engine, it seems like I designed the damned thing!" Herring said, as he reached for a tin funnel. "The battery's fresh, and I've already set the spark timing. Check the water level in the radiator, George. I'll fill the petrol tank."

When all was ready, Housam rolled up the pavilion's big door.

Grabbing the engine's crank handle in his right hand, Herring stooped to a position of leverage before making a preliminary pull. *Spark off*, he thought, mostly out of habit. "Let's see if she'll turn over."

A moment later, the engine clattered and wheezed before coasting to a halt. Flipping the switch, he positioned himself for another attempt.

"Switch on! Here goes nothin'!"

After another strong pull with similar results, Herring turned to Housam. "She's dry, George... it needs a 'choke.' Put your palm over the air intake for the next pull – we need to 'prime' the cylinders with a bit of gasoline."

Housam waited until Herring switched off the ignition before placing his hand on the engine's carburetor. High-voltage ignition shocks might be harmless, but he didn't appreciate being surprised. Presently, Herring pulled the engine through another clatter-wheeze-coast cycle.

"Now we'll try her again!" he said, flipping the switch to "on".

When he pulled hard on the crank, the engine experienced a powerful backfire, jerking the handle out of Herring's hand. As the iron lever whipped around, it whacked him squarely across the wrist.

"Damn it!" Gus hollered.

Rubbing his throbbing arm, Herring hobbled to the toolbox to retrieve an open-end wrench. Walking back to the engine, he glared at the innocent-looking crank mechanism. Noticing that the ignition switch was still on, he flipped it to the "off" position.

"George, these dry cells will only last about an hour; if I forget to switch off the ignition, you have my *permission* to do so! Otherwise, with carbon cells selling for 75 cents apiece... I'll go broke in no time!"

Loosening the machine screw that locked the distributor to the engine block, Herring twisted the head of the unit clockwise, as he explained the tactic to Housam. "By retarding the ignition timing, the spark will occur closer to top-dead-center on the compression stroke; cranking the engine should be safer without the damn backfiring!"

After a few minutes of tinkering, Herring's next pull on the crank produced a cascade of rifle-like discharges that emanated from the engine's exhaust. As he slowly twisted the distributor counterclockwise, the sporadic detonations turned into a deafening roar. Fifteen seconds later, he ended the engine's audition with a flip of the ignition switch.

"She doesn't sound half bad! What do you think, George?"

Herring looked around, but Housam was nowhere to be found.

After checking the torque on the engine's cylinder-head bolts, Herring strolled outside for a breath of fresh air. Within seconds he spied George trotting back from the south end of the pavilion.

"Where the hell did you disappear to?"

"Nature called... had to visit the outhouse."

Shaking his head, Herring walked to his toolbox and removed a contraption about the size of a grapefruit from the bottom cabinet drawer. It looked like a hollow wooden globe that had been sawed in half at the equator. Joined by a u-shaped bow of spring steel, the hemispheres were filled with Philippine mahogany shavings and held in place by a fine brass screen. The circumference of each bowl was fitted with a glued-into-place protective cuff of pliable cowhide.

After a brief examination of the homemade device, Herring turned and handed it to Housam. "Here, George, try these on for size."

Arms crossed, George looked defiant. "What the hell are they?"

"They're called *ear defenders*. Andy's father, Frank Vassallo, invented them."

As Housam slipped the band over his head and positioned the open end of the bowls over his ears, the leather seals pushed against his head and jaw. Yanking them off, Housam was skeptical. "What are these things supposed to do?"

"They'll keep the engine noise from damaging your eardrums – keep you from going deaf!" *And maybe you'll hang around after the engine starts*, Herring thought.

"What about *your* eardrums, Gus?"

"I'm already deaf! Too many engine tests *without* ear defenders for me!"

*

For its next trial, the engine's maximum power was determined by fine-tuning the ignition timing and fuel mixture.[2] To Herring's satisfaction, this time, Housam stayed to record the engine's speed by using the stopwatch and revolution counter.

At the end of the test, Gus asked, "What did you get for engine speed, George?"

Housam peered down at the chart. "Looks like 660 revolutions per minute."

Thinking for a moment, Herring nodded. He was heartened by the results. "That's good, but we're running without a *load*. The question is: how much will the engine slow down when the compressor's working?"

St. Joseph, Michigan

Sunday, September 25, 1898

For what seemed to be the tenth time, the coverall-clad Herring bent over his workbench, intent on dismantling the two-stage compressor. This time, his task involved the replacement of leather packing on the unit's two pistons. Hoping to identify possible shortcomings that might cost him additional time, he allowed his thoughts to drift through a systematic review of the compressor's components.

Each cylinder has one atmospherically controlled inlet valve and one pressure-actuated discharge valve. Other than the compression springs that control the opening and closing of these valves, the only two things that can go wrong are the pistons not sealing against their cylinders and the valves not sealing to their seats. Trouble is, the second stage still gets too damn hot... even with intercooling I keep scorching the leather—

At that instant, a loud banging jolted Herring from his thoughts; someone was at the door. Dropping his screwdriver onto the bench, he grabbed a rag to wipe his hands as he double-timed his way across the shop.

"Augustus Herring!" called out the visitor, extending his hand. "Your wife said I'd find you here!"

"Dr. Zahm," Herring said, shaking his hand, "I've been expecting you!"

Albert Francis Zahm, 36 years old, did not project a commanding image. A rather large head dwarfed his five-foot-eight, 130-pound frame. Although the disproportion implied superior intellect, Zahm's high forehead and closely spaced, cobalt-blue eyes suggested that he was a numbers man – an accountant, or perhaps a teacher of mathematics. When he stood to stretch, his gray, single-breasted jacket swung open, revealing a worn leather slide-rule case dangling from his belt.

"Albert, it's been three years since we last spoke! I was still working for Professor Langley at the Smithsonian."

"Yes! You were flying your two-surface rubber-powered model on the grounds of the Institution; I remember those wonderful flights as if they were yesterday!" Professor Zahm's voice exhibited an unmistakable Midwestern twang, revealing his Southern Ohio roots.

Gus said, "Let's not forget the Chicago Conference on Aerial Navigation back in '93, when we first met! I was attending the Columbian Exposition with Lillian, my wife to be, when I learned of the symposium; you helped me acquire the necessary credentials."[3]

Herring's entry into the aerial navigation fray had occurred during a period when disparaging songs and hurtful jokes were the public's reaction to the notion of human flight. As a result, many experimenters worked in anonymity and secrecy. On the other hand, Zahm had emerged from that period as a respected scholar and researcher in aerodynamics.[4] His recently won Ph.D.

and position as head of the department of Physics and Mechanics at Catholic University in Washington, D.C. allowed him to mingle with the scientific and engineering elite, including Bell, Langley, Walcott, and Barus.

"A lot of water has passed under the bridge since then," Herring said. "I hear that your new wind tunnel is up and running. Is it true?"

"The 40-foot machine is currently operating at wind speeds up to 25 miles an hour! However, a new variable-speed, direct-current generator/motor assembly from the General Electric Company will allow us to approach 50!"[5]

"I'm envious of your laboratory and resources. As you can see, I'm working on a shoestring budget, barely able to keep my experiments going—"

"Careful what you wish for, Augustus," Zahm interjected. "The problems associated with a university-controlled research facility are many – not to mention the egos and politics involved. However, I'm not here to burden you with my predicaments. I want to see what you're up to!"

Moving to the suspended two-surface machine, Zahm rocked back on his heels as he strained to take in all of the apparatus's elements, especially the power and propulsion systems.

"I'll lower the machine, Professor; no sense getting a stiff neck."

As Herring winched the flying machine to the floor, he described its brief history. "Four months ago, we flew this machine as a manned kite. The experiment enabled me to determine the propeller thrust and horsepower needed to fly it in a 24-mile-an-hour wind."

For more than an hour, Herring and Dr. Zahm discussed the many facets of the experiment, including the extended periods of gliding down the Lake Michigan shoreline.

"Well thought out, Augustus! Have you used these techniques before?"

"Yes. On a very windy day at the Indiana dunes, we determined the *drift* for one of my aspirating kites."

"I presume that you're referring to the Chanute glider trials of '96?"

"Yes. That reminds me... you wrote that Mr. Chanute had mentioned my experiments. I was somewhat taken aback, since he doesn't put any stock in the immediate future of dynamic flight."

"He mentioned just enough to arouse my curiosity," Zahm said. "That's why I decided to write you that letter. Be that as it may, Augustus, I'm very impressed with your efforts! Although you're struggling to provide adequate power, the physics related to the apparatus and its propulsion system have been impeccably applied. You are to be commended for your engineering efforts! In my opinion, your machine should perform as expected."

"I only wish that I could reliably charge the tank with air," Herring said. "I've struggled all summer – first with the stationary engine and now with this two-stage compressor."

"Any idea when you might be ready to attempt a trial?"

Shrugging, Herring turned to inspect the compressor components that were spread out on the workbench. "If everything goes well, I'll try to fly the machine this coming Saturday."

St. Joseph, Michigan

Wednesday, September 28, 1898

Some daylight still remained when George Housam ambled into the beachfront workshop. "Hey, Gus, did you see the big ol' full moon out there? You won't need the lantern to find yer way home tonight!"

As the two men stood next to the ready-to-run engine and air compressor, their talk turned to Professor Zahm and the flight trials scheduled for the weekend.

"Well, George, by the time we leave for home tonight... we'll know."

"Know what?"

"We'll know if the machine will be ready to fly on Saturday."

"What's so important about Saturday?"

"That's when Dr. Zahm will be back to watch us *try* to fly."

Kneeling to inspect the machinery, Housam shook his head. "Well, there's nothin' like a little extra pressure to liven up yer day... we better get crackin'!"

*

The first part of the evening was spent rolling out, cutting, and bending a 20-foot length of half-inch-diameter, annealed brass tubing. After emery cloth had been used to brighten the tubing ends, brass fittings were sweat-soldered into place. Next, the matching halves of those fittings were threaded into the compressor's outlet manifold and the inlet side of the tank's manually operated gate valve. That valve served to isolate the storage tank air from the compressor and allowed the fill tube to be removed.

Pulling out his pocket watch, Herring flicked open its cover. "Well, George, it's 9 o'clock. Can you think of any reason why we shouldn't try charging the tank?"

"Hell no! A little noise won't hurt nobody!"

Fifteen minutes later, the men had finished rolling the tailless flying machine out into the moonlit night air. The brass filling tube was reconnected, the gate valve opened, and the ball valve between the tank and air motor closed. Gus had previously adjusted the reducing valve to allow maximum tank pressure to be transferred to the motor.

All seemed to be in order.

Pleasantly surprised when the finicky Maybach-Daimler engine sputtered to life after only three pulls on the crank, Herring allowed it to warm up before rotating the distributor to its maximum horsepower position. Satisfied with the engine's performance, he waved his hand in the direction of Housam, signaling him to engage the air compressor.

Over the throbbing roar of the unmuffled four-cylinder engine, Housam, ear defenders in place, deftly levered the idler-pulley-controlled drive belts into position between the engine and compressor. With a piercing squeal and a puff of blue-gray smoke, the compressor clattered to life, pumping high-pressure air into the flying machine's storage tank.

Shielding his ears with the palms of his hands, Herring watched over the hot, vibrating machinery as Housam retreated to the relative solitude of the aeroplane to keep an eye on the tank's Bourdon-tube pressure gauge.

Ten minutes later, Housam ran up to Herring and shouted, "Pressure's at 595! You wanna shut her down?"

Abandoning his position, Herring dashed over to the tank and closed the gate valve, which required three complete turns to accomplish. Scurrying back to the compressor, he disengaged it from the engine by leveraging the idler pulleys away from the drive belts and flipped the engine's ignition switch to the "off" position.

With the exception of a high-pressure hissing sound, peace and quiet had been restored. Back at the flying machine, Herring quickly located the leak. "George, fetch a 3/4-inch open-end wrench! She's leakin' at the pressure gauge."

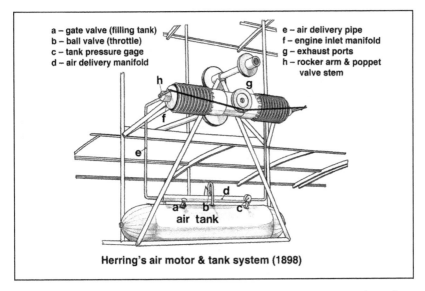

Closeup of compressed air motor, air storage tank, and transmission (1898); Richard Thompson illustration

As Herring gently tightened the hex fitting at the base of the gauge, the hissing gradually lessened and then stopped completely. "I thought that might be a problem when I installed the gauge," Herring said. "I wanted it to face me when I was in the operator's position, so I didn't tighten it down enough to seal the tank. Now it doesn't leak, but the gauge is facing sideways. I'll have to fix that before we make a flight trial. The pressure's at 590... should we give her a whirl?"

"Let's get her tied down first," George said, "then you can get aboard and open the ball valve."

By the time the motor and propellers were ready for a try, it was almost 11 o'clock. As Herring positioned himself within the framework, Housam elected to leave the kerosene lantern back in the pavilion; the full moon provided a surreal but workable light.

"Stand clear, George!"

An instant later, just as the propellers began to rotate, an uncontrolled blast of high-pressure air surged through the motor. By the time Herring had shielded his face and eyes in the crook of his elbow, the event had begun to diminish. Within seconds, the tank was empty and the propellers had yet to spin.

CHAPTER 66

Chicago, Illinois

Friday September 30, 1898

Chanute pressed his ear to the receiver as the long-distance operator switched him through to the aeronautical laboratory at Catholic University. Three minutes later, a voice crackled over the line.

"Zahm speaking."

"Professor, it's Chanute. Albert... do you have a moment?"

"Octave! What's on your mind this fine morning?"

"I just received a postcard from our friend Herring. He says he won't be testing the powered flying machine tomorrow. He also hinted that you had seen the apparatus—"

"Yes, yes," Zahm interrupted, "I had the pleasure of visiting with Augustus this past Sunday. I also received a communication from him yesterday, urging me not to make the return trip to St. Joseph. Apparently there's been some trouble with the machine's compressed air motor."

"So he says! I'm curious, what are your impressions of the contraption?"

Zahm, trying to sound casual, took a deep breath. "I was both surprised and pleased by Herring's proficiency. His testing and engineering skills are first class, and the workmanship is outstanding. I think there's a chance the machine will make a successful flight... although the compressed-air power system will limit its duration."

"Is he still clinging to Lilienthal's weight-shifting method of control?"

Zahm balked at the question. "Herring believes that the machine is compact enough to be controlled by shifting the operator's weight, with equilibrium assistance from his tail regulator... which I don't pretend to understand."

After a few moments of nervous silence, Chanute plunged into a condemnation of Herring's motives. "This may come as a shock to you, Albert, but it's clear to me that Herring's appetite for fame and fortune has warped his judgment. Based on our conversations, I believe that his sole objective is to extend the two-surface glider's time aloft, after which he will declare the problem of heavier-than-air flight to be solved!

"As you know, I have written extensively about automatic equilibrium and control. In my opinion, these problems must be resolved first! Afterward, artificial power may be employed in a thoughtful, methodical fashion, utilizing a foundation of experience and trustworthy data. I believe – as do most other *responsible* experimenters – that a practical flying machine will eventually be produced through the efforts of *many* practitioners, and not the egocentric meanderings of *pretenders* like Mr. Herring."

Zahm didn't answer immediately. Finally, he let out his breath. "Octave... Gus Herring is the most experienced operator of heavier-than-air gliding machines in the world. He has more time aloft than even the great Lilienthal – more than 2,000 flights! He has flown distances up to a quarter mile, has made sweeping turns and has performed extensive quartering flights into an ascending wind. Several of his glides have lasted for more than a minute! If Herring believes that it's time to add a motor to his apparatus, I, for one, must take him seriously!"

Undaunted, Chanute continued his harangue. "Albert, I know this man! If it weren't for pioneers such as Lilienthal, Langley and frankly... *myself*, Herring would still be fiddling with rubber-strip-powered toy flying machines! He's an intellectual *thief* who considers the technology to be his own. He pockets the ideas of others, claims them for himself, and then has the audacity to disrespect his mentors! Mention his name to Langley... if you dare!"

Zahm wasn't surprised by Chanute's outburst. "Octave, it's obvious you *hate* the man. Why do you continue to associate with him?"

After an uncomfortable silence where only the hum of the telephone line betrayed the older man's presence, Chanute finally spoke. "For the sake of aeronautics I hold my tongue and remain cordial... *someone* has to keep track of the scoundrel."

Rochester, New York

Monday, November 7, 1921

The unusually warm, shirtsleeve weather of the morning deteriorated during the lunch hour. As the litigants hurried back to the Monroe County Courthouse, a cold, wind-driven rain blew in off Lake Ontario. While Herring and O'Grady sat at the plaintiff's table awaiting the judge's return, the downpour drummed against the east-facing windows.

"Looks like we've got a Nor'easter blowin' in, Jim," Gus observed. "We could be in for a nasty afternoon."

Reaching into the inner pocket of his jacket, Herring pulled out a metal tube that resembled a container for an expensive cigar. As he unscrewed the cap, O'Grady leaned over and whispered. "What are you doing, Gus? You can't smoke in here."

"This isn't a cigar," Herring chuckled. "It's a pocket aneroid barometer."

Jiggling the two-piece instrument from its protective tube, Herring plugged the pieces together before holding the gadget at arm's length. "The scale shows 28.70 inches of mercury! Since we're sitting at 600 feet above sea level, this reading indicates a *very* depressed barometer. Hope you brought your umbrella, James!"

Judge Sawyer dashed into the courtroom, catching the bailiff unaware. Having witnessed this performance many times, none of the litigants seemed surprised. Thumping the battered oak block with his gavel, the judge abruptly jolted court back into session.

Herring disassembled his barometer as he walked to the witness stand.

"Mr. Herring," O'Grady said, "according to your previous testimony, the powered machine's first trial had to be postponed. What caused the delay?"

Slipping the instrument's protective container back into his pocket, Herring leaned back in his chair. "As fate would have it, a design *flaw* was discovered. The air motor's cam-actuated inlet valves couldn't withstand the high tank pressures. When the valves opened, their return springs were too weak to close them again, allowing air to escape through the exhaust ports without doing any actual work."[1]

"How did you correct this defect?"

"I designed and fabricated two new inlet valves – one for each cylinder."

"How long did that take?"

"More than a week. By Saturday, October 8, everything seemed to be working properly."

"What did you do next?

" I wired Dr. Zahm and Matthias, informing them the machine would probably be ready for a trial on Monday, October 10."

"Who else did you notify?"

"Nobody. Like I said earlier, until I was certain of success, the press would not be invited."

"However," O'Grady pressed, "you did invite Professor Zahm."

"Yes! We needed a credible witness."

"When did the professor arrive?"

Herring's explanation stalled in his throat. "He... he didn't. Albert was available the week before, but the air motor problem had pushed the schedule back. Then, due to university commitments, he couldn't take time off."

"What about Chanute?"

"A bad choice! Matthias harbored a clear-cut hatred for him."

O'Grady turned to a new page in his notepad. "Credible witnesses aside, what remained to be done on the machine before Monday's flight trial?"

"First, the thrust of the propellers had to be measured. Then the percentage of the machine's total weight, which the propellers were capable of lifting, had to be calculated."

"How was this done?"

"The measurement took place on the 'pendulum' machine."[2]

St. Joseph, Michigan

Sunday, October 9, 1898

Herring was perched atop his 10-foot stepladder when the pavilion's big overhead door rattled open, revealing George Housam's wiry figure. Slipping into the confines of the shop, he allowed the heavy gate to clang shut behind him. Leaning against the top rung, Herring peered down at his assistant, happy to take a break.

"No church for you today, eh, Gus?"

"Did you say a little prayer for me, George?"

"I always say a prayer for ya... Oh, by the way, Father Baker was askin' about ya!"

"Is that so! What did he say?"

"He said, 'George, I'm worried about that friend of yers. As a baptized Christian, why don't Mr. Herring go to church? If he meets his end in one of those Darius Green flying machines, I'm fearful his soul might end up in purgatory.'"

Clasping his hands together as if praying, Herring replied, "Ha!... church membership must be on the wane. But your Father Baker sure knows how to irritate me with that Darius Green hogwash. I've had to put up with that damned ditty all my life. Which reminds me, George, do you know the difference between Darius Green and a common housecat?"

Housam shook his head.

"A house cat only has nine lives!"

Sensing that his boss had finished discussing church matters, George scrambled over to hold the ladder. "I've got more important things to keep me occupied," Gus said. "Like getting this pendulum test set up and run!"

"Looks like you got a lotta work done after I left last night. Is the flyin' machine ready to get hooked up?"

"I've been waiting for you to get here, George. It's a two-man job to get the machine lifted into position."

Within minutes, the men had maneuvered the two-surface machine into place under the pendulum's support framework. By using the lift rope and pulley, Housam slowly hoisted the lightweight machine to a height where Herring, now stationed atop the ladder, could slip the pendulum arm's knife-edge pivot onto the steel rails of its stationary framework.

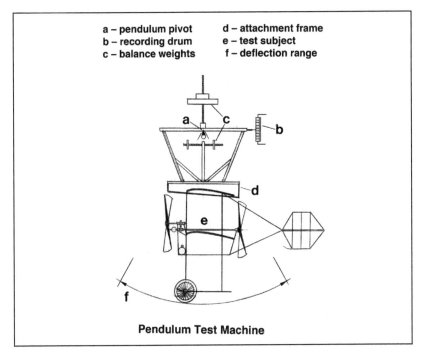

Herring's pendulum testing machine (1898); drawing by author

"Ha!" Herring grunted. "Perfect!"

Scrambling down the ladder, Gus squirmed out from beneath the contraption, which was now hanging only four feet above the floor.

"Aren't you gonna pull the ladder out?"

"Not yet. I still have two measurements to make. But first, I want to make sure the pendulum swings freely."³

Using his forefinger to push on the flying machine's strut, Gus nudged it about a foot to the rear. When he removed his finger, the apparatus swayed forward before settling into a rhythmic back and forth cycle.

"Reminds me of the pendulum on my old grandfather clock!" said Housam.

"That's the idea, George."

*

Using his tape to take measurements, Herring then installed a stationary recording device at the end of the pendulum tester's short indicator arm. The arm's lightly spring-loaded *Waterman recording fountain pen* was carefully positioned to press up against the pre-lined grid paper of the recorder's spring-powered drum. As the drum slowly rotated during a test, the pen traced the angular history of the pendulum arm on the grid's vertical axis and the seconds of operation on its horizontal scale. After careful adjustment, Herring was satisfied that the recorder delivered accurate results.

"Why do ya have ta know the arm's angle?" Housam asked.

Herring stopped what he was doing and turned to face his assistant. "Knowin' the pendulum arm's angle will tell me the propeller thrust, and a few other things. If I'm right, there will be some extra thrust at the beginning of the run. That's where the pressure-reducing valve comes into play. By regulating the pressure to where 40 pounds of thrust is delivered by the propellers, I can make the most of the air left in the tank. We'll find the valve's best setting by trial and error."

Glancing at his pocket watch, Herring decided it was time to take a break. "It's 4:30, George, time for supper. I'll meet you back here at 7 o'clock... and we'll run the first test."

Rochester, New York

Monday, November 7, 1921

Herring's testimony was temporarily interrupted as wind-driven hailstones began pelting the courtroom windows. Distracted by the meteorological phenomenon, a din of casual conversation arose in the gallery. Cracking his gavel, Judge Sawyer moved to restore order. "Mr. O'Grady, now that the novelty of our Rochester weather has ebbed, please resume your questioning."

"Mr. Herring, did you test for propeller thrust on the evening of Sunday, October 9, 1898?"

"Yes."

"Who was present for this test?"

"Matthias Arnot, George, and myself."

"When did Mr. Arnot arrive?"

"Just as we were about to dine. Matthias supped with us, and then we walked back to the pavilion to run the pendulum test."

"Did you experience any problems?"

"There were always problems, but this evening proved to be the exception. The stationary gasoline engine started and functioned flawlessly, and within 20 minutes, the compressor had fully charged the onboard tank."

"What happened next?"

"Early on, I had decided to run the pendulum test within the confines of the pavilion. Besides the difficulty of moving the equipment outdoors, the unpredictable wind might also have disrupted the results. Therefore, we had to rotate the entire works so that the wind generated by the propellers would blow through the overhead door opening. We also secured lightweight objects that might otherwise be blown about.

"Starting the motor required opening the ball valve between the supply tank and the motor. We needed to determine how to do this efficiently – especially during the flight trials, when the operator would have his hands full supporting the weight of the flying machine. After a prolonged discussion, it was decided to have an assistant open the valve remotely. To initiate this, we tied a length of light-gauge cord through a hole drilled in the ball valve's quarter-turn handle. The cord was then extended through the struts and piano-wire bracing to a position behind the lifting surfaces.

"Before starting the test, I switched on the recorder, making sure that the pen was delivering ink onto the recording paper's datum line. As I hollered to Matthias to pull the cord, I simultaneously hand-cranked the front propeller."

"Wouldn't the compressed-air motor start by itself?" O'Grady asked.

"Not necessarily. By helping the motor and two propellers to overcome their inertia, startup became much more reliable."

"How long did it take to run the test?"

"Less than 30 seconds."

"Please, Mr. Herring, describe the event."

Herring twisted in his seat before assuming a rigid military posture. "Immediately after startup, I thought: *This is a mistake!* Along with the shriek from the motor reverberating against the inner walls of the pavilion, a tornado-like flurry of wind from the machine's propellers hurled an instantaneous cloud of dust and dirt into the air. Despite the bedlam, the pendulum deflected as expected, equalized, and then made a gradual decline."

"Was the test a success?"

"It took a minute for the dust to settle before we could approach the machinery. With the flying machine still rocking at the end of the pendulum, I turned off the recorder and removed the grid paper from the drum. To my relief, the pen had worked flawlessly, and there was a clear-cut tracing of the pendulum arm angle for the duration of the run."

"What were you able to determine from the recording?"

"Calculations showed that the initial propeller thrust was 58 pounds; 18 more than was needed to fly the machine in a 20 mile an hour wind. However, only eight seconds later, the thrust had dropped below 40 pounds."

"Therefore," O'Grady said, "might we assume that in a brisk wind you would have eight seconds of adequate power at your disposal?"

"Not exactly. By adjusting the pressure reducing valve over the next two pendulum tests, the 40 pound minimum propeller thrust was subsequently extended to 12 seconds."

St. Joseph, Michigan

Monday, October 10, 1898

The Herring-Arnot machine sat trembling outside the pavilion's big door as the morning wind gusted to 18 miles an hour. The contraption faced into the wind, its vertical rudder sitting atop a wooden barrel, its nose relaxing on the sand. Two 50-pound sandbags were slumped over the wooden framework that supported the two spoked wheels, helping to steady the machine.

Wearing his dark gray coveralls and black captain's cap, Herring tinkered with the stationary engine's ignition system, while sipping from last night's pint bottle. *Can't stand warm milk*, he thought. Anticipating an early attempt at flying the new machine, Gus had asked for and received a week off from his job at Truscott's. Business being slow, old man Truscott welcomed the opportunity to save the company some money. George could only assist him after work and on weekends, but Matthias, who stayed at the Herring home, was on hand throughout the testing period.

"How does the old Daimler-Maybach engine look, Gus? Is it ready to go?" Arnot asked.

"You never know with this worn-out equipment. The thing might work fine one day and refuse to start the next. I'm hoping it runs like it did last night! However, I'm concerned about the four dry cells. If one or more of them gives up the ghost... the ignition system probably won't deliver the minimum voltage!"

"Do we have any of these cells in reserve?"

"I have two new Hellesen cells here at the shop and six more on order. They should arrive sometime this week... if we're lucky!"

Pacing nervously outside the empty pavilion, Herring decided to check the wind speed and direction again, something he had been doing every half hour for the last two hours. Picking up his pace, he hiked toward the shoreline 200 feet to the north, where the wind would not be as turbulent. Shuffling to a halt, Herring held the anemometer high above his head. Within seconds, he had his answer. The wind was still blowing out of the northeast, but its speed had increased to 22, gusting to 25.

Time to fly, he thought.

By 10 o'clock, the stationary engine and compressor had worked to pump up the tank to the required 600 psi. As he disconnected the long brass tube, a twinge of anticipation coursed through Herring's body.

This is it, Gus, he thought. *All these years of effort*!

Checking for leaks and finding none, he turned to Arnot and shrugged. "Well, Matthias, it's just you and me now...guess there's nothin' left to do but fly the damned thing!"

Fifteen minutes later, the Herring-Arnot machine sat on a compacted patch of sand near the Lake Michigan shoreline. While Arnot untied the rope that restrained the front propeller, Herring crawled into the operator's position and hoisted the 88-pound apparatus from its resting place. As the two propellers began to "windmill" at a furious rate, Gus altered his grip on the machine for proper balance. Checking his heading, he glanced back at Matthias, who had trotted to the rear, where he now gripped the starting rope. The moment of truth was at hand.[4]

CHAPTER 67

St. Joseph, Michigan

Monday, October 10, 1898

Crouched 10 feet behind the flying machine, Matthias Arnot yanked on the starting cord. The control valve snapped open, sending a surge of high-pressure air into the motor, setting it off with a primeval howl. As the machine's twin propellers spun up to speed, they hurled a veil of sand particles into the unsuspecting assistant's face.

Teetering forward in his runner's crouch, Herring lunged when he felt the tug – one, two, three strides, and off they went – man and machine thrust into a gusty, 25-mile-an-hour northeasterly breeze. As the contraption skimmed low across the Lake Michigan shoreline, Herring avoided grazing the sand by pulling his legs up into a fetal position beneath the air tank.

Flying at a pace equal to a fast walk, Herring's latest *Double Decker* settled into a steady, straight-ahead course. Without warning, an offshore gust walloped the machine on its port side, causing the regulator controlled cruciform tail to pivot. A fraction of a second later, Gus reacted by shifting his weight to the left, causing the apparatus to yaw into the disturbance while simultaneously leveling its lifting surfaces. In an instant, Herring instinctively transferred his weight back to the right and slightly forward, realigning the craft's heading into the prevailing wind.

No sooner than he had time to breathe – or so it seemed – the flight was ending. After only eight seconds, the Herring-Arnot machine glided back onto the sand, its twin propellers slowing to a wind-aided whirl. Dashing up from behind, Arnot latched onto a wingtip.

"Augustus! She flew... she flew!" Arnot shouted as Gus crawled out from beneath the conglomeration of steel tubing, wires, and varnished silk.

A moment later, a powerful gust lifted Arnot's wing, threatening to cartwheel the frail machine to destruction. "Hold on, Matthias! Tip the tail up!" Gus shouted. "Don't let her get away... and watch out for those spinning propellers!"

After a brief struggle, the twosome managed to plant the nose of the machine against the unyielding sand. Lashing the propellers to the frame, they lifted the contraption and carried it back to the pavilion for safekeeping. Returning to the shoreline, the men measured the distance flown. In less than 15 minutes, Herring's calculations were complete. Although the craft had flown only 50 feet over the sand, it had managed to progress 340 feet through the air during its eight- to 10-second flight... 25 miles an hour through the *air*, and four miles an hour over the *sand*, for an actual airspeed of 29. The men celebrated their achievement alone – there were no other witnesses.

Herring posing with compressed air-powered aeroplane (1898); Meiller/Herring Collection

*

As they relaxed back at the pavilion, Matthias tried to speak while devouring his egg salad sandwich. "For our next trial... we need a good witness... somebody to spread the word."

Herring swallowed a bite of his radish sandwich before answering. "You got somebody in mind? Zahm can't make it... he's got wind-tunnel tests goin' on—"

"What about Herr Dienstbach," Matthias interrupted, "the fellow from the New York *Times*? He's always been fair when writing about your exploits."

Herring's gaze turned to the flying machine. "I tried to telephone Carl last week. Someone from his office said he was in Germany and wouldn't be back until November."

A nervous silence descended over the conversation.

"What about Chanute? He could probably be here tomorrow!"

Scowling, Arnot wiped a smudge of mayonnaise from his lower lip with the back of his hand. "Christ, Gus, anybody but that old fart! He can't be trusted! By the way, have you seen the latest *Aeronautical Journal*? Just as I predicted... he's using *our* photographs!"[1]

Herring forged ahead. "Matthias, I agree with you, but we still need an expert to observe a successful flight trial before the snow flies."

Arnot shoved his hands into his pockets as he leaned against the peeling overhead door. "Do you think Chanute could be honest when he reports on our trials? This is the same person who belittled you for leaving him! The same person who ridiculed you for trying to solve the problem of powered flight on your own!"

Arms crossed, Herring patiently waited for Matthias to finish. "Despite our differences, we still correspond. The old man tells me about aeronautical activities happening around the world, and I keep him up to date on our flying machine."

Besides, thought Herring, *because of Michigan's ridiculous out of state check processing fee, I was losing 21 dollars a month on Matthias's stipend! Chanute offered to help by cashing the checks at his bank in Chicago and sending me the currency. Certainly I can't tell Matthias about this arrangement... he might think I'm being ungrateful.*

"One other thing," Herring continued, "as you know, my British patent was granted this summer, and the old man is listed as co-inventor. When he sees for himself that we have made progress toward producing a practical machine, he'll be more than happy to announce *our* breakthrough to the aeronautical community! Since Octave longs to be part of the winning team, I believe he'll provide a splendid review!"

Rochester, New York

Tuesday, November 8, 1921

As Herring trudged up the snow-clogged front steps of the courthouse, he thought, *Christ, where's the maintenance crew?* Entering through one of the massive glass-and-brass doors, Gus struggled into the building's vestibule before hesitating beside the statue of *lady justice* to stomp the snow from his buckle-up galoshes. Less than a minute later, he exited the otherwise vacant elevator and entered the familiar rear door of Judge Sawyer's federal courtroom. Aside from the elderly bailiff, who was replacing a burned-out bulb in the judge's bench lamp, Augustus was the first to arrive.

As the door creaked open, Vern turned from his task. "You're here early this morning, Mr. Herring. Where's that pretty daughter of yours today?"

"She decided to stay home today, Mr. Krehbiel," Herring said, removing his knee-length, black woolen coat. "It's much too stormy outside for her."

"Did you drive the *flivver* today?"

"Not today, I took the trolley... two of them, as a matter of fact. Before I left home this morning, the man on the radio reported that almost 18 inches of snow had fallen on Rochester since yesterday afternoon!"

The bailiff bristled. "First, there was that freak snowstorm in October that ruined six of my hardwood trees, and now this! My wife, Shirley, thinks we're headed for another ice age... but I just don't know."

*

Judge Sawyer delayed the beginning of testimony until the late arrival of the court's recorder and clerk. At 9:30, Herring was directed back to the witness stand.

"Mr. Herring," O'Grady began, "you concluded yesterday's testimony by describing a conversation that you had with Mr. Arnot. Please summarize that discussion for the court."

Herring crossed one leg over the other. "We needed an aeronautical expert to witness our next trial, but only Chanute was available."

"Was he your choice?"

"Yes... so I sent him a message over the wire."

"What did it say?"

Herring dropped his gaze, as if deep in thought. "I'll have to paraphrase, 'Will fly again on Tuesday October 11. Can you come? If so, come alone.' "

"Did you receive a reply?"

"That same day. Chanute said he would take the overnight lake steamer from Chicago."

"When did he arrive in St. Joseph?"

"His boat arrived on time at the Morrison Channel pier. Bringing only his camera and tripod, Chanute trudged the 700 feet through the sand to our pavilion shop, turning up a bit after 7 a.m."

"Was Mr. Arnot prepared for the encounter?"

"Matthias promised to remain calm. Nevertheless, he did mumble an occasional obscenity under his breath."

"When did the flight trial take place?" O'Grady inquired.

"By 8 o'clock, Matthias and I had moved the apparatus outside and reattached its tail. At that time, the wind was blowing a promising 15 miles an hour. In anticipation of an early attempt, I connected the air tube from the compressor to the onboard tank.

"As an omen of things to come, the stationary engine *refused* to start. A preliminary inspection revealed that the four non-rechargeable dry cells were indicating a low voltage on my galvanometer. Closer scrutiny identified the problem as a shorted ignition switch that had allowed the battery to discharge. Since there were only two fresh Hellesen cells on hand, there was nothing I could do but wire these in series with the four discharged units.

"Although the voltage was lower than desired, the stationary engine subsequently started, but then stalled when loaded by the compressor. Upon further investigation, I found that the breaker contact points in the ignition system's primary circuit had 'burned' from the excessive heat.

"After I had filed and reset the point clearance, the engine started and seemed to operate steadily. By this time it was almost noon, so we stopped to enjoy the basket lunch Lillian had prepared for us."

"What was Mr. Chanute doing while you were working on the engine problem?"

"He kept busy. I saw him mounting his camera to the tripod, preparing glass plates and tinkering with his light meter and lens settings. All the while, despite his vow to remain cordial, I sensed that Matthias was approaching his melting point."

"Did Mr. Arnot ask Chanute about his plans for taking photographs?"

"Toward the end of lunch, Matthias laid down the ground rules: there would be no photographs allowed to be taken of the machine in flight and only one with Matthias and me posing with the apparatus."

"What did Chanute have to say about these restrictions?"

"Not one word! He merely nodded and continued eating his pig's knuckles and sauerkraut. After lunch, he packed up his camera and set it aside for his trip home."

"Was Mr. Arnot satisfied?"

"Nothing short of Chanute's departure would satisfy Matthias!"

"When did you finally get the air tank pumped up?"

"About 3 o'clock. By the time the compressor had done its job, I was exhausted from all of the troubleshooting and fiddling. Then there was the other problem."

"What other problem?" O'Grady asked.

Rubbing the back of his neck, Herring slumped deeper into his chair. "The wind had died down! Although it wasn't calm by any means, 10 miles an hour wasn't *fast* enough to produce sufficient lift with our limited propeller thrust. In other words... the machine couldn't possibly fly!"

"What did you do?"

"When I told Chanute about the situation, he was unsympathetic, saying that he absolutely had to return to Chicago on the

7 o'clock boat. 'It's now or never if you want me to act as your witness,' he insisted."

"With conditions the way they were... what did you do?"

"In hindsight, I should have abandoned the trial! Instead, I foolishly decided to adjust the reducing valve so the motor would provide maximum propeller thrust at the instant of launch. I reasoned that the additional thrust might provide the machine with the necessary speed to help supplement the aero*curve's* lack of lifting power and keep it flying."

"Tell the court what happened."

"When Matthias pulled the cord to open the air valve, the motor shrieked and the propellers thrashed the air like never before. Before there was time to blink, I had already launched into the gentle breeze. At first the machine seemed to slowly pick up speed, but then the port side lifting surfaces began to settle, and no amount of weight shifting could change that.

"Nothing could be done. The tip of the lower lifting surface dug into the sand, causing the apparatus to rotate counterclockwise. After the wheeled alighting gear struck the sand sideways, kicking up a spray of the stuff, the portside framework collapsed to the sound of splintering spruce.

"Tipping further to port, the machine's front propeller beat into the sand, resulting in a cascade of black walnut shards flying about. When I regained my senses, I found myself trapped inside the piano-wire trussing, between the crumpled alighting gear and the lower lifting surface."

"Were you injured?" O'Grady asked.

"Other than bumping my head, which left me temporarily dazed, and a lacerated pinky finger on my left hand, I was fine."

"Where was Mr. Chanute while all of this was taking place?"

"Who knew? When Matthias and I finished dragging the wreckage back to the pavilion, he was perched atop one of the shop's wooden barrels."

"What was said?"

Before responding, Herring's eyes turned to the swirling snow outside the courtroom's windows. "It's been 23 years, but I still remember the old man's words: 'Herring... I knew that damned thing wouldn't fly! This is the *last time* you'll waste my time!' "

"What happened then?"

"Matthias, who had been bandaging my injured finger, lost his composure. Rising from his stool, he turned to face Chanute. Removing his pocket watch from his vest pocket, he flipped open its cover and peered at the dial. Looking up, he bellowed, 'Old man, you have exactly 10 seconds to vacate this facility... lest I chase you down and put my boot to your fat ass!' "

"Did Chanute comply?" O'Grady asked, as he labored to maintain his poise.

Herring tried to repress a smile. "It was the fastest I've ever seen Octave move! He rushed out the door shouting, 'Herring... we're through!' "

*

"Allow me to summarize," O'Grady continued. "The air-compressing equipment malfunctioned, the wind died, you crashed, and your expert witness disowned you! What else could possibly go wrong?"

Herring shook his head. "The local lumberyard disappointed me."

Confused, O'Grady glanced around the courtroom. "Please explain."

"I needed kiln-dried black walnut lumber to make a new propeller, and there was none to be had in St. Joseph."

"What did you do?"

"Bill Avery had what I needed at his East Chicago shop. Despite the risk of running into Chanute, Matthias volunteered to make the excursion."

"Besides wood for the propeller, what else did you need to make repairs?"

"Matthias also retrieved some cut-to-size Sitka spruce to replace the broken wing spar."

"While your partner was away," O'Grady said, "what were you doing?"

"I made new alighting *skids* from hickory that I had on hand."

"What about the cantankerous Maybach-Daimler stationary engine? Were you able to solve the unreliability problem?"

"Every lock has its key," Herring replied. "The Hellesen dry cell order arrived the following day – 24 hours too late!"

"How long did it take to repair the flying machine?"

Leaning forward, Herring clenched his fingers. "Including the propeller and alighting skids, which required an additional three days to fabricate, the machine was ready to go again by October 17."

"Why did you change from wheels to skids?"

"I thought that skids would hold up better to the rigors of landing on sand...we learned that wheels tended to dig in rather than roll."

"You had several days to reflect on Mr. Chanute's conduct. What conclusions did you reach?"

"Matthias reminded me of my biggest flaw."

O'Grady, raising his hands as if surrendering. "What might that be?"

"Besides being naïve, sometimes I don't see the forest for the trees."

"Would you care to elaborate?"

For a few seconds, Herring only blinked. "When I asked Chanute to become our expert witness, I failed to realize how much a successful powered flight would *threaten* his legacy. And when the machine failed to fly, no fault of its own, it provided the old man with yet another opportunity to slander me and diminish my work! Matthias had been right to oppose his invitation!"

O'Grady paged through his notes, signaling a probable change of topic. "Once the machine was repaired and ready to fly again, what were your plans?"

"We would try to fly again on the coming Saturday, and this time we would invite the press! Taking no chances, we explained the ground rules well in advance to the various newspaper editors."

"What were the rules?"

"We would *not* attempt to fly in substandard weather conditions, nor would we apologize for delays attributed to the flying machine's condition or that of its support equipment.

"In return for the journalists' cooperation and patience, we agreed to make ourselves available for questioning. In the event of a successful trial, all pertinent engineering details would be disclosed, along with a historical accounting of the venture."

"Weren't you worried about revealing your secrets to the public?"

Herring seemed tentative, but continued. "I was at a crossroads. I could try to conceal my ideas, or I could divulge them freely and hope the press would report them as mine. Here was the dilemma; three months earlier, Chanute and I had been granted a joint British patent for my manned, powered aeroplane. Back in February of that same year, the U.S. Patent Office had rejected my similar domestic application. While patent protection is always preferred, there are drawbacks. In this instance, all but the most minor details of my compressed-air-powered aeroplane were revealed in the claims of the British patent. Since this left me no protection in America, any alert individual or company in this country could tap into and use those ideas; there was no point in maintaining strict secrecy."

"With those rules and promises in mind," O'Grady said, "which newspapers sent a reporter?"

"Oh gad, let me think. There was the Chicago *Tribune*, the Chicago *Record* and the Chicago *News*, along with the Benton Harbor *Evening News*, our local paper."[2]

"Describe the setting on Saturday, October 22, 1898."

Herring paused, contemplating where to begin. "The day dawned clear, crisp, and windy – a promising start. By 9 o'clock the reporters began to arrive at the pavilion along with William Engberg, co-owner of a nearby machine shop that I had frequently patronized. Less than an hour later, as launch time neared, I could feel the nervous tension radiating from the observers as Matthias set up his tripod and camera on the stretch of beach directly upwind from the flying machine's starting point. Was this group about to witness a historic event, or yet another failure?"

"Describe what happened."

"It would be more authentic if I read from one of the newspaper articles that described the trial," Herring said. "The first account appeared in the Benton Harbor newspaper on October 24... two days after the fact.[3] The story didn't appear in Chicago until November 17... almost a month later. After that, papers in Detroit and New York City published an account of the flight, as well as the Elmira *Daily Advertiser*, that's in New York State... the paper you are currently holding."

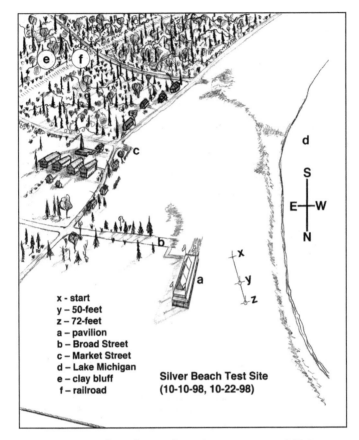

Silver Beach (1898). Site of Herring's two powered flights; drawing by author

Hoisting the old newspaper above his head, O'Grady addressed the judge. "Your Honor, the plaintiff wishes to place this newspaper into evidence."

After brief scrutiny by the judge and the defense, the court clerk assigned the document an exhibit number.

"Mr. Herring, please read from the article, titled, 'Herring's Powered Aeroplane,' that appeared in the Elmira *Daily Advertiser* on November 18, 1898. This article is an excerpt from a report in the Chicago *Record* newspaper, a day earlier."

Herring methodically arranged his reading glasses on the bridge of his nose before turning to the newspaper.

"Any day now, Mr. Herring—" Robbins squeaked, sarcastically.

"Hold your horses, Mr. Robbins," the judge directed. "The defense will get its opportunity to *procrastinate* soon enough!"

Peering in Sawyer's direction, Herring received the nod he was waiting for, and began to read:

" 'Perhaps fewer than half a dozen persons know that the Silver Beach bathing pavilion has been converted into an experimental shop, in which has been built a new flying machine; moving by reason of the power of its own engine and supporting not only the weight of the machine, but that of a full grown man besides.

'When the Record correspondent presented himself at the pavilion a few days ago he found the professor in overalls, working over a gasoline engine, which he explained, operated the air compressor that supplied the storage tank of the flying machine... in a few minutes everything was in readiness for a flight. The big door of the pavilion was raised and the machine, separated from its tail was carried to the outside at the end of the building. Here it was coupled to a brass tube from the air compressor and inside the gasoline engine was started... when the required pressure was obtained and the machine with its attached tail was moved out into the open stretch where it faced a 25 mph wind.

'The propellers were now turning at a furious rate by the force of the wind. Mr. Herring crawled underneath the apparatus and raised it so easily that it seemed to possess no weight at all. An assistant then moved behind the wings on the starboard side of the contraption and grasped a cord. After a signal from the professor, the cord was pulled, opening the air valve. As the engine shrieked and the propellers whirled furiously, Herring made a few forward steps and the machine leaped forward, an instant later flying in free air, with the skids nearly a yard above the sand and the operator's legs drawn up in a bunch near the tank.

'It was really flying – already the machine had covered a distance of 50 or 60-feet, when the speed perceptively slackened and a little further on the apparatus came gently to rest on the sand. The distance covered was afterward measured at 73-feet and the time of the flight was estimated by Mr. Herring at 8-10-seconds. He explained however, that though this represents a speed of only 5 or 6 miles per hour over the ground, the real speed of the machine was more nearly 30 mph, as it was advancing against a 25 mile wind... a second trial was attempted, but as something went wrong with the compressor, it was abandoned for the day and for the rest of the season.'"

Flipping through a second notebook, O'Grady pursued his next line of questioning.

"Mr. Herring, what did you say to reporters after your successful flight?"

"I emphasized that the *successful airborne condition of our experimental apparatus proved that the problem of manned, heavier-than-air flight was solvable and represented a step toward the development of a practical aeroplane.*"

"Did you suggest that the problem of manned, heavier-than-air flight had been entirely solved with your apparatus?"

"Absolutely not. Only that the problem was *solvable*."

"What progress did the flight of October 22nd represent?"

Then Gus emphatically stated: "*The apparatus was the first manned, powered aeroplane to rise from level ground into a prevailing wind, make steady progress along the ground, under control, and to land without breakage.*"

"In a letter to journalist Carl Dienstbach," O'Grady said, "you detailed some of the problems that needed to be addressed. What were they?"

"I'll mention a few that I remember," Herring said, rubbing his temple. "The power-to-weight ratio needed to be improved

– the early engines were still too heavy and too weak. Then there was the question of how to attain the necessary speed for take-off during calm conditions. Matching the propulsion system to a specific engine and airframe combination also presented a real engineering challenge. Of course, there was the question of how to ensure the safety of the operator in case of a mishap. I have always felt that the problems associated with flying machines should hold no particular difficulty if the principles of engineering were properly applied to their solution."

"Mr. Herring," O'Grady continued, "can you explain why only *one* of the three Chicago newspapers published the story about your flight – and why that article didn't appear until a *month later*?"

When Herring finally spoke there was a smile on his face but not in his voice. "I learned what had happened in a roundabout way. When Herr Dienstbach returned from Germany in early November, he telephoned me to discuss the October trials. He had obtained a copy of the Benton Harbor *Evening News* story and was surprised to learn that although three of the big Chicago papers had covered the event, none immediately published a report.

"Two weeks later I received another call from Carl – the same day that the Chicago *Record* finally ran the story."

"What did Herr Dienstbach have to say?" O'Grady prodded.

A wave of silence washed between them; eventually Herring spoke up. "Carl had spoken with his reporter friends at the *Tribune* and the *Record*, where similar accounts of the trials emerged. Inexplicably, these same reporters had their stories quashed by their editors... after they had contacted Octave Chanute for comment!"

"Did you learn what Chanute had said?"

"According to Carl—"

"Objection! Hearsay!" Robbins shouted.

"Overruled," Judge Sawyer chimed in. "I will allow the commentary."

"What did Chanute *allegedly* say," O'Grady said.

"Apparently his message was the same to all three editors; allow me to paraphrase: 'Report this so-called event if you must – it's your funeral! Herring has shown nothing that has not already been demonstrated by other competent experimenters, such as Professor Langley of the prestigious Smithsonian Institution. I'm afraid that Mr. Herring, a known fraud with questionable ethics, has set you up to satisfy his penchant for publicity in the pursuit of great wealth.'"

Addressing the judge, O'Grady asked for and received permission to read from the December Elmira *Daily Advertiser*, an exhibit that had been previously entered into evidence by the plaintiff. "Here's the pertinent part of this article, Your Honor:

'Monday, December 5, 1898

The Elmira Daily Adviser

Herring Called Fraud

'...Chanute returned to Chicago and, with a fanfare of publicity labeled Augustus Herring a fraud. Despite the subsequent press reports after his successful flight of October 22, the oft-heralded Father of Aviation has continued his vendetta against this erstwhile pioneer and his associate, Matthias C. Arnot, of this community.'"

"Objection, Your Honor!" Robbins cried. "Irrelevant! That article was merely an *opinion* piece! As we have grown to expect, the attorney for the plaintiff is introducing information that is not pertinent to this proceeding. The defense demands that Mr. O'Grady cease wasting the court's time and direct his attention to the appropriate issues at hand!"

"How do you respond, Mr. O'Grady?" the judge asked.

Flipping his notes to a previously bookmarked page, O'Grady strode out from behind his podium. "Your Honor, my reply to this objection *remains the same* as it has been since the beginning of this trial. Specifically, the plaintiff is determined to highlight any and all deliberate attempts to defame Mr. Herring's character. Mr. Chanute's reprehensible conduct subsequent to the experimental flying machine's misstep on October 11, 1898, and its subsequent success on October 22, represents but the latest example of a concerted effort to destroy this witness's reputation with the public and within the aeronautical community!"

An undercurrent rumbled through the spectator section of the courtroom. Banging his gavel, the judge held up his left hand, demanding silence. As the gallery grudgingly complied, Sawyer made his decision.

"Overruled. You may continue, Mr. O'Grady."

"Mr. Herring, do you know why the Chicago *Record* finally decided to break the story?"

"Because Herr Dienstbach's newspaper – the New York *Times* – threatened to pick up the Benton Harbor piece and scoop the other Chicago papers! Much to Mr. Chanute's annoyance, the *Record* reconsidered its position—"

"How did you learn of this development?" O'Grady interrupted.

"I learned of Chanute's maneuvering when I visited Bill Avery in December. According to him, the old man was still distraught because he had only *killed* two of the three Chicago stories."

CHAPTER 68

Rochester, New York

Tuesday, November 8, 1921

As the flood of employees and litigants rushed from the courthouse for lunch, Herring was lucky to find an unoccupied phone booth. After seven rings the operator was about to disconnect when a breathless Chloe finally picked up.

"Chloe, it's Augustus! I only have a minute... I'm on my way to lunch with O'Grady—"

"Daddy! I heard the phone ringing... I was outdoors... shoveling the Ford out of a snow drift—"

"Sorry to cut your fun short, my dear, but I want you to get gussied up. I'm taking you to a fancy restaurant tonight... there's a milestone to celebrate!"

*

An hour later, Herring was back on the witness stand.

"After the reporters left Silver Beach," O'Grady said, "what was next on your agenda?"

"Arnot and I stowed the flying machine inside the pavilion, locked up the shop, and walked back to my place to develop the glass-plate negatives. We were energized by the possibility of having an in-flight image to back up our powered-flight claims."

"Did you obtain a good photograph?"

Herring's skid-landing of compressed air machine (10-22-98); Meiller/Herring Collection

Herring seemed to stiffen. "Apparently, Matthias wasn't paying close attention. He snapped the picture after the machine had already touched down. The propellers were still spinning, and the alighting skids were kicking up sand. On the spur of the moment, we decided to try flying the machine again... to obtain a more persuasive photograph. While we waited in vain for the weather to cooperate, we formulated our plans for the upcoming season."

"What was decided?"

"I would design and build a larger version of the '98 two-surface apparatus and power it with a *steam* motor. I would also work at refining my lightweight gasoline engine, with an eye toward replacing the steam motor at a later date."[1]

"With the flying season over, what was your next project?"

"After Matthias left for home, I went back to my engineering job at the Truscott Boat Works. On evenings and weekends, I labored at disassembling the powered machine, placing it into two sturdy crates and moving them into my rented space at the boat works.

"On the last Saturday in October, I took a lake steamer to Chicago and returned the reconditioned Maybach-Daimler engine and air compressor to Art Pesch – an undertaking that raised my spirits! My lease at the Silver Beach bathing pavilion was about to expire, so I walked my metalworking tools and precision measuring instruments over to the rented shop space and carted my woodworking tools back to the shop behind my house. By

Thanksgiving, I had completed the drawings for the new flying machine, including a detailed list of specifications, copies of which I posted to Bill Avery for bid."

Herring paused to take a breath. "In December, I concentrated on converting the twin-cylinder compressed-air motor back into a gasoline engine – a single-cylinder job. I was anxious to improve the engine's performance for the express purpose of powering a *safety bicycle* – an idea that Matthias also liked. We hoped that selling a limited number of motor bicycles would help to offset the exorbitant costs associated with our flying machine work."

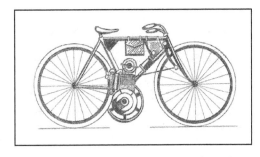

Herring's motorized safety bicycle (1899); Meiller/Herring Collection

"Objection, Your Honor!" Robbins shouted. "Are we discussing flying machines or bicycles? This whole line of testimony is irrelevant! The defense demands that Mr. O'Grady cease in his attempts to muddle the issues by encouraging the witness to meander through his every recollection."

Banging his gavel, Judge Sawyer held up his hand for silence.

"I'll take your objection under advisement, Mr. Robbins. However, this is a good time to end court proceedings for the day. Although the hour is early – my watch indicates that it's past 3 o'clock – I have other pressing court matters at hand; unavoidable issues that must be resolved. Court will resume promptly at 9 o'clock tomorrow morning."

With yet another crack of his gavel, court was adjourned. Turning to O'Grady, Herring grinned as he spoke. "Damn, that was unexpected... but welcome nonetheless! I'm taking my daughter out for dinner tonight, and this will allow me to beat the afternoon rush."

O'Grady leaned over and whispered into Herring's ear. "Tomorrow is an important day for us, Gus. It's time to testify about the happenings of '99... and you know what's coming."

"I have a pretty good idea," Herring said, grimacing.

"Good! I just want you to be prepared."

*

Gus surprised Chloe when he walked through the side door of his Rochester house.

"Daddy! What are you doing home this early?"

After a brief explanation, Herring hightailed it upstairs to freshen up, before changing into his Sunday best – a three-piece black suit, white, high-collar shirt and his favorite Kelly green bow tie. By 5 o'clock, he was back outdoors hand-crank-starting his Ford sedan. With the machine purring merrily, he set the idle speed and opened the heater vent before bounding back into the house to warm his hands. "Damn, it's cold out there – 20 degrees according to the thermometer—"

"Daddy! Please don't cuss. You know mother wouldn't like it," Chloe reminded him. "Where are we going... or is that a surprise, too?"

"We're headed to the Embassy Hotel, and might I add that you look wonderful! All of the young men there will be envious of me."

*

Finishing his dessert – a generous slice of apple pie washed down with a second glass of cold milk – Herring became talkative as they enjoyed the warmth of the dining room's fireplace. Sensing this to be an appropriate time, Chloe nudged her father into a discussion of the trial's direction. "Did Judge Sawyer appear sympathetic when you disclosed Mr. Arnot's failure to obtain an in-flight photograph?"

Leaning back, Herring thought for a moment. "I'm not sure about the judge's sympathies, but the photograph of our machine plowing through the sand paled in comparison to the image of Orville Wright's first hop, five years later. That snapshot has to be the most convincing image in the history of heavier-than-air flight, even though the machine was porpoising out of control and crashed seconds later!"

"Earlier," Chloe said, "you mentioned that tomorrow you're going to testify about the events of 1899. I'd like to hear the story so I can better follow the courtroom declarations. However, before you begin, *please* tell me... *what* are we celebrating?"

Herring gave his that's-fair-enough nod. "Sorry, my dear, how could I possibly forget? This morning, after finishing testimony about the powered flights, I experienced a feeling of joy and accomplishment, knowing that those modest hops back in '98 led directly to the problem's solution! At that instant, I knew we had to celebrate... aren't you glad I made that telephone call? Now, if you promise not to become bored to tears, I'll try to recall the events of 1899 – when you were only three years old."

As they settled back in their comfortable leather-clad chairs, the flickering candlelight danced off their faces, forming ghostly shadows on the walls and ceiling. Ignoring his aversion to strong stimulants, Herring followed Chloe's lead and ordered coffee.

"If I become incapacitated, my dear, you'll have to drive the flivver home," Herring said with a wink. "Where do you want me to begin?"

"Start with your *plans* for 1899."

When Herring did not immediately respond, Chloe added, "... Yours and Mr. Arnot's plans."

"After a lengthy discussion about our commitment to the development of a practical dynamic flying machine, we came to an agreement. Matthias would supply additional funds, and I would contribute the time necessary to bring a new apparatus to a flight trial."

"Do you remember how much money was spent on the flying-machine problem during your association with Mr. Arnot?"

Herring squared his shoulders. "I know roughly how much was spent between late '97 and September '99; I was thinking about it just the other day. Would you be shocked to learn that Matthias and I spent about $32,000?"

Chloe looked dumbfounded. "Where in heaven's name did all of that money come from? Was it all Mr. Arnot's?"

Reaching across the table, Herring gently took hold of his daughter's hand. "Chloe, I'm telling you this now because you've been my loyal and steadfast supporter. Truth is... I had spent all of my inheritance from your grandfather's estate. It was gone before I met Matthias. Then, I convinced your mother to let me use her $10,000 bequest for the two years in question. Arnot provided more than $20,000."

"All the while, you worked at Truscott's?"

"My salary from Truscott's paid for the house rental, food, and all of the other essentials to support our family. Anything left over went into the flying machine work."

Chloe slowly raised her free hand, a signal that she had something to say. "That explains why Grandmother Herring was so angry with you all the time. I remember her saying more than once, '... Augustus should have his head examined.' As for Mother, she was always sympathetic, but she often seemed sad when you went off to work on your experiments for long periods of time.

"I recall one particular Herring family gathering – I think I was about 10 years old. My aunts and uncles showed up with new automobiles and photographs of new homes, and my cousins told stories about tours of Paris and London – places I had only read about! That's when I realized that our lives were very different from theirs."

Wincing, Herring took a sip of his coffee. "Gad, this stuff tastes awful!"

Pushing his cup aside, Herring returned to the question of his sanity. "As far as Grandmother Chloe is concerned, she had every reason to be disappointed. If it weren't for my obsession to solve this damn problem, we might have lived a very different kind of life. Knowing how things have turned out, if I had it to do over again, I would have done them—"

Chloe interrupted her father with a squeeze of his hand. "Augustus... your drive and ambition make you who you are. You've always been happiest when working at solving technical problems. You could never have denied that part of yourself and been happy. Grandmother Herring didn't understand this... but Mother and I did. The idea thieves and legacy seekers have tried everything in their power to deny you the recognition and financial rewards you deserve! Because of your testimony, I now understand how these people have manipulated the courts and even the patent office. Because of that, you'll always be my champion!"

At that instant, their waiter reappeared.

"Aha!" Herring said. "Just the man I want to see. Take this evil-tasting stuff away and bring me another tall, frosty glass of milk!"

Chloe ordered more coffee, and the waiter hustled off into the shadows. For the next minute, the pair stared at the glowing embers in the fieldstone fireplace, its warmth both inviting and thought provoking.

"Let me see," Chloe said. "You finished the design of the new flying machine and sent its drawings off to Mr. Avery. What changes did you make?"

"I stretched the wingspan to what I thought was the controllable limit for a weight-shifting apparatus – 22 feet, with a chord of four and a half feet. The area of the lifting surfaces increased to 198 square feet – a nine percent boost from the previous machine. However, the biggest change was my choice of a powerplant. In order to achieve the long flights we had hoped for, I decided to use a steam motor."

"A steam motor! Wasn't that a step backwards? What about all of the hard work that you invested in the gasoline engine?"

Herring nodded. "As it turned out, my twin-cylinder gasoline engine developed only about two-thirds of the total power needed, and its action was somewhat irregular. I had estimated that at least 500 hours of 'bench time' would be needed to improve the engine... time that our team didn't have before the '99 season ended. In its place, I decided to use an *external* combustion design – a reciprocating, compound-action steam motor and its related systems. I felt confident with this choice because of my experiences at Stevens Technical Institute, and my successful refinement of Langley's steam motor for the unmanned aerodrome models."

"You mentioned that you were still working on a gasoline engine. How could you possibly find time to work on a new steam motor *and* a gasoline engine? Why would you waste your time on it in the first—"

"Hold your horses, my dear!" Herring interrupted. "I'll explain everything in due time. To begin with, I saved considerable time and effort by changing the twin-cylinder compressed-air motor back into a gasoline engine, while only using a single cylinder this time!

"After Bill Avery gave his word not to discuss our new flying machine with Chanute, I signed a contract for him to fabricate the wooden components. He promised to have the 'kit' ready by April of '99, so I had several months to concentrate my efforts on both the steam motor and the gasoline engine. I wanted to begin assembling the new flying machine in the shop behind our house no later than Memorial Day.

"Once I was able to get the gas engine to operate consistently on the test stand, I mounted it onto a safety bicycle to see how it would perform on paved streets and dirt roads. Before long, I was chugging around the streets of St. Joseph... but that didn't last. The neighbors complained about the racket, and the mayor demanded that I confine my testing to Lake Street."

"Besides aggravating your fellow citizens," Chloe remarked, "you worked a full-time job, designed and built a powerful and lightweight steam motor, designed and assembled a new and improved flying machine... and you still had time to experiment with gas engines and motor bicycles? Physically, how were all of these things possible?"

Herring grinned. "I was still a relatively young man with plenty of get up and go! For instance, if I wasn't working in my shop behind the house, I was spending long hours at various machine shops around town. During weekday evenings, rather than walking home for an hour or two of sleep, I often slept at places like Truscott's, Engberg's, or even the old Iron Works."

"Given that Mr. Arnot was footing much of the bill, what did he think of your agenda?"

Herring took a swig of milk. "Matthias thought that I was biting off more than I could chew until he *rode* that first motor bicycle; then he became excited about its commercial possibilities. In the 1890s, bicycles were all the rage in terms of personal transportation. I must admit that I briefly became distracted from the flying-machine work, while we investigated the potential for producing and marketing these motor bicycles. Shortly thereafter, I also built a new two-wheel machine that I called the *Mobike*; it outperformed the earlier motor bicycle by a country mile."[2]

Herring's single rider Mobike (1899); public domain

"How did the motor-bicycle rider get the engine running?"

Gesturing with his hands, Herring continued. "The machine was mounted from a back step after kicking along the road to get a start. After the operator reached the seat, the engine was thrown into gear by a pedal on the right side, after which it continued to run automatically until shut off. On long downgrades, the engine could be thrown out of gear, allowing the machine to coast. Then, near the bottom of the hill, the engine could be thrown back into gear."[3]

"Sounds like the motor bicycle wasn't intended for women to use," Chloe observed.

"Not at all!" Herring said, taking another sip of milk. "Your mother became quite an expert rider. Later on, we took many trips into the countryside with our tandem Mobike."

"Mother never mentioned that! I must ask her about it!

"Did you sell any motor bicycles? If so, how much did they cost?"

"The first few machines went on sale in June of '99... the retail price was $250, which included the bicycle. As a matter of interest, Wilbur Wright, the older of the two bicycle mechanic brothers from Ohio, built and tested his experimental kite about the same time. He used my two-surface planform, along with the trussing; Orville finally admitted to this in a sworn deposition back in 1908."

"Don't forget to mention that during your testimony tomorrow," Chloe said, as she wrote herself a note. "I'll remind you.

Are there any other details that you should consider mentioning concerning the Mobike?"

Herring shook his head. "Most of it isn't worth mentioning. I have the distinct impression that Judge Sawyer isn't paying much attention to engineering details."

"I think he hears everything," Chloe said. "Otherwise, why does he overrule most of the defense's objections?"

Herring raised an eyebrow. "Aren't you tired of hearing about these things yet? It's getting past your—"

"Oh no you don't, Augustus!" Chloe broke in. "I want to hear everything about 1899... I don't care how long it takes. Tell me why you decided to use the steam motor!"

Herring could do little more than marvel at her persistence. "In my pre-trial deposition, I stated that my business was the design and construction of gasoline engines, with an eye toward producing a powerful, lightweight, and reliable example that would sustain a heavier-than-air flying machine for minutes... if not hours. However, as I said earlier, developmental problems only allowed my gasoline engine to produce less than three horsepower – barely adequate to fly the apparatus, let alone provide for a satisfactory takeoff.

"Within a few years, improved gasoline engines would be powering perfected heavier-than-air machines. But back in '99, compressed air or steam motors were a better choice for early experimental machines such as mine. After converting the twin-cylinder gas engine into a compressed-air motor, I could adjust the cylinder pressure to achieve the necessary thrust for takeoff; unfortunately, there was little energy left over for sustained flight. As you can see, my dear, everything is a compromise when it comes to engineering. Which leads me to your question about the steam motor!

"I decided to use a steam motor because the technology would allow me to obtain the power and the duration necessary to extend my flights. However, these attributes came at a price; mechanical complexity!"

"Did your steam motor perform as expected?"

Hesitating, Herring thought, *How might I describe this without getting too involved?* "The motor and its boiler, the system's two main components, produced five and a half horsepower and weighed only 14 pounds. That's about one horsepower for every two and a half pounds of motor weight. For the day, that was a record power-to-weight ratio. The weight also included the gasoline burner, piping, and sheet-metal boiler covers; everything except water, fuel, and a condenser, which I didn't plan to use for the early flights."[4]

Herring's steam motor (1899); Meiller/Herring Collection

"Sounds challenging. Did you have many problems to overcome?"

"There were the usual troubles, but by the autumn of '99, most of them had been resolved, including one very strange difficulty with the coils of the 'flash' boiler. At startup, as the boiler was coming up to temperature, the coils tried to straighten out under the influence of the elevating steam pressure – as they moved, they *looked like* live human fingers!"

"Did Mr. Avery deliver the new flying-machine parts on time?"

"The crates arrived during the first week in May. When I telephoned Bill to confirm the shipment, he mentioned in passing that old man Chanute had been a constant thorn in his side. I prodded him to divulge specific information, but he was reluctant, not wanting to become embroiled in the conflict between his backyard neighbor and me. After assuring Bill that I would not divulge any of his comments, he reluctantly revealed the fundamental nature of the old man's odd behavior."

"I can't wait to hear this!"

Herring leaned close to Chloe and spoke in a lowered voice. "My dear, you are the only person that I have ever mentioned this to – I have kept silent for better than 20 years—"

Thinking how asinine this must have sounded, Herring abruptly sat back and resumed using his normal voice. "What the hell, the old bastard has been dead and buried for more than a decade – so, why am I whispering?"

"Please, Daddy, what did Mr. Chanute do?"

"Over a period of three months, Chanute engaged in *espionage*... and I believe that it was linked to the St. Joseph calamity that occurred later. I know this sounds farfetched, but hear me out. As it turned out, Bill didn't get started on the parts for the new machine until almost February. Chanute had just returned from a vacation in California. He stopped by the shop to say hello. His cheerful demeanor changed when he laid eyes on the plans for my new machine. When the old man stopped socializing and began to scrutinize the drawings and specifications, Avery pulled him aside, saying that I had instructed him to keep the machine's technical information confidential. Although Chanute tried to laugh it off, Avery said that he departed the shop in a foul mood."

"Something tells me that Chanute's snooping didn't end with that first encounter," Chloe coaxed.

"By no means! Chanute made a point of coming by the shop at least twice a week over the next three months – an unusual number of visits, according to Bill. Although he fastidiously avoided mentioning my new machine, Chanute took special interest in the grade and type of wood that was being used, as well as the component sizes that were produced. This took place whenever Bill was working on projects in another part of the shop or was speaking with a customer. More than once, the old man was seen slipping a six-inch steel ruler back into his pocket.

"At the beginning, when the plans were left in plain view, Chanute would always manage to sneak a peek at them on his way out of the shop. Before too many days had passed, Avery began stashing the drawings out of sight.

"Then there was the break-in—"

"Someone broke into Mr. Avery's shop?" Chloe interrupted.

"Yes, but there didn't seem to be anything missing; only paperwork was out of place – scattered all over the premises—"

"Let me guess... the flying-machine drawings were missing!"

"Not quite! That's because Bill brought them into his house at the end of each day. He thought that's what *they* were after. But there's more! Chanute had the nerve to complain when Bill kept the shop door locked during business hours. 'A damnable inconvenience!' he called it."

"Did Chanute ever criticize Mr. Avery for accepting your flying-machine work?"

Herring burst out laughing. "Very insightful, my dear! According to Bill, the old man only referred to *me* once, but it was a dandy!" At that very moment, Gus decided to stop talking and sample the after-dinner mints.

"Father, please don't tease! What did he say?"

"As Bill was crating the flying machine parts for shipping, Chanute made a last-ditch effort to derail the project. First, he chastised his former assistant for accepting work from a 'known charlatan,' calling it 'blood-money work.' Then the old man offered Avery double our agreed price – $1,200 – *not* to ship the parts! *Twice* the money, mind you! However, Bill was a man of his word, and wasn't swayed by the offer; I received the flying machine 'kit' right on time."

"What was Chanute's problem?" Chloe asked, puzzled.

Herring placed both palms of his hands flat on the table before him, as if to steady himself against some unseen force. "The son of a bitch thought I might succeed!"

Chloe's mood turned cold as she contemplated Chanute's desperate attempt to scuttle her father's plans. "Do you think that Chanute had anything to do with that awful night in September?"

Loosening the knot in his tie, Herring next undid the top button of his shirt. "That's better!" he sighed, leaning back. "Although it was mostly circumstantial evidence, at the time I believed he *was* involved, and I haven't had reason to change my mind. When I tell you who I *saw* that night, you'll understand why."

Chloe nodded, silently urging Augustus to continue. "How long did it take you to build the new flying machine?"

"First, we moved Matthias's crated '97 glider from the shop behind our house to my space at Truscott's, but there was still barely enough room to perform the work. George and I took nearly three months to complete the job. With the exception of the propellers, the apparatus was finished by August 10, 1899.

"After admiring the completed machine for a couple of days, George and I removed the lifting surfaces and set them aside while we carted the steel frame over to Truscott's in preparation for installing the steam motor and its accessories. At dawn the following Saturday, when the wind was calm and few people were up and about, we carted the lifting surfaces and cruciform tail across the sands of Silver Beach to Truscott's. By the middle of August, I began carving the black-walnut propellers in the shop behind our house."

After what seemed to be a long silence, Chloe, finished with her coffee, summarized their conversation.

"As fall approached with its ideal flying conditions, you and your assistants had completed construction of a new flying machine, a new steam motor, and most of the required accessories. You were also tinkering with a revised single-cylinder engine, had successfully used it to power a standard safety bicycle, and were in the process of building a prototype motorcycle, the *first* of its kind in the United States.

"Have I missed anything?"

Herring looked into his daughter's eyes and smiled. "You've got it, my dear! Now you know the circumstances that led to the most devastating day of my life."

CHAPTER 69

St. Joseph, Michigan

Sunday, September 10, 1899

Pushing on the oil-stained wooden lever, George Housam disengaged the two-inch leather belt from the overhead power shaft, causing the bench lathe he was operating to coast to a stop. Stepping back from the machine, he removed his thick safety glasses before turning to observe Herring, who was busy sweeping up brass shavings. Leaning his broom against a bench, Gus reached inside his full body coveralls and pulled out his pocket watch, snapping the cover open with a single, practiced motion.

"Damn, it's already 10:30! I have to quit. I promised Lillian I'd be home before 11 o'clock. If we get a move on, maybe I'll make it by midnight!"

The men had been working at Engberg's for the last four hours, but despite their best efforts, the steam control valve remained unfinished.

"The valve needs a little more work," George said. "If it's okay with the boss, I'll finish it after work tomorrow. In the meantime, let's get the shop back in order. You know how fussy the old man is about cleaning up."

By 11 o'clock, Herring was seated on the passenger side of Engberg's horse-drawn delivery cart as George struggled to lock up the shop in the dark. "Did you turn down all of the lamps?" Herring nagged. "You left one burning last Sunday, and Mike threatened to kick me out. Without the old man's help, I'd be up shit's creek without a —"

"Everything's off!" Housam snapped. "I know what I'm—" Housam stopped in mid-sentence as he noticed a curious orange glow off in the direction of Silver Beach. "Look over yonder, Gus. What do ya see? That flickering light reminds me of the driftwood fires that kids sometime set on the beach."

"Too damned cold and windy for any such shenanigans," Herring said, as he pulled up the collar of his overalls. Maybe the volunteer fire department is training new recruits; sometimes they burn stuff down at the west end of—"

From a half mile away, the men heard the muffled clang of the firehouse bell atop its tower. Listening intently, they counted the peals – two, pause, four, long pause – then the sequence repeated.

"That's box 24," Herring said. "Where's box twenty— "

"Somethin's goin' on over at Truscott's!" Housam cried, as he slapped the reins.

A few minutes later, their cart drew near the outskirts of the boat works. Herring seized Housam's arm, urging him to stop. As the rickety wagon skidded to a halt at Truscott's northernmost outbuilding, Gus pointed to movement among the backlit trees.

"George, look over there," Herring said in a restrained voice. "There's a man runnin' toward Morrison Channel... looks like he's tryin' to get away from somebody or something!" Seconds later, a pursuer hobbled across the same territory.

"That looks like old Phil, Truscott's night watchman," Herring shouted, as George urged the mare onto a dirt road that accessed the docks. "I'd recognize that limp anywhere!"

As they reached the edge of the channel, both Herring and Housam heard the unmistakable splash of oars and the labored exertions of an individual intent on putting as much distance between himself and his pursuer as possible. Seconds later, 67-year-old polio survivor Phil Graskowiak stumbled onto the dock waving his single-action Colt .45 revolver. As Gus and George ran to join him, the night watchman rasped, "Stop that son of a bitch! He just—"

At that instant a massive flash of white light accompanied a thunderous explosion that obliterated an outbuilding less than 300 feet to the south. Herring recognized the structure as Truscott's *paint storage* shed, where hundreds of gallons of highly volatile liquids were kept. The ensuing inferno sent flames 100 feet high, bathing the entire area in an otherworldly orange-white brightness, fully illuminating the would-be arsonist. During these five or six seconds of daylight-like clarity, the men had an unobscured view of one another. The arsonist, dressed in black, wore a tight fitting knit cap that framed his chiseled face and black handlebar moustache – features that Herring recognized at once. Stunned and rendered momentarily speechless, Herring leaned forward as he crouched, staring into the distant eyes of a loathed nemesis. With little hope of hitting his target, Graskowiak unloaded his "Peacemaker" in the general direction of the criminal, who had reached the safety of the heavily overgrown northern shore. Seconds later, like a fleeing rodent, he had vanished into the brush.

"Phil!" Herring shouted. "What in hell is going on?"

Struggling to catch his breath, Graskowiak made a pained effort to tell his story. "That bastard set fire to the machine shop!" he wheezed. "Then he ran down the line throwing torches or somethin' through as many windows as he could! Everything seemed to go up at once. I couldn't catch up to him! I'm gonna lose my job for sure!"

"Did ya pull the hook on the alarm box?" Housam asked.

Hyperventilating, Phil managed to nod before sinking to the ground.

After making sure the watchman was in no medical danger, Gus and George ran back to Engberg's cart. With Truscott's main complex still a quarter mile away, George urged the horse to a near gallop. When they came within 200 feet of the intense flames springing from the machine shop, the mare reared and would venture no closer. Jumping from the buckboard, George grabbed the beast by its harness and led her away from the mayhem. In the meantime, Herring edged closer to the blaze.

After tying the mare safely to a distant tree, Housam trotted up behind Herring, who was solemnly watching the flames burst through the shop's shattered windows. The radiating heat was so intense that the men felt the skin on their faces begin to blister. As they were forced back from the bedlam, Herring lost his composure. "All of my research is in there!" he cried. "The flying machines, the engines, the tooling for the Mobike... everything!"

Using his hands to shield his eyes against the intense heat, Herring stared through a familiar window, fixing his gaze on the shimmering image of his steam motor as it rested forlornly atop its steel assembly bench.

"Where the hell are the fire engines?" Herring screamed. "George, we've got to get some water on these flames! There still might be a chance to save the tooling, and maybe the steam motor!" Frustrated, Herring took off running. Following the railroad tracks, he headed toward Truscott's main woodworking shop and showroom to the southwest. Although smoke had begun to issue from the eaves of the building's gable roof, no flames were visible. As Gus, closely followed by George, slowed to a trot near the structure's front entrance, he came upon several other Truscott employees, including his engine department supervisor, Henry Nelson.

"Gus! Help us with the big slidin' door! We gotta save some of these showroom boats! Grab yer friend – he can help too!"

After the balky door had been successfully forced open, the crew of volunteers began hauling out boats, literally dragging them to an area beyond the railroad tracks. One notable 30-footer contained the latest version of Truscott's two-stroke vapor motor, a sunshade canopy and a fully functioning group of electric running lanterns – the company's latest innovation. As Herring struggled to help move the boats, he noticed that the first horse-drawn, steam fire engine had rumbled by on the red brick access road that led to the lake. He thought that he saw only two men aboard the hulking wagon, which meant that the volunteer fire department was short-handed. The big rig featured four, five-foot-diameter wooden spoke wheels that supported a vertical cast iron, 2,000-pound coal burner and boiler topped by a smoke-, steam- and cinder-belching stack. Directly behind the burner-boiler, a double-acting steam cylinder drove a piston-type force pump that discharged massive quantities of water through a flexible four-inch brass fitting.

Torn between assisting with the rescue of his employer's launches and helping the firemen extinguish the machine shop blaze 200 feet to the northeast, Herring's decision was made for him when three blasts from the engine's steam whistle called for assistance. Breaking away from the boat rescue, Herring and Housam sprinted the 100 yards to where the fire engine was being set up adjacent to its water supply – the Morrison Channel. As they approached the engine, a two-wheeled *hose cart* pulled by a lone stallion skidded to a halt beside them, its two-man crew scrambling to attach its flexible hose to the fire engine's pump outlet. Unhitching the horse, one of the cart men trotted the beast across the dock area to join its stablemates. Red-hot cinders from the fire engine's stack constantly threatened both man and beast.

As the two "hose men" struggled to push their cart toward the blazing machine shop, Herring and Housam joined in, careful not to step on the rubberized canvas hose that was unreeling between them. When they were within 50 feet of the conflagration, the helpers were waved back to the safety of the fire engine, while the cart men prepared the 20-pound brass nozzle to accept an imminent surge of water. Back at the engine, Herring and Housam pitched in to assist one of the firemen in joining 15-foot lengths of rigid, six-inch water pipe, which would be coupled to the inlet side of the pump; the pipe's opposite end was dipped deep into the channel.

The remaining member of the fire engine team, Chief Henry Hughson, struggled to shovel coal, monitor the steam pressure gauge, control the release of saturated steam to the motor, and man the whistle. Two blasts were the universal call for more coal, which George readily obliged by taking over the shoveling. Within five minutes, the boiler pressure had risen to 150 pounds, and the Chief, also acting as the throttle man, began directing steam into the engine. At first, it turned over reluctantly, sputtering a few times before coming up to speed. A single blast from the whistle was Hughson's signal that water was on its way as he cracked open the valve on the pump's outlet. Within seconds, the nozzle men braced themselves for the thrust of rushing water that shot a full 200 feet into the air.

Unfortunately for Herring, his area of the building was at a point farthest away from where the water put down. Hughson turned to face Herring, seizing him by the shoulder. "You're Herring, the flyin' machine man, ain't ya?"

Not waiting for a response, he continued. "The men tell me that you're an expert engine man, skilled with both steam and gas motors. Is that true?"

Not knowing where the conversation was headed, Herring nodded. "Good!" the Chief shouted "I'm puttin' you in charge of this here engine! Don't let the pressure get above 150 pounds, keep her rods lubricated and don't hydraulic the damn pistons with boiler water! Your friend here can act as your 'fireman' – that's *coal man* to you! Can you handle it?"

"Yes sir!" Herring and Housam shouted in unison. "But I need a favor!" Herring added.

"Make it quick, lad! As you can see... I'm shorthanded!"

"All of my equipment is located in the machine shop, at the other end of the building! Can you get a hose aimed through a window over there?"

"I'll see what I can do!" Hughson hollered, as he and his assistant hurried off toward a second engine that had just raced up the roadway.

*

As minutes turned into hours, it became obvious that the valiant effort to save Truscott's would come to naught. One building after another fell victim to the raging flames fanned by gusting 25-mile-an-hour winds. Well before daybreak, most observers conceded that the only remains of the company would be the 10 or 12 boats that had been pulled out of the showroom before it too went up in flames.

"Too bad the town only had three fire engines," Housam said.

Tired and grimy, Herring slowly shook his head as he leaned up against the now-silent fire engine. "It wouldn't have mattered, George. During the Great Chicago Fire, fewer than half of their 112 engines were in working order; even if they had had 200, the city still would have burned to the ground – just like here."

To survey the damage, the men remained until well after dawn. Although the main building's front and exterior walls were still standing along with its prominent brick smokestack, the structure's interior was a total loss; machinery, tools, and an inventory of partially finished boats that resided in the woodworking

Aftermath of the Truscott Boat Co. fire (1899); Meiller/Herring Collection

shop... all ruined. The machine shop had been completely destroyed – a forgone conclusion since it had been the first building to become fully involved. Its brick walls were still intact but had buckled badly by the intense heat, and only the massive smokestack from the adjacent powerhouse still stood as a solemn reminder to Herring that this had been the location of his rented space. As he walked through the gaping cavity that once housed a three-inch-thick oak man-door, Herring plodded through an ankle-deep slurry of ash and water, in the place that he had often referred to as his second home. Now he was hard-pressed to identify any of his possessions. The '98 powered flying machine, the new '99 powered flying machine, two motor bicycles, the prototype "Mobike," his tools, sand-casting patterns, prototype engines including his steam motor and accessories, jigs, fixtures, materials, working drawings, models, static and flying – all were gone except the '97 two-surface glider that Arnot had shipped to Elmira the previous week.

Herring's sobering thoughts were interrupted by Phil's gravelly voice. "There you are, Herring! I got somebody who wants to talk to ya!"

Strutting alongside Graskowiak was Chief of Police Charles Sauerbier, easily recognized by his blue parade uniform, gold-plated buttons and ceremonial hat. The portly 5-foot-6-inch 45-year-old had just completed his second year as chief, having previously been employed as a patrol officer under the former Chief C.H. Stuckey. Sauerbier's pronounced hooknose, black handlebar moustache and steel-gray eyes highlighted his otherwise pallid complexion.

As Herring trudged out of the sludge that inundated his shop's floor, he extended his hand to each of the men. Graskowiak obliged, but Sauerbier, the stench of whiskey on his breath, ignored the offering.

"Well Mr. Herring," the Chief said, "you disturb the peace with your damnable motors, and now you're party to an arson!"

Stung by the callous remark, Herring took a short step backward. "Pardon my frankness, Chief, but I've just lost more than $20,000 in machinery, tooling and materials in this fire. Furthermore, I've spent the entire evening assisting the fire department in trying to extinguish the inferno, while you were undoubtedly home... snoring! So please, don't give me any twaddle about being involved in an arson!"

Tilting his chin toward the overcast morning sky, Sauerbier teetered on the toes of his black leather shoes as his hand instinctively jerked to his holstered Billy club. "That's all well and good, *professor*, but I've got some questions that need answerin'. You can cooperate and give me what I want... or we can head back to the stationhouse. Which would you prefer, smartass?"

"Either way is fine with me," Herring said, staring down at the shorter man, "but my soiled shoes might foul your interrogation room."

Tiring of the verbal sparring, Sauerbier got to the point. "Look, Herring, I want the name of the guy who escaped across the Morrison Channel last night. Phil seems to think you recognized him."

Sauerbier waited with restrained impatience, his hands on his hips, his stubby fingers tapping.

"That's right... his name is Butusov. William Paul Butusov."

"How do you spell that?" Sauerbier asked, as he scribbled on his notepad.

"B-U-T-U-S-O-V," Herring replied.

"What's your relationship with this Butusov character?"

"We worked as assistants for the same man back in '96."

"Where?"

"Chicago."

"What type of work were you doing?"

"Experimenting with heavier-than-air flying machine gliders. Mine flew, his didn't."

"So, Butusov is a crackpot, too," the chief muttered under his breath.

Turning a not-so-subtle shade of red, Herring nonetheless kept his mouth shut.

"What's the name of the guy you worked for?"

"Chanute... Octave Chanute."

Looking up from his notepad, Sauerbier nodded. "The retired civil engineer? Bridges and railroads... yeah, I've heard of him. Before last night, when was the last time you saw this Butusov character?"

Hesitating, Herring shrugged. "Back in the fall of '96."

"What were the circumstances?"

"I was flight-testing a machine at the Indiana dunes, and he tried to kill me, but I thwarted the—"

"Did you file a report with the authorities?" Sauerbier cut in.

"No, but I made sure that he wouldn't try to pull anything again. I had my family's safety to worry about."

"Herring," the Chief said, "what the hell are you talking about?"

"Butusov emigrated to this country under false pretenses. As an enlisted man in the Russian Navy, he stabbed a fellow sailor to death and then deserted."

"How do you know this?"

Herring shrugged again. "I paid an investigator to look into his past. I still have the report."

"How was this information supposed to protect you and your family?"

"Mr. Chanute delivered a copy of the report to the Russian. An accompanying letter stated that if anything should *ever* happen to members of my family or me, the document would be forwarded to the U.S. Department of Justice and the Russian embassy in Washington. I knew that Russian officials would be anxious to have him extradited. Until last night, I hadn't heard or seen anything of him for almost three years."

Sauerbier removed his hat and scratched his bald pate.

"Tell me, Herring: why would Butusov want to burn down this here boatyard?"

Staring beyond the Morrison Channel, Herring gazed out at the endless waters of Lake Michigan. "In my *opinion*, Butusov was hired to destroy my flying machine work. To cover it up, he torched as many buildings as—"

Interrupting, Sauerbier raised his tinny voice. "Who in their right mind would risk going up the river over a foolish flying machine?"

Herring fired back his response. "If you expect me to name the people who want me to fail, you'll have a long wait! I will not subject myself to defamation lawsuits! However, I'll say that the Russian's previous employer would be an excellent place to start an investigation."

"I presume that you're referring to Mr. Chanute?"

"You said it, Chief… not me."

During the early afternoon of Monday, September 11, Chief Sauerbier placed a telephone call to the Chicago Police Department. In a candid conversation, he requested that Octave Chanute, a retired civil engineer and flying machine advocate, be interviewed regarding the whereabouts of William Paul Butusov, a former employee. Sauerbier identified the Russian as a suspect in his city's day-old arson.

Sauerbier also requested that a separate interview be conducted with William Avery, a neighbor and sometime assistant of Chanute. Avery was purported to have information concerning a dispute between Butusov and another former assistant, A.M. Herring. Sauerbier inferred that Avery might be able to provide investigators with a possible motive for Butusov setting the fire that destroyed the Truscott Boat Works, where Herring kept his flying machines and related materials.

Rochester, New York
Wednesday, November 9, 1921

It had taken Herring all of the court's morning session and half of the afternoon's to relate the events of 1899, culminating with his recollection of the devastating fire and his identification of Butusov as the alleged arsonist. Gus was exhausted.

As courtroom spectators caught their collective breath following Herring's dramatic testimony, Attorney O'Grady pressed on. "Mr. Herring, what was learned about the status and whereabouts of William Paul Butusov from the interviews of Octave Chanute and William Avery?"

Herring took a deep breath as he struggled to maintain his composure. "According to Chief Sauerbier, Chanute claimed that he hadn't seen or heard from the Russian since the winter of

'97. After some arm twisting by the Chicago police, Avery reluctantly recounted the sordid events of Chanute's espionage-like actions that occurred in his shop during February, March and April of '99."

"What reaction did Chief Sauerbier have to these interviews?"

"He was skeptical of my allegation that Butusov was hired by Chanute to torch the Truscott complex for the sole purpose of causing a 'temporary setback' in my research. Sauerbier concluded that Chanute's connection to the fire was entirely circumstantial, and that until the Russian could be found and interrogated, the arson would remain an open, unsolved crime."

"How did Chanute and Avery react to being involved in a criminal investigation?"

"Outwardly, Chanute laughed off the entire episode, but Avery felt that I had betrayed his trust in divulging Chanute's jealous behavior concerning my latest flying machine."

"Did your personal and professional relationship suffer because of this breach in trust?"

"To my regret, Bill Avery avoided speaking to me for almost five years."

"What, if anything, was done to apprehend Butusov?" O'Grady asked.

Herring shook his head. "There was an intensive local search, and Sauerbier sent a copy of my private investigator's report to the Assistant Attorney General at the Department of Justice, who reviewed its contents and turned it over to the Office of the Superintendent of Immigration, then a division of the Treasury Department. The Immigration people subsequently forwarded the report to their Board of Special Inquiry, who eventually got around to conferring with the Russians at their Washington, D.C. embassy.

"A month later, Russian Embassy officials confirmed that Butusov had indeed falsified his U.S. immigration papers, having fled his native Russia to avoid the capital offense charges of murder and desertion from the Imperial Navy. The Russians then demanded that our Justice Department act promptly to capture Butusov and his family for deportation."

"With the highest levels of the U.S. government now aware of this fugitive," said O'Grady, "what steps were taken to apprehend him?"

Herring slumped in his seat. "In addition to the Justice Department assigning agents to track him down, the National Bureau of Criminal Investigation printed 'Wanted' posters featuring Butusov's photograph; these were displayed on post office bulletin boards throughout the country!"

"When was he taken into custody?" O'Grady asked.

"To the best of my knowledge, Butusov avoided the nationwide dragnet and wasn't heard from again... until the summer of 1904."

CHAPTER 70

After the conflagration, Herring's life went from bad to worse. Old Man Truscott, infuriated by the possibility that Herring's presence may have had something to do with the company's $60,000 loss, fired him without further consideration.

Jobless and devastated by the destruction of his experimental flying machine and all of the related technology, Herring looked to Arnot for a financially sound course of action. Reluctantly, the men agreed to curtail all aeronautical work in favor of rekindling the manufacture and sale of their motor bicycles. With Arnot concentrating on marketing, Herring focused all of his energy on resuscitating the all-important single-cylinder gasoline engine.[1]

Rochester, New York

Thursday, November 10, 1921

As Judge Sawyer gaveled his court into session, Attorney O'Grady picked up where he had left off the previous day.

"Mr. Herring, how did you manage to resurrect the single-cylinder engine?"

O'Grady waited for an answer, but none came. "Mr. Herring, shall I repeat the question?"

"Oh... sorry, I was just thinking of something. Yes, yes, the single-cylinder engine. Working in the shop behind my rented house, I labored to recreate wooden 'patterns' using dimensions retrieved from the only engine that survived the fire. Making an accurate pattern is the first step in producing a cast-iron part.[2]

"The rough castings were hauled over to Engberg's shop for machining. Back at Church Street, George Housam and I assembled and mounted the engines onto their custom-built frames before carting them back to Engberg's barn, where the wheels, tires, and accessories were added. The finished machines were checked for defects by powering them up and down Lake Street. Of course, a final inspection was required before they were cleaned and crated for shipment to their direct-sales customers."

"Besides the motor bicycle," O'Grady continued, "was anything done to bring back the Mobike?"

Looking pale, Herring wiped his eyes with a handkerchief. "The partnership built and sold 226 Mobike motorcycles between June 1900 and August 1901. The Mobike's push-start, 1-1/2 horsepower engine could drive the single-rider machine up a nine-percent grade at 16 miles an hour, while the tandem could traverse 300 miles on a single tank of gasoline. The price of a new tandem, with puncture-proof tires and soft saddle seats, was $275. The single sold for $25 less."

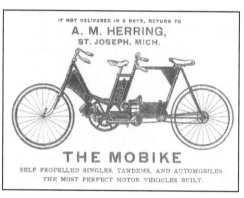

Herring's Mobike tandem motorcycle (1899); Meiller/Herring Collection

"Describe your burgeoning career as an *author* and *editor*."

"Besides providing company advertising for magazines such as *Horseless Age*, I also began writing technical articles for those same publications, for the sole purpose of promoting our motorcycle business. Eventually, I established myself as an engine authority and published author. As a result, I was offered a one-third partnership in a new magazine called *Gas Power*. Serving as its first technical editor, I presided over a niche market that evaluated new products, offered practical solutions to engine-related problems, and provided advertising opportunities for manufacturers and suppliers."

"Did the motorcycle business prosper?"

"Between the motor bicycles and the Mobikes, the company grossed almost $80,000 in 18 months. Our net profit, something less than $15,000, was used to satisfy outstanding vendor bills, labor, and production costs. All things considered, we felt relieved to be back in the black."

"With business looking up, were you able to resume your flying-machine work?"

Herring leaned back. "Time was our enemy. I recall writing a letter to Chanute in January 1900 that summarized my position. Although Matthias was anxious to restart the flying machine work, privately, I was planning to give it up. I had been corresponding with a Mr. Whitney, the owner of an automobile company, about taking charge of some experimental work."

"What became of this opportunity?"

"Nothing! Although the offer had been dangled in front of me for the better part of 1900, I eventually withdrew my name from consideration when Whitney reneged on his offer to sign a three-year contract for my services. It was simply too risky for me to uproot my family and move to Detroit without a guarantee of job security."

"During the summer of that same year, another distraction cropped up in the form of a magazine article. Can you elaborate on this piece of writing?"

"Octave Chanute's article in the June 1900 issue of *McClure's* magazine, 'Experiments in Flying,' was heavily illustrated with unethically obtained photographs of Mr. Arnot's '97 two-surface glider. Incredibly, the old man referred to that glider as 'my 1896 two-surface machine.' To make matters worse, the actual inventor of the machine was given only passing mention."

"Who was the inventor?"

Turning his hands palms up, Herring grinned feebly. "Yours truly!"

"How would you characterize Chanute's article?"

"Inaccurate... as usual! Besides the omissions, the icing on the cake was his endorsement of 'Langley's Law', the get-something-for-nothing dictum that states the flying machine allegedly uses *less* power the *faster* it flies!"

"How did Matthias Arnot react to the article and Chanute's unauthorized use of his photographs?"

Herring grimaced. "Chanute's piracy of Arnot's photographs triggered memories of the old man's appalling behavior during our unsuccessful trial of the compressed-air machine back in October of '98. To my surprise, despite his growing irritation, Matthias had nothing further to say."

"Then there was the tizzy over Chanute's article for the London *Times*. Can you comment on that?"

Herring nodded. "Four months after the *McClure's* debacle hit the newsstands, I had occasion to visit our Chicago-based supplier of bicycle frames. The consultation required me to lug a 22-pound engine to their facility. Afterward, with almost three hours to kill before my scheduled departure back to St. Joseph, I decided to take a hansom cab over to Chanute's East Huron Street home. Although my unannounced arrival broke the code for acceptable manners, I was nonetheless eager to show the old man the latest iteration of my lightweight engine.

"Barely acknowledging the engine's existence during this short, awkward visit, Chanute made best of the intrusion by delving into a discussion about automatic stability. Octave also informed me that *I* had been replaced by *two* new protégés: Edward C. Huffaker of Tennessee, my former co-worker at the Smithsonian, and Dr. George A. Spratt, an amateur experimentalist from Pennsylvania.

"Almost in passing, the old man mentioned that he had contracted with the London *Times* to author an article tentatively titled 'Flying Machines,' to be published in a supplement to their *Encyclopedia Britannica* for the 1901 calendar year."

"What happened next?" O'Grady asked.

Herring slid around in his chair, facing the judge directly. "Knowing that Matthias had developed a deep hatred for Chanute, I simply relayed the *Britannica* news without further comment or embellishment."

"What did Mr. Arnot have to say?"

"Nothing! I thought the issue had been forgotten… until a few months later. It was March 1901 when I learned that Matthias had written a passionate letter to the editor of the *Britannica*, suggesting that *I* be allowed to review Chanute's forthcoming essay to ensure against further injustices."[3]

"Did Arnot tell you that he had written this letter?" O'Grady asked.

"No, but when I received a copy of the correspondence from Chanute, who was vacationing in San Diego, I realized that the resentment was far from forgotten. Then, to make matters worse, the old man demanded to know if I *endorsed* Arnot's sentiments!"

"Had you had the opportunity to read the *Britannica* article?"

"That was the problem! The supplement had yet to be released in America, so I was obliged to take Chanute's word as to its contents."

"Did Chanute say he had mentioned your contributions?"

Herring looked astonished. "No! In fact, the old man was emphatic that he had not mentioned my two-surface glider, my automatic regulator, my power machine, or me. He claimed that there were 10 principal experimenters in the world, and my work didn't measure up to their lofty standards. Later, after obtaining a copy of the *Britannica* article, I saw that Chanute had included himself in that select group. It was obvious Mr. Chanute would not risk depriving humanity of anything as important as himself!"

"Did you respond to this disclosure?"

"After considerable thought, I replied to Chanute with a letter dated March 17, 1901, stating that he had treated me unfairly in both his articles and speeches. I vented many of my pent-up concerns, especially his declaration of having invented the two-surface gliding machine, as well as his malicious habit of not acknowledging my most important contributions.

"I emphasized that although I had exhausted two inheritances in the pursuit of developing a practical flying machine, unlike him, I had never worked exclusively for the fame or notoriety. This was a slap at Chanute's not-so-hidden agenda – the unapologetic, continuing quest for recognition in the field of aeronautics."

"Did Chanute contradict your grievances?"

"A week later, I received a list of 'facts' from Chanute that included inaccuracies, mistruths, and out-and-out lies concerning my work during the time that I labored as his assistant and afterward."

"Was there anything new in this letter?

"It contained the same threadbare drivel: that the two-surface machine was *his* design; that *I* had pleaded with him to join me in pursuing a British patent; that Mr. Avery had suggested the changes necessary to make my tail regulator work, et cetera, et cetera."

"Did you respond?"

"Not for almost a year. Business problems, due in part to Matthias's health issues, demanded most of my time."

"Tell the court what happened next."

Herring righted himself, as if preparing for an assault. "For more than a year, Matthias had become increasingly agitated and nervous. He could not get over Chanute's actions: the fraudulent use of our photographs, his repulsive conduct at our powered flight trials, his connection to the arsonist Butusov, and his spiteful articles in *McClure's* and the *Encyclopedia Britannica*.

"The first day of July was memorable because of the full moon. Returning to his Elmira home from an evening gathering of church elders, Arnot had to hand over the reins of his horse and buggy to his new wife. He had taken ill with abdominal pain, a condition presumably brought on by nervous prostration and a general breakdown in his physical condition. The Arnot family physician prescribed total rest, an explicit dietary regimen and a cessation of all activities, both physical and intellectual. The pain lingered almost a month, necessitating a further evaluation by a prominent internist called in from New York City.

"On Saturday, July 27," Herring said, his voice growing hoarse, "Matthias dictated a letter to me as he lay on his back prior to an operation for *appendicitis*. He stated his intention to set aside enough money to complete the flying machine work. As fate would have it, this would be his last correspondence. Four days later, with the emergence of yet another full moon – a rare *blue moon* – he died of peritonitis. Matthias Arnot was only 33 years old."

A stunned silence fell over the courtroom for the better part of 15 seconds. Herring had offered a powerful rendition of his friend and partner's final days.

"This must have come as a terrible shock," O'Grady said. "Did Mr. Arnot's letter contain any other disclosures?"

"After declaring his intent, Matthias offered instruction and advice as to the use of the $35,000 life insurance policy that he had carried, with me as beneficiary. Regrettably, the insurance company informed me that since they hadn't received the latest premium, the policy had been terminated on the day of Arnot's correspondence. In addition, Matthias's mother immediately withdrew the monthly stipend that her son had allocated to me."

O'Grady asked, "With the loss of the insurance policy money and the monthly stipend of $300, what became of the motor vehicle business?"

"Without financial backing and a partner to look after the marketing side of the company, I was forced to let the business go."

"Were you able to sell the company's assets?"

"There were two, 22-year-old upstarts who had heard about the company's dilemma and were anxious to purchase its unfinished rolling stock, tooling, and materials. These monies helped me to satisfy creditors and kept me out of bankruptcy court."

"Who were these men?"

"William S. Harley and Arthur Davidson of Milwaukee, Wisconsin."

Rochester, New York

Thursday, November 10, 1921

Returning to the courtroom, Herring looked even more disheveled and exhausted than he had in the morning. He had complained of various aches and pains during lunch as he passed up his customary salad, settling instead for a small glass of milk.

When O'Grady inquired about his condition, he claimed that lack of sleep contributed to his lethargy. Refusing to dwell on his "temporary circumstance," Gus said that he was anxious to continue with his afternoon testimony.

Without fanfare, Judge Sawyer directed Herring back to the witness stand. In a decidedly slow gait, Herring massaged his neck and shoulders while slowly making his way from the plaintiff's table.

"Mr. Herring," O'Grady said, "after the liquidation of your company, what did you do next?"

Herring looked down at the floor. "I took odd jobs. I worked as a railroad surveyor, a carpenter and a handyman. I did anything that I could to survive during the latter half of 1901."

"What did you think about during this disheartening period?" O'Grady asked.

Bending forward, Herring placed his left elbow on his thigh before resting his chin in the palm of his hand. "To begin with, I dwelled on Chanute's spiteful actions and the probable role he played in Matthias's death."

"Objection, calls for speculation!" Robbins shouted, rising from behind the defense table.

"Counselor O'Grady," Sawyer said, "where are you heading with this line of questioning?"

"Requesting the court's indulgence, Your Honor. Mr. Herring's recollections from this challenging period of his life offer insight into the very nature of this civil action. I promise the court that I'll keep this phase of my questioning brief."

" I'll hold you to that promise, Mr. O'Grady... objection overruled! Continue, Counselor."

"Mr. Herring, what led you to believe that Mr. Chanute had something to do with your partner's demise?"

"As I said this morning, Matthias Arnot had become increasingly agitated and nervous over Mr. Chanute's actions... the photographs, the fire, the magazine articles—"

"Objection, Your Honor!" Robbins interrupted. "At best, Mr. Herring's opinions rely upon unprecedented and circumstantial evidence."

"Unprecedented? I don't believe so!" O'Grady shot back.

Peering down from his bench, Sawyer pointed a skeletal index finger at Herring. "Mr. Herring, can you provide an instance of this situation happening in the past?"

"Certainly," Gus answered with a tongue in cheek smile. "I remembered how Orville Wright reacted after the death of his brother Wilbur."

"Continue," urged the judge.

"Orville believed that *Glenn Curtiss* was responsible for Wilbur's premature death."

"Objection! Irrelevant and argumentative!" Robbins shouted.

"Hold on, Mr. Robbins!" Sawyer intervened. "Continue, Mr. Herring."

"Orville made no bones about it! He said that Wilbur's death was due to the exhausting travels, mental anguish, and stress caused by his legal battles with Glenn Hammond Curtiss!"

The judge turned to O'Grady. "Counselor, have your witness substantiate his claim!"

"Very well, Your Honor."

O'Grady turned to his witness. "Mr. Herring, how did you learn of Orville Wright's claim against Mr. Curtiss?"

Herring rolled his eyes. "After Wilbur Wright's death in the spring of 1912, Orville made his feelings known in the *newspapers*. He stated that Curtiss should be held account—"

"Yes, yes," the judge interrupted. "Now that you mention it... I do remember that particular controversy. Objection overruled." Without further comment, the judge waved his hand for O'Grady to continue.

"Mr. Herring, assuming that Octave Chanute had played a role in Mr. Arnot's death, how did this unfortunate occurrence affect your long-term goal of inventing the world's first manned, dynamic, heavier-than-air flying machine?"

Herring hesitated as he thought about an era lost. "Chanute made sure I wouldn't encroach on his legacy anytime soon. My partner was *dead*, my business *lost*, my research *destroyed*, my inheritances *gone*, and my aeronautical achievements *disparaged* by his every written and spoken word."

Exhausted, Herring slumped in his chair.

Noting that the time was late, Judge Sawyer adjourned court until the following morning.

Rochester, New York

Thursday, November 10, 1921

James O'Grady was relaxing in his overstuffed easy chair when the ringer on his telephone sounded twice in rapid succession – the code for his neighborhood extension. Before lifting the receiver, he turned down the volume on Mozart's "Requiem Mass in D Minor." Speaking directly into the mouthpiece, he greeted the caller formally.

"Good evening... James O'Grady speaking."

"Mr. O'Grady?" the defeated voice whimpered. "It's Chloe... Daddy's back in the hospital!"

CHAPTER 71

Rochester, New York

Thursday, November 10, 1921

"My God, Chloe," O'Grady groaned, "is... is he all right?"

"I don't know... they won't let me see him."

"I knew something was wrong! He wasn't himself during lunch today... but you know your father, he wasn't about to let anything get in the way of his testimony. What happened?"

After taking a long moment to blow her nose, Chloe replied. "We arrived home about 5:30. Augustus said that he was exhausted and was going to take a nap on the divan. I began to prepare supper, but a little later when I checked on him... he was foaming at the mouth and unable to speak."

"What did you do?"

"I picked up the phone and pleaded with the operator to call for an ambulance!"

"How long did you have to wait?"

"Almost 20 minutes! Honestly, I didn't think they would arrive in time."

"Where are you now?"

"Rochester General... the same place as before. I think Daddy's had another apoplexy!"

*

O'Grady spent the next hour trying in vain to reach the hospital's attending physician. Beyond being told that Herring's condition was serious, there would be no further communication until morning. By the time he had gotten around to phoning Maloney, it was after 10 p.m., and his partner was in a foul mood.

"Yeah! Who the hell is—"

"Calm down Bill... we've got a *big* problem!"

"Another problem? Jesus Christ Jim, I just got out of the *cooler* an hour ago!"

After O'Grady finished explaining the predicament, the men methodically sifted through their options before agreeing on the only tactic that seemed workable: they had to find another witness – and find him fast! The most likely candidate on their list lived in Pennsylvania, not too far from Rochester; they would try to track him down first. A second lived in New York City and a third in Tennessee, a more distant and unlikely choice. Maloney would attempt to contact these men and persuade at least one of them to travel to western New York.

O'Grady had the unenviable task of convincing Judge Sawyer that another interruption of the plaintiff's train of evidence would not adversely affect the proceedings. The only thing he had to offer in return was the promise that their next witness would be ready to begin testifying on Monday morning. Both men knew that their civil action suit might well end in a mistrial if satisfactory arrangements could not be made.

Rochester, New York

Friday, November 11, 1921

"Court is hereby in session!" Vern the bailiff croaked, in his odd nasal voice.

After settling himself behind the bench, Judge Sawyer was about to address the litigants when he spotted O'Grady's raised hand.

"Yes, Counselor?"

"Permission to approach the bench, Your Honor."

With his normal routine broken, Sawyer paused briefly before waving both attorneys forward. "Mr. O'Grady, I fail to see Mr. Herring in the courtroom this morning. Might I assume there is a problem?"

"Unfortunately there is, Your Honor. Mr. Herring has suffered yet another apoplexy. I became aware of the emergency at 9 o'clock last night, when his daughter telephoned me from the hospital."

Sawyer cursed softly under his breath. "What is Mr. Herring's condition?"

"I just came from Rochester General, where I spoke with his attending physician. He explained that although Mr. Herring's latest apoplectic episode has left him weakened on his left side, with slightly slurred speech, he is expected to again make a full recovery. However, as before... there will be a recuperation period—"

The judge interrupted with a question. "Did he say when Mr. Herring could return to the courtroom?"

O'Grady shook his head. "Your Honor, despite my prodding, the doctor would not commit to a specific date. All he said was, '... the patient's rate of recovery will dictate when he can resume normal activities.' "

Having remained silent up to this point, Attorney Robbins chimed in with a predictable request. "Judge, because Mr. Herring will be away from the proceedings for an indeterminate period of time, the defense requests a *mistrial*."

Judge Sawyer cocked his head toward O'Grady. "Counselor, does the plaintiff have a counterproposal?"

O'Grady's face was beginning to flush. "Your Honor, if a mistrial is declared at this time, many weeks of productive testimony will be squandered, not to mention the court's valuable time. Although the plaintiff expected Mr. Herring to occupy the witness chair for at least another week, we are indeed fortunate to have two scheduled witnesses available to us on short notice. Therefore, in deference to Mr. Herring's regrettable affliction, the plaintiff respectfully requests a one-day continuance. Judge, I personally guarantee that our witness will be ready for swearing in at the court's appointed time on Monday, November 14—"

"Poppycock, Your Honor!" Robbins interrupted, "Mr. Herring's testimony, with all its stops and starts, has already produced more than its share of confusion! Now we must contend with *another* placeholder witness?"

Watching the judge scribble on his notepad did nothing to reduce O'Grady's anxiety as he subconsciously prepared for bad news. Pushing his chair away from the bench, Sawyer waved the attorneys back to their respective tables. "Judicial practice seldom allows for repeated interruption of testimony due to a witness's physical condition. However, I have decided *not* to declare a mistrial at this time. Court will adjourn until 9 a.m. on Monday morning."

Judge Sawyer caught the attention of both attorneys before they could leave the courtroom. "Mr. O'Grady, Mr. Robbins... I need a moment."

Caught unawares, both lawyers again marched to the judge's bench. "Counselor O'Grady, the court will again keep track of Mr. Herring's progress during his hospital stay. You are hereby required to deliver a brief written report to my chambers *each* morning – no less than 15 minutes prior to the scheduled start of the day's proceedings – beginning on Monday... is that clear?"

"Certainly, Your Honor."

*

Maloney was having his own difficulties. George Spratt and his son had been away from their Pennsylvania farm for more than a month; his wife said they were off testing a new aeroplane somewhere along the coast of Massachusetts.

With the pressure on, Maloney tried to contact Carl Dienstbach at the New York *Times*. To the attorney's dismay, he learned that Carl was currently on assignment in Europe and wouldn't return until January. Last on his list, Maloney tracked Edward Huffaker from his hometown in Chuckey City, Tennessee, to New York, where he had been the featured speaker at the Aero Club of America's monthly, Friday night meeting.

After five rings, Huffaker fumbled in the dark to locate the telephone. "Who's this?" he muttered.

"Attorney William Maloney. Is this Edward C. Huffaker?"

"Speaking."

"Sorry for the early call, Mr. Huffaker, but this is an emergency. I spoke with you earlier this year, and you agreed to be a witness for the plaintiff in the Herring-Curtiss Company versus Glenn Curtiss trial."

"I remember," Huffaker said, sweeping away the cobwebs, "but must you call at five in the morning?"

"My apologies, sir... but like I said, it's an emergency. We need you to testify first thing Monday morning."

"What's the emergency?"

"Mr. Herring, who was testifying in the trial, has suffered an apoplexy, and is currently in the hospital! We only have 52 hours to get you here, or the presiding judge will declare a mistrial!"

For several seconds there was silence. "Refresh my memory, Mr. Maloney... where's the trial being held?"

"For Christ's sake man... Rochester, New York! Listen, Huffaker, we need you here, and we need you here *soon*!"

Rochester, New York

Monday, November 14, 1921

As Judge Sawyer made his customary entrance into the courtroom, Edward C. Huffaker stared at him from behind the plaintiff's table. Maloney, having served his seven-day jail sentence for contempt of court, sat beside the plaintiff's latest witness with his arms crossed tightly against his chest.

As his first order of business, the judge called all three attorneys to the bench. Maloney eyed Robbins with disdain, and the judge sensed the big man's intense, lingering hostility.

"Gentlemen, as a result of our most satisfactory meeting in chambers this morning, Mr. Maloney understands that he may now act as *consultant* to the plaintiff... and sit quietly at the plaintiff's table. Mr. Robbins, you are directed not to interact with or tender any sort of exchange with Counselor Maloney.

"Gentlemen, this will be my last warning. As the court's chief officer, I will not hesitate to expel either or both of you from these proceedings! The court is not taking questions, or hearing any further comments on the matter... return to your stations."

Allowing a few moments for the attorneys to gather their notes, Sawyer addressed the plaintiff's table. "Mr. O'Grady, I am pleased to learn that Mr. Herring's condition has improved over the weekend. The court is looking forward to his rapid recovery and return to these proceedings."

"Thank you, Your Honor! So is the plaintiff. As a matter of fact, I spoke with Mr. Herring just—"

"Fine, fine," Sawyer said, cutting O'Grady short. "Who is your next witness?"

"Mr. Edward Chalmers Huffaker, Your Honor."

"Very well, Mr. Huffaker, step forward and be sworn in."

*

As was to be expected, the 65-year-old Huffaker's physical appearance had changed since his days as an assistant at the Smithsonian Institution, where he had worked with Augustus Herring. His five-foot ten-inch frame had filled out over the years to the point where he now weighed 190 pounds. His once dark-brown hair (what was left of it) had turned completely white, as had his bushy mustache. The wiry look of his youth was gone, giving Huffaker a more grandfatherly appearance that emphasized his deep-set, blue eyes. Gone were the tobacco stains on his teeth, the embedded grime under his fingernails, and the disheveled guise that had earmarked his earlier years. Dressed in a crisp dark-blue suit, white shirt and blue string tie, Huffaker looked more the part of a Southern gentleman; his toothy grin conveyed experience and confidence.

"Mr. Huffaker," O'Grady said, "what is your current occupation?"

"I'm a semi-retired surveyor and aeronautical experimenter," Huffaker said in a rolling Southern drawl.

"How long have you been interested in things that fly, sir?"

Taking a moment to think, Huffaker rubbed his chin between his thumb and forefinger. "Back in Seclusion Bend, Tennessee, where I was born, I can still recall watching turkey vultures soar overhead... and that was before Abraham Lincoln was elected president!"

"Tell the court about your education, Mr. Huffaker."

"In 1880, I graduated with honors from Emory and Henry College in Virginia, where I excelled in natural science and mathematics. Then I completed my master's degree in mathematics at the University of Virginia. I was offered a fellowship at Johns Hopkins University, in Baltimore, for doctoral studies in mathematics, but I demurred in favor of a public school teaching career."

"How long did you teach?"

"About six years, starting in 1886, then I turned to civil engineering in the '90s."

"Why did you give up teaching?" O'Grady asked.

Huffaker replied reluctantly. "My love of nature and the outdoors convinced me to forsake the classroom."

O'Grady continued, "After four years of laboring in Tennessee, you quit civil engineering and traveled to the nation's capital, where you worked for Professor Langley at the Smithsonian Institution. How did that opportunity come about?"

"Based on my observation of birds and air currents, I wrote a scientific paper titled 'The Value of Curved Surfaces in Flight.' After writing a number of letters to Langley, where I attempted to interest him in my theory of lift and how it's produced on a bird's wing, he finally agreed to read my manuscript. Afterward, although the professor was skeptical about my conclusions, he suggested that I submit the paper to Octave Chanute for use at the 1893 International Congress on Aerial Navigation being held concurrently with Chicago's Columbian Exposition. I changed the title to 'Soaring Flight' and mailed it to Mr. Chanute for his consideration."

"What was the core of your paper's hypothesis?"

Huffaker's eyebrows shot up. "As air follows the upper side of a bird's curved cross-section wing, its velocity increases relative to the straight flow across its lower side. Bernoulli's Principle predicts that increased velocity produces reduced pressure or a partial vacuum, which explains how the bird's wing generates most of its lift."

"What was Chanute's response to your work?"

"After the aeronautical congress, Mr. Chanute and I began exchanging letters. In August of '94, he visited my home in Tennessee, to join me in watching the buzzards soar. That fall, he wrote to Professor Langley, recommending me for employment at the Smithsonian... where I was eventually hired!"

"What were the terms of your employment?"

"I signed a one-year contract as an aide in the aerodromics laboratory. The pay was $100 a month, an excellent salary during the country's economic depression! I started work on January 2, 1895."

"What were your duties in the aerodromics laboratory?" O'Grady asked.

"I was put in charge of operating the small whirling table apparatus in a building adjacent to the Castle's south laboratory. It was there that I tested alternative wing configurations for aerodrome models *Number 5* and *6*. I also helped to construct and flight-test a series of small models that used twisted rubber strips to spin the propellers."

"When did Mr. Herring first come to your attention?"

Staring up at the courtroom's crystal chandelier, Huffaker's mouth dropped open as he searched his memory. "Professor Langley decided to *visit* Herring after reading about his flights with a Lilienthal-type glider in New York City. Later, he rambled on and on about Herring's superb flights with a two-surface model."

"When did you first meet Mr. Herring?"

"I had already started working at the Smithsonian... if memory serves me, it was sometime in May of '95. He was hired to supervise the men who worked in the aerodromics laboratory."

"As your boss, was Herring friendly toward you?"

Huffaker scratched the bridge of his nose. "By the time Herring was hired, Professor Langley had already rejected curved lifting surfaces for use on his steam engine-powered aerodromes. The Secretary asserted that a *flat plane* lifted equally well and was simple to construct, making it his hands-down choice. However, within a week of his arrival, Herring was already lobbying to replace them with what he called aero*curves*. When I learned of his preference, I congratulated him on his insight and gave him a copy of my manuscript. The next day he reciprocated with a draft of an article he was preparing for the *Aeronautical Annual*, in which he contended that a partial vacuum was responsible for the aero*curve's* lift at shallow angles. From that point on, we became friends and confidantes."

"Did Mr. Herring explain how he knew that a partial vacuum was at least partly responsible for an aero*curve's* lift?"

Huffaker nodded. "While experimenting with aero*curves* back in '93. In fact, he showed me a simple demonstration... one that proved the *existence* of a partial vacuum—"

"Mr. Huffaker," O'Grady interrupted, "can you describe this demonstration?"

Hesitating for an instant, Huffaker glanced up at the judge and then back to the attorney. "Mr. O'Grady, it's much easier to *show* than it is to *tell*. If acceptable to the court, I'll be glad to show—"

"Objection! Relevance!" Robbins cut in, scrambling to his feet.

"Your Honor," O'Grady countered, "if Mr. Huffaker believes that Herring's simple demonstration will help to clarify a fundamental aspect of flight, the court might consider—"

"Objection overruled!" Sawyer interrupted. "You may perform your demonstration, Mr. Huffaker."

Standing, Huffaker asked O'Grady for a sheet of paper from his legal pad. Glancing around the courtroom for a suitable flat surface, Huffaker pointed to the plaintiff's table.

"Over there, Your Honor?"

As the judge, defendants, and attorneys gathered around the highly polished oak-top table, Huffaker folded the paper crisply in half before crouching. "Gentlemen," he said, as he laid the paper flat with the fold facing him at the table's edge. "Notice that my opposed index fingers have been slipped between the upper and lower sheets and are pressing downward at the fold, securing its lower half to the tabletop.

"Also note that the upper portion of the folded sheet is free to move and slumps to the rear. In fact, if you view this uncomplicated arrangement from the side, it will bear an uncanny resemblance to a wing's curved cross-section... today we call it an *airfoil*. Hence, the question becomes: what will happen to the movable top flap of paper when I blow a sustained puff of air across this paper from the front?

"Would you care to speculate, Your Honor?"

Judge Sawyer shifted his weight from one foot to the other. "I am not a man of science; however, I would assume the top piece of paper would be flattened against the table top."

"Very well, Your Honor," Huffaker said, "we shall see."

Puckering his lips, Huffaker positioned them six inches in front of the paper and puffed a long, uniform stream of air. Rather than flattening against the tabletop, the top sheet flapped *upward*.

"Aha!" Huffaker beamed. "The laws of physics haven't changed since I last performed this little demonstration!"

Returning to his bench, Judge Sawyer took the opportunity to break for lunch.

"Court will resume promptly at 1 o'clock, at which time Mr. Huffaker will undoubtedly explain the ramifications of his demonstration. Court is hereby recessed."

*

Returning from lunch, O'Grady directed Huffaker to summarize the results of his demonstration.

"Bernoulli's principle predicts that when air passing over a curved surface speeds up, a partial vacuum will form over that surface, thereby allowing the slower moving, higher pressure air from below to push the surface *toward* the partial vacuum – a process known as *lift*.

"In the demonstration, when the top sheet of paper flapped upward, rather than flattening against the tabletop, visual evidence of a Bernoulli principle partial vacuum was confirmed."[1]

"Mr. Huffaker," O'Grady said, "did Herring's demonstration convince Professor Langley about the true nature of lift?"

Huffaker hesitated before answering. "No... I don't believe that it did!"

"How is it possible that the most famous American scientist of the day, Samuel Pierpont Langley, did not understand the theoretical nature and advantage of the curved surface... or *aerocurve*, as Mr. Herring called it?"

"I believe that Langley rejected curved surfaces because Otto Lilienthal had promoted their use. Out of spite, Langley refrained from switching—"

"Objection!" Robbins interrupted. "Speculation! The witness doesn't *know* what the professor was thinking!"

"Sustained," Judge Sawyer said, "keep your answers factual, Mr. Huffaker."

"Begging the court's pardon, Your Honor."

O'Grady restructured the question. "Knowing Langley's hostility toward curved surfaces, why were the lifting surfaces of aerodromes *Number 5* and *6* converted from flat planes to aero*curves*?"

Huffaker rolled his eyes. "It was the performance of Herring's two-surface model! In all my experience, I have *never* witnessed a model that came as close to perfection. Shortly after Langley saw that model fly, aerodrome *Number 5* was being fitted with aero*curve* lifting surfaces."

"Prior to Herring's arrival at the Smithsonian, how many successful flights had been flown by Langley's aerodrome models?"

Huffaker hiked a single bushy eyebrow. "None that I am aware of."

"Why had all of these flights failed?"

"There were several key components missing."

"Such as?"

"Besides the curved lifting surfaces, Herring's tail *regulator* helped with stability in gusty conditions. In addition, Gus's *gasoline* burner helped to boost boiler pressures, which improved the steam engine's power!"

"Mr. Huffaker, who in your opinion was *most* responsible for identifying the problems, creating the solutions, and implementing them into the aerodrome models?"

"Objection, Your Honor! Counsel is leading the witness!"

"Overruled, the witness may answer."

"Mr. Herring was most responsible for improving aerodrome models *Number 5* and *6*."

O'Grady referred to his notes before continuing. "Did pioneer aerodrome *Number 5* make its first successful flight while Mr. Herring was still employed at the Smithsonian?"

"No. Although Herring had completed all of the necessary changes by September of 1895, Langley insisted that both machines be completely dismantled."

"For what purpose?" O'Grady asked.

"Polishing... or what he referred to as 'cosmetic refinement.'"

"Mr. Huffaker, here in North America, isn't autumn the *best* time of the year for flying?"

"Yes... but Professor Langley had a reason for delaying the aerodrome trials until December."

"What was Langley's reason?"

"Among the men who worked in the aerodromics laboratory, rumor had Mr. Herring being fired or forced to resign."

"Objection!" Robbins hollered. "Rumors are just another form of speculation!"

"Overruled."

Undaunted, Huffaker continued. "From their years toiling within the confines of the Smithsonian, workers had become quite perceptive... firings had happened before. It was common knowledge that Professor Langley *never* allowed anyone but himself to accept praise for successes within his model aerodrome program. Because Herring was often outspoken regarding his contributions, the Secretary perceived his young supervisor as a threat to his sphere of influence. As a result, Langley's obsession with power and control forced Herring to resign his position."

O'Grady remained silent for several seconds, allowing the significance of Huffaker's statement to resonate throughout the courtroom. "Can you think of any other reason why Langley would want to get rid of Mr. Herring?"

Once again, Huffaker stopped to think.

"Actually, I believe that there were two additional reasons. First, Mr. Herring had difficulty adapting to Langley's authoritarian methods, especially the old man's refusal to delegate. Everything had to be approved in writing before any phase of the project could proceed. As you can imagine, this had a very deleterious effect on the staff, including Herring, who endured much stress and apprehension. Most of the time the men stood around waiting for approval on some mundane aspect of their work. There were occasions when I was so unsettled by the Professor's intense scrutiny of my work that I could barely perform my calculations!

"Second, there was Herring's unwavering support and admiration for Herr Lilienthal. Augustus often said that the 'birdman' was his mentor and guiding light for all things aeronautical, much to the Professor's chagrin, who seethed at the suggestion. Around the halls of the Smithsonian, it was no secret that Professor Langley despised Lilienthal, since most of the German's ideas about flight were in direct conflict with his own."

"How did Langley get rid of Herring, after only seven months on the job?"

Huffaker sighed. "In early December, after squandering the best flying weather of the season, Langley ordered Herring back to the Potomac for a series of flight trials with aerodrome *Number 5*. When Herring protested because a recent snow and cold spell had left ice on the river, the ensuing argument resulted in his resignation."

"When did the aerodrome models finally fly?" O'Grady asked.

"The following spring. If memory serves me it was May 6, 1896."

"Who was in charge of these flights?"

"Charles Manly, Herring's replacement in the aerodromics laboratory... and myself."

"Your position at the Smithsonian lasted *four* years, until December of 1898. Why did you leave?"

"The lack of job security and a growing disdain for Langley's domineering managerial style. The final indignity came when he began to attack my partial vacuum lift hypothesis. The Professor embarrassed me by assigning a number of tedious and meaningless field experiments that were designed to discredit the notion that... well anyway, on December 16, 1898, I tendered my resignation."[2]

Since it was nearly 5 o'clock, Judge Sawyer adjourned testimony for the day.

Rochester, New York
Tuesday, November 15, 1921

Huffaker returned to the witness stand, and O'Grady moved in for the kill.

"Mr. Huffaker, numerous accounts have been written about your stay at the Wright brothers' camp in North Carolina during the summer of 1901. Most of these stories have been unflattering towards you. What was the problem between you and the Wrights?"[3]

Taken aback by the question, Huffaker held out his hands, palms up. "There were several points of contention, but I thought that everyone had parted as friends. On the other hand, Mr. Chanute did inform me that Wilbur Wright had written him a letter of complaint."

"What was the nature of his complaint?"

"Objection! Hearsay!" Robbins shouted.

"Overruled," Sawyer said. "The witness may answer the question."

Huffaker slouched in his chair. "Wilbur said that I didn't help with the dinner dishes—"

Laughter cascaded through the courtroom... even Judge Sawyer had to smile.

"Besides the everyday chores," O'Grady persisted, "were there any legitimate complaints?"

"Legitimate, now *that's* an interesting word," Huffaker drawled, scratching his ear. "Wilbur accused me of endangering the mechanical integrity of his *stopwatch* and wind speed *anemometer* by setting them on the sand at Kill Devil Hills. Naturally, I took exception to his charge. I've been using precision instruments all of my working life, and I honestly cannot recall a single instance where I damaged one by setting it on a parcel of God's good earth.

"However, I do believe Wilbur had what Dr. Freud would call an *anal* character! Back where I come from, folks have a more earthy term for Mr. Wright's affliction... but this is polite company—"

Another round of laughter flashed through the courtroom. This time, Judge Sawyer's gavel interrupted the merriment.

"Order in the court. Gentlemen, stifle your laughter! Continue, Mr. O'Grady."

"Mr. Huffaker, tell the court about the *aeroplane* that you brought to the Outer Banks of North Carolina."

Kill Devil Hills, North Carolina

Thursday, July 18, 1901

The donkey cart ground to a halt directly in front of the Wrights' partially built wooden frame shed. Ed Huffaker, Octave Chanute's 45-year-old assistant, hopped down from his seat next to Dan Tate, his driver-escort. The four-mile cart ride over the barren coastal sands from the village of Kitty Hawk represented the end of a circuitous journey that had begun in Chucky City, Tennessee, two days earlier.

Having spied the cart almost a mile away, Wilbur and Orville Wright wasted little time in introducing themselves to their guest. Wilbur, the older, taller, and balder of the two, offered his hand. "You are Mr. Huffaker, I presume? My name is Wilbur Wright, and this is my brother, Orville. I trust your journey here was tolerable? When, pray tell, will Mr. Chanute be joining us?"

Huffaker looked a bit uncomfortable. "My train arrived at Elizabeth City on schedule, but there was a delay in getting my big crate hauled down to the docks. I was fortunate to catch the steamer, *Neuse*, before it departed for Roanoke Island, where I spent last night. From there, as you suggested in your letter, I arranged passage on the coastal mail boat, which delivered me to the Kitty Hawk docks about 9 a.m. this morning. It was quite a roundabout trip! The worst part of the excursion was hauling this crate on and off trains, boats, and carts! As for Mr. Chanute's date of arrival, I can only report that he said, 'I'll be along soon.'"

Over the next few days, Huffaker helped the Wrights complete their shed, which included the construction and installation of two top-hinged doors at either end of the structure. When swung into position and propped open with two-by-fours, the doors also acted as wide awnings that permitted a cooling breeze to blow through the aeroplane shed's interior.

Huffaker also assisted the Wrights while they reassembled their two-surface, tail-first glider. Although the brothers preferred to perform the work on the glider themselves, Huffaker helped by fetching tools, taking measurements, and tending to the hot glue pot. Occasionally, he was asked to hold some part while it was being fastened or clamped into place.

The hours that the men spent working together inevitably led to conversations relating their backgrounds, preferences, and positions regarding everything from food to politics. Huffaker soon learned that the Wrights were especially sensitive about two things: their lack of a *higher education* and their association with the *opposite sex*. One memorable exchange resulted after Huffaker had mentioned his university degrees.

"How have these college degrees advanced your position in life, Mr. Huffaker?" Wilbur asked. "Schoolteachers and field surveyors aren't positions one *strives* for to acquire great notoriety or wealth."

This is an interesting question, coming from a bicycle mechanic, Huffaker thought. Maintaining his friendly disposition, Edward answered honestly. "Had classes been held in the open air, I would have remained a public school teacher. However, my love of the great outdoors finally led me to surveying – a discipline vital to civil engineering. Loving what you do, that's the important issue, Mr. Wright. As a safety bicycle mechanic, certainly you understand what I'm saying."

Huffaker, whose father and uncles fought for the South during the Civil War, sensed a subtle Northern bias from occasional remarks made by the brothers. This realization, along with his status as an invited guest, convinced him not to tender unsolicited comments; instead, he decided to respond only when his Yankee hosts asked for his opinion.

On the morning of July 27, 1901, the Wrights, Huffaker, and Dr. George A. Spratt, another of Chanute's invited assistants, and the most recent to arrive, helped carry the newly finished glider to the Kill Devil Hills' practice grounds, 1,000 feet to the south. The area consisted of a level plain of barren sand from which a group of detached hills projected above the horizon. Because of the wind's force and direction, these sand hills continually changed in their height and slope. The brothers referred to the three hills as the 100-foot-high *Big Hill*, the 30-foot *Little Hill*, and the 60-foot *West Hill*.

After completing several short glides on the Little Hill, Wilbur, although concerned with the machine's pitch instability, decided that it was performing well enough to try his hand at gliding down the Big Hill. With the wind gusting to 25 miles an hour, Wilbur climbed aboard the center section of his double-decker's lower surface. He gingerly positioned himself in a fore-aft prone position within the specially designed hip cradle, a device whose side-to-side motion actuated the warping control for the glider's two superposed lifting surfaces. Wilbur's right hand extended out front to a control lever that actuated the forward horizontal rudder.

When all was thought to be ready, Wilbur signaled to Orville and Huffaker to lift the machine by its wingtips in preparation for a launch down the hill's 10-degree, northeast-facing slope. Waiting several seconds for the wind to stabilize, Wilbur then nodded, his signal for the men to proceed with the launch. After several quick steps down the slope, the machine was released to the whims of the wind and the fledgling skills of the elder brother. Within seconds of flying straight out from the hill, the fore-and-aft bucking became so pronounced that Wilbur feared he was losing control. It was all he could do to keep the glider from diving into the sand or climbing uncontrollably, at which point the apparatus would eventually stop flying – exactly what happened a few seconds later. As the glider reached the alarming altitude of 30 feet, Orville shouted a warning above the howling wind.

"Throw the rudder down! It's stopping!"

The call came too late to change the outcome. Expecting the worst – a Lilienthal-type nosedive to destruction – the men were dumbfounded when the contraption simply mushed slowly to the sand with no injury to itself or its operator.[4]

Believing that his poor piloting had produced the oscillations, Wilbur insisted on making a second flight, which ended as a duplication of his first effort.

That's enough of this madness, Wilbur thought. "Take the machine back to camp while we're both still in one piece!"

With the machine safely back in its shed, the brothers tried to identify the problem that was causing their glider to pitch out of control.

"I think the horizontal rudder may be too large," Orville opined. "It's making you over-control."

"Orv, I don't think so. If you will recall, last year's machine had a larger horizontal rudder, and it was quite stable in pitch... nothing like this."

"Perhaps we made a mistake in locating the center of gravity—"

"We already checked that twice!" Wilbur cut in. "It's correct according to Lilienthal's method. No... something else is happening."

After considerable speculation without determining a cause or identifying a solution, Wilbur asked George Spratt if he had any ideas. "According to my own experiments with curved lifting surfaces," Spratt replied, "I believe that the center of pressure is *reversing* its direction as the machine nears level flight – whereas it moves quickly aft. When the angle increases, it will reverse again and move forward, depending on the type of curvature—"

"I agree," Huffaker chimed in. "Center of pressure reversal is a problem with *all* curved surfaces. It happens when the faster airflow over the top of the surface *reattaches*, and the negative pressure lift moves to the—"

"Dr. Spratt," Orville interrupted, "how would you suggest we test your theory?"

Spratt laughed. "If this were my contraption, I'd remove the top lifting surface and fly it as a kite. By managing the restraining lanyards, the surface can be forced to fly at a shallow angle... where we can observe its actions."

Wilbur turned his head and winked at Orville. "After lunch, we'll get started removing the upper surface."

*

By 3 o'clock, the men were back atop little Kill Devil Hill with the glider's upper lifting surface. Within a few minutes, they had their answer. When forced to fly at shallow angles of attack, the wing began to buck, proving the center of pressure was indeed moving fore and aft... and was causing the problem.[5]

Back at camp, after discussing the situation with Spratt, the Wrights decided to change the curvature of the machine's two lifting surfaces. By adding a third spruce spar between the front and rear units, they would use the existing piano-wire trussing to flatten the curves while exaggerating their front edges.

Unable to remain in the background, Huffaker questioned the Wrights' decision. "Why are you placing all of the curvature within the forward 10 percent of the surface's chord?"

Wilbur nodded to his brother. "Do you want to explain it, Orv?"

After a pause and a sigh, Orville began his explanation. "Rather than using the arc of a circle for our lifting surfaces, we made the curve very abrupt at the front. This reduces the *area* on the topside of the curve that the air *sees* at small angles of attack. Less area means there will be less downward pressure, resulting in a less pronounced rearward movement of the center of pressure... thus helping to reduce the pitching problem."[6]

Huffaker grimaced. "Do you honestly believe that a flying machine is totally supported by air that strikes the bottom side of its lifting surface?"

Wilbur responded as if indulging a clueless child.

"Of course! What we know as the center of pressure, sometimes referred to as lift, is simply the *reaction* of a tilted surface to air molecules that are impacting it from the *bottom*. Think of it this way: the air acts like a *wedge* that is forcing the surface upward."

Unimpressed by Orville's simplistic explanation, Huffaker continued to grimace. "Mr. Wright, have you considered alternate explanations for why the center of pressure of a curved surface moves to the rear at shallow angles of attack?"

"And just what might that alternate explanation be, Mr. Huffaker?" Wilbur scoffed.

As Huffaker endeavored to explain how a partial vacuum was responsible for generating lift on a curved surface, the impatient Wilbur soon interrupted. "Mr. Huffaker, it seems to me that your explanation is best suited as a mathematics exercise in a college classroom. If you can't clarify your ideas to those working within the field, how do you expect them to be translated into an engineering solution?"

Feeling his face begin to flush, Huffaker nodded politely. "The problem is not mine, Mr. Wright. It's challenging to explain concepts without using mathematics – a skill you seem to lack."

Wilbur glanced over at his brother, who was shaking his head. "My brother and I have read everything we could find in the literature that was sent along to us by the Smithsonian and Mr. Chanute. However, Mr. Huffaker, I must say that your manuscript was impossible to read... let alone comprehend."

"Are you referring to my Smithsonian paper number 1135, 'Soaring Flight'?"

"Yes, yes, that's the one."

"If that article was too involved, perhaps you had the opportunity to read Mr. Herring's discourse on the topic of curved surfaces and partial vacuums in the '97 *Aeronautical Annual*. He was quite explicit in his—"

"Are you referring to the *bluffer*?" Wilbur interrupted, with a chuckle. "The same Augustus Herring who passes himself off as an expert by playing with *toy* flying machines?"

Orville chimed in, "Herring, the thief who makes off with other experimenters' ideas... such as his claim to have been the originator of Mr. Chanute's *Double Decker* glider?"

Huffaker held out his hands in a questioning manner.

"I've known Augustus Herring since '95, when we were both hired by Professor Langley for his model aerodrome project. He understood and appreciated my theory concerning lift, and I admired his creativity and mechanical talent; I even helped him find an apartment in Washington. Because our wives were both pregnant at the time, we spent a lot of time together when not working at the 'Castle.' "

"And we haven't forgotten," Orville retorted, "that *both* of you were born south of the Mason-Dixon Line."

Kill Devil Hills, North Carolina

Thursday, August 1, 1901[7]

Huffaker had been with the Wrights for two weeks when he finally decided to uncrate and assemble the *Chanute-Huffaker* glider. The day had dawned clear and sunny, and Edward hurried to complete his assigned chores around the campsite. Removing the screws in the shipping crate that housed his machine, his mind drifted back to his hometown... to the trials and tribulations he had endured in the 2-1/2 years since the death of his beloved Carrie Sue.

All of the glider models, all of the flight tests, all of the evaluation, and all of the changes made to this full-size, man-carrying apparatus, he thought. *Mr. Chanute and I have spent hundreds of hours designing and constructing this third method of automatic stability. And now we need assistance from these strangers to take the next step? Where is Octave? I can wait no longer! God willing, before day's end, the machine will have made its first trial.*

By 11 a.m., the component pieces of the glider had been removed from the crate and laid on the sand next to the shed. From time to time, the brothers would wander over to where Huffaker was working to silently ponder the idiosyncrasies of the unusual assembly process, before walking back to their own tasks. By noon, everyone had stopped for a lunch of smoked beef, hard-crusted Italian bread and hot coffee. Before long, the conversation drifted to the *Chanute-Huffaker* glider.

"Edward," Wilbur said, "how are you going to hold all of those framework tubes together?"

"Wire in tension, we don't need any fasteners or adhesives, just piano-wire and a few small turnbuckles! The idea came from Mr. Chanute's bridge building-experience—"

"What are the tubes made of?" Orville broke in.

"Rolled paper and hide glue... coated with shellac."

"Shellac!" Wilbur shouted, almost spilling his coffee. "How are those tubes expected to survive when they get wet?"

"Not well, I'm afraid, but the apparatus was only intended for the short term – it's not expected to last forever. For instance, the covering for the lifting surfaces is made from *Lawnsdale cambric muslin,* and it's finished with brushed-on coats of starch paste. That won't stand up to the rain either, but it shrinks and folds without creasing!"

"Why didn't you coat the tubes with something waterproof – like spar varnish?" Wilbur asked.

"I tried to have a special varnish mixed for me back in Chucky City, but they didn't have all the ingredients that Mr. Chanute specified. Therefore, I was obliged to use shellac and starch paste."

"I notice that you're using these paper tubes as struts between the upper and lower lifting surfaces," Orville observed.

"Yes, since fused paper tubes work well in compression, Mr. Chanute used them in combination with piano-wire to form a Pratt Truss, which holds everything rigid."

"Why didn't you simply use Sitka spruce for the struts?" Orville persisted.

"We were seeking a lightweight, inexpensive solution for the method of construction. You must admit that it's a clever design. By releasing the tension on only a few of the lengths of piano-wire, the entire contraption can be quickly folded in case of an emergency... like inclement weather. Another interesting feature is the method used for fastening the cambric cotton to the lifting surfaces; you'll see how that has been achieved after the machine is fully assembled. By the way, the cloth is allowed limited fore and aft movement – in step with the center of pressure movement. This was part of Mr. Chanute's scheme to provide automatic stability in both pitch and roll."

Wilbur gestured toward his own glider. "Orv and I design our machines to be durable. Do you honestly believe that paper tubes will stand up to the rigors of flying from these sand hills?"

Taking a long gulp of his coffee, Huffaker pondered how to best answer Wilbur's pointed question. "Well, sir, Mr. Chanute and I have been working on this idea for quite some time. The assembly technique works well on models, so we believe it will be successful on the full-size machine too. We'll soon find out!"

"Yes, we will," Orville chuckled. "When do you think the machine will be ready for a trial?"

"Once assembly is complete, I must locate the center of gravity with Wilbur aboard. After a few adjustments, it should be ready to try!"

When lunch was over, Dr. Spratt offered to assist Edward.

A few hours later, Spratt pointed beyond the Big Kill Devil Hill to the south. "Mr. Huffaker, look at those storm clouds; I believe rain is coming—"

An instant later, a lightning flash streaked between the distant dark sky and a point offshore, followed by a deep and prolonged rumble.

"An 11-second delay!" Huffaker shouted. "That's less than three miles away! The wind's picking up, too! We've got to get the machine folded and moved into the shed! Come on, George, I'll show you how it's done!"

In less than 20 seconds, Huffaker had released the tension on one of the turnbuckles. "Loosen those two on the left side, I'll get the last one on the right!"

A minute later, just as the wing's piano-wire rigging had been slackened, there was another burst of lightning followed six seconds later by a piercing clap of thunder. For some obscure meteorological reason, the entire area surrounding the campsite became eerily calm, as if the campsite were in the eye of a hurricane.

"Doctor, grab the top surface and set it down on the sand! It folds into four lengthwise pieces—"

At that instant the calm morphed into a gusty wind that seemed to spring directly from the ocean, a quarter mile to the east. "Hold on!" cried Huffaker over the growing howl of a freshening gale, "don't let the surfaces get caught up in—"

Just as the first raindrops thumped onto the exposed panels, Huffaker glanced toward the shed, desperately hoping to see the brothers hurrying to the rescue.

"Wil! Orv!" he shouted. "Help us move the machine inside!"

Instead of helping hands, what Huffaker witnessed was the methodical closing of the shed's overhead door. While he and Spratt battled to keep the glider from being blown away, the rain had intensified to a point at which the paper and shellac tubes were being pounded to pulp. After several minutes, a thoroughly drenched Huffaker and Spratt stumbled into the shed, having abandoned what remained of the glider.

"Why didn't you help us save the machine?" Huffaker bawled. "Wilbur, you promised there would be enough room in the shed if it rained! So did Mr. Chanute."

"Calm down, Edward!" Wilbur insisted. "There wasn't enough time – the roof was about to blow off the shed! We had our hands full just to get both doors closed!"

"When I looked over, at least two minutes before the first strong gust or raindrops, the two of you were standing there with your arms folded! That's not what I would call helping—"

"Call it what you like!" Orville interrupted. "This is our camp, and we're here to serve our own best interests! Face the facts, Mr. Huffaker; Mother Nature did you a favor."

"What do you mean by that?" Huffaker demanded, wiping his brow.

"By destroying your feeble apparatus, the squall has spared you the embarrassment of witnessing its imminent failure. If I were you, I'd consider myself *blessed!*"

Staggered by Orville's heartless remark, Huffaker struggled to regain his poise. "Mr. Chanute will certainly be interested in hearing of your opinion, Mr. Wright!"

Orville stood firm, legs spread and his arms folded. "Mr. Chanute is quite aware of your futile efforts, Mr. Huffaker... he told us as much in a letter more than a month ago."

"Edward," Wilbur said, trying to reduce the tension "you have two young children back home who need your attention. They would best be served by you scurrying back to them before they forget who their father—"

Huffaker interrupted, his palm facing forward "My first wife, Cora, died of consumption in '88... we had no children. Carrie Sue, my second wife, died of typhoid less than two years ago. We have two children, a boy named Moreland, who is five years old, and a girl, Mary Ada, who is almost three. I find it odd, Mr. Wright, that although you have never had a marriage or a family, you have the cheek to suggest how I should raise *my* children!

"Is it possible that your *spinster* sister, to whom you both write daily, has proven to be a suitable substitute for a conventional marriage partner?"[8]

Chanute-Huffaker glider. Wreck of the $1,000 beauty (1901); Special Collections and Archives, Wright State University

Without warning, the wind and rain stopped, allowing the brothers to shift their attention from Huffaker's insinuation to propping open the shed door.

"Orv, shall we take our walk to the village?" Wilbur asked. "It's time to retrieve our mail and procure supplies. You may wish to tag along, Dr. Spratt. We'll show you around Kitty Hawk."

As the three left camp, no attention was paid to the waterlogged remains of the Chanute-Huffaker glider that lay sodden on the sand.

Rochester, New York

Tuesday, November 15, 1921

After Huffaker finished recalling the Kill Devil Hills events of 20 years earlier, O'Grady summarized his life as of 1901. "Mr. Huffaker, at the time of your gliding machine's ruinous demise, you were 45 years old, you had lost two wives due to disease and were left with two young children to raise. Furthermore, you had sacrificed a lucrative career in civil engineering, all in the name of aviation. Is that correct, sir?"

Huffaker nodded.

"Answer the question verbally, Mr. Huffaker," the judge directed.

"Sorry, Your Honor. Yes, that is correct."

"What did you think when all of your efforts were laid to ruin during that rainstorm at Kill Devil Hills?"

Huffaker's face drained of color. "Everything seemed so futile. My years at the Smithsonian had yielded nothing in terms of recognition for my original insights, especially those concerning the theory of lift. Then the Wrights deliberately allowed my machine to be ruined, an act that allowed them to proceed unimpeded with their own experiments."

"What did Mr. Chanute have to say about the destruction of your machine when he finally arrived?"

Huffaker lowered his eyes, disheartened. "Mr. Chanute seemed more interested in erecting his six-man army tent than in hearing about the loss of *our* flying machine. Later he said, 'Perhaps it is best that we are not distracted from our goal of achieving automatic stability by a machine of questionable composition.'"

"What did Chanute's announcement suggest?" O'Grady urged.

"I was astonished to learn that he could write off our apparatus in such a cavalier manner – a machine that had consumed so much of *my* time and energy! It was then that I began to accept what Orville had claimed a few days prior; Chanute had written off the glider a long time ago.

"Later, in a letter from Dr. Spratt, I learned that Wilbur had taken a photograph of the remains and laughingly referred to it as 'the wreck of the $1,000 beauty.' "

"Thousand-dollar beauty?"

"Allegedly, $1,000 was the amount of money that Mr. Chanute had invested in our glider. Spratt said that Wilbur took the machine's destruction as a *joke* on me, but afterward it struck him that the joke was rather on Mr. Chanute, since the whole loss was his."

"Were there any other disagreements that may have led to the Wrights' intense dislike of you?"

"In hindsight, I believe that the Wrights' resentment toward me and Augustus Herring was *their* feeling of inadequacy, due to their lack of a formal education—"

"Objection!" Robbins intruded. "Calls for speculation!"

"Overruled. Continue, Mr. O'Grady."

"You were saying, Mr. Huffaker?"

"Early in my stay at the Kill Devil Hills campsite, it was evident that the brothers hadn't the slightest idea *why* their glider flew! Although I made efforts to enlighten them, including tutoring them in mathematics, they rejected my efforts, often with sarcasm and attempts at joviality."

"How did you respond to the Wrights' disrespect?"

"I simply smiled and chalked it up to ignorance. However, I must admit that I felt better after reminding myself that the brothers were simply common mechanics."

"Objection, Your Honor!" Robbins shouted, scrambling to his feet. "Although the defense certainly has no love for the Wrights, Mr. Huffaker's testimony flies in the face of what historians have been writing about them for the past decade. In fact, it's Mr. Huffaker who has been taken to task for being a sour miscreant, an overall negative influence, and a common *thief*—"

Judge Sawyer slammed his gavel down twice.

"That is quite enough, Mr. Robbins! Your prattle is mere hearsay! Mr. Huffaker is currently on the witness stand, not a group of *authors* and faceless *biographers*! The objection is overruled! You may continue, Counselor."

"Mr. Huffaker, tell the court what the Wrights said about Augustus Herring."

Huffaker seemed glad to have the spotlight turned away from him. "During one of my discussions with the brothers, I proffered that Herring was the first to identify the advantage of curved lifting surfaces over helicopter designs that relied upon direct propeller thrust to produce lift. Herring argued that the aero*curve's* favorable lift-to-drift ratio allowed it to raise greater weight than merely the thrust generated by its propeller.

"Wilbur pooh-poohed my allegation, saying that the lift-to-drift idea was known long before Herring had latched onto it. But when pressed... he couldn't recall the originator."

"Did the brothers admit to patterning their superposed lifting surfaces, their spacing, and structural truss work to Herring's *Double Decker* glider from 1896?"

"Objection!' Robbins cried. "Mr. O'Grady is leading the witness... again!"

"Sustained. Restate the question, Counselor."

"Mr. Huffaker, in your opinion, did the Wrights use *any* of Mr. Herring's ideas in their 1901 glider?"

"The Wrights would only say that they were *slightly* influenced by Chanute's *Double Decker* design. They remained steadfast in their refusal to acknowledge that Herring had anything to do with the origin of that machine.

"Eventually, the brothers resorted to name calling. Wilbur said that Herring was 'nothing more than an opinionated ingrate, a person who had infuriated both Langley and Chanute, before being summarily dismissed as their assistant.' Orville was irritated by Herring's assertion that he had used curved surfaces on his models back in the early '90s, that the 'forerunners' he used on his '94 Lilienthal glider were used for lateral control experiments, and of course, his alleged powered flights of 1898."

"Mr. Huffaker, how does today's scientific community view partial vacuums and lift associated with the curved surface?"

Huffaker smiled as he clasped his hands. "They agree with us – Herring and me, that is! However, it's important to understand that the Wrights didn't understand how lift was generated. Furthermore, nowhere in their writings, speeches, or patents does an acceptable explanation exist!"

"Since the defense felt it necessary to introduce the subject, Mr. Huffaker, which of the brothers accused you of being a thief?"

Huffaker mumbled something under his breath before answering. "Wilbur... but not to my face! Dr. Spratt spilled the beans in a letter. Old sourpuss noted that one of his blankets, of all things, had disappeared, and I was the prime suspect! The sad thing was, Wilbur's circumstantial recollections have been preserved in the form of letters written to Spratt and Chanute. Those falsehoods have been fertile ground for researchers and writers ever since, and probably will be for years to come."

"Have you drawn any conclusions concerning your month-long stay at the Wrights' Kill Devil Hills campsite?"

A wry smile spread across Huffaker's face. "The Wrights misunderstood me as much as they misunderstood the *Chanute-Huffaker* glider's design. It was Wilbur who delayed the testing and was *fearful* about flying my apparatus. He and Orville chose to abandon the machine during a thunderstorm... a despicable way to break a promise, but an effective way to eliminate distractions.

"I was also uneasy with Wilbur's and Orville's lack of formal training in either science or engineering. Although they were thorough *cut-and-try* experimenters, the Wrights owe most of their early success to beginner's luck. By comparison, Augustus Herring failed due to an incredible streak of *bad* luck, and his association with glory seekers cost him dearly."

"In 1915, you were called upon to testify on behalf of Orville Wright in the nation's most publicized trial of the era: the 'Wrights versus Glenn Curtiss.' Please summarize your testimony."

"I described the construction of Langley's *Great Aerodrome*, a machine that Curtiss and the Smithsonian Institution contended was the first successful manned, powered, heavier-than-air flying machine. Curtiss said that if the *Great Aerodrome* had been launched differently, it would have flown successfully before the Wrights' machine. The Wrights' lawyers questioned me as to the structural integrity of the big machine when compared to the model aerodrome of 1896. I'm afraid that my testimony hurt Curtiss' case."

"I understand," O'Grady continued, "that toward the end of your testimony, you tried to offer advice to the defendant?"

"Objection!" Robbins hollered. "Irrelevant!"

"Overruled," Sawyer said. "The court wishes to hear Mr. Huffaker's advice. Continue."

"I suggested that Mr. Curtiss should have chosen Augustus Herring's compressed-air powered two-surface machine of 1898... it would have made a stronger case for a successful design. Unfortunately, the judge used his gavel to shut me up... and the lawyers from *both* sides demanded that my observation be struck from the record."

"Why did you suggest Herring's machine offered a better defense strategy, if you were testifying on behalf of the Wrights?"

Huffaker grinned. "Although Wilbur was dead – he had died three years earlier – Orville was no friend; merely someone who wanted to use my expertise for his own selfish purposes. I agreed to testify on his behalf, but I quickly turned into a hostile witness. It's too bad that Curtiss cheated Herring of his rightful place in the Herring-Curtiss Company. Little did he realize that Herring could have been a valuable asset—"

"Objection!" Robbins squealed.

CHAPTER 72

Rochester, New York

Wednesday, November 16, 1921

Having endured the most stressful portion of his testimony, Huffaker returned to the courtroom confident that he could provide additional insight to his Kill Devil Hills experience.

"Mr. Huffaker," O'Grady said, "after the Wrights refused to lend a hand in saving the *Chanute-Huffaker* glider, and despite your subsequent quarrel with Orville... you didn't pack up and leave. Why, sir, did you stay on at the Wrights' campsite, and how did you manage to cope with the hostility that existed there?"

Huffaker began to fidget. "A few days after the incident, Mr. Chanute arrived in camp, which initially helped to boost my sagging spirits! Most of all, I looked forward to my mentor helping to defend our revolutionary flying machine."

"Is that what he did?"

"Sadly... no! To my dismay, he seemed reluctant to disagree with the brothers on virtually any topic! Confident they could resolve problems on their own, Wilbur in particular wasted little time or energy considering the ideas of others. When threatened by the elder Wright's passive form of arrogance, Chanute took the path of least resistance – he simply smiled and nodded politely!

"It became evident that aviation's elder statesman was reluctant to jeopardize this new and fragile relationship!"

"Can you give the court a specific example of the Wrights' arrogance?"

Tugging on his earlobe, Huffaker leaned forward in the witness chair. "Back in January 1920 – not quite two years ago – Orville Wright was called as a witness in the *Montgomery versus the United States* federal lawsuit. Part of his testimony related to the years 1899 through 1902 – a time when he and Wilbur had been experimenting with *model* flying machines. In his testimony, Orville said that after testing a particular model – one that was fitted with a fixed tail vane – they were disappointed when its performance was markedly less than its similarly shaped, full-sized counterpart.

"With Orville's testimony still fresh in my mind, I happened to reread an old article in the 1896 *Aeronautical Annual*, by Augustus Herring... where I discovered a curious coincidence! Mr. Herring had written about the performance differences between a small surface and that of a similar but larger surface. He maintained that the larger surface almost always produced more *lift* and less *drift* – a term we now refer to as *drag* – making the larger surface inherently more efficient."

Huffaker raised a forefinger for added emphasis. "I distinctly recall a campfire conversation back in 1901, where Wilbur Wright made specific references to Mr. Herring's articles in the '96 and '97 *Aeronautical Annuals*. At the time, I remember being mildly concerned when he dismissed the writing as, 'merely a mass of conjecture, dreams, pseudo science, and enthusiasm... of which the latter was most valuable.' Nineteen years later, in his sworn testimony, Orville claimed that he and his brother were the *first* to make the size difference observation!

"Frankly, I was struck by the out-and-out hypocrisy! Where was Orville's acknowledgement that, 'the labors of others had been of great benefit to us' – so often quoted by authors and others?"

Leaping to his feet, Attorney Robbins had heard enough. "Your Honor, the defense requests a sidebar!"

Gathering in front of Judge Sawyer's bench, everyone turned to the defense attorney.

"Your concern, Mr. Robbins?" the judge asked.

"Your Honor, the defense objects to this entire line of questioning! The ploy is obvious... counsel for the plaintiff is attempting to elevate Augustus Herring's status by discrediting the achievements of aviation's most famous and successful champions. First it was Langley, then Chanute, and now it's the Wright brothers! Although Mr. Curtiss and the defense have no fondness for the Wrights, these tactics are despicable and, quite frankly, are beneath the dignity of this honorable—"

"Nonsense, Your Honor!" O'Grady interrupted. "As attorney for the plaintiff, I am trying to make use of this witness's recollections—"

Tapping his gavel lightly, Sawyer cut O'Grady short. "The court needs to hear more about the plaintiff's direction in this case, Mr. O'Grady. Therefore, I suggest we withdraw to chambers."

Standing, Sawyer checked his pocket watch before addressing the litigants and courtroom personnel. "Court will recess for one hour. Testimony will resume promptly at 11:15."

Banging his gavel once, Sawyer turned and dashed toward the swinging door that led to his private chambers. Struggling to keep up, the attorneys looked vaudevillian as they chased after the judge and his flapping black robe.

*

Sawyer tossed his black satin robe onto the ancient cast-iron coat rack that stood in the corner of his Victorian-era chamber. Three of the room's windowless walls were burdened by wooden shelving that extended from floor to ceiling and contained hundreds of legal volumes, stacks of briefs from past cases, sundry court records, and memorabilia from the judge's 20-odd years of service to the State of New York's legal system.

Settling onto the easy chair behind his seen-better-days walnut work desk, Sawyer slouched to peek around the opaque red-and-gold-fringed shade that hung from a wobbly gooseneck floor lamp. "Find a chair that suits you, gentlemen," the judge sighed. "O'Grady, the defense has a point regarding your tactics. It's time for you to 'fess up mister! What's your game?"

O'Grady had been warned about Sawyer's bluntness when away from the bench, but his decidedly unprofessional demeanor momentarily confounded the attorney.

"Although I identified the plaintiff's direction in my opening statement, I'll be happy to retrace—"

"Opening statement my ass, Counselor!" the judge bellowed. "Belittling the integrity of dead American heroes is quite another story! Are you trying to turn my court into an arena for amending history?"

Momentarily lost for words, O'Grady glanced over at Maloney, who was grinning broadly. "Not at all, Your Honor! Allow me to review the plaintiff's two-phase strategy—"

"Maloney," Sawyer butted in, "wipe that shit-eatin' grin off your face! This ain't no laughin' matter!"

Robbins, delighted by the sudden turn of events, sat meekly with his arms folded, eyes staring up at the judge's dusty brass chandelier.

"First of all..." O'Grady continued, picking up where he had left off, "there are the *people*, and then there's the *money*. People have contributed ideas, interpreted the science and applied engineering principles... some have actually dared to experiment! As a consequence, toward the latter part of the nineteenth century, when it became apparent that some practitioner was about to create a practical powered aeroplane, Octave Chanute, among others, made concerted efforts to improve their historical positions."

"There he goes again," Robbins groaned. "O'Grady thinks he's a goddamn mind reader! The word 'speculation' isn't even in his vocabulary!"

"The testimony speaks for itself, Averill," O'Grady chirped, "or perhaps you haven't been paying attention!"

"Let's face it, Counselor," the judge said with a yawn, "you tend to speculate."

"Indeed!' O'Grady blustered. "Well, consider this, Your Honor: had it not been for a fire that destroyed his experimental flying machine and engines back in '99, Herring – not the Wrights – would be hailed as the first to fly!"

"Bullshit!" Robbins mumbled, slouching in his chair. "By the way, my name is *Frederick*... you asshole."

With no reprimand forthcoming from Judge Sawyer, O'Grady continued. "After the Wrights sold an aeroplane to the U.S. government in '09, they went on to manufacture machines for domestic sale... cashing in on their name recognition! The Herring-Curtiss Company began operations that same year but wasn't so fortunate. Although Glenn Curtiss was a household name to most aviation-minded Americans, Herring remained a virtual unknown. This being said, one of the plaintiff's *objectives* has been to establish Augustus Herring as an authentic aviation *pioneer* – one who deserves better treatment from the aeronautical community and historians alike.

"We believe that the witnesses thus far have substantiated our claim that Herring was a major contributor to the invention of the aeroplane, and we have also chronicled the injustices that were inflicted upon him. Testimony already heard by the court clearly shows that Mr. Herring was the originator of many innovations that led to the first successful flight, yet the Wrights and Chanute, in particular, have attempted to muddy the waters by their acts of omission and deception."

"O'Grady has a vivid imagination, doesn't he, Your Honor?" Robbins mocked. "Too bad his imagination doesn't coincide with the letter of the law! There has been no connection proven between the testimony of his witnesses and the Counselor's outlandish conclusions! For instance, where are Herring's patents? Oh, that's right... there ain't any!"

"Problem was, Your Honor," O'Grady said, "Herring let his enemies write his biography. That was his strategic blunder."

Swiveling his chair to the side, Sawyer sprawled before swinging his feet up onto the desktop. "Anything else you wanna throw in there, O'Grady?"

"As a matter of fact – I've just gotten started! The second phase of our suit is more pragmatic. To be frank, the glory is nice, but it's the *money* that separates the winners from the losers... and that phase of our case has yet to be addressed. However, since the court insists that we reveal our plans, I'll reluctantly comply... under protest, of course!"

With a backhanded wave, Sawyer signaled him to continue.

"Although we are currently providing testimony concerning a little-known aspect of the Wright brothers' notoriety, namely their misconceptions, mistakes, and wholesale theft of others' *ideas*, the plaintiff intends to point out that the Wrights became multimillionaires from being credited as the inventors of the aeroplane! Their all-encompassing patent for lateral control, which was generously granted to them by the United States Patent Office, has given them unprecedented leverage in the courts regarding infringement suits – but that's all been documented elsewhere.

"Your Honor, the point is this: the Wrights undermined the fledgling aeroplane manufacturing industry in the United States between 1906 and 1917. During that period, attorneys for the brothers filed multiple lawsuits against alleged infringers, most of whom terminated their manufacture of aeroplanes or accessories, rather than fight a losing battle in the courts. The lone exception was Curtiss's new company that willingly engaged the brothers. Through a series of legal maneuvers that delayed the final outcome, Curtiss was able to continue building aeroplanes for our allies in Europe. Finally, in 1917, the U.S. Congress, under relentless pressure from suppressed manufacturers, was forced to consolidate all of the U.S. aeroplane patents for the benefit of the war effort. This solution helped Curtiss's company garner more than $15 million in profits by the war's end!"

O'Grady pressed on. "A previous court case established that Judge Monroe Wheeler, Curtiss's attorney back in 1912, failed to file papers in New York State to dissolve the bankrupt Herring-Curtiss Company, so the organization remained viable throughout the war years... with many of Mr. Herring's engine and airframe innovations integrated into Curtiss's designs. As everyone now knows, because of our team's diligent research and Wheeler's oversight, the Herring-Curtiss Company was reactivated, allowing it legal standing to pursue this civil action.

"By establishing Mr. Herring as a legitimate pioneer and contributor to the invention of the aeroplane, we will show that he and his company are legally entitled to compensation from the profits of its successor, the Curtiss Aeroplane and Motor Company."

The judge swung his feet to the floor and stood. "That remains to be seen, Counselor. In the meantime, what do you propose to do in terms of toning down your courtroom rhetoric?"

"Rhetoric!" Maloney bellowed, rising from his seat like an erupting volcano. "What the hell does rhetoric have to do with exposing the truth about these thieves and charlatans? I think you've been paying way too much attention to Robbins, Your Honor! When Jim started talkin' about money just now, you could see the runt start to sweat! The cat's out of the bag, Judge! Robbins' damn well knows it's the plaintiff's dough, and his client stole—"

"Maloney!" Sawyer shouted, sliding out from behind his desk. "Have you forgotten about our agreement? For Christ's sake, sit down and shut up!"

As he recoiled, Maloney reminded the cowering Robbins of a growling bulldog whose head had just been jerked back by its owner's choke chain.

"Your Honor," O'Grady continued, in a calming voice, "As representatives of the plaintiff, Mr. Maloney and I are only interested in unearthing the facts through the testimony of witnesses who have intimate, first-hand knowledge of the circumstances. We have no desire to desecrate the public's perception of aviation's *sacred cows*. Our intent is to see that Mr. Herring receives his due *recognition*... and collects the financial compensation *owed* to him."

"It's come down to this, has it O'Grady?" Sawyer said. "Sacred cows and money."

"That's the way we see it, Your Honor."

"Very well then! Because you have clarified the plaintiff's strategy to my satisfaction, the court sees no reason to restrict your present line of questioning! You have to agree, Mr. Robbins, if the plaintiff can't question the *legitimacy* of Chanute and the Wright brothers in a civil action suit... then where? The defense's objections are summarily overruled! Let's get back to the court and place all this into the official record."

*

With less than an hour left before the court's regularly scheduled lunch break, O'Grady was anxious to have Huffaker conclude his testimony.

"Mr. Huffaker, was there anything else of significance that occurred during your stay at the Wrights' camp?"

Taking several seconds to contemplate, Huffaker turned his eyes toward O'Grady. "Two things stand out: after one of Wilbur's glides, Chanute pulled me aside and proclaimed that the brothers were, 'the next in the line of great gliding experimenters, carrying forward the work of Lilienthal, Pilcher, and *myself*.' Then he backed off a bit, saying that he was 'merely emulating the other two pioneers.' Chanute's grandiose description was repeated word for word in his *McClure's* magazine article."

"What did you think of your mentor placing himself in the company of such aeronautical greats?"

"Frankly... I was appalled! Mr. Chanute had never designed a successful machine, nor had he actually flown. Other than the possibility that he suggested using a common bridge truss for Mr. Herring's 1896 *Double Decker* glider, there were no other technical innovations that could be attributed to him. From my experiences at Kill Devil Hills, I was beginning to see Octave in a new light."

"Describe that new light," O'Grady urged.

"I saw a sly advocate, whose goal was to enhance his fame at the expense of any aeronautical experimentalist who showed promise."[1]

"From your observations, how did Chanute go about exploiting promising newcomers?"

"By offering them money. Chanute hired the most promising candidates to build and fly machines that incorporated threads of his own ideas. The results of these seemingly innocent transactions would then be used in his speeches and magazine articles, to claim partial or total responsibility for any advancement that may have—"

"Objection, argumentative!" Robbins croaked. "A rather harsh description of a man who was universally loved and declared the 'Father of Aviation.'"

"Overruled. Continue, Mr. O'Grady."

"What's your second item of significance, Mr. Huffaker?"

"Wrecked gliders and glory-seeking individuals aside, my month-long stay at the Wrights' camp allowed me time to reflect on the status of manned, heavier-than-air flight. I concluded that the bulk of the theory and successful techniques associated with aeronautics had already been well established before the brothers began their quest."

"What were some of those theories and techniques?"

"The knowledgeable use of aero*curves*, Pénaud and Herring's automatic stability methods, Wenham's multiple lifting surfaces, Herring's two-surface refinement, and Chanute's adaptation of the Pratt truss for bracing multiple lifting surfaces. These advancements, along with others, were available to the Wrights *before* they started experimenting."

"Realizing that many of the scientific and engineering problems had already been solved, did this change your opinion of the Wrights' ultimate accomplishment?"

Huffaker scowled. "In my opinion, the Wrights were talented mechanics with a weak background in science and mathematics. As I mentioned, their flying machine work was greatly assisted by the labors of earlier experimenters, whose ideas about design, construction, and a host of other topics, the brothers eagerly embraced."

Being noon, Judge Sawyer called a lunchtime halt to the proceedings.

*

All of the litigants retired to the Empire State Restaurant for their meal. Seating themselves at their customary table at the rear of the establishment, the threesome of Curtiss, Wheeler, and Baldwin had already ordered their lunches when the haggard figure of Robbins barged through the entranceway. Ignoring his adversaries, four of whom were seated at a table directly to his left, the diminutive attorney made a beeline for the defendant's table.

"Where the hell have you been, Averill?" Baldwin inquired. "You've only got an hour for lunch, ya know."

"Don't call me that!" Robbins shushed, as he unbuttoned his woolen, ankle-length winter coat. "Tom, how many times do I have to tell you? Call me Fred! Christ, even that asshole O'Grady called me the *A-name* this morning – in the judge's chambers no less!" Conversation tailed off when the waiter arrived with the men's food, providing Robbins the opportunity to order an open-face liverwurst sandwich on buttered toast, with horseradish mustard and dill pickles on the side, which he proclaimed as... "a man's meal!"

As the others enjoyed their food, Robbins began to think out loud about the morning's court proceedings. "Sawyer is starting to piss me off. He refuses to sustain any of my objections, and he keeps letting their witnesses express their *opinions*. I just tried to have a word with him about it—"

"Fred, are you nuts?" Wheeler interrupted. "You can't talk to the judge without their attorney being present – that could result in a mistrial, and our side would have to pay *all* of the court costs!"

Robbins finished slurping his coffee before he responded. "Monroe, do I look like a fool? I had O'Grady in tow when I tried to catch Sawyer on his way out of court! No luck though... he was gone with the wind!"

"Don't get excited, Fred," Curtiss said, without looking up from his plate of spaghetti. "Things are under control."

For one of the few times in his life, Robbins was at a loss for words. "What are you talking about, Glenn?"

Contemplating his second meatball, Curtiss waved his fork in a dismissive manner. "Nothin' in particular, but I hate to see you lose sleep over this crap. Everything will turn out okay. By the way... who's that guy sittin' with Huffaker and the two lawyers? He looks familiar, but I can't place him."

Robbins glanced over his shoulder. "That's probably Spratt, their next witness."

Curtiss sat up straight, wiping his mouth on his sleeve. "Ah yes, George Spratt! He used to be friendly with the Wrights, but they had a fallin' out over somethin'. He was also an admirer of old man Chanute – why, I'll never know! Maybe it was all the cash Octave threw at him over the years."

*

Uncharacteristically, Judge Sawyer was back at his bench 15 minutes early. At precisely 1 o'clock, he stopped writing and summoned Huffaker back to the witness chair.

"Has the attorney for the plaintiff finished questioning this witness?"

"Yes, Your Honor," O'Grady said.

"Very well, does the attorney for the defense wish to cross-examine?"

"Briefly, Your Honor," Robbins said, standing.

"Carry on, Counselor."

As Robbins slipped behind the podium, he used the toe of his shoe to maneuver a wooden block into position before stepping up to deliver his first question. Maloney, observing the action from the plaintiff's table, shook his head in disgust.

"Mr. Huffaker, you testified that back in 1901, Augustus Herring and you were the *only* two investigators who understood the true nature of lift as produced by a curved surface. Is that correct, sir?"

"To the best of my knowledge, yes, that's true."

"You also testified that a partial vacuum was responsible for a lift on a curved surface at shallow angles of attack. Is that correct?"

"Yes."

"Was your conclusion based on experimentation?"

"No. Only theory."

"If that was the case, how do you expect this court to believe that Dr. Spratt, who had spent the better part of five years *experimenting* with curved surfaces – the person who *identified* the mysterious center of pressure reversal at shallow angles – didn't understand the nature of the lift?"

"Spratt is an *analyst*, not a theorist – he had it wrong! And by the way, he wasn't the only person to have identified the center of pressure reversal—"

"Do *not* elaborate!" Robbins scolded. "Simply answer the question!... You also testified that the Wright brothers were *ignorant* as to the true nature of lift, is that correct?"

"Yes. They believed in the impact theory... as did Spratt."

"So we have you and Herring on one side of the issue, with Dr. Spratt and the Wrights on the other. How do you think Herring determined that partial vacuums were the source of lift on curved surfaces at small angles?"

"Objection!" O'Grady hollered. "The question calls for a conclusion."

"Overruled. The witness may answer."

"Mr. Herring used *steam* to observe the action of air blown over a curved surface. This knowledge coupled with an understanding of fluid dynamics, an engineering discipline, led him to conclude that the partial vacuum was playing a role in the lift phenomenon."

"Are you aware that Mr. Herring is an unreliable source of scientific or technical information—"

"Objection, Your Honor!" O'Grady interrupted. "Argumentative!"

"Sustained. Keep your questions based on facts, Counselor."

"Very well, Your Honor. Mr. Huffaker, are you aware that Horatio Phillips and Otto Lilienthal both used steam to observe the action of the air flowing over a curved surface... and that they both used the technique *before* Augustus Herring?"

"Objection!" O'Grady shouted. "Relevance! Where is the proof that either of these experimenters came to the same conclusion as Mr. Herring?"

"Sustained. Continue, Mr. Robbins."

"No further questions."

*

It was almost 2 o'clock before the plaintiff's next witness was directed to approach the bench and be sworn in.

"State your full name, age, and place of residence," the clerk ordered.

"Dr. George Alexander Spratt, 51, Coatesville, Pennsylvania."

At six-feet tall, and weighing 180 pounds, Dr. Spratt's appearance belied his age – he looked at least 10 years younger. His dark brown hair and his narrow face was highlighted by closely spaced blue eyes and a long Roman nose.

Anxious to get started, O'Grady skipped many of the preliminaries that usually dominate the beginning of a new witness's testimony. "Doctor Spratt, as the plaintiff's lead attorney, I have been told that although you graduated from medical school more than a quarter century ago, you decided *against* becoming a physician! If that is true, sir... what was your reason?"

"I had been diagnosed with a heart condition. Since physicians are required to work under stressful circumstances, I decided to take up farming."

"Isn't farming a physically demanding occupation?"

"Certainly, but I didn't work the fields. My father owns the homestead, and I *manage* the business."

"How did you become interested in aeronautics?"

"As an amateur naturalist, I was always fascinated by birds, especially the red-tail hawks and turkey vultures that soared the skies of our farmland in southeastern Pennsylvania. My university studies provided me with an understanding of analytical technique and instrumentation, which I later transferred to the study of aerodynamics. In 1895, fascinated by the potential of mechanical flight, I began devoting all of my free time to the study of curved surfaces."

"When and how did you become acquainted with Octave Chanute of Chicago?"

"In 1898. After three years of scientific investigations, I wrote a paper that explored the possibility of manned, heavier-than-air flight. Thinking that several of my conclusions were new and original, I began to seek an expert to critique my work. After reading one of Mr. Chanute's soaring articles, I decided to write him a letter of introduction, while simultaneously including a copy of my paper."

Spratt crossed one leg over the other, showing off his highly polished but decidedly old-fashioned high-top leather shoes. Casually dressed in a beige herringbone sport jacket with chocolate-colored accents and elbow patches, dark brown woolen slacks, and an open collar long-sleeved white shirt, the Pennsylvanian projected the image of an Ivy League professor.

"What did your scientific investigations involve," O'Grady asked, "and what was your most noteworthy discovery?"

"I developed an open-air *wind tunnel* and an instrument for measuring a surface's *lift*. Although my hilltop tunnel was cumbersome to use due to the constantly fluctuating wind speed, I was nevertheless able to collect and plot data that confirmed the center of pressure reversal on a circular arc lifting surface at shallow angles of attack."

"What did Mr. Chanute think of your work?"

"He offered me financial assistance to expand my experiments – Octave even suggested that I construct a full-size glider! I declined on the glider offer – thinking I would bungle the job! Besides, I was mostly interested in analysis, not construction projects."

"How long did you associate with Mr. Chanute, and what did you think of his work?"

"My friendship with Mr. Chanute continued for 12 years, until his death in 1910. Throughout, he was devoted to securing solid advancements for the science of flight. To this end, he provided helpful encouragement and financial assistance to promising young aeronautical experimenters – several, including myself, became his protégés. In the final analysis, his work commanded success – if not always fame."

"Objection! Conclusion!" Robbins shouted.

"Sustained," Sawyer said. "Doctor Spratt, keep your answers factual in nature. Continue, Counselor."

"These are interesting conclusions, Doctor Spratt; I will revisit some of them later. For the time being, let's move to your involvement with the Wright brothers. How did you make their acquaintance?"

Spratt felt the gears in his head began to whirl. "Mr. Chanute secured an invitation for me to attend their 1901 experiments at Kill Devil Hills."

"Why were you invited?"

"Mr. Chanute had written to the brothers about my *method* of measuring pressures on a curved surface... apparently they were intrigued."

George A. Spratt, Kill Devil Hills (1901); cropped from water damaged Library of Congress slide

"When did you first meet the brothers?"

"I arrived at their campsite on the morning of July 25, 1901."

"How did you get along with Wilbur and Orville?"

"Very well, but I had the advantage of being compared with Mr. Huffaker – who the Wrights' apparently despised. The brothers seemed partial to me because I was congenial, enthusiastic, a careful worker, and... I told funny stories. In other words, I was everything that Huffaker wasn't. Later in my stay that first year, Wilbur confided that he thought I was a kindred spirit who shared his and Orville's interest in aeronautics."

"Was there anything that you didn't like about the Wright's 1901 glider?" O'Grady asked.

Spratt hesitated briefly. "I didn't like their control system or the silly forward-mounted horizontal rudder. After all, have you ever seen a fixed-wing bird... or a bird with a rudder out front?"

A few chuckles emanated from the visitors' gallery.

"However, I kept my opinions to myself."

"You didn't agree with the manual wing-warping and horizontal rudder?"

"Manual control was and is... too dangerous. I was a proponent of Chanute's automatic stability concepts. Furthermore, the horizontal rudder should have been mounted behind the lifting surfaces, where it might have provided a degree of automatic stability – similar to Pénaud's force arrangement, or Herring's spring-loaded cruciform tail."[2]

"The Wrights were having problems getting their 1901 glider to perform up to their expectations. Did they ask you for advice?"

Spratt leaned back in his chair. "They were at a loss to explain why their machine bucked in level flight – more than could be corrected by manipulating the horizontal rudder's control. In desperation, they asked if I had any ideas. From my experimentation, I suspected that they were experiencing center-of-pressure oscillations. I suggested a simple experiment that allowed them to isolate the problem, which resulted in re-trussing the lifting surfaces to an entirely new contour."

"Did your suggestion reduce the bucking problem?"

"Markedly! From that point on, Wil and Orv wanted to know everything about my experimentation – especially my method for measuring the drift and lift of a curved surface."

"Why was that of particular interest to them?" O'Grady prodded.

"Serious experimenters knew that having accurate lift data for a given curved surface was valuable for determining the correct wing area needed to overcome the machine's weight. The Wrights became concerned when their glider from the previous year produced less lift than Herr Lilienthal's tables had predicted. Their 1901 glider had a wing area of 315 square feet, 70 percent larger than their previous effort, but it still didn't lift as forecasted. The Wrights were convinced that the German's lift numbers were erroneous and that they would require better data."

"What was the solution?"

Spratt's face darkened as he leaned forward, seizing the chair's wooden arms. "I provided the Wrights with the *method* for accurately measuring a curved surface's *drift* and *lift*. In 1909, Wilbur wrote to me... he said that *my* technique provided the key that unlocked the secret of powered flight! To this day, I have yet to receive a lick of credit!"[3]

Abruptly, there was a commotion in the visitors' section of the courtroom as a reporter, hurrying from the chamber, let the rear door slam loudly behind him. Moments later, the newspaperman shouted into a telephone's mouthpiece: "Hold the presses! We've got a scoop!"

CHAPTER 73

Rochester, New York

Thursday, November 17, 1921

O'Grady and Maloney sat at the plaintiff's table reading the front page story in the Rochester *Herald:*

Wright Brothers Accused of Stealing the Secret of Flight

Rochester, N.Y. – Almost 18 years after the Wrights' monumental first flight at Kitty Hawk, North Carolina, Dr. George A. Spratt, a friend and colleague of the inventors, dropped a bombshell during his testimony in the month-long trial of the Herring-Curtiss Co. v. Curtiss, civil action lawsuit that is being prosecuted at the Monroe County Courthouse in downtown Rochester, New York.

Dr. Spratt, a pioneer aeronautical analyst and valued assistant at the Wright brothers' Kill Devil Hills camp during the years 1901, '02, and '03, disclosed yesterday that he supplied the key idea that permitted the inventors to unlock the secret of manned, powered flight.

"...Wilbur's long dead, and Orville won't tell the truth!" Spratt shouted in response to a question from the plaintiff's attorney, James M. E. O'Grady. Defense attorney Frederick A. Robbins objected to Spratt's shocking assertion, declaring it "...argumentative and irrelevant." In a dramatic follow-up, the Coatesville, Pennsylvania scientist reached into his satchel and pulled out a handful of letters, waving them triumphantly above his head.

"The Wrights stole my idea, and here's the proof!"

A 20-minute legal squabble ensued, ending only when Judge S. Nelson Sawyer allowed one of the documents – a letter from Wilbur Wright to Dr. Spratt – dated October 16, 1909, to be placed into evidence. Henceforth, Spratt was permitted to read from its contents.

"It's quite true, that you told us your idea of balancing the lift of a surface against its drift and determining their relationship directly, instead of measuring each independently... we have not wished to deprive you of the credit for the idea..."[1]

Spratt then explained how the Wrights used their newfound knowledge. "By using my method for acquiring accurate 'drift to lift' data, and fabricating their own measuring instrument based upon my original design, the brothers tested a variety of small curved surfaces – today they're called airfoils – in a fan-driven wind tunnel. The results garnered from these tests enabled them to design efficient curved surfaces for their record-setting glider of 1902 and the world's first successful powered machine, a year later."[2]

Dr. Spratt's testimony will continue this morning, starting promptly at 9 o'clock.

*

As Dr. Spratt took the witness stand, a murmur of recognition washed through the courtroom's packed visitors' section.

"Order, order in the court!" the judge barked, banging his gavel. "Mr. O'Grady, you may resume questioning the witness."

"Doctor Spratt, based on your testimony from yesterday, is it safe to say that *you* unlocked the secret of powered flight for the Wright bro—"

"Objection!" Robbins broke in, scrambling to his feet. "Counsel is leading the witness!"

"Sustained. Restate your question, Mr. O'Grady."

Flipping a page on his notepad, O'Grady walked out from behind the podium. "Doctor Spratt, what was the single most important factor that allowed the Wrights to invent the manned, powered aeroplane?"

Spratt smirked. "Now *that's* a leading question! Generally, I'd say *all* of the knowledge the brothers absorbed from their aeronautical predecessors. Specifically, it was their ability to obtain accurate and usable *drift to lift* ratios from a large sampling of curved-surface wind tunnel models. The models that demonstrated the best performances were then enlarged and used on their full-size machines."

"Which machines are you referring to?"

"Their glider of 1902 and the powered *Flyer* of 1903."

"Doctor Spratt, do you know for certain that the Wrights *stole* the design for your wind tunnel measuring instrument?"

George A. Spratt's drift to lift balance (1901); Special Collections and Archives, Wright State University

"Yes. There were a series of events that took place between November 21, 1901, and December 15, of that same year."

"What is the significance of November 21, 1901?"

"That was when I *sent* Wilbur the drawings, specifications and a photograph of my latest drift-to-lift ratio measuring instrument – also known as the *tangential angle instrument*," Spratt said.[3]

"What is the significance of December 15, 1901?"

"That was when the Wrights declared that they had completed their wind tunnel testing – less than a month after I had mailed them the drawings and specifications."

"The Wrights were your friends... what raised your suspicions concerning their intentions?"

"In February of '02, two months after the brothers had finished their testing, Mr. Chanute, who had been helping with some of the monotonous calculations, wrote to me concerning a letter he had received from Wilbur. Apparently, Wilbur went out of his way to point out that *he* had designed and built a 'special machine for tangentials' – *their* drift to lift ratio measuring instrument – *before* beginning their final testing on November 22, 1901."

"Why was this suspicious?" O'Grady asked.

"As it turned out, their data was *faulty*! According to Chanute, three pages of tangent angle data had to be thrown out because of inaccuracies at low angles of attack. Chanute said that they crossed out 392 measurements – about 28 hours' worth of experimental work! This told me two things – the *design* of the Wrights' original measuring instrument was *defective*, and the discarded data required the brothers to build a *new* instrument later in December. So you see, Wilber had lied!"

"Did Wilbur respond to your November letter?"

"He answered on December 15, telling me that my new instrument was 'ingeniously designed and should give good results.' He failed to mention that they had had a problem in measuring tangent angles, requiring them to build a new instrument of their own."[4]

"Why was Wilbur Wright being secretive?" O'Grady asked.

"Objection! Calls for a conclusion!" Robbins barked.

"Overruled," Sawyer said. "The witness may answer."

"I think Wilbur was afraid that I would see through his deception. He was trying to cover his tracks by also lying to Mr. Chanute when he *backdated* his new instrument's time of construction."

"Let me get this straight," O'Grady said. "Wilbur Wright secretly copied *your* instrument for measuring tangent angles... and then he *lied* about when it was made?"

"Exactly... and that's not all! I now know why Mr. Huffaker pestered the Wrights with his 'character building' lectures – Edward had the brothers figured out long before me... and that's why they hated him—"

"—Objection! Speculation!" Robbins shouted.

"Sustained. Strike that last response. Continue, Mr. O'Grady."

"Let's get back to the hardware, Doctor. What were the similarities between your instrument and the Wrights'?"

Spratt threw up his hands in frustration. "To this day, Orville refuses to reveal the details of their so-called 'special machine for tangentials.' "

"What does that suggest?" O'Grady asked.

"He's hiding something!"

"Objection! Speculation!" Robbins hollered, scrambling to his feet again.

"Sustained. Strike. Continue, Mr. O'Grady."

"Describe *your* 'drift-to-lift' instrument, Doctor."

"My instrument used a parallelogram made from two discs and a connecting belt. An arm on one of the discs held the lifting-surface model in a vertical position. After setting this surface to a predetermined angle of attack, a steady wind would rotate the surface to a point where its lift balanced exactly against its drift, from which the surface's tangent angle was measured."

"Doctor Spratt, to the best of your knowledge, did anyone other than the brothers see their new measuring instrument?"

"Mr. Chanute described it to me just before his death in 1910 – their instrument was another parallelogram design, made from linked and pivoted hacksaw blades—"

"Objection! Hearsay!" Robbins shouted.

"Overruled. I will allow the witness's response."

"Very well, Your Honor," Robbins moaned, clearly upset. "The defense hereby requests 'a continuing objection.' "

By invoking a continuing objection, Robbins signaled Judge Sawyer that his rulings might be used as grounds for a future *appeal* by the defense. More immediately, Robbins would not have to stand and object to every follow-up question.

"So noted, Counselor," Sawyer growled, writing himself a memo. "Continue, Mr. O'Grady."

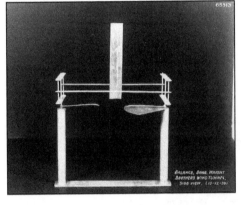

Wright brothers' version of George A. Spratt's drift balance (1901); Special Collections and Archives, Wright State University

"Doctor Spratt, did Mr. Chanute compare the two measuring instruments?"

"Yes. He said there wasn't much difference in the way of *design* and that the two instruments operated in a similar fashion."

"How did you interpret Chanute's analysis?"

"It was clear to me that the Wrights had *studied my design* before building their own instrument."

"When the Wrights invited you to the 1902 trials, were you aware of your contributions to their future success?"

"No, the brothers never discussed the results of their wind tunnel experiments with me. Naive as this may sound, I didn't put two and two together until six years later, when I read an article by Orville in the September 1908 issue of *Century* magazine."

"What was the article about?"

"How he and Wilbur *invented* the aeroplane."

"Was the story enlightening?" O'Grady asked.

"I learned," Spratt deadpanned, "that the brothers had analyzed the instability problem in their 1901 glider *all by themselves*. That they alone had figured out how to eliminate the wind-tunnel measurement errors that had dogged other experimenters. Orville tidied up his story by saying that the aeroplane's invention had been the direct result of their careful laboratory experiments, 'which alone made an early solution of the flying problem possible.' That's a direct quote! I'll never forget it!"

"Did you confront the Wrights about their memory loss?"

"Yes, but I waited for more than a year before writing to Wilbur. Having become totally disillusioned by the brothers, I didn't mince words. I wrote that I was astonished that my name wasn't mentioned in the articles or interviews that they granted. I said that they had taken my ideas and methods and had completely claimed credit for themselves, while ignoring me."[5]

"What was Wilbur's response?"

"He claimed that he and Orville had paid me back in spades... with data and ideas that many times outweighed the original 'loan.' To the contrary, I felt that their wind tunnel data benefited them... not me.

"To support their argument, Wilbur claimed that his 'special machine for tangentials' contained design changes that eliminated any debt that might have been owed to me in the first place... a statement that confirmed my original suspicion – that he had studied my instrument prior to building his own.

"A year later, in 1910, Wilbur again demonstrated his lack of character by *suing* the Herring-Curtiss Company for its use of interplane control surfaces. While 'design changes' exempted Wilbur from crediting me with the invention of my measuring instrument, he readily reversed his stance when it came to wing warping – claiming that *any device that accomplished the same outcome infringed on their patent!*" Spratt's voice grew louder with each word.

A tense moment passed before O'Grady resumed his questioning. "Doctor Spratt, there were many letters exchanged between yourself and the Wrights during and after those hectic years of experimentation. Can you describe the tenor of these letters?"

Spratt slumped in his chair. "As I recall, Wilbur's letters – he wrote most of them – were friendly, encouraging and informative. But most of all, I recall a degree of restlessness, especially when discussing the scientific underpinnings of curved lifting surfaces. The number of letters written during those first three years reflected Wilbur's burning desire to acquire information. In 1901 there were three. In '02 there were eight, and in '03 there were 15."

"Correct me if I'm wrong, Doctor, but weren't those the years when you were supplying the brothers with new and important scientific information concerning lifting surfaces?"

"Yes."

"How many letters did Wilbur write in subsequent years?"

Spratt pulled out a small notepad and flipped it open. "Let me see, there were five letters from him in '04, three in '05, two in '06, two in '08, and one in '09."

"What does this suggest to you?" O'Grady asked.

"The Wrights got what they wanted... then they abandoned me. Of course," Spratt said with a smirk, "there's that letter that I recently received from Orville."

"What did he want?"

"He wanted to *borrow* all 39 letters that Wilbur had written to me. He said that he was compiling a history of the aeroplane's development and needed to brush up on what was being discussed at the time."

"Will you honor the request?"

"First chance I get... I'll tell him to pound salt!"[6]

*

The court's lunch break found the litigants back at their usual eatery, where O'Grady and Maloney were ecstatic over Spratt's masterful, unscripted morning testimony.

"Doctor," Maloney said, "how do you feel after finally getting the chance to set the record straight after all of these years?"

"Vindicated! I hope you could tell from my testimony, the Wrights were unapologetic liars, and I was a sheep being coaxed to slaughter."

"Here's something that will make your day," O'Grady said. "We just received word that Orville Wright has telegraphed an offer to testify on Curtiss's behalf. Apparently, the Dayton *Daily News* picked up the Rochester *Herald* story and contacted him about it."

Surprised, Spratt lowered his pickled herring on kummelweck sandwich back onto its plate. "Is that something that's allowable?" he inquired, swiping a paper napkin across his lips.

"The defense can call anyone it wants," O'Grady said, "as long as they inform us by placing him on the court's witness list. But here's the kicker... Robbins told me that Curtiss said to, '... tell that little shit Orville to go to hell!' Apparently, Glenn enjoyed your testimony – especially the part where Wilbur got caught stealing the idea for your wind tunnel instrument."

Appreciating the irony of the situation, the three men had a good laugh before getting down to the business of the afternoon's strategy.

*

Judge Sawyer walked into court 10 minutes late for the start of the afternoon session. "Apologies to the litigants. There was an unforeseen bit of court business that required my immediate attention. Now that it has been taken care of, we may proceed accordingly. Mr. O'Grady, you may continue questioning your witness. Doctor Spratt, please return to the witness stand. A reminder, sir – you are still under oath."

"Doctor Spratt," O'Grady began, "I direct your attention to the months directly following the Kill Devil Hills experiments of 1901. Tell the court about your ongoing aeronautical research."

Spratt cleared his throat. "I took the months of September and October to finish and test my latest wind tunnel balance. Then I sent a photograph, description, and other materials to Mr. Chanute and Wilbur Wright."

"In that regard, you testified that you didn't hear from Wilbur Wright again until receiving his letter of December 15. During that interval, did you correspond with any of the other participants?"

"Yes, there were several letters exchanged between myself and Mr. Chanute."

"What was discussed?"

"Mr. Chanute was obsessed with the Wrights' manual control system... he believed it to be very dangerous. He told me that although the Wrights would probably succeed in their attempt at powered flight, he was afraid that it, 'would not be the best way.' "

"Besides arguing the deficiencies of manual control with the Wrights, how else did Chanute demonstrate his concern?"

Spratt stared down at his high-top shoes before raising his head. "He designed a 'rocking wing' glider and contracted to have it built for him by Charles H. Lamson, the California kite master. In October 1902, the finished machine was crated and sent to Kitty Hawk, where Chanute envisioned the Wrights flying the contraption. By experiencing the automatic stability features of his triple-decker machine, the old man hoped the brothers would change their minds about using manual control. As a hedge against the possible failure of the Lamson machine, Chanute hired Augustus Herring to rebuild his old *Katydid* multi-wing glider, which was also shipped to Kitty Hawk."

"With two additional gliders in their camp, weren't the Wrights being imposed upon?"

"Chanute avoided this possibility," Spratt said, "by bestowing the Lamson and multi-wing machines to the Wrights... they were *gifts*!"

"Why?" O'Grady asked.

Spratt massaged his right earlobe. "Chanute was pressuring the Wrights to change their minds about manual control. Also, by presenting the brothers with a significant gift, he was ingratiating himself to—"

"Objection! Opinion!" Robbins butted in.

"Sustained. Please refrain from making judgments, Doctor. Continue, Counselor."

"Did Mr. Chanute discuss the Wrights' wind tunnel experiments of November and December?"

"He said that he was pleased that they had undertaken the lift and tangentials measurements, but he was disappointed that the brothers had *stopped* experimenting to concentrate on their bicycle business. Chanute offered to offset their expenses if they would let him do so."

"Did they accept his offer?"

"Apparently not. So around Christmas of '01, Octave made another bold offer: he suggested that if the brothers so desired, he would introduce them to an old acquaintance, Andrew Carnegie, the so-called 'king of steel'—"

"For what reason?" O'Grady interrupted.

"To secure funds for their aeronautical research. Chanute suggested it would be possible to acquire up to $10,000."

"How did the Wrights respond to these offers?"

"Objection, hearsay!" Robbins shouted.

"Overruled," the judge said. "The witness may answer."

"The brothers turned the offer down as diplomatically as possible, saying that their investigations were only part of their ongoing *hobby*."

"Was there anything else that Chanute did to ingratiate himself with the Wrights?"

"Octave insisted he be allowed to help with the large number of wind tunnel calculations that needed to be carried out."

"You were one of Mr. Chanute's loyal assistants; what do you think these magnanimous gestures suggested?" O'Grady asked.

"Objection!" Robbins shouted, scrambling to his feet. "This calls for yet another conclusion!"

"Overruled!" Sawyer rumbled. "The court wishes to hear the witness's answer. Continue, Doctor Spratt."

"Back in '01, I thought that Chanute was just being his usual generous self. A few years later, I came to believe that he had hidden motives."

"Such as?"

"Although the Wrights wisely refused Chanute's offer to allocate money for their continued wind tunnel experiments, they became indebted to him in other ways. It began when Chanute supplied them with a list of aeronautical publications to read, along with suggesting potential locations for them to perform their experiments. It was also at his urging that I divulged my concept for determining the accurate drift force for a lifting surface. Soon afterward, Chanute convinced me to explain the inner workings of my tangential balance. Then he helped the Wrights with their calculations, followed by his gift of two full-size gliding machines. A few years later, these particulars and other gestures of Chanute's benevolence would allow him to claim partial credit for having *invented* the aeroplane."

*

"As an invited guest to the Wright's 1902 flight trials," O'Grady said, "when did you arrive at their Kill Devil Hills campsite?"

Spratt consulted a small notepad. "Early on the afternoon of Wednesday, October 1, 1902. That morning, I boarded the steamer, *Lou Willis,* for an uneventful trip from Elizabeth City to Kitty Hawk. Then I rented a chauffeured pony cart for the four-mile jaunt south along the beach to the Wrights' campsite."

"Was anyone there to greet you?"

"No. Wilbur and Orville were on the west slope of the Big Hill. Bill Tate, their hired helper, was also there, along with their brother Lorin, who had arrived the day before."

"What were they doing?"

"They were test-flying the new glider. I watched Wilbur and Orville make about a dozen flights. Later, I learned that all of these had been made with the lateral control mechanism deactivated."

"Were they having problems?"

"Wilbur said that the lift was much improved over the previous year's machine and that the bucking had subsided to the point where it could be managed by the horizontal front rudder—"

"Therefore, the machine appeared to be a success," O'Grady butted in.

Spratt shook his head. "Not exactly! The machine's wing warping – its lateral control – wasn't working as expected. Orville, the less experienced of the brothers, had recently crashed into what he described as, 'a heap of flying machine, cloth and sticks, with me in the center without a bruise or scratch.' "

"When did this crash take place?"

"On September 23, 1902... eight days before I arrived."

"What caused the accident?"

Sighing, Spratt leaned back in his chair. "Orville said he was sailing along straight and level when he noticed that one wing was getting a little too high and that the machine was sliding off in the opposite direction. When he moved the warping control an inch or so to bring the wing back down, the surface moved unexpectedly to a position that was higher than before. Assuring himself that he had moved the control in the correct direction, Orville then threw the wingtips to their greatest angle. By this time he found that he was descending backwards – toward the lower wing from a height of 25 to 30 feet – as a result of the machine having turned up at an angle of nearly 45 degrees in front – a condition he had not noticed while he was busy manipulating the warping."

"How long did it take to make the repairs?"

"Three days of working around the clock. However, because of unusually calm winds, they didn't get a chance to fly again until the following Monday, two days before I arrived. Unfortunately, the warping or 'end control,' as they called it, continued to produce unpredictable results. As Orville had discovered the hard way, a stray gust of wind could be downright dangerous to the operator."

"What did they do next?" O'Grady asked.

"On the evening of my second day in camp, the brothers engaged me in a long debate concerning the idiosyncrasies of certain lifting surfaces and how they might be responsible for the glider's strange reaction to warping. They were clearly unable to understand what had happened and were grasping at straws."

"When did they figure out what was wrong?"

"That same evening. While arguing the merits of twisting the wing ends in opposite directions, Wilbur suddenly made the connection. He said, 'What about the drift? Warping causes a drift imbalance between the left and right wings! This makes the machine slew toward the downward twisted wingtip – the one that's operating at the higher angle of attack!' "[7]

"Now that they understood what caused their problem," O'Grady said, "how long did it take to find a solution?"

"Less than 24 hours! Orville said an idea came to him in the middle of the night. I didn't completely understand the problem or the solution, but when Chanute and Herring arrived on Sunday, two days later, there was a rousing debate over the situation."[8]

"Do you remember what was said?"

"As I recall," Spratt said, "Chanute didn't have much to say... other than manual control was too dangerous. When Herring offered his view as to why wing warping worked perfectly while the glider was flown as a kite but failed to respond properly during free flight, Wilbur disagreed. The ensuing discussion of wind speed and its effect on a surface's lift and drift prompted Herring to question the validity of various engineering points that Wilbur had made in his speech to the *Western Society of Engineers*, a year earlier."

"Was this a friendly discussion?"

"The Wrights didn't like being corrected about anything, but they were in their element. They enjoyed a rough and tumble argument, especially if they decided to change sides halfway through. I never appreciated that tactic... it always struck me as being dishonest. However, Herring was very sure of himself, and he knew mechanical engineering. It turned out to be a very entertaining evening."

"Tell the court about the changes the Wrights were in the process of making to the glider's tail," O'Grady said.

"They started building the new tail the day before Chanute and Herring arrived. The old vertical tail consisted of two surfaces, and it didn't move. The new rudder was movable and had only a single surface."

"Did the discussions on the evening of October 5 have any bearing on the changes the Wrights made to their glider?"

Spratt frowned at the suggestion. "The brothers had already settled on their changes. They were simply exploiting Herring's knowledge and experience to confirm their decision. The Wrights always insisted that they had invented the aeroplane by themselves. They rejected any suggestion that anyone else may have contributed to their effort."[9]

"Having previously documented your criticism of the Wrights in this regard, are there any other instances where they exhibited a similar behavior?"

Biting his lower lip, Spratt lowered his head before turning his attention back to O'Grady. "Augustus Herring is a good example. Back in '01, Wilbur told me that he was annoyed by Herring's claims, his exaggerations, and his assertion that their glider was, 'nothing new'. When Herring arrived in camp the following year, there was more going on than met the eye.

"Thanks to Chanute's yarns about what he called Herring's 'jealous nature and tendency to claim more for himself than he rightfully deserved', all the men disliked him before he ever arrived — even Dan Tate, the Wrights' hired help! I rejected

Herring because Chanute said that he failed to acknowledge aeroplanes designed strictly with automatic control. The Wrights also embraced this flawed idea, but I trusted they would eventually see the light. Also, Herring subscribed to the partial vacuum theory of lift at low angles of attack, while everyone else in camp had faith in the *impact theory* – another good reason not to trust the interloper."

"It sounds as though Herring had the deck stacked against him. What happened to him first?"

"On Monday, while the brothers were busy working on their new rudder, Herring began setting up Chanute's refurbished *Katydid* outside the shed. Almost two weeks earlier, it had arrived by spritsail boat from Manteo on Roanoke Island. Wilbur and Dan had lugged the crate over from Roanoke Sound. That evening we carried the assembled machine to the big hill for a few glides, because Herring said he wanted to locate the center of pressure. On his second attempt down a 12-1/2-degree inclination, he alighted on the lower starboard wing and broke a spar. The apparatus was returned to camp where Herring had it repaired within an hour."

"Then what happened?"

"On Tuesday, when unfavorable winds prevented any gliding, Herring asked me to explain my testing machine. He was so enthusiastic about its possibilities that I decided to demonstrate how it worked, so we carried the equipment to the top of the West Hill. After a short period of instruction with the open-air wind tunnel, anemometer and balance instruments, Herring began taking data for the center of pressure and the ratio of drift to lift on several different curved surfaces.

"After lunch, I left him to work on his own while Lorin and I went drum fishing. Later that evening, Herring asked why I had attached the test curves to the balance instrument from the surface's *topside*. He tried to convince me that the top of a curved surface was more important than its bottom at low angles of attack. He claimed that the topside attachments would interfere with the airflow that produced partial vacuum lift. Since we didn't have any means of accurately soldering, Herring encouraged me to perform comparative tests once I got back home."

"Did you do this?" O'Grady asked.

"During the winter of '03, I took time to make the comparison—"

"What did you find?" O'Grady interrupted.

"Herring's suspicions were correct! At low angles of attack, all of the curved surfaces performed better with *bottom-side* attachment points."

"What did this determination tell you?"

Spratt's face lit up. "To question my belief in the impact theory of lift! I began looking for ways to test the partial vacuum theory. Herring and I rarely corresponded after our alliance at Kill Devil Hills, but he changed my understanding of how curved surfaces worked."

"Did you relay your findings to the Wrights?"

"Yes, but they said they weren't interested in what they called 'cockamamie theories.' Even after Wilbur's death in 1912, Orville refused to acknowledge that the partial vacuum had anything to do with a curved surface's ability to generate lift at low angles of attack."

CHAPTER 74

Rochester, New York

Sunday, November 20, 1921

Jim O'Grady honked the horn of his factory-fresh 1922 Buick Coupe before backing the shiny black machine into its garage space. Driving his new automobile back and forth to church each Sunday morning was O'Grady's reward for enduring yet another lackluster celebration of Mass by the apathetic Father Boobotch, the long-embedded priest at St. Andrew's Roman Catholic Church.

After hanging his black woolen coat in the hallway closet of their modest colonial-style home, O'Grady turned to Joanne, his wife of 30 years and waved a dog-eared yellow legal pad in her direction. "At least I had time to scribble down a few good questions for Herring! He'll be back on the witness stand tomorrow... hope he's up to it!"

Joanne's cheerful demeanor had little to do with church and everything to do with her passion for cooking. Anxious to get started on their Sunday dinner, she playfully nudged her husband into the parlor with the weekend edition of the Rochester *Herald*. Complying – as he always did – O'Grady first tossed a log onto the dying embers of his Franklin stove before ambling over to his favorite chair: a well-worn leather recliner that basked in the midday sunshine of the room's south-facing window. Casually slumping into his chair, O'Grady perused the newspaper's front-page headline.

Harding Taps Taft as Next Supreme Court Chief Justice

As a licensed, practicing attorney in the State of New York, O'Grady was impressed with Republican President Warren G. Harding and his "Ohio Gang" that ran the White House. The nomination and rubber-stamp confirmation of former President William Howard Taft was another sign that the country would continue its conservative ways well into the future. O'Grady smiled with satisfaction as he turned the page... where another interesting headline caught his attention:

Conflict at Kill Devil Hills

Dr. Spratt Tells of Wrights' Mistakes

ROCHESTER, N.Y. – On Friday, November 18, Dr. George A. Spratt, scientist, collaborator and valued assistant to the Wright brothers – inventors of the aeroplane – continued his outspoken testimony in the month-long trial of the Herring-Curtiss Co. v. Curtiss lawsuit being prosecuted at the Monroe County Courthouse.

"In 1902, the Wrights more than met their match in Augustus Herring!" Spratt proclaimed. "Herring proved them to be wrong in the most fundamental facets of flight!"

Seizing the opportunity, James M. E. O'Grady, attorney for the plaintiff, drew a roll of laughter from the court's packed visitors' section as he questioned Spratt.

"Specifically Dr. Spratt... where did the Wrights go wrong?"

A single whack from Judge S. Nelson Sawyer's gavel swiftly restored order, allowing the witness the opportunity to detail the Wrights' alleged mistakes.

"Wilbur Wright had the connection between air speed and a wing's lift and drag all wrong. His method for determining the power needed to fly an aeroplane was also wrong!"

"How do you know this to be true?" O'Grady asked.

"Mr. Herring pointed to a number of mistakes Wilbur had made in a lecture to the Western Society of Engineers, the year before."

"Doctor, to the best of your knowledge, was Wilbur Wright in error about anything else?"

"There were many things. For instance, the Wrights never did understand how a curved surface generated lift. We now know, twenty years later, that Herring was right about partial vacuums, and how they created lift at small angles of attack. The Wrights clung to their outmoded impact theory of lift – and that has been proven to be wrong! This faulty information led the brothers to design and employ queerly shaped curved surfaces that were dismal failures in their 1900 and 1901 gliders."

Aviation experts contacted by the Herald have confirmed that Dr. Spratt's testimony goes a long way in establishing Augustus M. Herring as the preeminent American authority on heavier-than-air flight at the turn of the century, although Mr. Curtiss' defense associates have hotly disputed the contention. At stake in this lawsuit are the multi-million dollar assets of the Curtiss aviation empire.

Rochester, New York

Monday, November 21, 1921

To everyone's surprise, after Dr. Spratt had completed his testimony late on Friday afternoon, the defense decided *not* to cross-examine. With the start of a new week, the sight of Augustus Herring sitting at the plaintiff's table in a wheelchair caused a stir within the court's filled-to-capacity visitors' section. Making his usual stormy entrance, Judge Sawyer addressed the plaintiff's star witness. "Welcome back, Mr. Herring! Once again, the court embraces the opportunity to examine your testimony and prays that your health has genuinely improved. Without further delay, counsel for the plaintiff will call its next witness."

Dressed in a new, charcoal-gray three-piece suit that daughter Chloe had selected for him, along with his favorite kelly green bow tie, Herring slowly stood. Carefully placing his cane, he shuffled to the witness chair. All eyes were focused on the thin, ashen-faced figure as Herring placed his left hand on the *Bible* and raised his right, again pledging to *tell the whole truth and nothing but the truth*. Favoring his left side, Herring pivoted in place and sat down somewhat heavily. Paying little attention to what he later described as his "temporary circumstance," Herring gingerly hooked his cane over the oak railing in front of him. With a nod to O'Grady, he was ready to proceed.

"Mr. Herring, I am going to review two statements that you made on Thursday, November 10 – the last day of your previous sworn testimony. You were speaking of the hard times you and your family had experienced after the death of Mr. Arnot – your business partner and aeronautical benefactor. You said, and I quote '…my partner was dead, my business lost, my research destroyed, my inheritances gone, and my aeronautical achievements disparaged by the written and spoken word.' "

"You continued by saying that to survive you worked a variety of odd jobs, including that of a railroad surveyor, carpenter, and handyman. As you said, '…I did anything that I could to earn a living during the latter half of 1901.' Using this difficult period of your life as a starting point, what happened next?"

Herring slowly shook his head. "Early in January of 1902, I swallowed my pride and wrote Mr. Chanute a letter of apology—"

"Apology? You apologized to the man who was attempting to destroy your life?"

"Objection!" Robbins cried. "Argumentative!"

"Sustained," the judge said. "Restate the question without the theatrics, Mr. O'Grady."

"Tell the court about your letter of apology."

"It was a difficult letter to write, but the bills were piling up and I had my family to think about. After explaining my economic predicament, I targeted the old man's soft spot: his legacy. I predicted that the powered aeroplane was about to become a reality, and its success would be a fitting climax and a lasting monument to his long study of the subject. I concluded by saying, 'If you would care to take the matter up again, I would willingly bury all differences as to credit, both past and future, in the production of a jointly invented machine.' "

"What about your differences, such as manual control versus automatic stability?"

"If we were going to work together again, I knew that I would have to fully embrace automatic stability. Unfortunately, I soon learned that all of my good intentions had come to naught. The old man didn't need me... he had hitched his chariot to two new mounts – the Wright brothers."

"Then what did you do?" O'Grady asked.

"When I didn't hear back from him, I concluded that the old man wasn't interested in a reconciliation. Therefore, I tried to interest other well-to-do people in helping to build a practical aeroplane—"

"Whom did you contact?"

"My best bet was the newspaper publisher William Randolph Hearst, an acquaintance from my Lilienthal glider days in New York City... but he was consumed with other projects. Then there was publisher James Gordon Bennett, another flying machine enthusiast. He begged off when he learned about Professor Langley's government-sponsored *Great Aerodrome* project. Bennett claimed he didn't want to play second fiddle to any taxpayer-subsidized boondoggle.

"One of my inquiries did seem to pay off when expatriate Sir Hiram Maxim, inventor of the machine gun, paid my way to England to explain my vision for a powered machine. He was impressed, and we came to an agreement. Maxim was to raise $100,000, and I was to receive a salary of between $300 and $500 per month, plus a one-sixth interest in the finished product."

"Then what happened?"

"Lady Maxim queered the deal! She said that Hiram had already squandered one fortune on flying machines and she wasn't going to let him do it again... and that was that!"[1]

"What did you do next?"

"By the third week of May, I was back working for the railroad during the day and writing more letters at night. Depressed and discouraged, I again wrote to Chanute, this time asking for work. To my relief, he directed me to estimate the cost of rebuilding his *Katydid* glider."

"What did you come up with?"

"A detailed breakdown of the tasks that needed to be performed, along with the hours and materials required—"

"What did the total come to?" O'Grady interrupted.

"Four hundred hours of work at 34 cents an hour... about $184. There was also the question of materials—"

"Did he accept your offer?" O'Grady asked, again cutting him off.

"He countered with $150, which I accepted... I needed the work, and he would supply the old *Katydid* frame and the materials. In passing, Chanute mentioned that he needed the rebuilt machine for some summertime experiments in North Carolina."

"Is that how you learned about the Wrights?"

Appearing uncomfortable, Gus shifted in his chair. "Actually, I had received a letter from Ed Huffaker six months earlier. He told me about the Wrights' abortive '01 trials, and tattled about the nasty things said about me. He also included a copy of Wilbur's lecture to the Western Society of Engineers... a very enlightening document!"

"With all that you learned," O'Grady pressed, "why would you still pursue a relationship with Chanute?"

"Besides the much-needed pay envelope, the old man kept me abreast of the latest developments in the world of aerial navigation."

O'Grady moved on. "What did you change on the old *Katydid* glider?"

"The size of the wings, and I made the machine lighter – much lighter. When Chanute visited my shop in July of '02, he approved my changes and praised the workmanship."

"What did the two of you talk about?"

"Chanute informed me that Mr. Lamson, a Californian, was constructing a *second* glider for him. It was a triple-decker with fore and aft rocking wings for automatic stability. Then the old man questioned the availability of Matthias Arnot's *Double Decker* glider from '97... said he wanted a *grand fly-off* between 'his gliders' and the Wrights' new machine; after the competition, he planned to *give* all three machines to the brothers so, in his words 'they could further observe and learn from their operation!' "

"What did you tell Mr. Chanute about the old Arnot glider?"

"That it was in need of some minor repairs. When he asked if I would consider selling the machine, I said that I would have to contact the Arnot family, since it was their property. I didn't tell him that the machine had been sent to Elmira, shortly before the Truscott fire."

"Why were you stringing the old man along?"

"I wanted him to bankroll the construction of a replica *Double Decker* glider. Unfortunately, he didn't think there was enough time to complete the job."

"Was anything else discussed?" O'Grady asked.

"Before leaving for Chicago, Chanute asked if I would be interested in accompanying him to Kill Devil Hills in October. He needed me to serve as his 'expert,' someone who would not only set up the machines but also fly them. When I consented, he stopped short of offering me a verbal commitment, saying that he would have to check with the Wrights, since we would both be guests at their camp."

With that bit of information, Judge Sawyer called for a break in the testimony. It was time for lunch.

*

Despite the best intentions of William Maloney, Chloe took charge of pushing her father's wheelchair out of the courthouse. However, when they approached the curb, the big Irishman stepped in and lifted the apparatus down to street level. Once they had crossed Main Street, the process was reversed, and Chloe was in charge again until they reached the sanctuary of the restaurant.

After placing their lunch orders, Maloney asked Herring why he hadn't indulged in his customary glass of whole milk. "I'm not allowed to drink the damn stuff anymore! The doctors say there's too much fat in milk – and it probably contributed to my apoplexy."

"Other than being weak, how do you feel, Gus?" O'Grady asked.

"I'm tired, but I'm glad to be back! I would have gone crazy had it not been for Chloe filling me in every day. I'm sure you saw her sitting in court taking notes."

At that point in the conversation, Chloe excused herself and retired to the ladies' room.

"What did you think of Huffaker's testimony?" O'Grady asked.

"The Wrights treated him shabbily," Herring said. "They should have been ashamed... and why didn't Chanute come to his rescue? In addition, I think that Huffaker's allusion to the brothers' masculinity and possible questionable behavior with regard to their sister was perceptive – I know that I felt jittery around both of them – especially Orville."

"What did you think about Spratt's testimony?" Maloney chimed in.

Herring shifted in his chair. "Back in '02, during my 10-day visit to the Wrights' camp, I honestly didn't associate Spratt's method of analyzing the performance of a curved surface with the success of the brothers' glider. It may have taken him several years, but the good doctor finally put two and two together... the Wrights had stolen both his method and the design for his tangential balance.

"However, Huffaker had the brothers figured out. As a result, he preached to them relentlessly about the importance of maintaining a high level of personal character – a big reason why the Wrights hated him. A decade later, Doc Spratt confirmed Huffaker's suspicions when the brothers sued our company for the use of interplane control surfaces."[2]

"Are you aware that Orville offered to testify for Curtiss after reading about Spratt's testimony in the Dayton newspaper?" Maloney said.

Herring grinned. "The Wrights never could handle the truth... that will become obvious in my upcoming testimony."

"That reminds me, Gus," O'Grady said, "how did the Wrights react to Chanute's request for you to be his 'expert' back in '02?"

Polishing his teaspoon with the corner of his cotton napkin, Herring hiked an eyebrow. "After I returned home from the Kill Devil Hills trials, there was a letter waiting for me – it was from Spratt. Among other tidbits of information, the good doctor informed me that Wilbur had been upset at the possibility that I would become Chanute's 'expert' at the trials. Wil then informed the old man that he and Orv preferred Bill Avery to me. According to Spratt, Wilbur said that '...Herring has a jealous disposition and claimed more credit than he deserved.' "

"Why, then, didn't Chanute bring Avery to the trials?" Maloney asked.

"Years later, Bill told me that Chanute had inquired about his availability, but he couldn't take time off from his booming electrical contracting business."

"Sounds like you were everybody's second choice," Maloney laughed.

At that moment, their waiter reappeared with their lunches, with Chloe close behind. Herring had ordered grilled blue pike from nearby Lake Ontario, with a small chef salad and a glass of sweet apple cider. As she helped to carve his filleted fish into smaller portions, Chloe could no longer hold her tongue. "As this trial moves along, I find myself more and more disgusted with Mr. Chanute's campaign to slander Daddy. His contrived, malicious conduct has been most shameful! The personal letters, the newspaper reports, the magazine articles and the word-of-mouth notions that he planted – all in the name of shameless self-promotion!

"The 1902 Kill Devil Hills excursion is a good example. Although Daddy had never met the Wright brothers or Dr. Spratt, they all had a negative impression of him – all because of Chanute's backstabbing. If that reprehensible old man were alive today, I'd walk right up to him and slap his face!"

Fits of laughter erupted from their table.

Kill Devil Hills, North Carolina

Sunday, October 5, 1902

Chanute and Herring arrived at the Kill Devil Hills campsite amid a steady downpour. Their six-mile voyage by spritsail boat from the town of Manteo, on Roanoke Island, had taken almost two hours. Knowing their approximate time of arrival, Wilbur had dispatched Dan Tate and his donkey cart to carry their guests and their luggage from Baum's landing on Roanoke Sound to their campsite a mile and a half to the north.

Retreating to the protection of the shed, Chanute introduced Herring to Wilbur, Orville, and Dr. Spratt. Wilbur then presented their older brother. "Although Lorin is here on vacation, he has graciously agreed to act as our photographer."

Anticipating their arrival, Wilbur and Orville had prepared sandwiches and a large pot of hot coffee. As the men sat to eat, Wil led with a prayer: "Bless us, Oh Lord, and these thy gifts, which we are about to receive, from thy bounty, through Christ, Our Lord. Amen."

Having eaten their fill, the men were sipping their coffee and talking quietly when Wilbur abruptly interrupted. "Gentlemen, Orville and I have few rules here, except for one... Sundays are to be used exclusively for reading, preferably the *Bible*, and for rest. There will be no exception to 'God's order' tolerated at *this* camp."

(Later, when Dr. Spratt and Herring were alone on West Kill Devil Hill, Spratt enlightened the newcomer about the reason for Wilbur's statement: Chanute's stories of Herring's *Sunday* flying sessions in New York City and his *Sunday* repair sessions at the Indiana dunes.)

With that bit of nasty goings-on out of the way, Wilbur changed the topic. "Gentlemen, tell us about your trip."

Chanute readily obliged. "Other than a rough ride aboard the spritsail, the trip was actually quite pleasant! We left Chicago on Thursday at 1 o'clock in the afternoon, on the Big Four railroad that connects with the Chesapeake & Ohio, in Cincinnati. That evening we took advantage of our Pullman car reservations and slept on the train. Friday's daylight hours were spent taking in the scenery as we railed toward the Atlantic; we arrived in Norfolk at 7 o'clock that evening. After a satisfying supper at a downtown restaurant, we retired to one of the city's finest hotels – the Calumet. Early Saturday morning, we continued our journey by railing the short 30 miles to Elizabeth City, where we purchased supplies and booked passage aboard a steamer headed for Roanoke Island. Upon arriving there, we commandeered a hack to carry our goods and ourselves to the Manteo Hotel, where we endured somewhat less than first-class food and accommodations. This morning, Mr. Tate and the little sailboat did the rest... three days later, here we are!"

"I hope that you took our advice and procured heavy canvas or 16-ounce sailcloth while you were in Elizabeth City," Orville said. "You'll need it for your cots—"

"Yes," Herring interrupted, "but the sailcloth and blankets are waiting for us at Baum's. We couldn't carry everything."

"That's quite all right, Mr. Herring," Wilbur said. "It's stopped raining. Why don't we walk down there and retrieve the rest of your gear? Besides, it'll give us a chance to get acquainted. Later, if you decide to sleep in the second story as we do, suitable cots can be prepared quickly."

As the men, both 35 years old, traipsed off to the south, Chanute peered after them with trepidation, thinking... *I don't trust that bastard to be alone with Wilbur.*

Walking in silence, they faced the specter of Kill Devil Hills looming straight ahead.

"This must be the experiment site," the wide-eyed Herring said.

"When we get a little closer," Wilbur promised, "you will see there are actually three hills. Each has good and bad features. Although the West Hill is the second highest, we don't use it very often… only when the wind blows out of the west. We prefer the Little Hill or the Big Hill; they're best suited for our trials. How were the Indiana dunes for gliding?"

Gus shrugged his shoulders. "Unlike this place, the wind direction was critical. It was best when it blew out of the northwest, directly off Lake Michigan. Unfortunately, we often had to wait days for that condition to materialize. Then there were the storms! Did Mr. Chanute tell you about the catastrophe of '96?"

Wilbur nodded as they trudged past the 60-foot-high West Hill.

"Have you ever heard of a nor'easter, Mr. Herring?"

"Certainly! My family was originally from Conyers, Georgia, a town named after my grandfather. I know all too well the destructive power associated with hurricane-force winds!"

"Let us pray that we do not encounter any nor'easters while you're here, lest there are no machines *left* to fly afterwards!"

"I'm curious, Mr. Wright, how did you become acquainted with our friend Chanute?"

Wilbur hesitated, deciding how much he should divulge. "Back in '99, I wrote to Mr. Chanute, asking for information about various facets of aerial navigation. After he accommodated my request, we continued to exchange letters. A year later, my brother and I invited him and two of his assistants to last year's trials. Afterward, he insisted upon helping with our wind tunnel calculations, and now… here he is again."

Wilbur's deportment changed. "Originally, Mr. Chanute wanted us to operate his two old gliders, but we thought otherwise, instead suggesting that he supply his own 'expert' for that task. Now, Mr. Herring, it's your job to manage his machines. I can't begin to tell you what a relief it is to Orv and me that—"

Herring interrupted. "Why didn't you simply tell him to leave his machines in Chicago? After all... it's your camp."

"Listen, Herring... how do you tell the *Father of Aviation* that we prefer having our experiment time to ourselves? And that's not all. In addition to the nuisance, Mr. Chanute seems oblivious to the workings of our machines, and no amount of explaining seems to help. Orv and I think he's a bit senile."

Wilbur quickly recovered his poise. "I have a pointed question for you, Mr. Herring... what is the *real* reason for you being here?"

Kicking up a shower of sand in front of him, Herring, taken aback by Wilbur's frankness, decided he had nothing to lose by being honest. "I don't know what the old man told you, but I'm only here for the *money* he's paying me to set up and demonstrate these machines... I have a family to feed back in St. Joseph, and times have been difficult."

Herring's eyes were fixed on Wilbur's. "Now that we're clear as to our intents and purposes, Mr. Wright, I have some hard-earned *advice* for you and your brother... that is, if you're interested."

"Advice?"

"It concerns your future in the field of aerial navigation."

Stopping, Wilbur glanced behind him, as if someone were following. When Herring also paused, the two came face to face.

"Advice is like opinion," Wilbur said, "It's usually not worth the time it takes for the telling. However, because we are being so candid, I'll try to listen. What do you have to say, Mr. Herring?"

Turning back toward the direction of Baum's, Herring began to pick up the pace as he gathered his thoughts. *Should I tell him the* truth *about Chanute? No one took the time to advise me when it counted!*

"Please listen carefully, Mr. Wright; I'll say this but once.... if you become indebted to Mr. Chanute, *your* ideas will become *his* ideas, which will then be used to enhance his—"

"*Indebted?*" Wilbur interrupted. "What *exactly* do you mean?"

"If you accept support from Mr. Chanute, whether it is money to help with your experiments, the gift of his obsolete gliders, or even assistance with your calculations... the old man will be keeping score!"

"Are you suggesting that he will somehow use his generosity against us?"

As if to purge his mouth of a sour taste, Gus paused to spit on the sand. "Chanute will spread the word of your successes in his speeches, magazine articles, and newspaper interviews. As he continues to report about you and Orville over the months and years, the old man will insert *himself* into the story, to be perceived as a confidante who had actually participated in your work—"

"Mr. Herring," Wilbur broke in, "that's a far-fetched analysis if I ever—"

"Do you think so?" Gus interrupted. "Well, Mr. Wright, you have heard but half of my advice! As you continue your association with the 'great man,' little by little, people will come to view him as your *mentor*, while the two of you will be regarded as his subservient *assistants*. As a final offense to your self-respect, Chanute will make *no* effort to correct these impressions!"

"Surely you are exaggerating!"

"Hear me out, Wilbur... this old man is very, very clever. Mr. Chanute may have you convinced that he doesn't understand the operation of your machinery, but I'll guarantee that he has made detailed sketches and notes concerning everything that you and your brother have ever shown or discussed with him."

"Are you accusing him of stealing our ideas?"

Herring laughed. "Not directly! Octave doesn't need to pilfer your ideas... he only needs to *sound* knowledgeable in order to snatch some of the credit for developing a practical flying machine! Chanute has all of the money he needs – what he really *wants* is the glory!"

"Give me an example of his so-called cleverness."

Herring stopped in place, taking time to remove his captain's cap and run his fingers through his hair. "Two examples come to mind. Back in '95, shortly after I moved to Chicago from my position at the Smithsonian, I wrote an article entitled 'Dynamic Flight.' Mr. Chanute, my new boss, encouraged me to submit the manuscript to James Means for possible inclusion in the 1896 *Aeronautical Annual*. Means praised the content, but the article was almost not published. Upon reviewing the proofs that Means had sent to me, I discovered that the drawings and description of my 1892 two-surface rubber-powered model had been *deleted*. I was told this was done to meet the issue's space requirements. Because it was my most successful flying model design, I was disappointed and resisted its being edited out of my article... but the allure of earning a five-dollar author's fee and seeing my name in print caused me to relent. Then in the fall of '97, at Chanute's invitation, Means traveled to the Indiana dunes to watch my manned glider flights. After considerable prodding on my part, he admitted that *both* Langley and Chanute had requested that he cut the two-surface design from my article! When I pressed for details, Means confessed that both men had voiced concerns as to the model's origin and designer—"

"What is the point?" Wilbur cut in.

"Chanute had led me to believe that Professor Langley was behind the *Aeronautical Annual* debacle... but Mr. Means convinced me otherwise. Here's my argument: if the two-surface model had appeared in the '96 *Annual*, it would have established me as its sole designer, and Chanute would not have been able to claim the design of the manned glider of 1896 as his own... as he now does with impunity. As a matter of fact, *his* man-carrying *Double Decker* is nothing more than a scaled-up version of my 1892 flying model – and that includes the wing trussing!"

Wilbur thought, *A good reason for Orville and me to keep our experiments to ourselves.*

Wilbur said, "Mr. Chanute told us that the *Double Decker* glider was *his* design and that your only role was to make the drawing according to his specifications."

"I rest my case!" Herring said before continuing, "My second example of Chanute's cunning involved the photographs he volunteered to take of the late Matthias Arnot's *Double Decker* glider back in the summer of '97. To make a long story short, the old man confiscated the photographic plates and used the prints to illustrate his many published articles, claiming that the images were of *his* 1896 machine, which, of course, they weren't."

They were almost to Baum's before Wilbur spoke again. "Why did you stop experimenting after those powered hops back in the fall of '98?"

"Didn't Chanute tell you? All of my experimental machines, engines and tooling were lost in a horrendous fire! Then my partner died, and I was forced to liquidate my motor-bicycle and motorcycle business!"

Wilbur grimaced. "What are you working on now?"

Herring's disposition brightened with the change of subject. "I've been spending most of my free time building a small gasoline engine for a model flying machine. It's about half done, but I'm having trouble making some of the castings – that new aluminum alloy is a *bitch* to work with. If the engine shows the same relative power for its size as my bicycle engine, it should give between 0.25 and 0.32 brake horsepower, and its weight will be something less than two pounds with accessories and supplies for the short run."

For the next several minutes the men engaged in a spirited discussion related to the lack of suitable lightweight engines for flying machines and the problems associated with their design and construction. Then, as suddenly as their conversation started, it ended, not to be rekindled for the remainder of their walk to Baum's place.

On the way back, with each man carrying a load of bulky but relatively lightweight blankets and sailcloth, Wilbur asked another question. "What does Mr. Chanute think about your fondness for dynamic flight?"

Herring couldn't help but laugh. "Chanute believes it was premature for me to have tried manned, powered flight back in '98. I'm certain you know... he's a loud and persistent proponent of automatic stability. On the other hand, like Lilienthal and presumably yourself, I have always gravitated toward operator-controlled flight. That said, I must admit that my tail 'regulator' mechanism does provide a surprising degree of automatic stability."

"Speaking of automatic stability," Wilbur said, "the crate containing Mr. Chanute's multi-wing arrived more than a week ago, but there has been no word concerning the Lamson machine."

"Maybe we'll both get lucky and the goddamn thing will get lost in transit."

Wilbur stopped in his tracks. "Mr. Herring... would you mind *not* using profanity in my presence? My siblings and I were raised in a Christian home where my father, a bishop in the Church of the United Brethren in Christ, did not tolerate any form of blasphemy or cursing."

Gus shrugged... it was no skin off his nose.

"With that out of the way," Wilbur continued, "I suggest that after you and Mr. Chanute have tended to your sleeping arrangements, the two of you must take a look at our new machine. Orv and I welcome the opportunity to answer any of your questions."

CHAPTER 75

Kill Devil Hills, North Carolina

Sunday, October 5, 1902

After the evening cleanup, the men gathered around the Wrights' most recent – and somewhat battered – glider. When Wilbur had finished explaining the basic operation of their control system, including some of the problems they had experienced leading up to Orville's crash, he paused to answer questions.

"Let's see if I have this straight," Herring said. "With last year's glider, as you activated the wing-warping mechanism to raise a downward-tilted surface, the machine began an unexpected rotation around its vertical axis in the direction of that surface. Although the surface initially began to rise... the warping controls seemed to reverse, causing the wing to drop even lower than before."

Wilbur nodded.

"At the conclusion of last year's trials, did you understand why the wing-warping system wasn't working properly?"

Wilbur shook his head. "Not then... but we knew something had to be corrected. After all, our sole objective has been to fly a *straight path* down the hill and to bring the machine back to level when tilted by the wind."

"I can see where attempting a *turn* would be risky," Herring observed.

"When our intended direction is disrupted by a gust," Orville added, "the glide most often ends in a spinning turn around the low wing, or a side-hill collision!"

"For this new glider," Herring continued, "you added a stationary, two-surface vertical rudder to help nullify the unwanted rotation. Although its weathervane-like surfaces were intended to keep the machine headed straight down the hill... you say they were ineffective?"

Wilbur nodded again.

"Gentlemen, gentlemen," Chanute interrupted, "the answer is as plain as the nose on my face... it's called automatic stability! All of these controls are unnecessary when aerodynamic forces are allowed to make the requisite corrections; besides, automatic control is safer! If Otto Lilienthal and Percy Pilcher were alive, I'm sure they would agree with me!"

Politely waiting for Chanute to finish, Herring addressed Orville. "Question, Mr. Wright: have you tried flying any of your machines as a kite?"

"Of course," Orville responded. "At one time or another, all of them have been flown in a ballasted, tethered condition."

"Did your wing warping produce the desired results?"

"Always!" Wilbur chimed in. "It has always worked flawlessly when actuated from the ground."

"Did you notice any difference in *pull* between the left and right tether lines while the warping mechanism was being activated?"

Wilbur hesitated. "We use an upper and lower tether line for both the port and starboard wingtips. Because the drift force is significant at high angles of attack for a machine this large, Orv takes one side and I, the other. In that way, we can coordinate the warping and still restrain the apparatus."

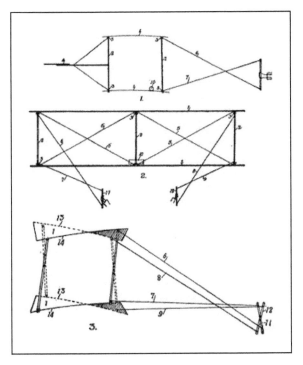

Wright kite showing tethered wing-warping and trailing horizontal surface (1899); Library of Congress

"Therefore," Herring continued, "you didn't notice the difference in pull between the port and starboard tether lines during warping... and I suspect that you didn't use spring scales to measure the—"

"There was no reason!" Orville interrupted.

"To the contrary," Herring admonished, "the pull will *always* be greatest on the tether line that controls the downward twisted wingtip."

There was silence for several seconds before Wilbur spoke. "Pray tell, why is that?"

Herring waved his hand in a dismissive manner. "You said so yourself... the downward twisted wingtip is flying at a higher angle of attack than its counterpart. Its *drift* and, therefore, the *pull* will always be greater on that side."

The brothers glanced at each other.

"How did you come to that conclusion?" Wilbur inquired.

Gus began his explanation. "I experienced this peculiar phenomenon almost a decade ago, when testing my Lilienthal-type glider with two movable auxiliary rudders that I had installed on the port and starboard wings."[1,2]

"Well," Orville broke in, "perhaps you should enlighten us as to the remedy for this drift imbalance!"

Sensing hostility in Orville's voice, Gus considered dropping the matter altogether and turning in for the evening. "I'm not certain that you need or want enlightenment, Mr. Wright. Even an amateur can see that you are in the process of making changes that are intended to alleviate the—"

"Try not to be offended by my brother," Wilbur interrupted. "He thinks that our solutions are the only ones worth considering. Nevertheless, I am intrigued by your analysis. Please, continue."

"If you wish," Herring said. "I'll describe the controls that I used on my '94 Lilienthal-type glider."

"Please do."

"The two auxiliary rudders, hinge-mounted to the *front edge* of the port and starboard lifting surfaces – one on either side of the operator – moved up and down in opposite directions to preserve the machine's roll equilibrium when tilted by a gust. The rear vertical rudder counteracted the yawing tendency about the machine's vertical axis, caused by the inherent drift imbalance between the auxiliary surfaces as they operate at various angles of attack and airspeed—"

"How did these controls work together?" Wilbur asked, cutting him short.

"I found it necessary to use the vertical rudder to preserve a straight course when I actuated the auxiliary surfaces. I suspect that your wing-warping system will also require the coordinated application of vertical rudder."

"Where did you learn these things?" Wilbur asked.

Herring hesitated before answering. "I take no pleasure in offending you or your brother, Mr. Wright, but you may not welcome some of the things I have to say."

"Don't concern yourself, Augustus... I have a thick skin."

"My friends call me Gus or A.M. – either will do! Let's begin with your speech to the Western Society of Engineers, back in September of last year, where you alleged that an aeroplane would fly more than *twice* as fast on only *twice* the power. In reality, doubling an aeroplane's airspeed will result in almost four times the drift – it's a *square* function. This calls for four times the thrust and eight times the motive power to overcome the drift – it's a *cube* function. The power required to fly that fast can be obtained by multiplying the thrust in pounds times the flight speed in feet-per-second. These concepts were all verified in my 1898 manned kite experiments at St. Joseph's Silver Beach."

"For the sake of argument," Wilbur offered, "let's say that you are correct. What does this have to do with our side-to-side equilibrium problem?"

"No offense intended," Gus continued, "but your lack of experience in dealing with well-established engineering principles seems to have hampered your ability to deal with this problem."

Unconvinced, Wilbur frowned as Herring walked over to their glider's starboard wingtip, where he used his hands to simulate the twisting motion of what the brothers called their "end control."

"This is what I'm referring to, Mr. Wright: within certain angular limits, a wing's lift will increase *directly* with its angle of attack; double the angle, double the lift. The wing's drift, however, will increase as the *square* of its angle of attack; double the angle, *quadruple* the drift. Therefore, when your 'end control' warps the starboard-side wing downward at its rear, increased lift will *initially* cause the surface to rise. In other words, the wing will begin to roll upward, on the machine's longitudinal axis. However, an instant later, the stronger drift force takes over, slowing the starboard wing's advance and causing the machine to yaw about its vertical axis. As you might guess, this slowing reduces the surface's lift and halts its upward roll. At this point, the operator might notice that the controls aren't working properly—"

"Get to the point!" Orville demanded.

"The point is... the *drift force* dominates the *lift force* at all but the most shallow angles of attack. But, that's only part of the story!"

Gus was in his element. "As the machine begins to yaw, the *airspeed* of the portside wing begins to exceed that of the starboard-side wing... which brings into play another important relationship—"

Waving his hand, Wilbur called for a halt to Herring's monologue. "How does knowing about all of these relationships make the lateral control problem any easier to solve?"

Herring extended his hands, palms up. "It has to do with understanding the *magnitude* of the forces involved. Perhaps an example will help... picture an aeroplane flying straight ahead at 40 miles per hour. If you increase the thrust until the speed has *doubled* to 80 miles per hour, the wing will generate four times the lift and drift, a *square* relationship for them both – two squared equals four. For the time being, we're going to concentrate our attention on lift. If the thrust is reduced to a point where the machine slows to 20 miles per hour – half its original speed – the lift is only one quarter of what it was at 40 miles per hour; one half squared equals one quarter. As you can see, *lift* is very sensitive to small changes in airspeed."

"Go on," Wilbur said.

"Here's how *lift* and *airspeed* apply to the control reversal problem: if your 32-foot-wingspan glider has an airspeed of 25 miles per hour, and a drift imbalance due to wing warping allows it to yaw around the vertical axis at 30 degrees per second... the higher, portside wing will travel about 50 percent faster than the lower, starboard-side wing! If you square this ratio and allow the faster portside wing only *half* the angle of attack of the slower starboard-side wing, the higher wing will still develop about 20 percent more lift, and the machine will begin rolling toward the lower wing. Surprise, surprise... the controls will act as if they were *strongly reversed*!"

After an uncomfortable period of silence, Orville spoke up. "That's absurd! There couldn't be that much difference in lift between the yawing wings!"

"It can be proven quantitatively," Gus said, "but I don't want to get caught up in a numbers game. For the sake of discussion, I have been speaking from a qualitative perspective. With the engineering concepts related to the problem identified, it's obvious that your vertical vanes couldn't generate *enough* force, *fast* enough, to restrain the machine's yaw about its vertical axis during warping."

Again there was silence.

"So... what's the solution?" Wilbur questioned, extending an upraised palm toward Herring.

"Any number of things. You might eliminate the stationary rudders and make one surface with *twice* the effective area. If

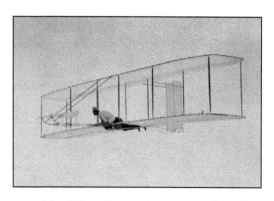

Wright glider with dual fixed rudders (1902); 1903 Journal of the Western Society of Engineers

"that doesn't hold the machine against yawing, I would consider moving the rudder to a position farther behind the wing, giving it more purchase against the rotation."

"And if those fixes don't work?" Orville scoffed.

"Then I'd try what you're doing now – I'd make the rudder movable—"

"I'm curious," Wilbur broke in, "Why do you think a movable rudder will work, as opposed to your earlier suggestions?"

Gus turned toward the disabled flying machine. "I just explained how my '94 Lilienthal machine worked! You asked for my opinion, and I have tried to explain the scientific and engineering concepts involved. You may choose not to believe what I say, but you're *not* the first to have delved into the intricacies of aerodynamic roll control—"

"Is that so?" Orville interrupted. "Is that why you still advocate weight shifting to control your machines... because it's a *better* method?"

After the ensuing laughter had subsided, Chanute decided it was time to add his two cents to the discussion. "Gentlemen, all this bickering is very entertaining; however, I must agree with Mr. Herring on one point – wing warping is nothing new. Louis-Pierre Mouillard, the French engineer who studied the flight of buzzards in the Egyptian desert back in the '70s, later patented a wing-warping roll-control system. As a matter of fact, I have a copy of the patent in my files."

Staring down at his high-top shoes, Orville offered no further rebuttal.

Seemingly a world away, Herring paid little attention to Chanute's ramblings; he was busy formulating a response to Orville's cynicism. "Mr. Wright, like Lilienthal, I knew that weight-shifting control would allow me to get into the air quickly and provide the opportunity to learn the idiosyncrasies of the wind and its gusts. While I was accumulating more than 2,000 flights – almost 11 hours in the air – the art of weight shifting gave me the opportunity to learn and practice without the constraints of counterintuitive systems of control, such as your end-control method. In my opinion, the *only* limitation attributed to weight shifting is the machine's wingspan. Through trial and error, I have found that the span should not exceed 20 feet."

"Wonderful!" Orville shouted. "But a practical machine will require much larger wings to carry the additional weight of a motor."

"Lilienthal found a way to overcome the wingspan limitation, Mr. Wright... it's called wing stacking. By the way, it seems as though you have taken full advantage of a particularly successful *Double Decker* machine from the not-so-distant past. When compared, I find that your cambered, trussed wings have the same construction, shape, and spacing of certain machines previously flown at the Indiana dunes—"

"Come, come, Augustus," Chanute scolded, "certainly you're not suggesting that *you* originated the *Double Decker's* layout? Are you forgetting that it was invented by Wenham and further developed by Stringfellow and others... *myself* included? Historically speaking, your two-surface design held no more promise of *practical* flight than a teakettle held promise of becoming a steam locomotive!"

After another round of laughter at Gus's expense, a grinning Wilbur urged Herring to continue his analysis. "All jesting aside, Mr. Herring, repeat your reasons for making the rudder movable."

Stung by Chanute's latest barb, Gus struggled to concentrate on Wilbur's request. "The idea is to introduce a restraining torsional force *before* the downward warped wing has had a chance to slow down. Actually, it's all about controlling various aspects of the machine's *inertia*!"

"Fact of the matter is," Wilbur said, "we believe that the stationary vertical vane has made the problem worse. As the machine tilts toward the low, downward warped wing, it also begins to *slip* sideways in that direction, creating a *flank* wind that pushes on the rudders, making the rotation worse. This combination of events result in the low wing auguring into the sand – we call it *well digging*. Then, two nights ago, Orv had a brainstorm. He envisioned a *movable* vertical rudder that could be controlled by the operator in opposing the yaw—"

"Then Wil came up with an improvement!" Orville chimed in. "Yesterday, he suggested we mechanically link the rudder's action to that of the warping mechanism. When the wings are warped, the rudder immediately turns in the proper direction to oppose the drift-induced yaw!"

"Sounds reasonable to me," Gus agreed. "These changes should produce a *straight-line* course down the hill."

*

Sensing that the topic of lateral stability had run its course, Dr. Spratt, who had been silent until then, decided to join the conversation. "Mr. Herring, I have a question concerning an aeroplane's power requirement. You stated that to double the speed of an apparatus, eight times the power would be required. As I'm sure you are aware, Professor Langley still holds fast to his claim that as an aeroplane's speed increases, the amount of power needed to fly... decreases."

Gus slowly shook his head. "It's clear that the *special circumstance* of very slow flight has clouded the Professor's vision. Like the unequal lift produced by yawing wings in our lateral control discussion, Langley has convinced himself that this circumstance is common at *all* airspeeds—"

"What 'special circumstance' are you referring to?" Chanute interrupted.

Knowing that Octave would object to his portrayal of *Langley's Law*, Herring prepared to defend his position.

"Almost a decade ago I attended *your* International Conference on Aerial Navigation, in Chicago – a meeting that was sponsored by the 1893 Columbian Exposition. When Professor Langley finished reading his treatise on 'The Economy of Speed', he solicited questions and rebuttals from the audience. It was then that I provided an alternative to his conclusion that *higher aeroplane speeds were more economical of power than at lower speeds—*"

"Perhaps you should explain your *motives* for opposing the Professor's theory," Chanute interrupted.

"Motives?" Gus replied, raising an eyebrow. "Unlike your unconditional endorsement of Langley's Law in *McClure's* last year, I have simply rejected the contention that getting something for nothing is sound science."

Clearly annoyed, Chanute grumbled, "Just what is *so* important about this so-called special circumstance?"

At this point, Gus plunged into another detailed technical explanation, similar to his rebuttal in Chicago. When he finished there was a strained silence before Chanute spoke up. "If what you say is true, why wouldn't this reduction of power happen at all other doubling speeds – such as from 20 to 40 miles per hour?"

"I can answer that," Dr. Spratt broke in. "At higher airspeeds, the forces of lift and drift are no longer feeble; they become much more significant when exposed to square and cube factors."

Gus beamed. "You've got it, Doctor! Langley's Law only applies to the special circumstance of *very low airspeeds*."

"Say what you will," Chanute opined, "but the good Professor has contributed more to the knowledge base of aeronautics than any other experimenter of our—"

Disagreeing, Gus cut the old man short. "Unless you disagreed with him, like Lilienthal did. Langley stonewalled the *Smithsonian Press* when they wanted to translate and print an English version of Otto's book, *Bird Flight, the Basis of the Flying Art*. Although he quashed the proposal, Langley had a dozen chapters of the manuscript translated for his *personal* study!"

Taking a moment to blow his nose, Chanute seemed unwilling to continue the quarrel.

"I find it interesting," Spratt continued, "that Langley says an aeroplane's power requirements diminish with speed, while Wilbur maintains that it increases linearly, and you suggest that power will increase exponentially!"

"The exponential increase in power idea didn't originate with me," Gus was quick to confirm. "Since I'm fluent in German, I first read about this phenomenon in Lilienthal's book. Do you think that Langley had something to protect by not authorizing a translation and reprinting of Otto's manuscript?"

*

"There's one more thing, Mr. Herring," Spratt persisted. "During one of last year's discussions, Mr. Huffaker mentioned that you are a proponent of the partial vacuum theory of lift for a curved surface. Is that correct?"

"Absolutely! In his book, Lilienthal hinted that there might be other *influences* at work – some unidentified condition that acted on curved surfaces, allowing them to perform more efficiently than a traditional flat plane. To investigate his hypothesis, I used my whirling arm device, aero*curve*-type kites and an elementary wind tunnel to test various shapes and camber depths."

"How did you learn that the partial vacuum was Lilienthal's 'unidentified condition'?" Spratt asked.

"Because an aero*curve's* lift-to-drift ratio was so much better than that of a flat plane at shallow angles of attack, I knew there had to be something more going on than a reaction to air particles striking their lower surfaces. The breakthrough came when I directed steam from a teakettle over an aero*curve* model that I had mounted in my wind tunnel. Before the steam fogged the viewing window, I was able to see the unmistakable Bernoulli flow patterns that formed over the top of the surface. Several years later, Mr. Chanute informed me that the Englishman, Horatio Phillips, had also used steam in a wind tunnel of his own design but *not* for the purpose of observing flow over a surface. Although Phillips identified the superiority of the curved surface, his explanation for why it worked better than a flat plane was awkward."[3]

The others had been listening in silence to Spratt and Herring's aero*curve* dialog, when Wilbur suddenly spoke up. "Everyone agrees that the cambered surface works better than the flat plane, Mr. Herring. But why does the cambered section's center of pressure move suddenly to the rear at shallow angles? We certainly don't want to repeat Lilienthal's lethal nosedive."

Gus corrected Wilbur. "Otto's standard monoplane didn't nosedive into the ground – it *slipped* sideways to its destruction – a severe example of your well-digging experience. But that's a discussion for another time. At low angles of attack, the aero-*curve* allows air to flow in a smooth, arched path over its topside, which eventually attaches near its rear edge. To arrive at the same time as air passing beneath the surface, the curving air on top must *speed up*. According to Bernoulli, this higher velocity air generates reduced pressure... a partial vacuum that we call *lift*. The center of pressure rushes rearward when the curved airflow attaches to the back of the surface."

The brothers and Spratt all squinted simultaneously. Such talk made them uncomfortable.

"We believe that a cambered section allows part of its airflow to impact on the front of its upper surface, thereby keeping it from flipping over, but it also contributes to the pressure reversal at shallow angles!" Wilbur insisted. "Furthermore, no amount of theorizing on your part will convince us that lift is anything more than air particles impacting the bottom of any surface, forcing it to function on a principle similar to that of an inclined plane."

"Is that so?" Herring chuckled. "Tell me, how well did your hump-nose aero*curve* work on last year's glider?"

Wilbur cocked his head and grimaced. "Not so well... that's why we spent last winter testing dozens of cambered sections in our new wind tunnel."

"Without an understanding of *how* or *why* they work!" Herring cackled.

"I wouldn't be so cocksure, Mr. Herring," Orville spat. "There's no proof that curved surfaces work the way you—"

"Thanks for the warning, Orv," Herring cut in, "but I'll take my chances. In the meantime, you might occasionally consider other points of view. You know the old saying... 'It's a wise man who listens.' "

Ignoring his brother's growing agitation with their brash guest, Wilbur pressed on. "Don't you think that air impacting the bottom of a cambered surface produces lift?"

"Certainly... but most aero*curve* surfaces produce their best lift-to-drift ratios at angles less than about 20 degrees, and that's when upper airflow attaches, the partial vacuum is formed, and the center of pressure rushes to the rear. I believe that an aero-*curve* produces its lift from a *combination* of air particle impact on the bottom and partial vacuum lift on the top – just as Bernoulli had predicted!"

At that moment, Chanute pulled out his pocket watch and flipped open its solid-gold cover. "It's after midnight! Time to turn in gentlemen; there's flying-machine work to be done tomorrow!"

CHAPTER 76

Rochester, New York

Tuesday, November 22, 1921

Following yet another of his flamboyant entrances, Judge Sawyer addressed the assembled litigants. "Due to the impending Thanksgiving holiday, the Supreme Court of New York will adjourn at the conclusion of tomorrow's session... and will reconvene at 9 a.m. on the morning of Monday, November 28, 1921."

After a satisfying shuffle of his papers, Sawyer peered down at O'Grady. "Counsel for the plaintiff... you may recall your witness."

Formalities complete, O'Grady posed his first question to Augustus Herring. "Acting as Old Man Chanute's designated expert, how would you rate the upgraded *Katydid* glider's performance at the Kill Devil Hills trials?"

"Objection!" Robbins clamored. "The defense objects to Mr. Chanute being referred to as the 'old man.' "

"Your Honor," O'Grady said, "this is a clear example of 'hearsay exclusion.' Our witnesses have *all* referred to Chanute as the 'old man.' "

Dropping his chin, Sawyer stared over his reading glasses. "I'll overrule. However, referring to an individual in derogatory terms does little to demean that person in the eyes of the court. As you may have noticed, Counselor... this is *not* a jury trial!"

"Sorry, Your Honor... I'll rephrase the question. How did the rebuilt multiwing glider perform?"

Herring's eyes narrowed. "It was a complete disaster. The damn machine refused to fly!"

"How was that possible?"

Herring shook his head as he recalled the facts. "Before making any changes to the original *Katydid*, I reviewed its design. My stress analysis calculations suggested that I could lighten many of the machine's wooden components and reduce its mass by almost 20 percent. In hindsight, the weight reduction was responsible for the contraption's appalling failure. Unfortunately, I didn't discover the error of my ways until there was a malfunction."

"What failed?"

"The vertical wing struts were bending under load, allowing the wings to twist. At first I thought the piano-wire trussing was at fault... but that was not the case."

"What did the Wrights have to say about the multiwing's problems?"

"Not much. They had their hands full with their own machine. However, after watching an early attempt to fly the *Katydid*, Orville thought the trouble might have resulted from structural weakness – as it turned out, he was correct."

"How did Chanute take the failure of his pride and joy?"

"Objection, Your Honor!" Robbins shouted. "Another unsubstantiated characterization. There's no record of this machine being referred to as Mr. Chanute's 'pride and joy.' "

"Sustained," said the judge, rolling his eyes.

"Very well," O'Grady responded, "I'll restate the question. Mr. Herring, how did Chanute react to the failure of the *Katydid*?"

Herring smiled at the memory. "Saying that he was outraged would be an understatement."

Kill Devil Hills, North Carolina
Friday, October 10, 1902

Gus was up and dressed at the break of dawn. While the others slept, he began preparations for breakfast – eggs, bacon, a loaf of Italian bread, a tub of butter and a pot of steaming hot coffee. Stepping into the cool, damp morning, Gus watched as a bald eagle flapped overhead, clutching a hapless drum fish in its talons. Turning to the eastern horizon, he lifted his hand to shield his eyes as the sun poked through a distant layer of stratus clouds. He thought... *A fine day on the way! With reinforced struts and re-trussed surfaces – everything should be stiff enough. Today's the day – either the damn thing flies, or I'll wreck it in the try*!

Two days earlier, Herring had attempted several glides down the Big Hill's 13-degree slope in winds ranging from 20 to 28 miles an hour. In each instance, the *Katydid's* lifting surfaces had distorted so badly that it was unable to sustain flight. Dan Tate helped Gus haul the machine back to the shed, where he devoted the remainder of Wednesday and all of Thursday making the necessary changes. As he gazed at the placid, gray Atlantic, Herring's thoughts returned to the shores of Lake Michigan, where the *Katydid* had also stumbled in its attempts to glide the slopes of the Indiana dunes. In an instant, he was nudged back to the present by a tug on his shoulder.

"Augustus! We need to talk... in private!" Clutching him by the arm, Chanute steered his assistant toward the shoreline. When they were within a stone's throw of the water, the old man stopped. "I'll get right to the point: why are you trying to make me look like a fool?"

The old man's candor took Gus by surprise.

"What are you talking about?" he asked, taking a short step backward.

Chanute's neck had turned crimson. "You know damn well what I mean! I hired you to *improve* the multiwing so it would perform better – not worse! Now you act like its failure was a forgone conclusion!"

Herring slowly folded his arms in front of his chest. "Where have you been the last two days? I've spent all my waking hours modifying the *Katydid's* struts and trussing."

Chanute shook his head. He was having none of it. "You claim to be an engineer? How could you foul up a straightforward stress-analysis calculation? It's downright embarrassing... especially since it came on the heels of *your* recommendation to rebuild the multiwing in the first place—"

"My recommendation?" Herring interrupted. "Your memory is—"

Disregarding Herring's rebuttal, Chanute continued. "You said you wanted to 'beat Mr. Wright'... a bullshit story if there ever was one! Admit it, you used me to wrangle your way into the Wrights' camp!"

Ignoring the allegation, Gus addressed the engineering *faux pas*. "If the multiwing's lifting surfaces were stationary and not pivoted, I would agree with you. However, the calculations became much more complicated with their fore and aft movement. If you recall... I mailed you a copy of my stress analysis and asked for your input."

"That's a lie! I never received any calculations from you!"

Herring turned and spit on the sand. "Another convenient loss of memory! Okay, since you're accusing me of incompetence and deception, here's another problem for you to get lathered up about." Herring decided to be vague. "Wilbur mentioned that the *Katydid's* past flight distances and angles of descent were *exaggerated*."

Chanute extended his hands, palms up. "What distances, what angles?" he demanded.

Herring forced a scornful smile. "The brothers are shrewd. They used flight data from your June 1900 *McClure's* article to test your conclusions in their wind tunnel last fall... didn't they tell you? Wilbur said the multiwing's 1896 glides were better than his measurements indicated they should have been—"

"I don't believe you!" Chanute butted in. "Why would Wilbur Wright confide in you – someone he despises?"

Herring shrugged his shoulders. "Wilbur said the *Katydid's* glides might not have been made under *real* test conditions—"

"More contrived bullshit!"

"Don't take my word for it, Octave, ask him yourself... it's no skin off my nose."

Herring knew that Chanute would never press Wilbur on the issue. The last thing the old man wanted was to create waves between himself and his latest aeronautical protégé.

"As a matter of fact," Gus continued, "I remember that *Katydid's* best flight was only about 80 feet... how did that distance get stretched to 300, like you wrote in the *McClure's* article?"

Chanute's face was now firebox red. "You were so mesmerized by *your Double Decker*," he stammered, "that you didn't pay any attention to what was going on around—"

Herring interrupted, "Have you forgotten that *I* was the operator for most of *Katydid's* flights back then?"

Chanute had heard enough. "Never mind that! I've made a decision! Since you haven't lived up to your end of our agreement... you're not going to be paid!"

Gus wasn't surprised the old man had gone back on his word. "Our agreement called for me to make *mutually* agreed-upon changes to the multiwing glider – which I have done. I was hired to set up and test the machine – which I am doing, along with the Lamson, which just arrived in camp. Are you suggesting that I have to satisfy some additional unstated requirement before I receive my fee?"

"Stop trying to put words in my mouth!" Chanute shouted.

"Then you had better spell out what you mean by 'my end of the agreement.' "

"Listen, you nincompoop," Chanute sputtered, "I've already spent a king's ransom on these trials! The very least you could do is act interested!"

Herring raised his eyebrows. "Fascinating.... Mr. Huffaker said that you walked away from your $1,000 paper-tube glider last year... with nary a regret."

"Damn you, Herring, last year has nothing to do with the here and now!"

At that moment, a faint voice interrupted the men's bickering. "Ahoy there!" Spratt shouted, as he jogged toward them.

"It must be time for breakfast," Chanute growled. "I'll finish with you later!"

*

By 9 a.m., seven men carrying two gliding machines and related support equipment, trudged through the damp sand toward the Kill Devil Hills test site. A few minutes later, Wilbur, Orville, and Dr. Spratt, accompanied by Lorin, carrying his tripod-mounted camera, plodded to the top of the Little Hill. Slogging on to the Big Hill, Herring and Dan Tate carried the *Katydid*, followed by Chanute with his anemometer, stopwatch, and notepad. At the summit, they stopped long enough to witness Wilbur's first glide of the day – what looked to be a hop of about 150 feet in 8 seconds

"We're in luck," Gus said, "the wind is blowing straight up this 15-degree slope."

Paying little attention to Herring, Chanute stared at the anemometer's metric gauge as the instrument's three-cup rotor spun above his head. "Nine meters per second," he grumbled, peering at the yellowed conversion chart taped to the instrument's wooden storage box. "Let me see... that's 20 miles an hour. Pretty windy for this early in...."

"Perfect for gliding," Herring weighed in, as he wiggled into the forest of spruce sticks and piano-wires. Lifting the apparatus by its parallel wooden rails, he shifted his hands fore and aft on the smooth, varnished surfaces until satisfied with the balance. Then he tilted the contraption's nose down in front, allowing the breeze to hold it fast against the sand.

"Steady the wingtips, men... I'll give her a try!"

With his legs protruding below the glider's framework, Herring assumed his practiced sprinter's crouch before glancing to the left and right. Nodding, he lunged forward, hurling himself and the machine down the slope. With his feet kicking up a spray of sand, Herring struggled to achieve the minimum airspeed needed to ....

*

"Mr. Herring, are you all right?" Tate panted. "Gus... can ya hear me?"

Herring had crashed the *Katydid* in spectacular fashion. Unresponsive, his bleeding body lay face down in the sand, 75 feet from the crest of the hill. In contrast, the apparently undamaged *Katydid* rested serenely in an upright position, 20 feet farther up the slope.

Several seconds passed before Chanute was able to crab-crawl down the hill to his fallen assistant. Removing a clean white handkerchief from his vest pocket, he pressed it against Herring's partially severed, bleeding left ear. With his free hand, he began sweeping sand away from the younger man's nostrils. Swiveling his head toward the frozen form of Dan Tate, he barked an order.

"Run, Mr. Tate! Fetch Dr. Spratt!"

Minutes crept by before Spratt, Tate, and Lorin Wright arrived. Kneeling beside Herring, Dr. Spratt quickly determined the extent of the fallen birdman's injuries. He asked the inevitable questions: "Was he conscious? Did you move him? Has he moved his limbs? Besides the ear, have you noticed any other injuries?"

A little later, Chanute talked about the wreck. "That was one hell of a smash, Doctor. Might he have broken his neck when he crashed through the machine's framework?"

Having secured their glider back at the Little Hill, Wilbur and Orville scuttled up the Big Hill's northern slope to join the other members of the expedition. "Is he going to be all right, Doctor?" Wilbur inquired, out of breath.

"That remains to be seen. Since Mr. Herring is still unconscious, I believe he has sustained a concussion. I've checked for bone displacement and fractures in his head, neck, and spine; everything appears to be in place, but we must move him back to camp. Once there, I can stitch up his ear and perform an in-depth evaluation."

Within half an hour, the still-comatose Herring had been loaded onto a makeshift stretcher made from sailcloth and two spare wing struts. Next, the men began the physically demanding task of carrying Herring 1,000 feet back to the aeroplane shed.

After his ear had been sutured and bandaged, Dr. Spratt soaked cotton cloths in the Wrights' cool well water and applied them to Herring's forehead, in the hope he would emerge from his unconscious state. Shortly before noon, Herring's eyelids suddenly snapped open. "Where's Butusov?" he asked, delirious. "He's got my measuring tape! Mr. Avery... how far did I fly? Did I make it to the lake?"

"Mr. Herring, lie still and try not to talk," Spratt ordered. "You've had a little mishap with your glider – but you'll be okay."

A few minutes later, Herring rolled over and sat up. "What the hell happened?" he asked, gingerly fingering the cotton bandage that covered his reattached ear.

*

By 1 o'clock, everyone but Gus had eaten lunch. A pounding headache, the result of his concussion and ear injury, had left him bewildered and somewhat nauseous. Although a headache powder had helped, Herring could do little more than listen as the others endeavored to learn more about his misfortune.

"Mr. Chanute," Wilbur inquired, "what seems to have caused the crash?"

Chanute held a single palm upward. "The start was like hundreds of others I have seen him make over the years, with one exception: the machine refused to rise!"

"What do you mean by that?" Orville asked.

Chanute shook his head. "The damned thing *refused* to fly, and that's when Herring tripped up—"

"I couldn't keep up," Herring rasped, wincing in pain. "I remember the machine wasn't lifting and that I had shifted my weight slightly to the rear... that's when my feet got tangled."

Wilbur stood and walked to the shed's open door, peering out at the ocean's roiling surface.

"What happened next?"

"I don't remember," Herring said, shrugging.

Dan Tate, who usually remained silent during aeroplane discussions, spoke up. "After Mr. Herring tripped, the front of the machine dug into the sand, which threw him through the wood and wires. He did a forward summersault and landed way out in front of the contraption, on his stomach. Then the machine reared up in front like a big old goose comin' in for a landin'; then it did a backward flip and flopped down... right side up!"

"That's a pretty good description," Chanute said approvingly, writing furiously into his ever-present pocket notebook. "There's nothing I could add that would help to clarify the event."

Rochester, New York

Tuesday, November 22, 1921

Returning from the lunch break, the judge dispensed with the usual formalities and ordered Herring back to the witness stand. O'Grady accommodated the court's promptness by immediately delving into the details of the Kill Devil Hills crash.

"Mr. Herring, did you learn what caused your accident?"

Herring slouched in his chair. "As I mentioned earlier, I had unintentionally weakened the multiwing's structure by over-lightening some of its key components, and it nearly cost me my life; I never made that mistake again."

"From this morning's testimony, you said that the *Katydid* escaped the incident undamaged. Did Mr. Chanute want you to try flying it again?"

Herring shook his head. "Other than a few trials as a kite, my crash spelled the end of the *Katydid's* sad existence! Even the old man had to admit that the apparatus was too dangerous to fly."

O'Grady flipped to the next page in his notepad. "After the accident, you steadfastly refused to take any time off. Tell the court about that decision."

"With favorable gliding weather, Orville and Wilbur decided to return to the Big Hill that afternoon and resume testing their machine. Their older brother, Lorin, was scheduled to return home in a few days, and they wanted to obtain some photographs of the machine flying. Although I felt under the weather and lacked energy, I wanted to see how the Wrights' latest control system performed. Over the objections of Dr. Spratt, I carefully donned my cap and slowly hiked back to the sand hills, where I sat and served as the official timer."

"What happened next?"

"Two of Wilbur's glides frightened me. In both cases, the machine came to a complete stop high in the air. Today, unlike then, we know that the machine's lifting surfaces were on the verge of an *aerodynamic stall*. In both instances, the glider pivoted with one wing high and landed roughly, facing sideways on the hill.

"The brothers were ecstatic that their glider neither fluttered back to earth, as it had the previous year, nor slipped sideways into the hill in 'well-digging' fashion. Although Wilbur admitted that he hadn't directed the machine to a side hill landing, the coordination of the vertical rudder and the wing's end-controls seemed to provide better management than he had previously experienced."

"Then there is this remarkable photograph," O'Grady said, holding a large black-and-white print above his head. "Your Honor, the plaintiff wishes to enter this into evidence."

"What do you have there, Mr. O'Grady?" the judge asked, motioning with his finger for both attorneys to approach the bench.

Kill Devil Hills. L to R-Chanute, Orville, Wilbur, Herring, Spratt, Tate (1902); Revue générales des sciences pures et appliquées, vol. 14, Nov. 1903, p. 1136

"It's a group photograph," O'Grady said, "showing six men sitting on the sand in front of the Wrights' 1902 glider. As you can see, the machine's upper wing is providing them with shade—"

"Objection, Your Honor," Robbins butted in. "There's nothing on this photo that identifies when it was taken."

"Look again, Counselor. A date has been scratched into the emulsion of the glass plate negative. If you look closely, you'll see it in the print's lower-left corner. It says 10-10-02."

"Objection overruled," Sawyer said. "Mr. O'Grady, can these men be identified?"

"Yes, sir. From left to right, you have Mr. Chanute, Orville Wright, and Wilbur Wright; they're sitting on the left side of the structure that's carrying the machine's horizontal rudder. On the other side, you have Augustus Herring, George Spratt, and Dan Tate. With the exception of Herring and Tate, the men are all wearing starched, high-collar white shirts with ties. Dr. Spratt is wearing a dark-colored suit coat and bowler hat, while Chanute, Herring, and Wilbur appear to be sporting captain's hats. If you look closely, you can see the cotton bandage protruding from under Herring's cap. Wilbur is wearing suspenders—"

"Objection, Your Honor!" Robbins shouted, cutting short O'Grady's description. "What does this have to do with Mr. Herring's testimony?"

O'Grady continued, "After the exhibit has been allowed into evidence, I'll explain its significance."

"Objection overruled," Sawyer said, cracking his gavel. "The court will allow entry of the plaintiff's exhibit."

After the usual delay for entering the item as exhibit number 26, O'Grady was finally allowed to continue with his line of questioning. "Mr. Herring, why is this photograph important?"

"This print *proves* that I was present at the 1902 Kill Devil Hills trials."

O'Grady moved from behind his podium. "Why is it necessary to prove that you were there?"

"Later, after my departure, certain individuals tried to refute my presence there."

"Objection! Argumentative!" Robbins shouted.

"Overruled."

"Can you identify the persons who perpetrated this story?"

Herring stared out the window for several seconds before turning his attention back to his attorney. "Wilbur and Orville Wright along with Octave Chanute!"

A burst of whispers erupted from the spectators' area.

"Order in the court!" the judge shouted, thrice cracking his gavel. "The bailiff is hereby ordered to maintain a watchful eye on our visitors! Further outbursts will result in an immediate eviction from these proceedings! Continue, Mr. O'Grady."

Looking up from his notes, O'Grady pushed his wire-rimmed bifocals farther up the bridge of his nose. "How did you learn about your 'exclusion' from these trials?"

"Dr. Spratt spilled the beans during his testimony in the Herring-Curtiss Company's bankruptcy trial – years after the fact."

"Why would these people try to perpetrate such a boldfaced lie?" O'Grady queried.

"Objection!" Robbins said, leaping to his feet. "Calls for speculation."

"Sustained."

"I'll restate the question, Your Honor... Mr. Herring, did Dr. Spratt tell you why these men felt threatened by your presence back in 1902?"

Herring looked astonished. "Spratt testified that the Wrights, spurred on by Chanute's paranoia, were worried that my verbal analysis and suggestions for improving their control system may have jeopardized their chances for obtaining a U.S. patent!"[1]

"Objection! Hearsay!" Robbins puffed.

"Your Honor!" O'Grady said, raising his voice, "Dr. Spratt's testimony is all a matter of public record."

"Objection overruled. Continue, Counselor."

"Mr. Herring, what was your reaction to the Doctor's sworn testimony?"

Herring leaned forward in his chair. "At first, I was confused. I didn't understand why they considered me to be unethical. To be specific, why did the Wrights believe I would steal their ideas concerning lateral control? Back in '02, I was only interested in securing a good paying job! The brothers didn't apply for their patent until 1903, and it wasn't granted until '06. Apparently, they sweated for better than three years while the application was pending, hoping that I wouldn't submit a similar assemblage of claims... which I didn't!"

"Objection, speculation!" Robbins shouted.

"Sustained. Strike from the record."

"Did Dr. Spratt explain the Wrights' strategy during his testimony to the bankruptcy court?" O'Grady asked.

"Yes, he did. They would deny that I was *present* at their '02 camp, or that I had made *suggestions* for improving their control system."

Rubbing his temple with his forefinger, O'Grady contemplated his next question. "What about the other attendees at the '02 trials? Weren't they present when you analyzed the problem and suggested improvements?"

Herring raised his eyebrows. "They were all there, but the three Wright brothers clammed up, and Chanute denied that I had accompanied him to North Carolina! Spratt was a steadfast friend of both Chanute and the Wrights until well after their patent had been granted. Mr. Tate knew that I was there, but he died shortly thereafter."

"What about Chanute's contacts?' O'Grady asked. "The old man was in touch with everyone."

"I knew that Chanute had talked with Langley, but the professor died back in '06. Chanute died in 1910, Wilbur in 1912. None of them ever revealed their devious secret, including Orville – the hypocrite. I was completely in the dark during those chaotic years, unaware that my presence at the Kill Devil Hills trials had been kept secret... today, some would call this historical revisionism."

Edging back to his podium, O'Grady glanced at his notes. "Mr. Herring, on the evening of October 5, 1902, as an invited guest at the Wrights' Kill Devil Hills campsite, you were asked to analyze the problems associated with their machine's lateral-control system. Afterward, you answered questions and offered suggestions related to the improvement of said system. Assuming this statement is correct, why did the Wrights go out of their way to conceal your role in the development of active control systems?"

"Objection! The attorney for the plaintiff is calling for speculation!" Robbins shouted.

"Overruled!" Judge Sawyer said. "The court wishes to hear the witness's response."

Surprised by the unexpected opportunity to state his opinion, Herring looked contemptuous. "The Wrights' method of control – the system that I had amended in both theory and practice – proved to be the key ingredient in solving the problem of *practical* heavier-than-air flight. Although their landmark patent of 1906 had been challenged in the courts, it was eventually upheld in the broadest terms imaginable, giving the Wrights precedent over most methods of lateral control. Their all-encompassing patent allowed the brothers to hold sway over America's aeroplane industry for 11 years, until Congress enacted the *Cross-License Agreement of 1917*. Prior to that legislation, the domestic aeroplane industry had atrophied due to the Wrights' exorbitant licensing fees and stifling business practices. As a result, our country's aviation industry was unprepared to assist Europe in The Great War – the same war that made the Curtiss Airplane Company rich!"

Rochester, New York

Wednesday, November 23, 1921

Testimony began again at 9 a.m.

"Mr. Herring," O'Grady said, "when did you fly the Lamson glider?"[2]

"After taking Saturday and Sunday to rest, I was up early Monday morning, anxious to uncrate the Lamson machine and start its assembly. Since Wilbur had accompanied Lorin to Kitty Hawk to meet his steamer, Orville and Dr. Spratt lent me a hand. Late in the afternoon, we hauled the glider to the Little Hill for its initial trial."

"Did the machine show promise?"

Herring flies Chanute-Lamson triple-decker at Kill Devil Hills (1902); Special Collections and Archives, Wright State University

Herring shook his head. "My best flight – the one where Orville took a photograph with me flying the contraption – was only about 50 feet in length, and that was after making a multitude of adjustments and changes."

"What were some of the challenges?" O'Grady probed.

Gus hesitated as he searched his memory. "One of its biggest problems was the forward-mounted horizontal rudder. Although it was similar in appearance to the Wrights' forward surface – it was *fixed*. When a gust of wind turned the nose upward, this rudder made the condition worse, and no amount of weight shifting could reverse this tendency."

"What did you decide to do?"

"I removed the damn thing; but that made the contraption tail heavy. My only solution, short of rebuilding the entire machine, was to add sandbags to the front framework. Since the glider was already overweight, things went from bad to worse."

"Where did Lamson get the idea to mount the horizontal rudder in front of the wings?" O'Grady asked.

Herring chuckled. "Chanute! Where else? The old man was such a two-faced so-and-so. Although he despised the Wrights' use of their manually controlled forward rudder, I believe he wanted to impress them with his design flexibility. Apparently, Lamson didn't seem to understand what Octave was attempting to do, and made the surface stationary."

"Objection! Speculation!" Robbins cried.

"Sustained."

"Can you imagine," Gus continued, "the old man's fore and aft 'rocking wings' struggling to provide pitch stability to an apparatus whose forward-mounted horizontal rudder was fighting their every action? If I hadn't been risking my life in that conglomeration of mistakes, it might have been laughable. Unfortunately, it was an extremely dangerous contraption to fly, and I was leery of being injured or worse while trying to operate it."

"How did this end?" O'Grady continued.

"After trying everything in my bag of tricks, to no avail, I finally threw up my hands and refused to risk my neck any longer."

"What did Chanute have to say about your decision?"

"He demanded that I proceed with yet another series of flights! He had the nerve to say, '... show some courage, man!' I suggested that if he wanted to risk his own neck, he should step up and launch himself off the hill. I told him that I was finished."

"How did the Wrights react to your decision?"

"Neither of the brothers volunteered to carry on with the Lamson experiments! No matter, I was through arguing with Chanute about his pathetic gliders. As a matter of fact, save for *one* other flight four years later, my gliding career was over."

"How much longer did you stay at the Wrights' camp?" O'Grady asked.

"The old man and I left that very afternoon. Since we were both headed to Washington, D.C., we traveled together – nine hours of silence!"

"What route did you take to get there?"

"A donkey cart to Kitty Hawk, a steamer to Elizabeth City, and a train to Washington via Norfolk. We arrived in the capital about 10 p.m."

"Where did you stay?"

"I checked in at my old stomping grounds, the Willard Hotel, where I was to meet my mother and sister Elizabeth the following afternoon. The ladies were railroading their way back to New York City after a visit with another of my sisters who lived in Clarksville, Georgia. The layover provided us time for a leisurely dinner and to catch up on the latest family gossip.

"Since I had several hours to kill the following morning, I decided to walk over to the Smithsonian – it was like old times! I wanted to talk with Professor Langley, to see if I could gain an appointment to work on his *Great Aerodrome* project—"

"Wait a minute," O'Grady interrupted, tapping the side of his head with his fingers. "If I'm not mistaken, you and the Professor weren't exactly on speaking terms."

Herring took a deep breath. "In retrospect, I believe this was a last-ditch effort to secure employment before facing my mother's scrutiny. Besides, a lot had happened since my resignation from the Smithsonian's model aerodrome project, and Langley was under a great deal of pressure from his government sponsors to produce a successful man-carrying machine. Although the Spanish-American War still dragged on, his critics – and there were many – were impatient with his lack of progress. The Professor knew my work, so I hoped he might rehire me."

"How did your plan work out?"

Herring grimaced. "Aside from providing me with a bit of ammunition for my meeting with Mother – not so well. Langley wasn't available when I arrived unannounced, but I was fortunate to be interviewed by Mr. Watkins, the curator of the Smithsonian's mechanical collection. If you remember, he served as one of Langley's technical advisors."

"How did the interview go?" O'Grady asked.

"Very well – at first. Watkins was enthusiastic about getting me back into the fold. In fact, he stated that the Professor would be interested in discussing some of the new technologies that I had been working with since my departure – especially control systems."

O'Grady hesitated. "Did that send up a red flag?"

Herring squirmed in his chair. "I had a good idea what was coming next, and Watkins didn't disappoint. He said that the Professor was especially interested in learning about the Wrights' system of control. Apparently, Chanute had preceded me to the 'Castle' and mentioned that we had spent better than a week at the Outer Banks site."[3]

"How did you feel about divulging the Wrights' secrets to a competitor?"

"I told Watkins that since the brothers' system had yet to be patented, it wouldn't be ethical to discuss it."

"Did that affect the rest of the interview?"

"Let's just say that I didn't get the job! Later, I learned that Chanute had written Langley a damaging letter – saying that I had botched my assignments, was not to be trusted, and other less than complimentary remarks."[4]

O'Grady turned to address the judge. "As the attorney for the plaintiff, I must point out the irony of these circumstances! On one hand, Chanute and the Wrights considered Herring to be *unethical*. On the other, Mr. Herring was penalized for taking an *ethical* stand against Professor Langley's boldfaced attempt to bribe him!"

"The court so notes the contradiction, Mr. O'Grady. Please continue."

"Your good friend, Octave Chanute!" O'Grady boomed. "When did you rid yourself of that reprehensible old man?"

Herring took a moment to fashion his response. "One might think that Chanute's damning letter to Langley would be the final straw, but areas of mutual concern kept popping up! For instance, early in January of '03, I wrote to him requesting payment for expenditures that I had incurred while traveling home from Washington... if I remember correctly, the amount came to $35. Four months later, he still hadn't paid me. Then there were the assorted letters concerning a potential entry in the aeroplane competition at the *1904 St. Louis World's Fair*... but nothing ever came of that. There also was the chance conversation I had with him during that event, but I'm getting ahead of myself."

O'Grady winced as he returned to his notes. "You hadn't seen your mother for several years; was she pleased with the reunion?"

"With Mother, you never knew," Herring said. "We had a prolonged conversation about my damaged ear, followed by the usual candid *tête-à-tête* concerning my future. At the end of the evening, she handed me a sealed envelope, with a warning not to open it until after she and my sister had departed the following morning."

"Did she give a reason?"

"Apparently Mother didn't want there to be an argument."

"About what?"

"I understood... once I opened the envelope!"

CHAPTER 77

St. Joseph, Michigan

Saturday, October 18, 1902

Hauling his battered Gladstone bag and gray canvas duffle from the hansom cab, the weary traveler fished two dimes from his watch pocket and paid the fare. Although fatigued from the 18-hour train ride that had begun at 8 o'clock on Friday morning, Herring was still in high spirits as he tossed the duffle over his shoulder and marched up the concrete walkway to his home. All was quiet at this early morning hour.

As he fumbled to find the key, the door swung open to reveal Lillian's smiling face. After depositing his gear in the parlor, Gus directed his drowsy spouse to the kitchen table where, despite the hour, he began relating the events of the past two weeks. Lillian, wrapped in her warm bathrobe, winced when he described the crash. With barely enough light from the room's 25-watt bulb, she reached out and tilted his head to better judge the damage to his ear. Nodding, she leaned back.

"It's only a matter of time."

"Before what?"

"Before a flying machine puts you in your grave."

Gus reached for her hand across the table. "I've been thinking... it's time to quit."

"Quit? What are you going to quit?"

There was a pause. Gus glanced at Lillian, his confidence faltering. "I've decided to quit flying gliders."

Having heard similar declarations in the past, Lillian remained stoically unimpressed. Sensing her vexation, Gus reached into his shirt pocket and pulled out an envelope. "I have something to show you," he said, sliding the packet across the polished tabletop.

"What's this?"

"Open it."

Fumbling to extract the envelope's contents, the first thing Lillian noticed was the cashier's check from the Bank of Hartford.

"Gus!" she stammered, "Five thousand dollars? What on earth is this for?"

"Read the letter, my love."

Less than a minute later she looked at Gus and offered her response. "She misses the children... that's why she wants us to move back to the city. She wants her grandchildren to be *available*... especially during the holiday season."

Herring scratched his head and frowned. "I don't understand why she misses *us*... Mother has my older brother William and my younger sister Elizabeth to keep her company—"

"Augustus," she broke in, "your brother William and sister Elizabeth don't have children. That trip to Georgia to see your sister Chloe's newest baby must have started her thinking."

"About what?

"Of what she *thinks* she's missing! That's why she's bribing you!"

"A bribe? That's a bit strong, don't you think?"

"Perhaps... how about a carrot on a stick? A big, juicy carrot!"

"God, can you believe that Old Man Bleckley is still churning out kids with my sister Chloe? What is he, 75?"

Lillian blushed at the thought.

"She's only 37, Gus, two years older than you."

Shaking his head, Herring's thoughts shifted to another sister who lived in Baltimore.

"What about Sarah? She has three kids—"

Lillian cut him short. "Yes, but Sarah would never agree to move away from her much-loved Maryland Historical Society."

"That leaves us! What do you think, my dear? Would you like to move back to Long Island? You read the letter... Mother will buy us a home anywhere we want!"

Lillian lifted her eyebrows and smiled broadly. "You know I love Long Island... it's where I grew up!"

"Mother's offer is hard to pass up, but what about her conditions? We must return no later than June and must remain for the duration of *her* life."

"She's already 67, Gus; how much longer might she live... five or 10 years?"

Herring tapped his fingers on the tabletop. "She also demands that I find suitable employment. Do you think my old engineering consulting business qualifies? Then there are the flying machines... she doesn't like me dabbling with 'fool killers,' as she calls them. I suppose there's no way around that; I'll have to curtail my aerial navigation work – one of the reasons I've decided to quit gliding."

"I thought so!" Lillian said, parking her hands on her hips, "You had already made up your mind!"

"Great minds think alike, my love!"

Rochester, New York

Wednesday, November 23, 1921

Due to the impending Thanksgiving holiday, attorneys for both sides anticipated the judge's calling for an early end to the court's afternoon session. With that possibility in mind, O'Grady and Maloney spent most of their lunch hour discussing how best to direct Herring's remaining testimony, with the objective of leaving a lingering impression on the judge.

To the regret of all in attendance, Sawyer failed to declare a plan for an early adjournment. Instead, he carried on business as usual, by ordering Gus back to the witness stand.

"If I understand your mother's stipulations correctly," O'Grady said, resuming his interrogation, "you had about seven months to move your family back east. Is that right, Mr. Herring?"

"Yes."

"Concerning your new house on Long Island... did your mother purchase the dwelling outright, or did she simply arrange for the down payment, with you taking a mortgage?"

Gus looked rueful. "Because my father had been very successful in business, Mother became a millionaire upon his death. As to our new house... a few years after the fact, I learned from a relative that the property had been purchased from funds that were earmarked for *me* by my late Grandfather Conyers. Somehow, Mother had arranged to become the silent custodian of my inheritance."

"Therefore, your mother enticed you and your family to move back East with *your* own money."

"That would be an accurate assessment," Gus said.

"When did you move from St. Joseph to Long Island?"

Gus hesitated as he tried to remember the exact date. "It was May 1903... toward the middle of the month."

O'Grady turned to a new page in his notes. "Tell the court about your writing career."

Herring sat military straight. "As an internationally recognized combustion engine expert, I became a regular contributor to a number of magazines, including *Horseless Carriage* and *Horseless Age*. In December of '02, I was contacted by a group of investors from the Benton Harbor-St. Joseph region of Michigan."

"What did these investors want?"

"As publishers, they wanted me to become editor for their new magazine—"

"What was the name of this magazine?" O'Grady interrupted.

"*Gas Power*."

"To whom did the magazine cater?"

"Those who made, sold, or used gasoline-fueled engines."

"Because you had already decided to move back east, how did that decision sit with these publishers?"

Herring threw one leg over the other. "When I explained that I would be moving in the spring, I was offered the position of *managing editor* in the soon-to-be established New York office – complete with a one-third partnership in the venture!"

O'Grady closed his notepad. "Mr. Herring, prior to moving, did you continue with your aeronautical experiments?"

"In light of my mother's 'bequest,' I was free to complete my miniature gasoline engine and install it in my *Double Decker* model."

"How was this new engine different from the rest of your designs?"

Herring couldn't resist basking in his success. "In conjunction with using a castable, lightweight aluminum alloy, I incorporated all of my most successful innovations into this one design, hoping to shatter the power-to-weight barrier for internal combustion engines. I was certain that it was possible to make a gasoline-fueled engine that weighed only three or four pounds per horsepower. Earlier, I had calculated that if the new engine showed the same relative power for its cylinder displacement as my Mobike engine, it should realize between one-quarter and one-third brake horsepower and would weigh less than two pounds. By November 1902, I was ready to test this new engine."

St. Joseph, Michigan

Friday, November 7, 1902

It was already dark outside when Herring pushed open the door to his "mystery barn" workshop. Working his way around the perimeter of the shop, he systematically lit the wall-mounted gas lamps before walking over to the big, cast-iron coal-burning stove. Shoveling three scoops of bituminous into its belly, Herring painstakingly stoked the embers left over from the morning's burn before flaming the skinny end of a wood splinter. Closing and latching the heavy hinged door, he cupped his hand around the reluctant flame as he shuffled to the center of the room, where he lit the two remaining lamps that hung above his cluttered workbench. Extinguishing the splinter between his thumb and forefinger, Herring thrust one hand into the pocket of his gray canvas work pants as he leaned forward to inspect components of his latest masterpiece... the miniature engine. Staring intently through a magnifying glass, Gus was startled when the shop door thumped open behind him.

"George! I hoped you would drop by tonight; do you have something for me?"

As was his custom, his friend and sometime assistant, George Housam, streaked straight for the stove. Always cold during the fall and winter months, George stomped his feet and extended his hands over the iron behemoth's radiating warmth. "I don't know how you can work... it's freezin' in this dang place."

Here we go again, Herring thought. "I just lit the stove a few minutes ago, and it's already 55 degrees... how hot do you want it?"

Undoing the top button of his slush-gray, knee-length winter coat, Housam continued to shiver. "How about 80 degrees? I'm tellin' ya Gus, I'm movin' south," he said, dropping three oversized wedding-band-like metal rings onto the bench.

"Ah, you've done it!" Herring exclaimed. "What took you so long?" he continued, with a wink. Picking up his magnifier, Herring slid over to the nearest hissing lamp for a closer look. Pulling a one-inch micrometer from his apron pocket, he measured the thickness of a ring. "One hundred thousandths of an inch – right on the money! Did you have any trouble machining these things?"

Housam nodded as he glanced around for a place to sit. Spotting an old folding chair, he plopped himself down. "*Sintered iron is brittle stuff* – I broke two rings during the final grindin' operation." Housam hesitated before continuing. "But I still don't understand how you're gonna get the rings on the piston? When ya get ready to try, let me know... I'll head for home. I can't stand to see a grown man cry!"

"Relax, George," Gus said, grinning. "I've been sneakin' rings into piston grooves for a long time... since at least the eighth-grade picnic. There's a trick or two that I use, but I won't bore you with the details... you'll just have to hang around and watch."

"Speakin' of details, why are ya foolin' with this little bitty engine in the first place? What happened to the full-size flyin' machine engine?"

Gus sighed. "Money, George! Money! But engines are engines, no matter what their size. What I'm learning now will help when I build the big machine's engine."

"Why do ya think this motor will be better?"

Finding himself a chair, Herring sat facing his inquisitive assistant. "Less weight and more power! Five years ago, the lightest gasoline engine weighed 150 pounds per horsepower! I hope to reach four pounds per horsepower with this one."

"Hope?"

"Until I test her on the *prony brake dynamometer*, there's no way to know the exact horsepower. However, I do know it will weigh about two pounds, not counting the fuel and accessories."

"So, I reckon you're lookin' to get half a horse out of her."

"Very keen," Gus said. "One of the reasons I chose you for my helper."

Ignoring his boss's dry sense of humor, George continued. "Besides being lighter, why do ya think this motor is better?"

Gus removed his cap and scratched his head.

"It's a better design! I've learned a lot since the old opposed-twin-cylinder days."

"Are ya talkin' about the aluminum castings?"

"That's part of it... certain aluminum alloys are as strong as steel, yet they're almost three times lighter!"

Snatching a part off the bench, Herring continued. "This cylinder is cast from the stuff, and so are the front and rear crankcase halves and the bolt-on cylinder head."

George held out his hand in a questioning manner. "How do ya expect an iron piston ring to ride on that soft aluminum casting and not wear it out?"

"Ah... very insightful!" he said, handing Housam another delicate-looking cylinder. "Take a gander at this."

Taking a moment to adjust his wire-rim spectacles, Housam turned the piece over in his hand. "Beautiful! Where'd ya get this made?"

"Chicago... Engberg's was booked up. It's made from tempered tool steel that's been internally ground and honed to a tolerance of two thousandths of an inch."

"We couldn't produce anything near this good on our old machines. What's it for?"

"It's called a *drop-in liner*. After expanding the aluminum cylinder with a bit of heat, I'll push it into place, making a light interference fit. Tool steel makes a great mating surface for iron piston rings and the assembly is a lot lighter than my older steel cylinder with its integral cooling fins. But, there's a problem... I'm not happy with the quality of the cylinder casting, so the old steel cylinder will have to do for now."

Housam nodded. "What about the design? This engine looks a lot like your earlier motors, only smaller."

"The bicycle and Mobike engines taught me a lot, George. The twin-cylinder engine – the one that was destroyed in the fire – used a single pushrod-activated overhead poppet valve for *both* the inlet and exhaust operations. That was a mistake... I believe that it killed about half of the possible power.

"The new system uses two separate overhead poppet valves. The exhaust valve is opened and closed by a push rod that operates between the camshaft and the rocker arm. The inlet valve opens and closes automatically, and only allows the air/gasoline mixture to enter."[1]

"How the hell does that work?"

Herring handed him the partially assembled cylinder head. "Push down on the stem of the inlet valve – it's the one with the coil spring."

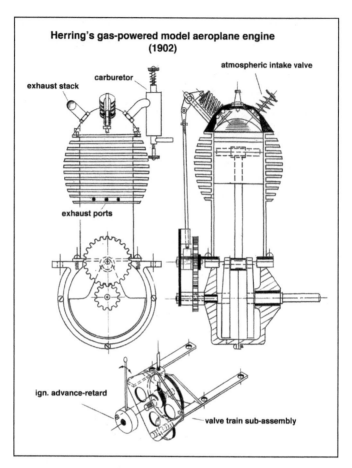

Herring's 1902, 4-stroke cycle model aeroplane engine; drawing by author

Obeying, George seemed surprised by the results. "There's hardly any resistance. I could almost blow it open with a good strong puff."

"That's the secret... it took me a couple of nights' work on the lathe to wind a compression spring that would barely close the valve against its seat. Here's the idea: with the exhaust poppet valve closed, the piston moves away from the cylinder head on the inlet stroke. This creates a partial vacuum in the cylinder, allowing atmospheric pressure to force air through the inlet pipe to the poppet valve, which is then pushed open against its weak spring."

George snorted, signaling that he understood. "How does the gas get mixed with the air? Ya gonna use the spoon and sponge—"

Gus hooted to the rafters as he handed Housam the engine's cylinder head. "No more spoons... that's what this inlet pipe is for. As air blows through the narrow part of the pipe, called a venturi, it speeds up, thus lowering the pressure. The partial vacuum allows the gasoline to be sucked into the pipe at that point. Actually, atmospheric pressure *pushes* the gas from the tank and into the pipe. It's Bernoulli's Principle all over again – the same law that explains how an aero*curve* surface generates lift."

Housam held up his hand in a "stop there" gesture. "How do ya mix the right amount of gas into the air?"

Gus pointed to a delicate-looking machine screw at the inlet pipe's venturi. "With this *needle valve*. Screw it in, and *less* gasoline sprays into the pipe... screw it out, and *more* sprays in. Simple, eh?"

George's questions kept coming. "How are you gonna oil her, Gus?"

"I thought you'd never ask! I've decided to mix mineral oil in with the gasoline."

Housam squinted. "How's that gonna oil the parts below the piston? The crank, the rod bushings, the piston pin?"

Herring cocked his head and grimaced. "To tell the truth... I'm not quite sure! That's where your piston rings come in. Usually, I would use three compression rings on the piston to get a good seal to the cylinder wall. This time I'm only going to use *one*."

"One ring?" George hollered. "Will that seal good enough? Besides losin' compression, you'll leak high-pressure gas into the crankcase."

"That's the *idea*," Herring said. "I'm pretty sure that piston ring 'blow by' will be enough to lubricate the engine's bottom end while still sealing well enough to give good power. As a side benefit, the friction due to ring drag will be much less!"

Housam grinned, shaking his head. "That's pretty smart... if it works. Come to think of it, how are you gonna get rid of all that oil that'll pool up in the crankcase?"

Herring pointed to a small hole at the bottom of the crankcase casting. "That's what this oil dump hole is for."

"What'll you do if your newfangled oiler don't work?"

"I'll add a second compression ring to the piston and squirt some mineral oil into the crankcase before each run – just like we did before."

George hesitated for a moment. "That's okay fer short runs, but not fer long ones."

"We'll just have to see what happens. In the meantime, let's get started on the final assembly... I wanna run her tomorrow."[2]

St. Joseph, Michigan

Saturday, November 8, 1902

Herring and Housam stood outside the mystery barn in the cold morning air, deciding their next move.

"I'm tellin' ya Gus, Old Lady Fahey's gonna ring up the coppers if you run the motor here. Remember what happened the last time?"

"I'll take my chances, George. I'm going to give this engine a try, and that's that!"

"You remember what the judge said, don't ya? If ya didn't take the noise down by the lake, he's gonna fine ya... maybe throw ya in the hoosegow."

"That was a couple of years ago," Gus said, as he rotated the propeller. "I've been good since then... hey, this thing's got great compression!"

Housam nodded, still worried about spending the rest of his weekend behind bars.

"Come on George," Gus said, handing him a clipboard, "help me run through this checklist."

For the next several minutes, the men worked on checking off all of the particulars associated with operating the little engine.[3] The second-to-last item listed was Herring's longtime nemesis: sparking plug leaks. Would this delicate, homemade ignition component seal against cylinder pressure? Using a glass pipette, he carefully deposited a drop of mineral oil at the intersection of the plug's ruby mica insulator and its steel body.

"Grab the propeller down by the hub," Herring directed. "Then slowly rotate the crankshaft until the piston passes through the compression stroke. That's when I'll use this magnifying glass to look for bubbles."

A critical step, the sparking plug had to pass this final test before the men could attempt to run the new engine. "Hold it! Hold it!" Herring hollered. "We're blowing bubbles between the brass cinch nut and the insulator!"

Back inside the shop, Gus disassembled the tiny component. *I must be crazy,* he thought, *trying to contain combustion pressure with something this tiny and fragile.*

A few minutes later, while inspecting the parts under the magnifying glass, Housam identified the problem. "There's a toolin' scratch runnin' the length of this-here insulator... it could be causin' your leak."

After inspecting the insulator himself, Herring stood, turned, and slapped his assistant on the back. "George, you've got the eyes of an eagle! Here's another mica insulator with the center electrode cemented in place... let's hope it works better!"

With his fingers crossed, Gus watched as his upgraded sparking plug withstood the rigors of the latest pressure test. Next, he pulled out his pocket watch. "Quarter past eleven. What do you think... should we give her a try or wait until after lunch?"

"Let's have lunch."

"It's a little early. What the hell, we might as well push her outside... just to see if she'll start."

"Then we'll shove everything back into the shop," George suggested, "lock the door and hightail it into your house! If the coppers show up, Lillian can say she doesn't know anything about it! One thing's fer sure, that old bag's gonna squeal... I saw them lace curtains of hers wiggle. She's probably got her hand on the 'phone already, just waitin' for the motor to make a 'pop!' "

Hands on hips, Gus could do little more than shake his head. "Come on, let's get this test stand outside!"

Within minutes, Herring had opened the gas shutoff valve and slipped on an old leather glove. Grabbing the propeller with his gloved hand, Gus slowly turned it counterclockwise through two complete revolutions of the engine's crankshaft. Bracing himself, he "flipped" the prop rapidly with his middle and index fingers. Hearing a slight gurgling sound, Herring knew that gasoline had entered the cylinder through the inlet poppet valve. Judgment time had arrived! Would she run?

Reaching behind the machinery, Herring toggled the ignition switch to "on" and rechecked the position of the ignition-advance lever. Grabbing the propeller, he again rotated the engine through its compression. This time, as he expected, there was a solid "thump" of resistance against his grip – the engine had fired its air/gas mixture!

Glancing up at Housam, who was standing safely behind the heavy wooden test stand, Herring nodded in preparation for the much-anticipated event. He rotated the propeller another turn and a half. It was again poised at the beginning of the compression event. Switching to his index and middle fingers, Gus braced himself and flipped with all his might.

*

By 11:30 a.m., the men were standing in the rear entranceway of the house, laughing and trying to catch their collective breaths. It had taken them less than five minutes to run the engine, shut it off, and push the equipment back into the shop before scrambling inside.

"What's all the commotion about?" Lillian asked. "I heard the little engine run... it's so loud!"

"That's why we're in here," Gus sniggered. "Old Lady Fahey probably called the coppers by now! You'll tell 'em I'm not here, won't you, dear?"

"Honestly, you men are acting like children. Helen's not even home. I saw her and Brian walking arm in arm down Church Street more than an hour ago... so you're safe."

Herring offered a mock glare in Housam's direction. "I thought you saw the old lady behind her curtains?"

"Must have been seein' things."

"Anyway, you're just in time," she said. "I was coming to call you two to lunch. Get washed – we're having spaghetti with meatballs with my special sauce."

George brightened. "Meatballs, my favorite!"

As the men finished their noontime feast, Housam wiped his chin on his white cotton napkin.

"She sure ran great, Gus. I've been thinkin' though, why are you usin' that old-fashioned-lookin' propeller with the steel tubes and cotton coverin'? Your hand-carved wooden propellers for the old twin-cylinder engine worked great. I thought I heard this one flutterin' a bit."

"It's all about getting enough surface area to load the engine. Remember, there were two props for the twin-cylinder engine – one pulling and one pushing. This engine only has a single two-bladed propeller, so I need more area to do the job. These fabric-covered units are quick to build, lightweight, and still give enough flywheel action to keep the engine running smoothly. If I have time, I'll carve a black walnut propeller before we fly the *Double Decker* model."

*

By late afternoon, Herring's latest masterpiece had been run five more times. The lubrication system seemed to be working exactly as he had envisioned, as evidenced by the streak of mineral oil that had deposited itself along the bottom of the test stand. Better yet, the tiny sparking plug had held its own against the engine's unrelenting combustion pressure.

During the runs, which lasted between two and three minutes each, Herring spent most of his time crouched behind the machinery adjusting the spark timing and fuel mixture, while enduring a steady blast of air from the tractor propeller. Each small adjustment was evaluated by listening to feedback from the engine's exhaust note; a higher frequency indicated that it was running faster, and vice versa. However, for the purpose of calculations, Herring relied on George and his stopwatch to acquire speed data from the crankshaft's revolution counter. This task was performed several times during each trial.

Flipping the ignition switch to the off position, Herring marveled at the instantaneous silence that enveloped the neighborhood as the 22-inch propeller windmilled to a stop. Sidling over to Housam and his clipboard, Herring was certain that this last run had been the best yet. "Let's see, 9.2 seconds to complete

400 revolutions. That works out to 2,608 revolutions per minute... almost 40 turns a second! Yes sir, that's the best yet!"

"Faster 'n anythin' I ever seen," George stammered. "Ya sure 'bout them numbers? Hard to believe anythin' can spin that fast!"

"That's pretty good for a gasoline engine," Gus said, "but I once had a single-acting steam engine that turned more than 5,000. Later, we'll find out what kind of horsepower she's putting out... that'll be the real test."

As Housam wiped the oil and exhaust residue from the engine's external components, Gus pored over data from the previous runs. After a minute, he pushed his pencil behind his good right ear and turned to Housam. "Let's give her one last run before calling it a day. Fill the tank while I change the ignition cells."

A few minutes later, they were ready to restart the engine when Herring had a thought. "George, I noticed that you're standing right next to the spinning propeller – remember, that's dangerous. If something flies off, you will be right in the line of fire. Stand to the front or the rear of the prop, but not directly in its plane of rotation."

Housam shook his head. "Trouble is... if I stand behind it, I can't see the counter because you're in the way."

"Then try standing out in front of the propeller."

Preparations complete, Herring flipped the propeller and the engine roared to life again – it was another one-flip start. Slipping back to his normal position behind the engine, he quickly advanced the spark lever and was beginning to adjust the needle valve when, as Herring would later describe it, "There was a tremendous thump" immediately followed by the sound of George screaming and cursing.

The tube-and-fabric propeller had failed at the hub, causing shrapnel to be thrown out by centrifugal force. Compounding the mechanical failure, a piece of debris ruptured the tinplate gasoline tank, showering its volatile contents onto the hot engine. This resulted in what was later described as a "huge orb of flame" that enveloped the test area.

With its propeller disintegrated, the engine shook violently as it sped up, howling out of control. In a desperate attempt to terminate the mayhem, Herring swiped at the ignition switch with his free right hand, resulting in the runaway machine's almost immediate stoppage. Unable to endure the scorching heat on his right side, Herring propelled himself out of harm's way by barrel rolling to his left before hitting the ground.

Moments later, after scrambling to his feet, Herring's first conscious image was that of his faithful assistant sitting spread-eagled on the ground, clutching at something above and to the side of his right knee. Upon closer inspection, the last three inches of a battered 10-inch length of quarter inch steel tubing protruded from his canvas work pants. Housam's condition appeared to be dire; he had been impaled, and the wound was gushing blood from what had to be a severed artery. To make matters worse, George seemed to be lapsing into shock.

Lillian, having heard the commotion from inside the house, rushed to the scene. After a quick appraisal of her husband's condition, she raced over to George, using his leather belt to apply a tourniquet above the wound.

"Gus!" she shouted. "Hurry! Call for an ambulance! Then bring me my scissors, some kitchen towels, and some blankets!"

CHAPTER 78

Rochester, New York

Wednesday, November 23, 1921

His head spinning, Herring paused to refill his water glass as the courtroom's gallery buzzed with chatter.

"Order in the court!" Sawyer barked, wielding his gavel.

With silence restored, the judge swiveled in his chair. "Are you prepared to continue, Mr. Herring?"

"Yes, Your Honor... thank you."

"Carry on, Mr. O'Grady."

"Mr. Herring, what are your recollections of that fateful day in November 1902?"

The courtroom became very still.

"After I telephoned for help, I ran back outside and busied myself with the test stand fire."

"What did you do?"

"With Lillian attending to George, I grabbed a shovel from inside the shop and pitched dirt onto the raging flames."

"Were you successful?"

"I had the fire smothered in less than a minute."

"What else do you remember about that dreadful day?"

"I remember the thundering black stallion, snorting steam from its nostrils. I remember the doctor's black topcoat flapping in the breeze as he knelt beside George. I remember the attendant, dressed all in white. I remember the ambulance, its cherry wood panels reflecting the fading afternoon sunlight – the gold leaf lettering that shouted ST. JOSEPH HOSPITAL. Most of all, I remember George being carted away on a stretcher... unconscious."

"Did they take you to the hospital?"

"No. After assessing my burns, the doctor instructed Lillian to apply cold water compresses to the affected areas. Afterword, she was to slather a generous layer of butter onto the damaged skin."

"Then what?"

"Lillian propped me up with pillows as I reclined on the sheet-covered divan in our parlor, where I eventually fell into a fitful sleep."

"Did you sleep the whole night through?" O'Grady asked.

A flicker of surprise showed in Herring's expression. "No... I was jarred awake by an outrageous but very realistic nightmare."

"Did you get back to sleep?"

"No... so I lit the kerosene lantern and wandered out into the cold, in my pajamas, to check the charred remains!'

"Why?"

"To make sure that the Russian, Butusov, had not made off with my experimental engine after all!"

O'Grady raised an eyebrow.

"When did you learn about George's condition?"

"While I picked at my lunch, Lillian told me that the hospital had upgraded George to 'serious but stable' condition. Apparently, the surgeon was able to successfully repair the torn artery in his leg, but his sliced sciatic nerve would probably cause a permanent numbness in his foot. The good news was... they were confident he would recover."

"Were you well enough to visit him in the hospital?"

"Yes, but the hospital wasn't allowing visitors for a few days."

"In the meantime," O'Grady continued, "did you have a chance to look at the damage to your engine and test stand?"

Herring gave a sigh of irritation. "Encapsulated in a dense layer of soot and dirt, the engine looked to be in terrible condition. However, after a careful cleaning with a stiff bristle brush and a bit of kerosene, only the wooden test stand and some of the accessories sustained significant damage; the engine came through the ordeal unscathed.

"The propeller was another matter. Since it had fragmented, finding the root cause of the problem was a very subjective undertaking. After inspecting all of the broken parts that I could find, the most likely culprit was the hub – a steel piece that attached the propeller to the engine's crankshaft. The gas torch brazing that fused the radial blade struts to the hub... had *all* fractured. However, it was impossible to determine whether these fractures had occurred prior to or after the propeller began to disintegrate—"

"Objection! Speculation!" Robbins broke in. "The witness is trying to deflect responsibility from himself!"

Judge Sawyer glared at Herring over the top of his glasses. "What does the witness have to say about this?"

"Your Honor," Herring said, beginning to blush, "I have always taken full responsibility for the safety of my assistants. This accident occurred with my machinery, and Mr. Housam suffered because of its failure. Afterward, my only concern was to make sure that a similar incident would never happen again."

Swiveling back in his chair, Sawyer tapped his gavel. "Overruled. Continue."

"What was the remedy?" O'Grady asked.

"After considerable thought, I concluded that my 'flip-start' method of cranking the engine had subjected the brazed parts to unacceptably high loads. If I was going to use a similar propeller design in future tests, a less stressful method of starting had to be found."

St. Joseph, Michigan

Tuesday, November 11, 1902

As he approached Housam's ward, Herring was confronted by the unmistakable stench of urine and disinfectant. *Another good reason to avoid hospitals,* he thought. Immediately inside the third floor's Recovery Room C, six iron-frame beds had been crammed into an area intended for five. At the far side of the space, light filtered through a curtainless double-hung window, revealing that its yellowed glass hadn't been cleaned in at least a decade. As Gus's eyes searched the cribs for George, he wondered, *Who designed this place? Those two little wall-hung gas lamps couldn't possibly illuminate this gloomy room.*

Rusty water stains further offended old whitewashed walls, chipped and gouged from countless gurney strikes. Twelve-foot-tall ceilings, whose dusty crown moldings projected pre-Victorian-era plasterwork, provided non-comatose patients with a shadowy visual distraction.

Of the six men who occupied Recovery Room C, Housam and three other men seemed to be in various stages of recuperation; two others could have passed for cadavers. Housam, lying next to the drafty window and a rattling steam radiator, wagged his hand at Herring.

"Damn, it's good to see a friendly face! Clara left for home a while ago... poor thing had been here for two days and nights."

Complaining that he was either too hot or too cold, George's sling-elevated leg prevented much in the way of movement. Dragging a rickety wooden stool over to the end of the bed, Herring stretched to shake his friend's hand, his own first- and second-degree burns still stinging. Glancing around the space, Gus slowly shook his head.

"Christ, are you ready to come back to work? This place is depressing."

Propping himself up on an elbow, Housam winced and dropped back onto his pillow. "Shit, I'm ready ta go right now – pain or no pain. Speakin' of pain... I've got this stiff neck and headache that won't quit. But, like I've been tellin' the docs, I gotta get back to work... this place is costin' me nine bucks a day... plus I'm not getting' paid for all the time I'm off—"

"You can relax about the money," Gus interrupted. "I'm paying for your hospital and doctor bills, and all of your back wages at Engberg's."

Touched by his part-time boss's generosity, George's lower lip began to quiver. "Gus, where *you* gonna get the money? You're not exactly rollin' in dough."

"Don't worry. It's what Lillian and I have decided. Your only concern should be how to get out of here and back home. Besides... there's a bunch of testing that needs to get done on our project."

'What ya got in mind?"

Herring started to shrug his shoulders, but his burns shouted otherwise. About to curse his bad luck, he thought the better of it. "Well, we gotta mount the engine on the model. Then we have to fly it as a kite, same as we did with the full-size machine back in '98, but without me aboard this time!"

"You gonna measure all the same stuff?"

Herring nodded. "Yep. The main thing we have to find is the machine's drift in a 20- to 22-mile-per-hour wind. That will tell us what the minimum propeller thrust needs to be. We'll also need to determine the lift-to-drift ratio. That will tell us something about the model's efficiency.

"We'll get all this information from the pull on the tether line, along with the angle it makes with the ground. Think of it this way: a powered flying machine can be likened to a kite being towed along by a propeller instead of a cord, where most of its weight is raised by its aero*curve* lifting surfaces. Because propeller thrust only needs to compensate for the model's drift, and not its lift, the engine's power requirement is a whole lot less than… say, a model helicopter, where *all* of its lift must be generated by the propeller."

"Sounds good ta me," Housam said, as he struggled to sit upright.

Herring helped by stuffing an extra pillow behind his friend's back. "After the kite test, we'll have to construct another pendulum-testing machine – a small one this time. Then we'll mount the model on the pendulum arm and run the engine with its new propeller. The trick is to find the *throttle setting* that will deliver the *minimum* amount of propeller thrust needed."

"You'll git that from the kite experiment. Right?"

Herring nodded.

"Gus, I've been thinkin' about all the testin' we've been doin'… why don't we just take the little machine out to a cow pasture and give her a try?"

Although Herring recoiled at Housam's suggestion, he didn't show his displeasure. "George, let me explain something. Before I can build a practical, powered flying machine, I have to understand why the damn things act the way they do. As you know, I've been working on this problem for more than 15 years and have spent more than $30,000 in the process... but I *still* don't have all the answers. And another thing... I don't like slippin', fallin', and rollin' in cow flop!" At that, both men had a good laugh.

"If we're going to succeed," Gus continued, "I'm convinced there's a *three-step process* that should be followed. We need to test-fly the experimental machine as a kite, tailor the propeller's thrust on the pendulum apparatus, and determine the engine's horsepower on the dynamometer... in that order."

"Explain that swingin' arm test again, will ya, Gus?" Housam asked, closing his eyes.

Herring took the opportunity to scrutinize Housam's leg bandages while he spoke. "Okay, let's say the engine/propeller combination needs to deliver one pound of thrust to overcome the model's drift. A simple calculation tells me that the pendulum arm needs to deflect nine degrees for that to happen. All we have to do is throttle the engine until the model moves to that angle."

As he spoke, George's lips twisted into a sardonic grin, and he opened his eyes. "Throttle the engine? How ya gonna do that?"

"With an air-inlet valve. I can control the engine's horsepower and therefore, the propeller thrust by metering how much air is allowed into the combustion chamber."

"That's somethin' new... what does the valve look like?"

Carefully reaching inside his coat pocket, Herring pulled out a mechanism about the size of an unhusked walnut. "I brought the carburetor along so you could see it for yourself."

As soon as he handed the polished brass and aluminum assemblage of parts to Housam, the bedridden man's demeanor perked up again.

"Look inside the inlet pipe."

Slipping on his reading glasses, Housam peered into one end of a half-inch brass tube. "Looks like a little stovepipe damper!"

"Exactly! That's where I got the idea... from my coal-burning stove. By turning the shaft on the outside of the inlet pipe, I can adjust how much air the engine is allowed to breathe. It will take some trial and error before we get the correct throttle setting and fuel mixture, but when we determine that, there's only one thing left to do."

"What's that?" Housam croaked, still grinning.

"The dynamometer test. We'll remove the engine from the model and mount it on my fan brake torque stand. We'll have to find out how much horsepower she produces at our newly found partial throttle setting. By keeping the new flight propeller as the load, it'll be a simple matter to retrieve the torque and shaft speed data before making the horsepower calculation."

"Then we'll be done?" Housam deadpanned.

"Not quite, but the last step won't take that long. By moving the throttle to the wide-open position, we'll be able to find the engine's maximum horsepower with the same propeller."

"Besides braggin' rights, why is that so important to know?"

Herring swung his hands up gingerly, locking his fingers behind his head. "Horsepower is like a yardstick, George. In the first dyno test, the horsepower required will be less than the maximum the engine is able to produce. That's a luxury for me! I've never had an engine that was powerful enough to meet the minimum requirements for aero*curve*-assisted flight! Nevertheless, simple arithmetic will give me the answer I'm looking for: by dividing the engine's *maximum horsepower* into its *throttled horsepower*, I can determine the *horsepower fraction* that's required to fly the model.

"On the other hand, knowing how many pounds of flying machine can be lifted per horsepower allows me to compare the performance of different machines; for instance, Langley's steam-powered aerodrome *Number 5*, the one I helped to develop, lifted only 29 pounds per horsepower.

"There's something else. For the machine to move forward along the ground, it will have to generate more thrust than simply being able to overcome its drift in a 22-mile-an-hour wind. That will only allow it to hover above the ground! Extra thrust is needed to move the machine forward against the wind."

"How much extra horsepower do you think the little engine will make?"

"I don't know... that's what the dynamometer tests are for."

"If we wanna fly two times faster, we'll need twice the power, right?" Housam said, glancing at Gus out of the corner of his eye.

Herring flinched... then grinned. "George, you old scoundrel! Are you tryin' to pull my leg? On second thought, if Wilbur Wright doesn't understanding the concept of power and speed..."

Housam chuckled through his smile.

"Okay, you asked for it!" Herring said as he began his power-versus-speed harangue. "If you double a flying machine's airspeed, its drift will increase as the *square* of the speed increase... otherwise, there will be four times the drift, in pounds. To overcome this drift, the engine will have to produce four times the thrust, in pounds, multiplied by twice the speed, in feet per second, which equals eight times the power, in foot-pounds per second."

"Eight times... why so much?" George interrupted, playing the perfect straight man.

Herring shook his head. "Because, doubling the airspeed is all about the *cubing* of the airspeed increase."

Housam thought about this for several seconds. "If that's the case, I'm bettin' your little flyin' machine will never see 40 miles an hour!"

Both men laughed. *George might be damaged goods*, thought Herring, *but he hasn't lost his sense of humor*. Aloud, Gus said, "I agree, but there are a few other things— "

"Like what?" Housam cut in, rolling his eyes.

"Like takeoff power. More power will be needed to get the model up to its minimum flying speed, where the aero*curve* surfaces can deliver sufficient lift to overcome the weight of the machine. Once the model reaches that break-even speed, any extra horsepower might be used to make the machine fly faster, until its rapidly increasing drift equals the propeller thrust... then the airspeed will stop increasing."

"What do you mean by *'might be used'*?"

Herring reached over and patted George on the hand. *Ice cold*, he thought. "That's another one of these compromises I've been talking about. For example, for the propeller to work its best when accelerating the model to its minimum flying speed, its twist or *pitch* has to be reduced."

"What do ya mean by that?"

"We have to take some twist out of the propeller to prevent it from *churning* the air as it tries to move the model from a standstill to its minimum flying speed. The problem is, once the flying machine reaches that speed, the engine is running pretty fast."

"Why is that?"

"Reducing propeller pitch takes *load* off the engine. Anyway, the engine can only run so fast before something breaks, and we don't want that to happen!"

As an afterthought, Herring added an analogy. "It's like trying to go fast on one of my Mobikes without slippin' the transmission belt to a larger pulley wheel – it just won't go any faster until you do."

Scratching his head, Housam considered the dilemma. "Seems to me you should invent a propeller that changes its twist while it's on the go!"

"Maybe so, my friend, but I've got my hands full as it is! By the way, George, it's good to see you smiling."

"Smilin'? I ain't smilin'!"

Not knowing what to make of his friend's peculiar comment regarding his mask-like grin, Herring chalked it up to the stress of his injury and soon took his leave, promising a return visit in two days.

*

Lillian reported the troubling news to Gus on Wednesday afternoon. "George has *lockjaw*, and it's gotten worse since your visit! He's having muscle weakness and cramps near the wound, and his back muscles have begun to contract and spasm. According to Clara, his facial and throat muscles were the first to be affected—"[1]

"That explains the rigid smile and arched eyebrows," Herring interrupted, "He was also complaining about a headache and stiff neck."

*

Word of George's death, attributed to acute respiratory failure, came late in the afternoon on Thursday, November 13. Upon hearing the news, Herring scrambled to send telegrams to the men who had known and worked with Housam on the flying machine trials of 1898. Bill Avery of Chicago and Henry Clarke of Philadelphia showed up for the funeral.

After the burial, a wake was held at George's home. Grief-stricken by the loss of his friend, Gus Herring delivered an envelope containing $750 to Clara, George's distraught wife. It was enough to pay the hospital bills, plus an additional six months of wages he would have earned at Engberg's.

For their part, the flying-machine men lamented the loss of another respected associate. Two of the original five were now dead, and the others were experiencing problems of their own, including Gus. On the evening of the funeral, Chief of Police Sauerbier paid an unannounced visit to Herring's mystery barn, where Gus and his out-of-town guests had gathered. "Herring, if you have the *cheek* to run another damn engine in a residential neighborhood, I will arrest you for disturbing the peace and disorderly conduct... along with anything else legal counsel can dream up!"

Sauerbier cited his concerns. "Besides your suspected involvement in the Truscott fire, you're now responsible for the death of a respected, lifelong citizen of St. Joseph!"

Stunned, Herring held his tongue. With no response forthcoming, the chief continued. "The residents of our little town have long since had enough of your foolishness! Most consider you to be either batty, a crank, or both!"

Sauerbier finished on a personal note. "I want you out of St. Joseph, Herring... the sooner the better!"

*

Henry Clarke, between jobs and financially depressed, readily accepted Gus's offer to work on the gasoline engine-powered model aeroplane project. Room and board plus 20 cents an hour sealed the deal. With Chief Sauerbier's threats still fresh in his mind, Herring again sought out the old Silver Beach Pavilion site for his base of operations. Located in St. Joseph's commercial-industrial zone, the pavilion was not subject to the town's residential noise ordinances. At $14 a week plus the cost of heating the place, Herring contracted for the remainder of the year, with an option for an additional month's occupancy.

Early on Thursday, November 20, Herring and Clarke began loading tools, equipment, and supplies onto a rented horse-drawn cart. After three trips to the pavilion, the majority of the move had been accomplished without incident, including the transfer of the disassembled model flying machine.

Friday dawned bright and sunny with a crisp 20 mph wind blowing from the southwest. With air temperatures in the mid-20s, conditions were far from comfortable on the frozen sands of Silver Beach, but quite favorable for kite flying. With hot potatoes stowed in their pockets for warmth, the men had little trouble obtaining the required pull force and angle data needed to determine the *Double Decker* model's lift and drift.

Back inside the heated pavilion, Clarke gobbled his bologna sandwiches while Herring, too occupied to eat, drank a glass of cold milk and worked on his calculations.

"Listen to these numbers, Henry! The drift in a 22-mile-per-hour wind is less than a pound! Eighty-six-hundredths, to be exact. That's all the thrust the propeller will have to produce to fly the machine. There won't be any ground speed, but she'll still be flyin'! There's more good news: the lift-to-drift ratio works out to be eight to one; that's better than my '98 power machine."

"What's our next step?" Clarke asked, as he sipped his coffee.

"We're gonna mount the ready-to-fly model on the pendulum machine, run up the engine, and determine the throttle opening where the propeller will deliver 0.86-pound of thrust... turns out, that's only a seven-degree deflection from the vertical. Then we'll lock the throttle arm in place and move the engine over to the dynamometer. We need to find how much power she's puttin' out."

"What'll that tell us?"

"Among other things, the efficiency of the propeller."

"Okay, where's the propeller, and how about the pendulum machine?" Clarke asked as he glanced around the pavilion.

Herring laughed and slapped his old friend on the back. "We have to build them! That's what we'll be doing for the next couple of weeks."

Rochester, New York
Wednesday, November 23, 1921

Suspending his gavel above the oak block, Judge Sawyer directed the litigants to "... enjoy your four-day holiday... but remember," and here, he paused, "court proceedings will continue at 9 a.m. on Monday, November 28."

The crack of Sawyer's gavel unleashed a frenzied rush from the courtroom. Catching the first train available, Gus and Chloe arrived at their Long Island home just before midnight. Reunited after many weeks of separation, Lillian listened intently as Gus described the hardships associated with the trial and overcoming his two mild apoplexies. To the regret of all, the long Thanksgiving weekend passed quickly. Early Sunday morning, both women accompanied Gus back to Rochester by train, where they would remain together until the end of the trial.

Rochester, New York
Monday, November 28, 1921

After an abbreviated "welcome back" from the judge, O'Grady was directed to recall the plaintiff's witness back to the stand.

"Mr. Herring, prior to the holiday adjournment, you told the court that a pendulum machine and propeller had to be constructed before engine testing could proceed. Is that correct?"

Herring shifted in his chair. "Yes... but there was a problem. The pendulum machine was ready in short order, but the propeller held me up because I wasn't able to get it brazed until after the holiday. By the time I retrieved it from Chicago, we were into December."

"Why was the work done in Chicago?"

Staring down at his high-top shoes, Herring hesitated a few seconds before answering. "After George Housam's regrettable death, I no longer trusted my high-temperature soldering – so I paid a professional to perform the work."

"What happened next?"

"Henry and I installed the engine and accessories into the model. While he was busy with the details, I finished work on the new propeller."

"What had to be done?"

"Besides covering the steel tubing blades with a double layer of Japanese silk and coating them with pyroxiline dope, the entire assembly had to be balanced both vertically and horizontally and then corrected for any fore and aft runout."

"Explain to the court how this new propeller was safer than the one that disintegrated."

Herring continued, "A grooved pulley wheel, known as a sheave, was incorporated into the propeller's hub—"

"Yes, yes," O'Grady interrupted, "but how did these changes make the propeller *safer*?"

Unperturbed at being cut short, Herring continued. "The hub was strengthened by brazing the steel spars – the ones that formed the forward and trailing edges of the blades – rigidly to the pulley's forward face. In addition, the pull-rope method of starting eliminated most of the harsh forces associated with the 'hand-flip' method."

O'Grady breathed deeply before launching into his next series of questions. "When did you finally run the pendulum trial?"

"On Monday, December 8, 1902."

O'Grady looked skeptical while extending an upturned palm. "How can you remember the exact date?"

"It was the same day I got the *eviction notice*."

"Eviction? From where?"

"The pavilion."

"But, didn't you sign a lease?"

"Yes... but the owner of the pavilion said he was under pressure to get rid of me."

"By whom?"

"After some prodding, he admitted it was the chief of police and a handful of concerned citizens. They convinced him that I was a menace to the community."

"What did you do next?"

Herring rolled his eyes. "I had an attorney talk to the man."

"And?"

"In return for my written promise not to sue him, the man refunded all but two weeks of my rent... but we had to be out before December 15, the following Monday. Later that week, I learned that Sauerbier had also approached my Church Street landlord, but the man had refused to throw us out. He told the chief that I had always paid my rent on time and took good care of his property."

"What about the unfinished tests?"

"With only six days left, Henry and I worked around the clock to wrap things up."

O'Grady turned to a new page in his notebook. "Tell the court about the pendulum test."

Herring's demeanor changed for the better as his mind shifted to the technical side of his story. "With Sauerbier looking for trouble, I decided to run all my engine tests *inside* the pavilion."

"With all the windows and doors closed?" O'Grady asked with a frown.

"Actually, the toxic accumulation of exhaust smoke turned out to be a minor inconvenience."

"How did you handle that problem?"

Herring grinned. "It was a three-ring circus! After shutting the engine down, I would run to open the big door while Henry pushed open a window. Then we both rushed outside until all the smoke and most of the building's heat had cleared out. We burned two cords of hardwood in the stove that week!"

"Other than the wasted heat, how did the testing turn out?"

"Besides Henry's understandable reluctance to approach the spinning propeller, the machinery performed as designed. Engine starting proved to be safe and predictable, while adjustments to the throttle, fuel mixture, and ignition advance were routine. Within the first hour of operation, I had established the throttle setting that delivered a steady seven degrees of deflection."

"What came next?"

"We removed the engine, propeller, and accessories from the model and mounted them to my small engine *dynamometer*. We had to determine the engine's power at the throttle setting that delivered 0.86 pounds of propeller thrust. After that, the only thing left to do was to run the wide-open-throttle test."

Momentarily distracted by the wind-driven snow and sleet pounding against the courtroom windows, O'Grady hesitated before asking his next question. "Mr. Herring, explain the purpose of a dynamometer."

Clasping his fingers together in front of his chest, Herring welcomed the opportunity to enlighten the court about this aspect of mechanical engineering. "A dynamometer is a machine that measures the *torque*, or twisting force, of any rotary-shaft-output motor or engine. It can handle windmills, steam motors, or anything in between."

"Why did you have to measure your engine's torque?"

"So I could calculate its *power*. Two things are needed to do this: the engine's torque, and its shaft *speed* with a given load... in our case, it was the flight propeller."[2]

"What did you learn from the dynamometer test?"

Pausing, Herring struggled to recall the specifics of an engine test performed almost two decades earlier. "If memory serves me, calculations showed that the propeller was operating at about 77 percent... not a bad efficiency for the day. All propellers waste a bit of power when converting the engine's power into *thrust*."[3]

"Did that test complete your work with the dynamometer?"

"There was only one more test to run... and I was in for a nasty surprise."

CHAPTER 79

St. Joseph, Michigan

Wednesday, December 10, 1902

Gale-force Arctic winds blew across the warm waters of Lake Michigan, creating winter mayhem in the town of St. Joseph.[1] With only four days remaining on his renegotiated pavilion lease, Herring paid little attention to the inclement weather. He was consumed by a more immediate problem: learning why his little engine wouldn't start, and how he could fix it.

Unable to see clearly within the shadowy confines of his improvised shop, Herring hung a second kerosene lantern close to the dynamometer-mounted engine. Adjusting the lantern's mantle to maximum brightness, he began his investigation by poking around the tank and carburetor, looking for blockages or leaks. Unable to find anything askew, he turned his attention to the finicky spark-ignition system. With the aid of his *d'Arsonval* voltmeter, Gus pinpointed the problem within seconds. As he rose from his wooden stool, the frustration in his voice was unmistakable. "These damn dry cells are low on voltage *again*! Henry, I need you to fetch a new set from the Church Street shop."

Because the big pavilion door faced directly into the storm's fury, Clarke opted for the seldom-used man door on the leeward side of the building. When he slammed the door shut behind him, he disturbed a wagonload of snow, which plummeted from the structure's steep gable roof. Leaning forward in a futile attempt to sweep the slush from his neck and shoulders, Clarke tried to maintain a positive attitude. *Slogging through drifts up to my knees will probably take two hours,* he thought. *Thank goodness for baked potatoes!*

*

As darkness settled in, a dog-tired Clarke stumbled back into the pavilion carrying an ice- and snow-covered pasteboard box.

"I was getting worried," Herring said. "Thought something might have happened to you!"

Draping his sodden woolen coat over the old wooden rack adjacent to the stove, Clarke's attention turned to the box and Lillian's roast beef on rye sandwiches that awaited him and his boss. Famished, he could almost taste the buttered baked potatoes and steaming hot coffee that surely would accompany the impending feast.

"Nasty out there," Clarke stammered, warming his hands over the stove's crackling wood fire. "I'd say there's two feet on the ground, with four-foot drifts in places... and the snow's not lettin' up! What you been doin'?"

Herring had already replaced the first of four depleted dry cells. "I assembled the regulating tail and got it mounted on the frame. Then I adjusted the strip rubber springs... but it's all guesswork until we fly the machine."

After supper, Herring began his pre-start routine by making certain that the throttle's "butterfly" was locked at its wide-open position. As he wrapped the cotton cord around the starting pulley, an atmosphere of tense anticipation filled the shop. With the first pull, the engine sputtered and then growled to life, its propeller blowing a cloud of dust and debris off the floor and into the air. Squatting safely behind the whirling blades, Herring – as was his practice – alternated between adjusting the ignition advance and the fuel mixture. With his head tilted toward the exhaust, he listened intently to its frequency – a sign that the engine's shaft speed was either increasing or decreasing.

With stopwatch and clipboard in hand, Clarke stood behind and to the right of Herring, where he could monitor the repositioned revolution counter. Given a nod from his boss, he started the watch on an even number and waited for 300 revolutions to roll off the mechanical counter before stopping the instrument with a flick of his thumb. Using the conversion chart, it was a simple matter to translate seconds into revolutions per minute. Convinced that the engine was running as fast as it could, Herring made a cat-like move to the rear of the dynamometer, where he noted the exact position of the torque-arm pointer on the stationary scale.

Suddenly, they heard a loud crash... as if someone had slammed a door. Startled, the experimenters turned toward the source of the noise, where three shadowy figures stood in the blue-gray fog of exhaust smoke. With his tuning-and-performance test complete, Herring toggled the ignition switch, triggering an abrupt halt to the engine's reverberating bark. Ignoring the uninvited guests, Herring and Clarke rushed to open the door and window before hustling outside for a breath of fresh air, closely pursued by Police Chief Sauerbier and two uniformed patrolmen.

Frowning, Sauerbier stated the reason for his visit. "Augustus Herring, I'm placing you and your assistant under arrest!"

Wheeling about to face his antagonist, Herring's unexpected movement prompted the two cops to reach for their Billy clubs. "What's the charge? This isn't a residential zone; besides, I operate my equipment behind closed doors!"

Dangling an official-looking document at arm's length, Sauerbier continued. "You've been charged with contaminating a public eatin' place – a possible felony. The judge will decide. I'm here to take ya in! Git yer coats!"

After stopping down the stove's damper and quenching the kerosene lanterns, Herring and Clarke were herded out of the building. With wind-whipped snow stinging his face and hands, one of the cops struggled to padlock the door.

"This is an official crime scene," Sauerbier snarled. "It's *illegal* to reenter the building without a signed release from the judge. Guess that puts a damper on yer bullshit, eh Herring?"

Locked within the squalid confines of the town's two-horse paddy wagon, Herring and Clarke huddled against the snow and cold, the wagon's three barred windows affording little protection. The chief and two patrolmen sat beneath an abbreviated canopy that also proved inadequate. By the time they arrived at the station house, everyone was frozen to the bone. It was 8 o'clock before the alleged polluters of public property had been formally charged. Placed in a 12x12-foot cell with two drunks and a shoplifter, Herring was eventually allowed his one telephone call.

Recovering from the initial shock of her husband's incarceration, Lillian phoned family attorney Samuel J. Morningstar, instructing him to arrange for the men's bail... but the storm delayed his appearance until noon the following day. Bail for the two had been set at $50. Upon their release, Morningstar explained the charge. "The warrant states that by operating your engine within the confines of the Silver Beach pavilion, you have contaminated a place where food is prepared and sold to the public – a third-degree misdemeanor in the State of Michigan."

"If convicted... what's the penalty?" Herring asked.

"Up to a year in prison and a thousand-dollar fine – to be determined by the judge."

"That's outrageous! It's a trumped-up charge, Sam! Who ever heard of contaminating a public place with exhaust fumes?"

The attorney shrugged. "I'll speak to the judge before we go to trial tomorrow. I'll see if I can get the charge reduced to a violation and a warning. However, I suspect the chief will use his influence to push for a conviction on the higher count. You know he's got it in for you."

*

The Berrien County courthouse underscored its separation from the police headquarters with two, 10-foot-tall, three-inch-thick white oak doors of medieval split-arch design. Used jointly by the town of St. Joseph and the county, the facility's Victorian theme, opulent materials, and old-world craftsmanship spoke of its significance to the community. The centerpiece of the room's 30x48-foot *golden rectangle* proportions was a splendid crystal chandelier, recently updated with clear-glass Edison bulbs for illumination. Wall-mounted kerosene lamps lighted the room's cherrywood appointments, which included the walls, flooring, pew-style benches, and 12-panel doors. Much of the woodwork, such as the judge's bench and the barrier railings, were embellished with polished-brass fittings.

It was 8 p.m. before Judge Frederick Durkee addressed the case of Augustus M. Herring and Henry V. Clarke. Flanked by two armed bailiffs, the accused, accompanied by Attorney Morningstar, were marched to the bar. Lillian watched nervously from her seat in the gallery, which had slowly filled to capacity during the course of the evening. The arrest and trial of Professor Herring was big news in St. Joseph.

One of the latecomers was the chief of police, who made a spectacle of himself by barging to the front of the gallery, where he stood menacingly, overseeing the contingent of litigators, courtroom personnel, and the accused. The presence of a reporter and photographer from the Benton Harbor *Herald* added to everyone's sense of anticipation.

After reading the charge, the judge asked, "Gentlemen, how do you plead?"

"Not guilty, Your Honor," Herring and Clarke said, in chorus.

Having previously decided on a non-jury trial, the men were anxious to get started.

"The prosecuting attorney may proceed," the judge directed.

Simon D. Greenleaf, the county's new assistant district attorney, was a 26-year-old recent graduate of Chicago's prestigious Northwestern University School of Law who was taking his first case to trial since having been recruited a month earlier. The fledgling attorney, a rail-thin young man of average height who parted his straight brown hair down the middle, peered fretfully through cumbersome bifocals.

"The prosecution calls Chief of Police Charles Sauerbier."

Instantly recognizable by the multitude of gold-plated buttons that adorned his navy blue uniform, Sauerbier endured the mandatory swearing-in process before taking his seat as a witness.

"Chief Sauerbier," Greenleaf stammered, "how were the deplorable conditions at the Silver Beach pavilion first brought to your attention?"

"Citizen complaints."

"Objection, Your Honor," Morningstar said. "Lack of specificity."

"Chief Sauerbier," the judge challenged, "who were the complainants?"

"I didn't write their names down! Besides, I checked the pavilion myself... it was a pig sty!"

"Objection sustained," Durkee said, glaring at Sauerbier. "The court expects you to provide the accusers' names... the defendants have the right to know! You will provide these names at the end of tonight's session or be held in contempt of court. Is that clear, Chief?

Sauerbier squirmed in his chair. "If you say so, *Your Honor*."

Turning toward the prosecutor, Durkee flicked his hand out. "Continue, Mr. Greenleaf."

"Chief, what did you observe while inside the Silver Beach pavilion?"

Sauerbier snorted. "Smoke! So thick you could have cut it with a knife!"

"Anything else?"

"Ain't that enough? If it was summer, the hamburgers and frankfurters would have been—"

"Objection! Argumentative!" Morningstar barked. "Since when does smoke constitute a pig sty?"

"Sustained. Please refrain from embellishing the facts, Mr. Sauerbier," the judge scolded.

"Is there anything else you want to add, Chief?" the prosecutor asked.

"Yeah! Saul Green, the town's chief medical man... he agrees with me—"

"Objection! Hearsay!" Morningstar interrupted. "If he's called as a witness, Dr. Green can speak for himself."

"Sustained. If the prosecution has no further questions, the defense may cross examine."

Paging through his notes, Morningstar meandered to within 10 feet of the prosecution's witness. "Chief Sauerbier, how long have you known the defendant, Augustus Herring?"

Sauerbier stared down at his faultlessly polished lift shoes before leaning back. "I don't remember."

The defense attorney took a step closer. "Allow me to refresh your memory. Didn't you interrogate Mr. Herring immediately after the Truscott boat factory blaze back in '99?"

Sauerbier inclined his head. "Lemme think... now that you mention it, maybe I did."

"Maybe? Three years ago you accused the defendant of having played a role in that fire! You insinuated that the arsonist might have been motivated because of Mr. Herring and his flying machine work... isn't that true, Chief?"

"Objection! Counsel is... well, he's leading the witness Your Honor!" Greenleaf stammered. "Besides, it's all hearsay!"

Morningstar stared up at the judge. "The defense withdraws the question, Your Honor."

No longer able to control his temper, Sauerbier slowly stood before he began shouting at Morningstar. "So what if it's true? As a matter of fact, the son of a bitch is *still* my prime suspect—"

The sudden "pop" of the newspaper photographer's flash powder added to the courtroom confusion. Rising from behind his bench, Durkee banged his gavel. "Order in the court! Chief Sauerbier, control yourself!"

Wheeling around to face the judge, his neck purple with rage, the chief continued his tirade. "This trial is supposed to be about Herring breakin' a town ordinance! It's about Herring puttin' the public in harm's way!"

Sweeping his arm toward the door, Durkee's black robe fluttered open, revealing fire-engine-red suspenders and a sleeveless white tee shirt. "Chief, you're out of order!" the judge bristled. "You are hereby excused from offering any further testimony! Vacate these chambers! Immediately!"

As Sauerbier stormed out of the courtroom, Durkee delivered a warning. "Rest assured, Chief, the court will not forget your contemptuous conduct... there *will* be repercussions!"

A prolonged murmur throughout the gallery prompted Durkee to bang his gavel several more times. "Call your next witness, Mr. Greenleaf!"

"The prosecution calls Dr. Saul Green."

Heads swiveled one way and then the other as the gallery awaited the arrival of Dr. Green. After several awkward moments, everyone realized that the old physician was indeed absent, causing more whispering from the spectators.

"Try the Pig's Ear," a male voice mumbled, referring to the doctor's favorite tavern.

"Order in the court!" the judge bellowed. "There's no need for comment!"

Turning to the young prosecutor, Durkee once again instructed him to call another witness. "Due to the unexpected absence of an important witness, the prosecution requests a week's adjournment—"

"Objection!" Morningstar cried, cutting Greenleaf short. "The defense is prepared to continue with its case! A continuance due to the prosecution's ineptitude is totally unacceptable."

The judge rubbed his cleanly shaven chin with an index finger. "Request for a continuance is denied. The defense may present its case."

Morningstar promptly called Herring to the witness stand. "Mr. Herring, did your little aeroplane engine contaminate the interior of the Silver Beach Pavilion?"

Herring looked peeved. "Absolutely not!"

"The prosecution contends that smoke from your engine's exhaust has the ability to contaminate food that will be prepared there next summer... a full six months from now! Do you have anything to say about this, Mr. Herring?"

Herring shook his head in denial of the charge. "Mr. Clarke and I have eaten lunch inside the pavilion for almost a month. If there were any health-related food issues, we'd know it by now!"

After prosecutor Greenleaf had cross-examined both Herring and Clarke, Judge Durkee opened the court to comment from the gathered citizenry. The first four individuals were unanimous in their condemnation of the "professor" and his experiments, which they said had already caused unrest and death within their town.

To Herring's astonishment, the following eight were in complete favor of his flying machine efforts, while expressing concern for the frivolous charges levied against him and his assistant. One was Michael Engberg; the seven others were complete strangers to the defense.

Noting that neither individual had a previous criminal record, Durkee reduced the charges to a misdemeanor and levied a token fine of two dollars. Afterward, in private, he offered sage advice to Herring. "Find someplace else to perform your experiments, young man. I'd suggest another township. Sorry to say, it seems as if Chief Sauerbier will be extra vigilant in regard to your activities."

*

With little more than 24 hours remaining on his lease, Herring faced the grueling task of moving his equipment and model flying machine back to his Church Street shop. Transport was difficult enough during ideal conditions, but the heavy accumulation of snow made travel by horse-drawn wagon next to impossible. Desperate, Herring endured the 90-minute slog to Engberg's shop, where he sought the advice of his friend and business associate. "Mike, I've got a problem, and I think you're just the man who can help. How can I haul my equipment up the hill from the pavilion to my shop when wagons are getting stuck all over town?"

Engberg looked put out. "Won't old man Williams give ya a few extra days on your lease?"

"I talked to him last night. He seemed sympathetic, but Sauerbier is holding him to today's midnight deadline. The chief said that if we're not out by then, he's gonna lock me up."

"That son of a bitch!" Engberg hissed. "I've got a way... follow me out to the barn!"

An hour later, Herring and Engberg slid to a halt beside the pavilion's front door – in a horse drawn sleigh. Wheeling around to face the newcomers was the snow-encrusted form of Henry Clarke, who had been shoveling for most of the morning.

"You're a sight for sore eyes!" he panted. "But, there's problem... I can't get inside the damn building! The police padlock is still—"

"I figured as much," Engberg cut in, jumping down from the buckboard. "I'll take care of that." Grabbing his long-handled bolt cutter from the back of the sleigh, Engberg made short work of the second-rate lock. "You're all set! Let's start loading your stuff!"

With the sleigh full of equipment and materials, Engberg accompanied the men to Herring's shop, where the shoveling routine was repeated. This time, both Lillian and young William helped, along with a sympathetic neighbor who had learned of Sauerbier's vendetta.

Heading back for another load, Herring first dropped off Engberg at his shop.

"If ya have any more trouble with Sauerbier, call me... I fer one am fed up with him oversteppin' his bounds." Hesitating at the entrance, Engberg watched his sleigh slip down Lake Street toward Silver Beach.

*

By nightfall, the only item that remained to be moved was the model *Double Decker*. Choosing not to risk any damage, Herring decided to break the model down into its major components. This required another two hours, which included packing the pieces into the sleigh. Lillian, an experienced hand at protecting delicate belongings from damage during the family's many moves, sent along two blankets and three old pillows.

With 10 minutes to spare before the deadline, the men prepared to vacate the pavilion. After closing the damper on the wood-burning stove and extinguishing all but two of the kerosene lanterns, Herring turned to his assistant.

"Henry... it's time. Let's get the hell out of here!"

Lanterns in hand and exhausted from their daylong efforts, the men shuffled toward the big door for the last time when they heard a shout. "Herring! You haven't heard the last of me! I'll be watching you... *professor!* Step out of line *once*, and you'll be behind bars with all the other thieves and scoundrels! Simpletons like Judge Durkee can't protect your sorry ass forever!"

The diminutive Sauerbier strutted about the premises clutching the remains of his police-issue padlock, which he tossed at Herring's feet. "Now yer destroyin' police property, eh Herring?" he snarled. "You owe the taxpayers of St. Joseph 50 cents for this here lock. Pay up... or I'll run you in here and now!"

Herring reached into his front pocket, pulled out a crumpled dollar bill, and thrust it toward his antagonist. Unfolding and inspecting the currency, Sauerbier slipped it into his watch pocket. "Hope you're not waitin' fer change, Herring, 'cuz you ain't gettin' any. Fifty cents is my fee fer havin' to come all the way over here in such shitty weather! Now get out and—"

"Buy *two* padlocks Chief!" Herring interrupted. "One to shit on and one to cover it up with!"

Ignoring the remark, Sauerbier sauntered over to Henry Clarke and stared up into his face. "As for you, Clarke... isn't it time you headed back to the 'city of brotherly love'? You've worn out your welcome here in St. Joseph."

*

Early the next day, determined to put the Sauerbier encounter behind him, Herring immersed himself in the reassembly of the gas model. Undeterred by the chief's stern warning, Henry stayed aboard to help with preparations for a flight trial. "With all the distractions last week," he said, "you didn't tell me how that last dyno test turned out. How much horsepower did she develop at wide-open throttle?"

Herring nodded as he continued to work on the *Double Decker's* motor mount. "I did the calculations while we were stuck in jail... you were sleeping."

Herring pointed to a crumpled sheet of paper lying on the workbench.

"We got 0.47 brake horsepower at 2,800 revolutions per minute. That means the engine is capable of lifting 108 pounds per horsepower.[2] There's an interesting comparison here," Herring continued. "My calculations show that only 12 percent of the model's total dead weight, in the form of propeller thrust, is needed to fly our model in a 20-mile-an-hour wind. I have to laugh when I think about Langley's strict policy. We couldn't try any of the aerodrome models unless the propellers were delivering half the total dead weight. Funny thing... the damn things barely flew anyway!"[3]

"Now that we've got all the facts," Henry said, "what's next on the agenda?"

Herring's latest revision to the airframe involved raising the engine. In its current condition, the propeller hung slightly *below* the central landing skid, where it stood a chance of striking

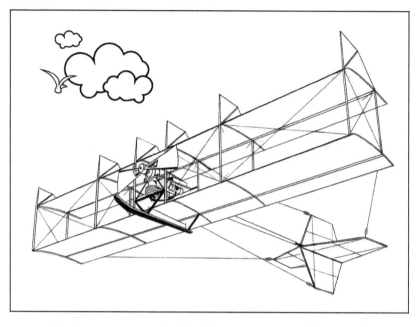

Herring's gas powered 'Flyin' Fish' model aeroplane (1-3-1903); drawing by author

the ground during alighting. To counteract the higher center of gravity, Gus decided to invert the engine, pointing the relatively heavy cylinder assembly downward. With all the requisite changes required, he figured that the total dead weight would increase from seven to nine pounds... a 22 percent increase.

"Two pounds!" Clarke hooted. "Why so much? Won't the extra weight affect the machine's performance?"

Herring paused a moment. "I don't think so. The machine has 11 square feet of lifting surface; that works out to a wing loading of only 10 ounces per square foot. By adding two pounds, the loading only increases to 13 ounces. The extra weight also includes the new *wing sails* I'm going to use—"

"What the hell are wing sails?" Clarke interrupted.

Striding over to his workbench, Herring flipped his wastebook open to a set of freehand sketches. "This is my latest idea for *passive* lateral control. Old Man Chanute would wet his pants if he saw this – I believe that wing sails could represent one of automatic stability's finest hours! The idea came to me while talking to Wilbur Wright a couple of months ago.

"Notice the six triangular sails lined up from wingtip to wingtip above the top lifting surface? In nautical terms, they're called *jib* sails—"

"They look like *fins* on a fish!" Clarke cut-in.

"So they do," Gus admitted. "The six jib sail *masts* are nothing more than extra-long front wing struts... I had to make six new ones.

"Next, a sleeve is sewn into the long adjacent side of each sail-cloth jib and is slipped over its mast and fastened in place. The rear corner of the jib is then fastened to an eye-fitting directly behind the mast at the rear wing spar."

"What are these things supposed to do?"

"Well, that's just it," Herring said with a tone of resignation. "Wing sails are *supposed* to add a degree of lateral stability by opposing any sideways slide the machine makes toward a downward tipped wing. Most often, this happens when a gust from the side disrupts the machine's equilibrium."

"I thought your regulating tail was supposed to take care of that."

"There are times when the regulator is overwhelmed, unable to prevent a sideways slip, especially when a big gust tips the machine. I believe these sails will oppose an otherwise inevitable slide, allowing my automatic regulator to do its job. The Wrights say that sideslipping leads to what they call 'well digging.' "[4]

Rochester, New York

Monday, November 28, 1921

Returning from lunch with Lillian and Chloe, Herring had barely reached the plaintiff's table when the judge ordered him back to the witness stand.

"Mr. Herring," O'Grady began, "when did you finally get to test the new gas model?"

Taking a deep breath, Herring leaned forward. "Not until after the first of the year – the actual date was Saturday, January 3, 1903."

"Why was there a delay?"

"Changes took longer than expected, and my assistant wasn't available... he had traveled back to Philadelphia for Christmas."

"Meanwhile, did you get anything accomplished?"

"As the new editor of *Gas Power* magazine, I was responsible for preparing the upcoming issue during the daytime. On evenings and weekends I worked at modifying and preparing the *Double Decker* gas model. Then there was the question of locating a suitable flying site; anywhere within Berrien County, which includes St. Joseph, was out of the question."

"Did you find a new site?"

"Yes, near the village of Stevensville, Michigan, which is in Lincoln Township. Heading toward Chicago, it's about five miles southwest of St. Joseph."

"Why did you decide on Stevensville for the trials?"

"Stevensville was outside Sauerbier's jurisdiction... and the wide-open spaces were ideal for our flight trials!"

"It was the middle of winter!" O'Grady said. "How did you transport the model and your equipment?"

Herring grinned as he rubbed his shoulder. "We modified Engberg's horse-drawn sleigh. We added hickory bows that supported a canvas cover over the cargo area. It looked like a little Conestoga wagon... with runners!"

"Once you arrived in Stevensville, did you have to work outside in the cold and wind?"

The mere thought of the ordeal caused Herring to rub his hands together. "Some of the time, but fortunately, we had use of an old barn."

"Tell the court about the flight trials."

Stevensville, Michigan

Saturday, January 3, 1903

Mike Engberg had accompanied Herring and Clarke to his friend's 200-acre farm, half a mile east of Stevensville. Harold Fose, an octogenarian former farmer, had struck it rich from a half-dozen crude-oil stripper wells scattered across his property. Deemed unproductive for commercial exploitation, the trial wells were abandoned during the mid-1880s. Disenchanted with the meager income provided by farming corn and soybeans, Fose used his homegrown mechanical skills to connect a worn-out steam engine to a force pump salvaged from an old Aermotor windmill. To his delight, the first of his "dry wells" began pumping *sweet crude* at the modest rate of two barrels a day. At 42 gallons to a barrel, Fose's lone well produced almost 2,500 gallons that first month, grossing him $60, at a dollar a barrel.

Finding a buyer for the stinking black fluid turned out to be easy. A small refinery south of Chicago promised to buy all he could pump. The Chicago and Western Michigan railroad ran across his land, so Fose convinced their management to lay down a siding long enough for a single tank car. Within a year, he was filling two tank cars a month. His most expensive problem involved moving the crude from his well sites to the railroad siding. After much hand wringing, he installed two 3,000-gallon holding tanks at each well, connecting them with three-inch iron pipe to another engine and pump located at the siding.

Two years later, Fose abandoned his habit of monitoring the weather. He also forgot about the growing seasons; after all, petroleum was grossing him the princely sum of $400 a month! To no one's surprise, Fose sold all his farm equipment, vowing never to spread another pile of manure. Instead, he hired three roustabouts to operate the engines and pumps, while he assumed the role of company president.

*

As the hybrid sleigh slid to a halt in front of the century-old limestone-faced farmhouse, an elderly man greeted the newcomers. Dressed in a tattered red woolen shirt and a pair of ancient bib overalls, Harold Fose ambled over to his friend. With all the subtlety of a barbed-wire fence, he delivered a gentle but well-placed blow to Engberg's ribs.

"Hear you've been playin' with toy flyin' machines! That true, Michael?" Fose asked, his unkempt mane of trailing white hair flapping in the mid-morning breeze.

"Not me!" Engberg barked, gesturing toward Herring. "This is the young genius I've been tellin' ya about. Augustus Herring, meet Harold Fose. When he's not around, we call him *Fosey.*"

"How long did it take ya to get here?"

"Two hours," Engberg said. "We took the path that ran next to the tracks – just like ya told us to."

"Pretty good time considerin'.... I gotta get me one of them there sleighs."

The men, although anxious to start assembling the model, were obliged to follow their host into his first-floor office for coffee.

"When I'm not out in the field, you'll find me in here, workin' on the books," Fose said, as he turned up the mantles on the two wall-mounted kerosene lamps. With coffee already brewed in anticipation of their arrival, Harold snatched three semi-clean cups from their wooden pegs... carefully eyeballing each, before dusting them with a sweep of his shirttail. The room smelled of mold and ancient dust, and its red-oak walls and 10-inch rough-hewn ceiling beams reminding Herring of his father's old hunting lodge back in Georgia – especially the wall-mounted animal heads, firearms, and memorabilia from bygone expeditions.

"You must keep your taxidermist busy," Herring said, as he surveyed the shadowy collection.

"Do the work myself," Fose said, puffing out his chest. "My only hobby, other than huntin'."

Herring nodded his approval.

"Tell me, Herring," he said, as the men poured their coffee, "is there any money to be made in the flyin' machine business?"

Herring slowly shook his head. "Not for a while, I'm afraid. First, somebody has to show that they're capable of flight with a man aboard."

Fose mimicked Herring by also shaking his head. "Lemme see if I got this straight, young fella. There's no money to be made in flyin' machines, and ya get to be called a fool for tryin'!"

Sensing that his host was only half serious, Herring grinned.

"Well..." Fose continued, "don't let the sons of bitches change your way of thinkin'! The naysayers told me my land couldn't produce oil more than 20 year ago... but now ya don't hear 'em crowin'! Matter of fact, hardly a week goes by that one of 'em don't ask me if there could be any of that stinkin' black stuff on their land!"

Stopping to guffaw and blow his nose, Fose had another thought. "I was 60 years old when I got into the oil business, so you're never too old to strike it rich! The important thing to remember is not to quit. Don't let bastards – like that chief of police – stop ya! When Michael told me 'bout the fix you were in, I decided right then and there to let you use my land... and the barn out back. To make things cozy, I even moved a coal-burnin' stove out there, too. The old barn's kinda leaky, but the stove will take the chill off while you're gettin' yer machine ready ta go."

After thanking Fose, Herring joined the others aboard the sleigh for the jaunt to the barn, about a thousand feet to the south. Skidding to a halt in front of the structure's massive wooden doors, Herring jumped from the buckboard, and with some difficulty, slid them open. Within seconds, the sleigh's steel runners scraped noisily as they passed onto the barn's hard-packed dirt floor. While Engberg unhitched the mare, moving her to an empty stall, Herring tended to the Franklin stove. Prior to closing the doors, Clarke lit four kerosene lanterns that he had retrieved from the wagon. Having completed these initial tasks, the threesome gathered around the sputtering stove to drink more coffee and plunge into their bag lunches.

"Will you look at this place?" Engberg marveled, shedding his heavy coat. "It must be over a hundred years old. Those beams were cut from first-growth timber. You don't find 30-footers with an 18-inch face anymore!"

As Herring gazed upward, the magnificence of the old structure's construction became apparent; it was a collection of pinned mortise and tenon, rabbet, pinned lap, and other fine joinery techniques. Although the barn had been maintained regularly throughout its long existence, a natural dark-brown patina had formed over its vertical red-oak siding, giving the construction a somewhat sinister appearance when viewed from a distance or at night.

A U-shaped hayloft surrounded the barn's central open area, providing a ceiling for the half dozen horse stalls and storage bins. The remainder of the 30x40-foot structure had been devoted to the storage of heavy farm equipment. With the success of Fose's crude oil business, all the machinery had long since been sold.

*

It was mid-afternoon before the gas-powered model, now dubbed the *Flyin' Fish*, was ready for its first trial. Outside, brilliant sunshine was mirrored by acres of snow-covered fields. Returning to the shadowy confines of the barn, Herring stomped the snow from his boots. "Temperature's holding at 20 and the wind's averaging about 19 out of the west. It's cold, but conditions are perfect... gentlemen, it's time to go flyin'!"

Engberg, burdened by a heavy wooden toolbox and a quart tin of gasoline-and-oil mix, slogged through ice-encrusted snowdrifts that crept up past his knees. Herring and Clarke followed in his footsteps, struggling to keep the unwieldy model from being torn from their grip by the freshening wind. Three hundred feet east of the barn, Herring shouted, "Far enough! Let's set up here!"

After stomping a clear spot in the virgin snow, Clarke turned the *Flyin' Fish*, into the wind before setting it down. To protect the model from being overturned by wayward gusts, Herring placed the heavy toolbox in front of it. Using their boots as flattening tools, the threesome tramped a two-foot-wide, 50-foot-long "runway" that extended from the model, directly into the wind. Finished with the preparations, Herring decided that a "dry run" engine start-up and hand-launch were in order. "While I'm trying to start the engine, I want both of you to steady the machine from behind... one at each wingtip. When it starts, hold on tight; it'll try to get away from you! I'll run around the backside to adjust the fuel and the spark. When I'm satisfied that everything is ready to go, I'll wave... that'll be the signal to lift her up to shoulder height."

Clarke and Engberg raised the model as directed.

"Next," Herring said, "I'll slide under and grab hold of the short skid and tail boom."

With his pre-launch instructions complete, Herring had his two assistants release their grip and step to the side. "The rest will be up to me," he said, as he rotated the dormant machine into the breeze. "I'll balance my grip, run down the path, launch her a little nose high, and hope for the best!"

Transferring three fluid ounces of the fuel mix to the tiny brass tank, Herring reminded himself to move the spark advance lever to the start position and to open the petcock valve between the tank and the engine's rudimentary carburetor. Turning the propeller over by hand several times, he listened for the first faint sound of liquid gasoline sloshing about within the engine's cylinder – a signal to close the petcock. Primed with fuel and with his assistants in place, Herring flipped the ignition switch to the "on" position.

Wrapping the starting cord around the pulley, Herring explained that he was leaving the fuel petcock closed – he wanted to see if the engine would "run-out" its prime. "With this inverted cylinder, it won't take much to 'flood' the sparking plug with too much fuel... and I don't want to remove the plug out here; I'd probably lose it in the snow!"

With that bit of explanation, Herring gripped the machine's motor mount with his left hand, hesitated long enough to wink at his comrades, and gave the starting cord a forceful pull.

"*Thump-thump-thump*"... then silence. The engine didn't fire. A second and third pull produced the same gloomy results.

Somewhat irritated, Herring flipped the ignition switch to the "off" position and ordered the machine to be taken back to the barn.

*

By the time Herring had cleaned the fuel-fouled sparking plug, it was 3 o'clock.

"The engine's too cold to start," he declared. "We'll have to warm up the cylinder so the gasoline vaporizes... or I don't think she's gonna run."

After a few seconds of silence, Clarke spoke up. "How 'bout warming it up over the stove and starting her inside the barn? Then we'll walk the machine outside to—"

"That's a possibility," Herring interrupted, "but it's a long way out to our launch site."

"Then let's launch her right outside the barn," Engberg said.

"Too risky, Mike! If the machine makes a big sweeping turn, it might come back and smash into the building! I don't wanna take the chance—"

"I've got it!" Clarke chirped. "Let's use the old hot-potato trick—"

"... to warm the cylinder!" Engberg chimed in.

Within seconds, Herring had three medium-sized spuds cooking atop the stove.

It was almost 4 o'clock before they were ready to try again. This time, before Herring began his start-up routine, he stuffed the three muslin-wrapped spuds around the engine's cylinder, tying them tightly in place with a strip of India rubber.

"There... we'll give her a minute to warm up."

In the meantime, Gus checked the spark-timing lever and ignition switch. With Clarke and Engberg in position, Herring plucked the potatoes free from the cylinder, snapped on the ignition, and opened the gas petcock. Grabbing the starting cord in his right hand and bracing the motor mount with the other, he gave a long, smooth pull.

The engine sputtered once, hesitated... then roared to life.

CHAPTER 80

Stevensville, Michigan

Saturday, January 3, 1903

Moving with deliberate caution, Herring trudged to his spot behind the trembling machinery and whirling propeller to make adjustments. First, he advanced the ignition timing to a predetermined point and then cracked open the throttle... just enough to increase the propeller speed to almost 1,800 revolutions per minute.[1] He spent 15 agonizing seconds fiddling with the carburetor's needle valve, monitoring the ever-changing exhaust frequency by ear. At last, Herring leaned back – all was ready!

Engberg and Clarke had hoisted the *Flyin' Fish* to shoulder height. An instant later, Herring ducked under the model from the rear, seizing it by the alighting skid and tail boom. Adjusting his grip to compensate for propeller thrust, Gus nodded... the signal for his assistants to move aside.

Striding forward, Herring soon picked up his pace. As the end of the tamped-down runway drew near, he began to sidestep... as if about to throw a javelin. In a single graceful motion, Gus pivoted on the ball of his left foot, released his left hand, and thrust the protesting machine slightly nose up into the wind. As the model accelerated from his right hand it began to nose over, making a beeline for one of the many tall snowdrifts that littered its path.

The men held their breath as the engine slowly picked up speed. Gradually, the *Flyin' Fish* reversed its descent and began a gradual climb. Moving into the wind at a pace comparable to a brisk walk, the cruciform tail was observed to rock in response to an errant gust. Unable to prevent the lifting surfaces from tipping to starboard, the tail regulator relinquished control authority to the billowing wing sails, which swung the contraption back into a yawing turn to port. With its engine continuing to bellow, the model was now flying downwind with increased ground speed. A dozen seconds later and 100 feet above the snow-swept field, it began to soar in wide, hawk-like circles.

When he thought to check, Henry Clarke's stopwatch indicated that the model had been flying for more than two minutes and was at an estimated altitude of 300 feet. "Gus!" he hollered, "I'm having trouble keeping her in sight! If the engine don't stop soon... we're gonna lose the damn thing!"

The white model was blending into the gray sky and snow-dominated landscape. Herring thought: *When we get her back, I'm gonna paint the wings orange!*

"Christ!" Engberg puffed. "With a gas tank smaller than a shoeblacking tin, how much longer can that motor run? Looks like she's tryin' to fly back to Silver Beach!"

Aware of the encroaching shadows of dusk, Herring shouted, "Henry... how long has she been up?"

Stopping in his tracks, Clarke stared at the stopwatch. "Four and a half minutes! Hey... it's startin' to snow!"

Suddenly... the sky went silent; the engine had finally stopped running.

"Fix yer eyes on her boys!" Gus shouted. "Don't look away and don't even blink... or you'll lose sight of her!"

Five hundred feet to the east of where the men stood, the model continued to glide, making leisurely 200-foot circles into the wintry abyss. The moment Clarke thought he saw the model touch down, he clicked off the stopwatch. It read: 6 minutes and 22 seconds. "Mark the spot!" Gus shouted. "Line up with something in the background, like that evergreen off in the distance! Otherwise, we might walk right by her! It's gonna be dark in half an hour, so we only have one chance at bringing her back today!"

Although he didn't say so, Herring was worried; worried they might not find the model that had performed beyond his expectations. *I shouldn't have filled the tank to the top*, he thought. *There's too much time and money invested in that engine.*

As the heavy snowfall continued, the sky darkened and the gusts strengthened. Using their gloved hands to shield their eyes from the onslaught, the men found it difficult to find the spot where they thought the model had touched down. After another 15 minutes, doubt was beginning to creep in. "According to my line of sight," Herring said, "we should have reached her by now!"

"I thought she alighted closer to that big tree off in the distance," Clarke said.

"No! No! It come down somewhere back there!" Engberg growled, pointing behind them.

As the men stopped to scan the surrounding territory, Herring threw up his hands, admitting to himself that his gas model was indeed – lost. "We have to quit looking. If we don't get back to the barn before dark, we could get lost out here and freeze to death! Come on, men, let's head back... tomorrow's another day."

As they began their long hike back, something caught Clarke's eye. "Wait a minute... what the hell is that?" he said, pointing to the southwest.

At first glance, it looked like just another distant snowdrift, except for one odd feature: the drift was headed north to south, rather than east to west like all the others. Raising a single frost-encrusted eyebrow, Herring slogged ahead for a closer look. As he approached the anomaly, he wheeled about and shouted, "She's here, Henry! You found her!"

As they gingerly brushed snow off the machine's two lifting surfaces, one thing became apparent: there was no damage. The *Flyin' Fish* had come through its first harrowing trial unscathed. "Henry Clarke," Herring crowed, "from now on we're callin' you *eagle eye!*"

*

Since the continuing snowfall had obliterated their tracks, finding Fose's barn in the dark became a challenge. As destiny would have it, a dim and misty partial moon occasionally broke through the cloud cover, providing the men with a guiding light. After an hour's hiking, Herring rattled the barn's big door shut, signaling the end to a long but productive day.

As the men sat sipping coffee, Mike Engberg spoke up. "Well, Professor, you did it... you're the first to fly a gas-powered, heavier-than-air machine—"[2]

He was interrupted by the barn door rattling open.

"Well I'll be damned... yer back," Fosey wheezed. "I was about to hitch up my team and come lookin' fer ya!"

Glancing around his barn, Harold spied the big *Flyin' Fish* model. "And ya found your machine! When I saw how high it was flyin', I thought it was a goner! Who's the fool now, Mr. Herring? If you're sellin' stock... I want some!"

CHAPTER 81

Rochester, New York

Monday, November 28, 1921

Herring sat alone at the plaintiff's table with his eyes closed. The courtroom, so quiet, he could hear the distant sounds of passing automobiles and trucks on Main Street. Within a few minutes, this unnatural tranquility would melt away as court personnel and litigants returned from their late afternoon recess.

"Mr. Herring," O'Grady said, once court was back in session, "after you and your team had returned to the safety and comfort of Fosey's barn, what were your thoughts concerning the spectacular success of your gas-engine-powered flying model?"

"Objection!" Robbins cried. "The term *spectacular success* is argumentative!"

"Sustained," Sawyer concurred. "Tone down the rhetoric, Counselor."

Promising to restate the question, O'Grady continued. "Mr. Herring, what were your thoughts concerning the performance of your gas-powered model?"

Herring leaned back in his chair, smiling at the recollection. "I was elated and relieved! I also felt vindicated—"

"Vindicated?" O'Grady interrupted.

"In 1896, my original U.S. Patent Office application was rejected because the examiner said I had failed to demonstrate a practical, operational machine. Apparently, the success of my rubber-powered two-surface model didn't count. In October 1898, my manned, powered flights should have resulted in a pioneering patent. Instead, the *revised* application was slapped with a notice of 'final rejection.' During January 1903, when the latest powered model flew for more than six minutes, my thoughts returned to those rejected applications. Then and there, I decided to try again... to revive my 1896 application!"

"Did you address this possibility with your patent attorney?"

"My poor health, and Mr. Whittlesey's schedule, prevented me from meeting with him in Chicago until February of '03."

"You were ill?"

"I returned home from Stevensville with a cough and a mild fever. The condition soon worsened, requiring me to remain bedridden for the better part of a week. The only work I managed to do during that time was to write a letter to Chanute."

O'Grady's response was sharp. "Why on earth would you write to him?"

"Octave still owed me $35... I reminded him of the debt."

"Did he finally reimburse you?"

"Not until May of '03... I didn't realize that the old man had taken his daughters on a four-and-a-half-month tour of Egypt and Europe. By the time he returned to Chicago, my family had already relocated to Freeport, Long Island—"

"Speaking of your move from St. Joseph, Michigan," O'Grady interrupted. "How did that come about?"

"During the third week in January, Lillian and I took the *Lake Shore Special* to New York City. Using Mother's Gramercy Park apartment as our base of operation, we borrowed her Cadillac touring car to motor around Long Island's south shore in search of available property."

"What was the rush?"

Herring cleared his throat. "Two reasons prompted us to move as soon as possible: my tenuous relationship with Police Chief Sauerbier, and Mother Chloe's limited-time offer to buy us a house."

*

On the second day of their search, while exploring the town of Freeport, the Herrings found a modest, 2-1/2-story inland house at 309 South Main Street. The dwelling was backed up to a canal, which connected to a series of Venetian-like channels that furrowed through the flat countryside before exiting into the Atlantic Ocean. Behind the main residence was a detached boathouse/shed where Gus could tinker with his engines.

Freeport attracted many of New York's well-to-do vaudeville actors and movie stars, who used the seashore location as a summer vacation retreat. Most important to Herring, a convenient commuter railway system connected Freeport with the city and *Gas Power* magazine's newly established Manhattan editorial office.

*

"Mr. Herring... concerning your letter to Chanute. Did you have another reason for writing to him?"

Herring considered the question for a moment. "I wanted to *document* the successful flight of my gas model."[1]

"Was there anything else on your mind?"

Herring smiled. "I couldn't resist shaking the old man's cage. With tongue in cheek, I feigned anguish over my inability to secure a job at the Smithsonian. This provided me the opportunity to inquire about his promised letter of recommendation to Professor Langley. As expected, Chanute ignored my query."[2]

O'Grady slid out from behind the podium. "Mr. Chanute often retreated to Europe during Chicago's long winter months. Was his '03 vacation any different?"

Herring shrugged... his answer was of no consequence. "Since I was preoccupied with my pending move to Freeport, I paid little attention to what Chanute was up to – other than to brood over the money he owed me."

*

Before returning to St. Joseph, Herring signed the papers for their new house and also finalized a three-year lease for the editorial office of *Gas Power* magazine.[3] On February 4, 1903, he traveled to Chicago, where he met with George P. Whittlesey, his patent attorney. Providing an enthusiastic update of his gas model's success, Herring posed the question of how to persuade the U.S. Patent Office to revisit his several-times-rejected 1896 patent application. To his disappointment, Whittlesey did not wish to pursue the issue and abruptly declined further involvement. Undeterred, Gus returned to St. Joseph, more determined than ever to find a patent attorney who would fight aggressively for his cause. In the meantime, Henry Clarke was hired to help the Herrings prepare for their move to Long Island and to ready the *Flyin' Fish* for additional flight trials. Between the second week of February and the end of March, Herring, accompanied by Clarke, made four more trips to Stevensville, where they realized more than a dozen successful flights.

The entire month of April was spent packing and making arrangements for the move back East. On Friday, May 8, the Herrings said goodbye to their friends and boarded the train for New York. Within a week of their arrival in Freeport, Herring again wrote to inform Chanute of his move, his further success with the *Flyin' Fish* gas model, and most notably... that his reimbursement check had yet to arrive.

By the end of the month, the Herrings had settled into their new home. William and Chloe had started attending classes at their new school, and Lillian began repainting the kitchen. Augustus dutifully boarded the shuttle train each morning for the 40-minute commute to Manhattan. By June, after a 10-year pause, Herring again hung out his shingle, trumpeting the resumption of his engineering consulting business. That same week, he contacted a young patent attorney by the name of S.W. Scherr Jr., whose legal office was conveniently located in Manhattan.

New York, New York

Monday, June 15, 1903

After searching New York's massive, multi-faceted directory, Herring eventually stumbled across Carl Dienstbach's telephone number. Finding him at home, Gus quickly arranged an impromptu meeting at the reporter's Greenwich Street apartment.

Dienstbach had recently returned to New York following an extensive journalistic assignment in Berlin, Paris, and London. After an enjoyable afternoon of flying-machine talk, Herring, anxious to continue their discussions, invited Dienstbach to supper the following Saturday.

Arriving in Freeport by commuter train and hansom cab, Dienstbach toured Herring's boathouse-shop, with its experimental apparatus and constructions scattered about like favorite old toys. After Lillian's wild-strawberry pie dessert had been enjoyed by all, the men slowly made their way back to the shop, where Herring ran his little half-horsepower gasoline engine. With summer solstice in full effect, darkness didn't prevail until almost 10 o'clock. As the flame from Gus's ancient oil lamp cast dancing shadows onto the blackened cedar walls, their conversation continued into the early morning hours.

"You haven't mentioned the Chanute lectures," Herring said, ignoring the croaking bullfrogs.

Dienstbach winced, as if pained by the memory. "The two of you have had issues... that's why I've been hesitant about discussing his lectures."

Herring grinned as he slouched on his wobbly oak chair. "Don't worry about me, Carl, I've already heard most of the old man's lies."

"If you insist... but I hate ending the evening on a sour note!" Standing, Dienstbach shuffled over to a south-facing window, raised it, and stared toward the humid gloom of the Atlantic. "Chanute spoke in Berlin three months ago – the third week in March. The hall held 700 and there wasn't an empty seat to be found. Using a translator, the old man informed the gathering that he and his daughters had recently completed an 11-week tour of Egypt and Italy. He said he had left his family to entertain themselves in Venice, while he dutifully forged onward to Vienna, where he delivered a speech promoting the aeronautical agenda for the upcoming St. Louis World's Fair. Then he claimed to have 'hopped' an overnight train to Berlin. Actually, I believe that Chanute divulged his itinerary simply to illustrate that he was well to do."

"Did he have anything interesting to say?"

"He talked about his experiments, and his assistants, including you. I believe he referred to you as a 'talented mercenary.' For the most part, he gossiped about the Wrights and how they were helping *him* to carry out *his* aeronautical plans. The biggest surprise was how Chanute downplayed the importance of European airships or 'gas bags,' as he called them."

A quizzical look flashed across Herring's face. "Why on earth was he concerned about airships?"

"That became more apparent two weeks later, when he delivered his Paris lecture."

"Did you get to speak with him in Berlin?"

"I tried, but he was continuously surrounded by well-wishers and admirers; apparently, he didn't recognize me, so I was consigned to the role of observer. Strangely, there was one man in particular who caught my attention. Not only was he always at the old man's side... he looked familiar, but I couldn't put a name or a place to him. Later, I saw Chanute and this bearded man in a popular Berlin restaurant... they were engaged in a heated discussion. I must admit this unlikely association pricked my interest, if not my ability to recollect."

"You mentioned Paris. I assume you are referring to Chanute's April 2nd Aéro Club de France lecture... did you attend?"

"Since I speak French, Mr. Moedebeck at *Illustrierte Aeronautische Mitteilungen* sent me there. He thought the old man, being of French descent, might divulge some fresh aeronautical tidbits to his former countrymen."

Herring stood and wandered over to his motorized rowboat, where he began tinkering with its engine. "Was there anything different about the Paris lecture?"

"He briefly described the Wrights' method of retaining lateral equilibrium by warping the wing while simultaneously moving the rear vertical rudder... but he declined to elaborate. Chanute also renewed his attack on 'airships,' saying that 'Frenchmen of talent were being distracted by powered gas bags, when they should be working to solve the problem of heavier-than-air flight.' "

Dienstbach waved his hand and continued, "It soon became apparent that Chanute's lectures were an exercise in self-promotion, where others – you know who I'm referring to – only assisted in carrying out *his* experiments."

"What about that mysterious man you described earlier? He didn't show up in—"

"Yes, I spotted him again!" Dienstbach said in a hushed tone. "He was sitting right there in the last row! After the lecture, he headed down to where Chanute was receiving best wishes, and they left the building together. I still can't put a name to the face... but it'll come to me eventually."

Later that year, Dienstbach would recall the mystery man's identity while attending the 1903 *Frankfurt Aeronautical Show*.

Rochester, New York

Tuesday, November 29, 1921

Judge Sawyer called Herring back to the witness stand.

"Mr. Herring," O'Grady quizzed, "besides Mr. Dienstbach's rendition of the events in Berlin and Paris, what evidence do you have that Mr. Chanute had been promoting himself during his lectures?"

Reaching into his document case, Herring rummaged about for a moment before pulling out a tattered magazine. "Just so happens I have a copy of the April, 1903 issue of *L'Aérophile*, France's leading aeronautical journal."

"Your Honor," O'Grady said, "the plaintiff wishes to place this publication into evidence."

Minutes later, without objection from the defense, the journal was processed and entered as plaintiff's exhibit number 48.

"Mr. Herring, please read the passage that best illustrates Mr. Chanute's lack of full disclosure."

Herring nodded slowly and turned to a page marked with a paperclip. "The title of the article '*La navigation aérienne aux États-Unis*', translates to 'Aeronautic Navigation in the U.S.' begins on page 223... where the author states:

> 'Chanute's speech at the Aero Club of France on April 2, spoke of the Wright brothers as his devoted collaborators, as young, intelligent and daring pupils who worked under his guidance. The brothers had written to Chicago for technical information, on the basis of which they built machines similar to those of Mr. Chanute, and were actively carrying his work to completion.'"

O'Grady turned to face Judge Sawyer. "Your Honor, all across Europe, Mr. Chanute portrayed the Wrights as his *apprentices*! This should not surprise any attentive observer... hadn't Chanute *stolen* Mr. Herring's intellectual property? His ideas concerning aeronautical—"

"Objection! Conclusion!" Robbins shouted.

O'Grady spun away from the bench with a flourish. "I withdraw the question, Your Honor."

"Mr. O'Grady," Sawyer admonished, "please save your trickery for a jury trial! Objection sustained."

"Thank you, Your Honor!" Robbins said, puffing out his chest.

Jury trial or not, O'Grady had placed an exclamation point on Chanute's ploy to ensure his legacy at any cost. He continued, "Besides Vienna, Berlin, and Paris, where else did Chanute address the European aeronautical community?"

Herring nodded again in the same labored way. "His last lecture was delivered in London. Then he and his brood took a steamship back to America. The European lectures provided Chanute with enough material for more than a dozen major magazine articles. I would be remiss not to mention the photographs he used for these articles... they were *pirated* from the late Matthias Arnot's negatives that illustrated our 1897 Indiana Dunes glider."

O'Grady stared at Herring for a moment. "Previous testimony has shown that Chanute had pilfered your ideas... did a similar fate befall the Wrights?"

"Objection! Conclusion! Calls for speculation!" Robbins cried out.

"Overruled. The court wishes to hear the witness's response."

Herring leaned forward. "The old man also acted against the Wrights' request not to use the in-flight photographs he had taken of their 1902 glider by including them in his speeches. After Chanute's European speaking tour, the Wrights no longer trusted him. According to Orville's recent testimony in the *Montgomery v. United States* lawsuit, Chanute wanted to publish the entire story of the brothers' 1902 glider, including their wing-warping method of lateral control. Although the Wrights had applied for their U.S. patent in March of '03, the foreign patent applications had yet to be addressed. Desiring to keep the details of their control system secret, Wilbur asked Chanute not to include any of this information in his lectures or articles. Not wanting his friend to feel slighted, Wilbur cited a provision in French and German patent law that rejected any claim previously disclosed in print or lectures."

"How did Chanute respond to this constraint?"

"According to Orville, Chanute said '[My] article would have proved quite harmless, as the construction is *ancient and well known...*'

"Astonished and hurt by Chanute's assertion, Wilbur let the remark pass for the time being. Nevertheless, early in 1903, Pandora's box had been pried open, and the 14-year dispute over the priority of lateral control was underway.

"To complicate matters, in Paris, Chanute stated that 'the amateurs Wilbur and Orville Wright introduced only three improvements: placing the horizontal rudder at the front, placing the operator in a prone position, and *warping the wings* to steer to the right and left.'"

"What effect did Chanute's disclosure have on the brothers' ability to obtain foreign patents?"

"Since Chanute didn't understand *how* the Wright's lateral-control system actually worked, his disclosures were of questionable value to the foreign experimenters. While he openly discussed wing warping in his Paris speech, he failed to adequately explain its link to the vertical rudder… the other component of the Wrights' lateral-control invention."

"Because Chanute lacked understanding," O'Grady said, "were the Wrights able to salvage their foreign patents?"

"Nine years later, the German Patent Office canceled any patent rights the brothers may have held to their invention in that country. Chanute's *faux pas* also caused impediments for Orville's infringement lawsuits in France… but both Wilbur and Chanute had died by then."

O'Grady pivoted on his heel and stared at Curtiss, whose head was slumped forward in probable sleep. "History shows," O'Grady shouted, "that Mr. Chanute took significant credit for the Wright brothers' aeronautical success!"

Curtiss exhaled loudly as he jerked to attention.

"How does that resonate with you, Mr. Herring?"

Herring inclined his head. "It was *déjà vu*. Chanute had used similar tactics against me! By insisting that other experimenters had demonstrated priority in most phases of flying-machine design, such as Wenham's rudimentary biplane configuration or Pénaud's stabilizing tail, Chanute actively undermined my claims, while promoting his role as facilitator. My Indiana Dunes glider of 1896 is a case in point. The old man was cunning when he spoke to newspaper reporters or wrote for the aeronautical journals. Chanute falsely implied that he had designed the machine, and provided me with a sketch and instructions for completing the job. Like the Wrights, I was credited with being little more than a 'paid assistant'... a mere mechanician who obediently followed his master's instructions."

On that note, Judge Sawyer recessed for lunch.

*

Court resumed at 1 o'clock.

"Mr. Herring," O'Grady continued, "in November 1903, you received a letter from Dr. Spratt. Included was a *copy* of another letter that Chanute had sent to Spratt. What news did these letters contain?"

Defense attorney Robbins objected, Herring produced the letters, and O'Grady had them entered into evidence. Ten minutes later, Herring was allowed to address O'Grady's inquiry.

"Doctor Spratt thought I might want to hear about Chanute chasing an old delusion—"

"Old delusion?" O'Grady cut in. "What do you mean?"

With his wire-rimmed reading glasses in place, Herring began reading from Spratt's letter.

> " 'I left Kill Devil Hills for home on the cold and windy morning of November 6. By chance I encountered Mr. Chanute on Manteo Island... he had just arrived from the mainland by steamship. After a brief discussion concerning the Wrights' chances for success with their powered machine, he agreed to keep me informed as to their progress.' "

In the other letter from Chanute, dated Wednesday, November 11, 1903, Herring read:

"'I have pushed the brothers rather vigorously to fly my oscillating-wing machine while they await the return of their repaired propeller shafts. My machine is still stored here from last year's fiasco that was due entirely to Herring being a bungler.

'I also broached the topic of the 1904 St. Louis World's Fair. I offered the boys an expense-paid trip to the exhibition, if they would agree to display and fly my Katydid glider.'"

"Spratt also mentioned that Chanute was trying to purchase the *Éole*, Clement Ader's old flying machine from the French government, so the brothers could learn and benefit from its construction methods and to perfect its design."

O'Grady slipped from behind the podium. "Doesn't it seem strange that Mr. Chanute couldn't see the aeronautical Promised Land ... when it stood shimmering, right before his eyes?"

Herring smirked. "During the Montgomery trial, Orville testified that Chanute wanted the brothers to modify and fly the old Ader machine for *him*! He didn't think the Wright machines were superior – only that the brothers were more skilled at handling them. Wilbur and Orville were of the opposite opinion."

*

As the court session droned on, Herring found himself daydreaming about the circumstances surrounding the Wrights' first powered flights in December of '03, and his own flights five years prior. He remembered reading Orville Wright's comments in *Century* magazine:

> *The first flight lasted only twelve seconds. A flight very modest when compared with that of birds, but it was, nevertheless, the first in the history of the world in which a machine carrying a man had raised itself by its own power into the air in free flight, had sailed forward on a level course without reducing its speed, and had finally landed without being wrecked.*[4]

Orville's recollections were similar to those of Herring's powered flights. The difference? Herring didn't claim his two Silver Beach flights of 1898 were historic – only that manned, powered flight, was *solvable*.

Herring also pondered the patent-related issues that he and the Wrights had faced back in 1903. The Patent Office had quickly rejected the Wrights' application of March 1903, suggesting that they submit a working model. As Herring had learned earlier, the agency still regarded heavier-than-air flight and perpetual motion as twin delusions, and they did their best to prevent "cranks" from squandering their money on fees. As it turned out, the Wrights had abundant time and cash to invest in the drawn-out process. In this regard, Herring remembered that on Christmas Eve 1903, he put the finishing touches on a letter he would soon mail to the brothers in Dayton.[5]

"Why did you write to the brothers?" O'Grady asked, bringing Herring out of his reverie.

"I learned of their Kill Devil Hills success from the newspapers and wanted to congratulate them."

"Was there anything else on your mind?"

Herring did not answer at once. "—I proposed that we join forces, acting as a single party to get the broadest patent claims, eliminate competition, and avoid future litigation."

"Join forces? Why?"

Herring leaned forward. "If the brothers were going to turn their invention into a company... I wanted to be part of it."

"How did you justify your proposal?"

"It was obvious... we had similar machines! There were bound to be infringement issues—"

"Which machines are you referring to?" O'Grady interrupted.

"My gas-powered *Flyin' Fish* model and their latest powered two-surface machine."

"Your letter suggested that you had patented claims that would prove costly to the brothers if future litigation became necessary."

Herring shifted in his chair. "At the time, I was confident the patent office would reverse its 'final decision' of 1898, and grant me a pioneering patent... especially in light of my highly successful powered model."

"The patent office had yet to act on your behalf – is that correct?"

"Your Honor!" Robbins shouted. "The witness has just admitted to misleading the Wright brothers about nonexistent patents! Five years later, he *lied* to Mr. Curtiss about the same phantom patents! Herring never had—"

Sawyer banged his gavel repeatedly, calling for order. "Order! Order in the court! The defense will have its opportunity to cross-examine the witness in due time! In the meanwhile Mr. Robbins, you are treading on very dangerous ground. Another such outburst, and I will cite you with a contempt of court penalty! Is that understood?"

"Very well, Your Honor," Robbins acquiesced as he slid into his chair.

"Very well indeed! Continue, Counselor."

Determined to maintain a straight face, O'Grady looked up from his notes. "Mr. Herring, tell the court about your continuing patent office problem. What were your thoughts back in 1903?"

"In my letter to the Wrights, everything but the U.S. Patent Office debacle was accurate. Since I already held a British patent for my invention, I assumed that obtaining a U.S. patent with similar claims would be a formality. The gas model's performance was spectacular, like that of the Dune Park two-surface glider before it and the rubber-powered, two-surface model before that.

"My compressed-air-powered machine performed comparably with the first three flights of the '03 Wright Flyer, but there were no accolades... nor did I seek them! If not for Matthias Arnot's slip-up in releasing the camera's shutter *after* the machine had touched down, credit for the world's first successful manned, powered flight would have rightfully belonged to me and the State of Michigan – a full five years before the Wrights' first feeble effort in North Carolina!"

"Objection!" Robbins cried. "The witness is speculating again!"

"Sustained. Move on, Counselor."

"Mr. Herring, how did you learn about the construction details of the *Wright Flyer*?"

Herring made an impatient, brushing-away gesture with his hand. "I didn't know the exact construction, but I *presumed* their glider from '02 had laid the groundwork for the new machine."

"Did you suggest how your proposed company should be formed?"

"They would share a two-thirds interest, with me taking the remaining third."[6,7,8]

New York, New York

Monday, December 28, 1903

Carl Dienstbach had returned from the Frankfurt Aeronautical Show just in time to celebrate Christmas at his New York apartment. The following Monday, he rode a hansom cab across town to meet with Herring at his mother's lavish Gramercy Park residence. Always on the lookout for new material to use in *Gas Power*, Herring listened attentively as Carl showed him photographs and relayed stories about the big German trade show, including tales of the latest European engines being developed for heavier-than-air flight.

Waiting until Herring's mother had left the parlor, Dienstbach leaned close to his friend and whispered into his ear. "I finally saw beyond the facial hair and fine clothing... and placed that mysterious fellow from the Chanute lectures. But I'm hesitant to disclose his name."

"Why?"

"He's certainly not a friend."

"Out with it, Carl... who is he?"

"Butusov."

CHAPTER 82

Rochester, New York

Tuesday, November 29, 1921

O'Grady and Maloney sacrificed their usual sojourn to the Empire State Restaurant to meet with Herring in the lawyers' tiny alcove adjacent to the judge's chambers. The liverwurst sandwiches and drinks that Maloney had ordered from the courthouse's German delicatessen were delivered soon after Judge Sawyer recessed for lunch.

Herring's afternoon testimony would concentrate on his activities leading up to the 1904 St. Louis World's Fair, more formally known as the *St. Louis Louisiana Purchase Exposition.*

Back in court an hour later, O'Grady got right to the point. "Mr. Herring, did moving to New York City allow you to meet and congregate with others who were involved in aeronautical research?"

Herring's demeanor brightened. "Unlike St. Joseph, Michigan, where I worked in relative anonymity, New York was teeming with like-minded individuals."

"If you please, provide the name of one such individual."

Herring hesitated, as he probed his memory. "Charles Matthews Manly would be a typical example. He served as Professor Langley's chief engineer after I left the model aerodrome project—"

"Yes," O'Grady interrupted, "and a few months later, the machine flew successfully. Tell the court: for what was Mr. Manly most remembered?"

"The *Great Aerodrome*... the manned, heavier-than-air machine that twice failed to fly – just prior to the Wrights' success in North Carolina."

"How did you become acquainted with Mr. Manly?" O'Grady prodded.

"Carl Dienstbach introduced us back in February of '04. Although Charles was nine years my junior, we became friends and eventually worked to help form the Aero Club of America."

"Did Manly help with your flying machine projects?"

Herring paused briefly before continuing. "Charles assisted with the design and eventual troubleshooting of the radial engines for my first Signal Corps aeroplane. As I recall, Charles also accompanied me and Wilbur Kimball to the 42nd Street Armory; where we flew the *Flyin' Fish* gas model."

O'Grady grimaced; this was something new. "How was it possible for a large, engine-powered model to be flown indoors?"

Herring cracked a smile as he shifted in his seat. "The armory was huge, and the administrator of the facility was a steadfast advocate of flying-machine research. Once a month, he cleared the drill floor of equipment and turned it over to me for experimentation!"

"How did you control the model's flight path?"

"By securing a 60-foot, #13 piano-wire tether to the model's starboard wingtip and fastening the opposite end to a freely rotating horizontal wheel at the top of a six-foot-tall central pylon. After a bit of tinkering, I was able to realize remarkably stable clockwise flight."[1]

"How long did these flights last?"

Herring hesitated, performing the mental calculations. "On several occasions, the model flew 70 tours around the pylon in about 12 minutes. That translates to flying five miles in 12 minutes, or 25 miles an hour... then the fuel ran out!"

"Why didn't you fly the model outdoors?"

"Stevensville had acres and acres of uninterrupted Michigan farmland... New York City didn't. I was also concerned that the machine could be lost or commandeered by some unscrupulous thief who might have chased a wayward flight. Whereas, a tethered model couldn't get away from me."

"What were you trying to accomplish during these armory flights?"

Herring considered this. "I wanted to determine an accurate relationship between the engine's power and the lift produced by its aero*curve*."

"Why was that important?"

Trying to be patient, Herring nodded... he was comfortable explaining points of technique. "The establishment of a reasonable *power-to-lift* ratio is imperative to any practical heavier-than-air flying machine. This ratio helps to determine the payload limits, including passengers, that the machine is capable of lifting."

"By the way, who was Wilbur Kimball?"

"Kimball experimented with what are commonly referred to as *helicopters*. These are heavier-than-air, vertical takeoff and landing machines. Two years after the armory experiments, we shared a beautifully outfitted shop on Broadway Avenue."

O'Grady stared down at his notes. "Aside from your friendship and Aero Club activities, did you and Mr. Manly participate in any other aeronautical endeavors?"

"We attended the 1904 St. Louis World's Fair together."

"Please elaborate."

"With the fair opening on April 30 and running through December 1, I approached Manly about us attending during the week of October 3 – five months distant."

"Why were you interested in that particular week?"

"The *International Aeronautical Congress* was holding its annual meeting there on October 4 and 5."

New York, New York

Thursday, April 7, 1904

As the men enjoyed their lunches in a Manhattan restaurant, Manly watched as Herring gulped down his second glass of milk.

"My daddy always said that milk clogged your arteries," Manly stated.

Herring flinched. "Maybe that's true, but my daddy always said that alcohol rotted your liver. What's worse... my apoplexy or your cirrhosis? I'll take my chances with milk!"

"*Touché*," Manly chuckled. "By the way, I've been thinking about your idea of attending the fair in October. As you know I originally decided not to attend when the fair's engineering committee *canceled* the engine-performance competition... they claimed a low number of entries—"

Herring interrupted with a laugh. "The word got out, Charles! When potential entrants learned that you were going to bring the *Great Aerodrome* engine, they decided not to waste their time and to save the entrance fee! By the way, was Professor Langley disappointed when the engine competition was canceled?"

Manly frowned and paused a moment. "I'm afraid it deepened his depression. On the brighter side, I received a letter from Mr.

Chanute. He invited me to speak at the Congress... but it's a long trip. What do you have in mind for travel arrangements?"

Herring raised his eyebrows. "I've already done some checking. We should go first class, and take the most famous train in the world... the *20th Century Limited*. After all, a man of your stature deserves nothing but the best!"

Manly tugged at his earlobe. "Kind of expensive... even if you could get tickets."

Herring persisted. "Not if we share some of the costs. Each ticket costs $34.70, and that includes the extra fare and Pullman charge.[2] Here's the best part: we'll make the trip to Chicago in only 18 hours; that's six hours faster than the *Exposition Special*, the train that Lillian and I took to Chicago's Columbian Exposition back in '93. Besides, when we're in St. Louis, we can share a room."

"What about getting to St. Louis... isn't there a more direct route?"

Herring winked. "There's a more direct route out of Cleveland or Toledo, but it's slower. Coach seating on the *Alton Limited* out of Chicago will get us to St. Louis in less than six hours."

Thinking ahead, Manly had a suggestion. "On the way home, we could save money by riding coach on the *CCC&St. L* back to Cleveland. Once there, we'll transfer to the *Lake Shore Limited*; she's slower than the *20th Century*, but less costly!"[3]

"And she's still part of the New York Central's Water Level Route," Herring grinned. "What do you say? I'm game if you are!"

The following morning, Herring hiked over from his 42nd Street commuter stop to the Grand Central Terminal. To his chagrin, the October 2 booking of the *20th Century Limited* had already sold out. After subtly bribing the reservations officer with a freshly minted one-dollar bill – and the promise of another

– Gus would be contacted if there happened to be a cancelation. The next day, he received a telephone call from the reservations office. Two tickets were available if he wanted them.

Rochester, New York

Tuesday, November 29, 1921

"Hence it was settled," O'Grady said, "you and Mr. Manly attended the World's Fair together in the fall of 1904."

"Correct," Herring confirmed, crossing his legs.

"Besides the Aeronautical Congress, was there anything else that captured your imagination?"

"The $200,000 in prize money! That alone was enough to lure 'cranks' out of the woodwork – both lighter- and heavier-than-air advocates. The organizers hinted that the Brazilian balloonist, Santos-Dumont, as well as the Wright brothers would be competing for the $100,000 Grand Prize."

"Were there any other reasons why you wanted to attend the fair?"

O'Grady was pressing him hard and Herring knew why. "Actually, I wanted to see what progress had been made with lightweight internal combustion engines... but then there was the Chanute situation."

"What situation are you referring to?" O'Grady asked, followed by a barely audible sigh of relief.

"I learned that the old man had promised the organizers to show off and fly a replica of my *Double Decker* glider—"

"Objection! Argumentative!" Robbins interrupted, scrambling to his feet. "Again, history shows that the original two-surface glider of 1896 was designed by Mr. Chanute, who financed its construction and testing. In this context, the machine has *always* been referred to as the *Chanute glider*. The witness continues to take undue credit for an achievement he had little to—"

"To the contrary, Your Honor!" O'Grady shouted, cutting Robbins off in mid-sentence. "Prior testimony has shown that Mr. Herring designed, built, and flew a rubber-spring-powered two-surface model back in 1892. While working for Mr. Chanute in 1896, Herring designed, built, and flew a full-size version of this same machine... as a glider. The facts in the matter have already been entered and are part of the court's record."

Consulting his notes for several seconds, the judge suddenly raised his eyes and glared over the top of his pince-nez spectacles. "Mr. Robbins, the court finds no reason to admonish Mr. Herring for calling the machine in question *his Double Decker* glider. Objection overruled. Continue, Mr. O'Grady."

"Mr. Herring, how did you learn that Chanute had built a replica of your glider?"

"Back in March of '04, I received an unexpected telephone call from Chicago – it was Bill Avery. Although we hadn't spoken for almost *four* years due to the Truscott boatyard fire fiasco, he nonetheless sought my opinion concerning parabolic aero*curves*, and how they might perform on my two-surface glider. I thought it was a strange question, and with a little prodding he spilled the beans."

"Tell the court what you learned."

"According to Bill, Chanute had contracted him to construct an 'improved' version of my two-surface glider for a Frenchman he met during his European lecture tour."

"When was Mr. Avery contracted?"

"Back in November of '03."

"Who was the Frenchman?"

"A would-be aviator by the name of Jacques Balsan."

"What if anything did Mr. Balsan have to do with Chanute's World's Fair entry?"

Herring uncrossed his legs and leaned forward. "After the glider was almost complete, Balsan reneged on the deal, forcing Chanute to absorb the cost of its construction. By then, the old man knew the Wrights weren't going to pursue the Grand Prize, so he decided to grab some of the limelight for himself."[4]

O'Grady shook his head. "Excuse my ignorance in such matters, Mr. Herring, but how did Mr. Chanute propose to fly an unpowered glider? To my knowledge, there aren't any sand dunes in St. Louis!"

A murmur of laughter rolled through the courtroom.

"Chanute spent the spring and summer developing an electric-motor-powered *winch* that would haul the glider into the air by winding up a 400-foot-long rope. Since Chanute had designated Avery to be the contraption's operator, Bill was nervous about the old man's selection of the aero*curve* and his method of launching the machine."

"What did you think of Chanute's new curved surface?"

"Objection!" Robbins hollered. "Calls for speculation!"

"Sustained. Continue."

"Why didn't Chanute use a more modern curved surface – perhaps one that was derived from the Wrights' wind tunnel testing?"

"Objection! Speculation!"

"Sustained."

Aggravated, O'Grady bit his lip. "Did Chanute justify his use of the unusual aero*curve* in any of his written articles or speeches?"

"Not that I am aware," Herring said. "On the other hand, neither he nor the Wrights had any idea of how an aero*curve* actually worked!"

Freeport, Long Island, New York
Saturday, April 30, 1904

Charles Manly pushed himself away from the dinner table after finishing his second slice of Lillian's homegrown-rhubarb pie. "A wonderful meal, Mrs. Herring. Your roast beef and mashed potatoes are second to none! And the pie was outstanding! Thank you for having this bachelor over for dinner!"

Minutes later, he and Gus removed themselves to the solitude of the parlor, where Manly, at the urging of his host, removed his shoes and stretched out on the divan before discussing the recent newspaper article about Herring's work in aeronautics.

"I just finished reading your captivating treatise from the March 13 edition of the New York *Daily News*. I found your recollections to be riveting, and the photography revealing. To be frank, Gus, I didn't realize that you had accomplished so much... especially during the last decade. Tell me about this article."

Herring settled deeper into his favorite overstuffed chair. "Apparently the *Daily News* reporters heard about my tethered-flight experiments at the 42nd Street armory. They tracked me down and one thing led to another – resulting in the first newspaper statement that I've made concerning my total body of work in aeronautics."

Manly propped himself up on one elbow. "I found your statements concerning Langley, and especially Chanute, to be... what should I say... fascinating."

"Indeed... in what regard?"

"You mention being sought out by other scientists... that most of Chanute's successful experiments were conducted along lines laid out by *you*; assertions such as that."

Several uneasy moments of silence followed. Then Gus spoke. "Those are true statements, Charles. I have tried to be diplomatic in my dealings with Chanute, but time and again he ignores my contributions regarding the origin and success of the two-surface glider... among other things."

Manly's head dropped onto a doily-decorated cushion. "But Augustus, by repeatedly kicking the anthill, you're focusing negative attention onto yourself. You know he corresponds with every enthusiast in the world—"

"—Has he spoken with you lately?"

"As a matter of fact, he has, and that's why I'm addressing the situation with you now. Apparently you wrote a short note to him back in February, asking if he had any information on the subject of flying machines for your *Gas Power* magazine."

"Yes, and true to form he didn't reply."

"That's what I'm trying to tell you! Chanute said, 'I am ashamed... almost... to say that I have not answered him.' He's really got it in for you, Gus!"

Herring was quiet for the better part of a minute. "What can I say, Charles? By the end of last year, Chanute had hitched his wagon to a fresh team... he thinks he can afford to be vindictive. Unfortunately for him, the Wrights are on to his self-serving tricks."

Removing a stack of newspapers and magazines from beneath an end table, Herring pulled out a three-month-old issue of the Chicago *Daily News*. "Here's an article my machinist friend sent to me from St. Joseph. It has an interview with the Wrights. Listen to this... "

Slipping on his reading glasses, Herring edged closer to the table lamp. "The title of the piece is 'New Principles of Control'. It gives a description of their flights of December 17. Anyway, the interesting part is when Wilbur says '... all the experiments have been conducted at our own expense, without assistance from any individual or institution.' "

"What's the date of the article?"

"January 6, 1904. The Wrights had heard about Chanute's European innuendos concerning *his* role in *their* success. According to George Spratt, it took Chanute only a few days to question Wilbur's quote, asking him '... what did you have in mind concerning myself when you formed that sentence that way?' "

"How did Mr. Wright respond?"

"According to Spratt, the old man didn't get a clear answer. However, the brothers knew that they were being referred to as his *pupils*, and most insiders were led to believe that he had also *financed* them. I believe that Chanute fully realized that Wilbur's *Daily News* statement was made in an attempt to correct these perceptions. Wilbur was intentionally vague, not wanting to offend his powerful friend; but it's apparent to me that the Wrights no longer trust the old man."

Manly raised an eyebrow. "What do you make of their squabble?"

"Even though Chanute contributed *nothing* in terms of science or technique, he yearns to be remembered as the unselfish mentor who guided his pupils, apprentices, and employees to the ultimate solution of the problem."

"Can you relate to the Wrights' predicament?"

Herring raised his head and winked. "The old man pulled all the same tricks on me... it's an old story!"

New York, New York

Sunday, October 2, 1904

The Long Island commuter train groaned to a halt in the heart of Manhattan. Scrambling to disembark, Herring – accompanied by Lillian, who had come to see him off – was about to ascend the Grand Central Station's limestone steps, when a red-capped attendant appeared from behind a nearby pillar. "Carry your luggage, sir?"

As the threesome entered through one of the dozen brass-framed glass doors, it was a short stroll through the bustling depot to Waiting Room B. As Herring dug into his pocket for a two-cent tip, the attendant held up his hand. "No thank you, sir! Compliments of the New York Central!"

From across the way, a familiar voice beckoned. "Gus, Lillian, over here!" Manly shouted. Once seated, Herring flipped open his pocket watch to compare its time with that of the colossal four-faced clock that loomed conspicuously in the Grand Hall. "1:15 p.m.... Plenty of time before our 2:45 departure."

"This is quite the terminal," Manly beamed. "There are 17 tracks, 11 platforms, and three waiting rooms."

"That's not the half of it," Herring said. "The place has grown from three to six stories, and the outside has a whole new French Renaissance *façade*, complete with domed towers—"

"Gentlemen, what do you think of that?" Lillian interrupted, pointing to a large poster hanging on the far wall. The illustration depicted a streamlined, propeller-driven, lighter-than-air gasbag flying above a train with the caption: 'You may fly someday, but the quickest way now is the 20th Century Limited.'

"That's a fascinating advertisement," Gus said, "but I'll bet it won't be a powered gasbag that ends up beating the train. What do you think, Charles?"

Manly grunted. "We'll be better prepared to discuss that after seeing the dirigibles perform later this week."

"Too bad Santos-Dumont won't be there," Gus replied.

*

After an hour of nervous anticipation, another "red cap" appeared at the entrance door and announced that all ticketed passengers for the *Century* should report to track number 34.

At 2:30 p.m., the restraining rope was removed, allowing 42 passengers to walk on the plush red carpeting that covered the concrete platform leading to their assigned cars. Herring noticed that the Pullmans, traditionally colored chocolate brown, had changed; they were now a dark shade of olive green. Inside, the sleeping berths were nowhere to be found; they had been transformed into coach seating for daylight riding.

"So... this is *Gonzalo*," Manly said with a grin. "Reminds me more of a luxury hotel than a railway car!"

"Thank you, sir!" George, their porter, said. "Mr. Pullman, God rest his soul, would be tickled to hear you say that!"

As their luggage was being stowed into furniture-quality Honduras mahogany compartments, the pleasant aroma of boiled linseed oil and carnauba wax wafted into the cabin.

"Don't feel too niggled about the name, Charles," Gus said, "the other sleeper car is named *Petrucio*."

Before settling into his high-back leather seat, Gus slid over to the car's arched side window, grinned, and waved to Lillian, who was standing on the platform blowing him kisses. Within seconds the conductor shouted, "All aboard the *20th Century Limited* for all points west... Albany, Utica, Syracuse, Rochester, and *Buff-lo*, New York... Cleveland, and *Tole-do*, Ohio... Elkhart, Indiana and *Chica-go*, Illinois. All passengers please be seated and prepare for departure!"

Within seconds, the locomotive's 20 tons of traction force jostled the cars slightly as they creaked and shuddered into motion.

"Now that we're rolling," Gus said, squinting to read his pocket watch, "God willing... we'll be in Chicago's LaSalle Street

Station at 9:45 a.m. tomorrow, just in time to catch the *Alton Limited* to St. Louis. With a bit of luck we'll be in our hotel room by 5 o'clock tomorrow afternoon!"

"I'm surprised," Manly said, "the *Century's* trainset only has five cars behind the locomotive and tender."

"Hmm," Herring pondered. "What she lacks in size, she makes up for in speed and amenities."

Manly picked up the *20th Century Limited* brochure and began browsing. "Listen to this. There's a 'combined baggage, buffet, smoking and library car' located behind the tender – the male passengers mostly use it. Next in line is the 'dining car.' The third and fourth cars are 12-section 'sleepers' that convert to coach seating during the day. Bringing up the rear is the 'observation car' – where most of the women and non-smoking males go to watch the countryside fly by. The observation car also has eight private staterooms.

"There's a maid, a valet, a manicurist, and a barber on board. If you need to dash off a letter, there's also a stenographer. George will shine your shoes while you sleep, and the morning newspaper, along with a fresh carnation, will be at your fingertips when you awake. Finally, if you don't feel like getting up in the morning, you can order breakfast in bed! Just like at home, eh Gus?"

*

Rolling out of Grand Central Station, the *Century* followed the Harlem River for a few miles to Spuyten Duyvil, where it began tracking the Hudson River toward Albany, 143 miles distant. As the men relaxed, they began to notice some of *Gonzalo's* amenities.

"I like the dark green carpets and leather upholstery," Manly commented. "Do you know how hard it is to keep a carpet like that looking clean?" A 28-year-old bachelor, Charles had tipped his hand. Not having a maid, he admitted to doing all his own housework. "I don't mind pushing the carpet sweeper around the apartment; besides, maids are expensive. I wouldn't be able to afford this trip if I had a maid to support."

Gus snickered. "You're beginning to sound like Wilbur and Orville... but they have their sister to do the housekeeping."

"I like the stained glass skylight windows," Charles observed, ignoring Gus's teasing. "When the sun shines through, the interior of the car changes color; even the mahogany carvings and ornamentation take on a different appearance. Pullman went all out with these new wooden cars; they spared no expense, that's for sure."

Although the Pullman Company exploited the full glory of a dying Victorian era, their marketing was aimed in a different direction. In 1904, the company motto was: "Travel and Sleep in Safety and Comfort."

"Let's go to the observation car while there's still daylight," Herring suggested. "I want to see how those new car-connecting vestibules work. The company's been bragging about passengers not having to step outside the cars anymore."

The flexible vestibule, a bellows-like innovation, worked to perfection as Gus and Charles strolled from their sleeper, through the car immediately to their rear and into the observation car. Proceeding to the back, the men were fortunate to find a pair of unoccupied overstuffed chairs, and they settled in to enjoy a panoramic view of the splendid scenery on the Hudson Valley route. A semicircle of floor-to-ceiling double-strength glass lined the rear of the car, providing observers with an unprecedented viewing experience as the countryside sped by at nearly 70 miles an hour. *Exceeded only*, Herring thought, *by the thrill of flying.*

"The sensation reminds me of flying the *Double Decker* glider... except we're pointing in the wrong direction!"

"I can't relate," Charles mused. "My two attempts at flying replicated the trajectory of a cliff diver. What was it that reporter from the Washington *Post* said? Oh yes, 'The *Great Aerodrome* entered the water like a handful of mortar!' " With that bit of self-deprecating humor, Charles laughed heartily as Gus slapped him on the back, sharing the lighthearted moment.

"It took a brave man to try flying that big aerodrome, Charles."

Manly flicked his hand, as if whisking an insect away from his face. "It really doesn't matter, Gus. History only remembers the first to reach success, and the Wrights have apparently claimed that honor. I must admit however, I feel sorry for Professor Langley... who is suffering mightily.

"I also regret leaving my position at the Smithsonian, but government funding had run out. Since I was about to be terminated, I felt it was in my best interest to resign; I submitted the paperwork the Monday after Christmas."

Surprised to hear the details of Manly's sudden departure from the Smithsonian, Gus nevertheless sympathized with his friend. "As you know, I also left the employment of Professor Langley, but for much different reasons. I resigned after a heated argument concerning his unyielding, authoritarian method of managing the model aerodrome program."

"I heard all about the feud from the artisans. To a man, they agreed that Langley had made a mistake by forcing your resignation. Although your departure provided me with work, the same constraints and restrictions were foisted upon me! To be honest, Gus, I learned from your experience and rolled with the punches. Of course, the sad demise of the *Great Aerodrome* is a story for another time but, I still believe we should have been the *first* to fly!"

As the *Century* approached Albany, the car's interior lights flashed on. The autumn sun had dropped behind the surrounding hills, darkening what had been a spectacular first hundred miles.

"Charles, what do you say we get back to *Gonzalo* before we stop in Albany for a load of water. We're only there for a few minutes, and I want to get a look at the prototype 12-wheel locomotive while there's still daylight."

As they shuffled back to their car, Gus's words had pricked Manly's curiosity. "Are you saying the New York Central is committing its premier long distance express to an unproven locomotive? What's so special about this engine?"

Herring waited until they were seated again before relaying some of the technical details. "To begin with, it's called a Pacific-type, built by the American Locomotive Company, right here in Schenectady, which is the old Brooks Company of Dunkirk, New York. Anyway, she's a 133-ton 4-6-2, which means *four* pilot wheels, *six* drivers, and *two* trailing wheels. Nothing special there except the layout has been optimized to suit the Central's water level route. But that's not all. The new locomotive generates 3,900 horsepower at 66 miles an hour! Add this to the fact that there are *no* engine changes between New York and Chicago, and the new engine translates into reduced fuel and maintenance costs. Eleven years ago, the *Exposition Flyer* needed *two* locomotive changes to make the 980-mile trip."

At that instant, the *Century* began to throttle back for its stop at the state capital. Three minutes later, with a warning from the conductor not to delay their return – or risk being left behind – the two mechanical engineers leapt from *Gonzalo's* vestibule door and jogged 150 feet to where the trembling, steam-huffing behemoth stood.

"Look at the size of those drivers!" Manly said, pointing to the imposing cast-iron wheels.

"Seventy-nine inches is only an inch bigger than the *Exposition Flyer's* drivers," Gus said, "but the *Century's* superior horsepower gives a higher top speed at level running conditions, and better acceleration coming out of coal and water stops."

Oblivious to the long filler pipe that swung out from the 35,000-gallon wooden water tower, the men continued their rapid-fire technical discussion of the engine's finer points.

"Stand back!" the fireman growled as he crouched atop the tender car, the pipe's spigot dangling above an open hatch. "Ya might get yer fancy suits wet!"

As the men scrambled to put distance between themselves and the tender car, Herring paused to check the wind direction. "Good... the wind's blowin' toward the engine. Wouldn't want a clinker to burn a hole in that *fancy* new suit, now would we, Charles?"

Reacting instinctively, Manly brushed an imaginary cinder from his shoulder. "Why are they just takin' on water?"

"Because the tender is sized to take on coal every other stop – they'll take on both in Utica."

Manly warmed to the topic. "Couldn't they eliminate the water stops by using *track pans*?"[5]

Gus patted Charles on the shoulder. "That's possible, but the New York Central has decided not to use them on the *Century*. Besides the water freezing in the pan during the winter months, the inevitable overflow from the tender tank sprays all over the sleeper cars, making a mess of their stained glass windows. Then there was the problem of unfortunate animals drinking from the pans, and getting scooped up... I prefer not think about that!"

Manly cringed and changed the subject. "I wonder how much coal and water the tender carries—"

"Fourteen ton of bituminous," a gravely voice butted in, "and 10,000 gallon of water!" Wheeling around, the men watched as an ancient-looking mechanic in oil-stained coveralls limped toward the locomotive with his oilcan.

Building upon this new information, Herring continued. "If you could look down at the tender from the front, the big central coalbunker is surrounded by a u-shaped water jacket. As we know, one pound of coal converts six pounds of water into 200 pounds of steam pressure."

"That rings a bell," Manly said. "I remember that ratio from my undergraduate days at Cornell."

As the fireman topped off the tank, Herring took it as a signal to head back to their Pullman. Whisking away any possibility of coal dust from his lapels, Gus slowly shook his head. "That fireman sure has his hands full! Besides filling the tank with water, it's his job to supervise refilling the coalbunker. Once underway, he shovels coal into the burner and keeps tabs on the steam pressure."

"One thing's for certain," Charles quipped, "you rarely find an overweight fireman!"

Their car suddenly creaked and launched forward as the *Century* began its parallel run along the Mohawk River toward Utica and Syracuse. Once the train was underway, George announced that supper was about to be served in the dining car. Ten minutes later, Gus and Charles were busy ordering squab and roast beef with all the trimmings.

It was almost 8 o'clock before they shuffled lethargically, drinks in hand, to the observation car. The dimly lit car was unexpectedly deserted, allowing the men to choose the prime seating, in the middle and directly behind the imposing bank of panoramic-view windows.

"Great meal, eh Gus?" Manly said, as he dropped into the chair next to Herring.

"I should hope so ... cost me three bucks!" Gus chuckled.

They stared into the darkness as the iron rail's wooden ties retreated into the void of an eerie red glow. The hypnotic spectacle reminded Herring of yet another aspect of the modern railroad. "Charles, have you noticed? The *Century* doesn't have a traditional kerosene lantern hanging out behind the car."

Manly turned his head from left to right as he leaned forward. "Maybe not... but that red glow is coming from somewhere."

"Two *electric* lanterns, one on either side of the car! They're powered by an axle-mounted direct-current generator... every car has one. In railroading, centralized dynamos are a thing of the past."

"If that's the case, they must be using lead-acid storage batteries to keep them lit when the train is stopped."

"That would be my guess. It's still easier to dim lights with direct current than it is with alternating current – like inside this car," Gus said, with a sweep of his hand.

"That should make Mr. Edison happy," Charles snorted. "He's lost the battle of the currents to Tesla and Westinghouse... but there are still plenty of applications for direct current."

As the *Century* charged toward Syracuse and the flatlands of the Lake Ontario watershed, Manly sipped the last of his Bourbon Manhattan, while Herring swallowed what remained of his milk.

"You know, Charles, the emerging flying machine business is similar to the battle of the currents... will dirigibles or aero*curves* become the dominant technology?"

Setting his glass down with a flourish, Manly turned to face his friend. "That reminds me, Gus, earlier you mentioned Alberto Santos-Dumont. What do you make of that mystery?"

Santos-Dumont, a Brazilian expatriate who resided in Paris, was the celebrated designer and operator of powered, lighter-than-air gasbags. Recently, he had stormed back to France following an incident of alleged sabotage at the site of the St. Louis flight trials.

"It's a strange story, that's for sure," Gus said. "An article in the New York *Times* said that Exposition organizers had bent over backwards to lure Dumont to St. Louis to compete for the $100,000 Grand Prize for flying. When he complained about the speed requirement, they reduced it. When he squealed about the layout of the course, they changed it... twice, as a matter of fact! Officials even offered him bonus money for an early success! According to the *Times*, when Dumont finally agreed to come to America, the consensus among experts was that he would probably win the big prize."

"Reputation has a long reach," Manly said. "Three years ago, Dumont won the $20,000 Deutsch Prize for traversing a seven-mile, closed-course flight that included circling the Eiffel Tower. He beat the 30-minute time requirement by only 30 seconds, traveling at an average speed of 14.5 miles an hour. Most newspaper accounts didn't mention that he failed *seven* times with *six* different gasbags before finally succeeding!"

Santos-Dumont called his St. Louis gasbag *The Racer*. It was 131 feet long, 23 feet in diameter, and held more than 44,000 cubic feet of flammable hydrogen gas. There were two, 15-foot propellers, one at each end of the pilot's gondola. Originally designed for a 60-horsepower Clement engine, Dumont projected the speed to be a fanciful 50 miles an hour.

"I understand that three months ago, officials from U.S. Customs broke the seals on each of the three big shipping crates – they weighed two tons apiece – and inspected the contents. At that time, the gasbag was found to be in perfect condition—"

"I heard that, too," Gus interrupted, "but all of Dumont's preparations and pandering came crashing down the next morning when one of his mechanics found four long slashes in the still-folded gasbag. Making matters worse, the slashes had penetrated three pleats, making 12 cuts in all. Oblivious to the problem, the pint-sized Dumont – he's only five foot four inches tall – dressed in his traditional high-collar shirt and signature Panama hat, pranced about the grounds lobbying officials for yet another change in the course's layout!"

Manly sank lower in his chair. "I don't understand why someone wasn't guarding their equipment."

"That's the scandalous part of the story, Charles. The organizers went to the expense of providing two Jefferson Guards; one for the early shift and one for the late. After questioning, the late-shift guard admitted to leaving his post. They say he went for coffee."

"Did they arrest a suspect?"

Herring flipped his hand. "The article said that a man in a *wheelchair* was arrested for loitering in the vicinity sometime after midnight. Officials finally let the cripple go, saying he was physically incapable of committing the crime. The organizers offered a $1,000 reward for information leading to the culprit's arrest and conviction, but nothing has come of it."

"I heard that Dumont had a similar problem back in Europe."

Herring nodded and handed Manly a newspaper. "There's an article in this morning's New York *Times* that claims it's the *third* time he's had one of his gasbags damaged."

"Three times?" Locating the article in the second section, Charles began to read aloud. "It says 'the first reported incident of sabotage occurred in Paris on Sunday, April 12, 1903. Inexplicably, less than a month later, on Sunday, May 3, another of Dumont's balloons was discovered slashed to ribbons in a London based warehouse, just hours before it was scheduled to be inflated—' "

Gus interrupted Manly's recitation. "Nothing for more than a year, then it happens again in this country! Very strange, wouldn't you agree, Charles?"

"Without doubt! In St. Louis, Dumont passed the damage off as the work of a vandal, not one of his rivals. He also steadfastly refused to cooperate with police or help to press the investigation. Instead, he and his mechanics left for Europe three days later, just prior to our Independence Day celebration. Upon departing, he told exposition officials that the gasbag could be repaired, and that the machine and his entourage would return sometime in September – a promise that he broke. Just the other day, he told a London reporter he would 'never return to St. Louis,' and the rumors have been flying ever since."

"That they have, Charles. For instance, the Jefferson Guards have suggested that Dumont lacerated the envelope himself... to save face. They say he feared being called a failure."

"That doesn't make sense, Gus! Dumont had the rules and the course changed to his advantage. Besides, why would he damage his own machine after spending so much money in its construction and transport from France?"

Suddenly, George appeared, his gold-plated company watch in hand. "Misters Herring and Manly, there's no rush sirs, but your sleepers are ready for when you fixin' on turnin' in."

"We'll be up front shortly, George," Herring said. "What time do you have?"

The porter glanced at his watch. "10:05, sir."

Reaching into his vest pocket, Herring pulled out the *Century's* schedule of stops. They were almost to Rochester. It was an ideal time to retire. While the train was stopped, they could avoid the destabilizing forces that made common tasks somewhat difficult. This was a trick practiced by his father during his many years of travel by rail, and Gus hadn't forgotten the advice.

Before climbing into his bunk and sliding the aisle curtains closed, Gus had one last suggestion for his traveling companion, who had already deposited himself in the upper bunk. "Charles, you might want to set your timepiece *back* one hour. Although it's technically inaccurate to do so now, the New York Central chooses to make the time zone change outside Buffalo, when the companion *Century* traveling in an easterly direction passes."[6]

*

From Buffalo through Cleveland to Toledo, a distance of almost 300 miles, the tracks ran along Lake Erie's southern shore. Next, they traversed the rolling prairies of Indiana, slipping by the pristine waters of Lake Michigan and headed toward Chicago.

After a restful eight-hour sleep, Herring awakened to wash, shave, and dress before ambling into the dining car for his breakfast. He ordered his traditional poached egg, bacon, toast, coffee and especially, his tall glass of cold milk. Before the order arrived, Manly joined him, looking a bit worse for wear, claiming that he had trouble sleeping due to the symphony of dueling snorers... Gus and the occupant of the compartment opposite them.

"Snore? Certainly I don't snore! You're the first person to claim that I snore! Not even my dear wife has ever accused me of snoring! Are you certain the dreadful noise wasn't emanating from someone else's compartment?"

Manly crossed his legs, squinted, and glared squarely at Herring. "I didn't take my recording *snore-o-meter* on this trip, so I can't replay the episode for your enjoyment. However, I can assure you that *you* do indeed snore... in fact, you snore almost as loud as Professor Langley!"

Herring jerked his head back. "You certainly know how to hurt a man's feelings, don't you, Charles?"

As the waiter returned with Gus's meal, Charles had decided on his breakfast. "I'm going to have French flat cakes, butter, with pure maple syrup, and an order of caviar toast with marmalade and Swiss cheese on the side... exactly what I would consume on a typical workday back in New York," he whispered.

"You certainly are in a pleasant mood, my friend, if not altogether truthful. What are you anticipating for today?"

Charles held his hand up high, as if asking for silence. "Other than a safe trip, I'm looking forward to riding the *Alton Limited*, Alton Railway's flagship service between Chicago and St. Louis! I'm looking forward to seeing what has been modestly referred to as 'the most *beautiful* train in America.' I can't wait to lay eyes on the *three* shades of maroon with gold-leaf trim and lettering!"

"Well," said Gus, emptying his glass, "you won't have long to wait; we're about to roll into Elkhart, Indiana. It's our last stop before the LaSalle Street Station."

CHAPTER 83

Somewhere south of Chicago, Illinois
Monday, October 3, 1904

Manly was disappointed. After enjoying the superb amenities aboard the *20th Century Limited*, the *Alton Limited* out of Chicago seemed anticlimactic... partly because they were riding in coach, in rather cramped conditions. On the positive side, this was the last leg of their 24-hour excursion, and the World's Fair loomed directly ahead.

Looking up from his Chicago *Times*, Herring tapped a finger on the article he had been reading. "Listen to this, Charles. A 29-year-old German physicist has made an exciting discovery. Ludwig Prandtl found that liquid flowing through a tube slows down near the walls. He refers to this phenomenon as a *boundary layer*."

Manly pondered for a moment. "What does this 'boundary layer' have to do with aeronautics?"

Herring patiently started again. "The boundary layer idea clarifies something that has baffled me for years! It *seems* to explain why my machines and those of the Wrights, especially the Wrights, stretch their glides when they're in close proximity to the ground."

"But, Gus—"

Herring rustled the pages. "Hear me out, Charles. Allow me to explain how I *think* this boundary layer business works: to begin with, it's linked to the *Bernoulli Effect*, which you know is the relationship of a fluid's pressure to its velocity; reduced velocity produces increased pressure, and vice versa.

"At altitude, the increased speed of the airflow across the topside of the aero*curve* produces lift due to the partial vacuum that's generated. The airflow across the bottom of the aero*curve* impacts that surface, resulting in a positive pressure. Together, the positive and negative pressures generate the aero*curve's* total lift at any given air speed.

"Here's where the boundary layer comes into play. When a glider flies into the wind, close to the ground, the bottom side of the aero*curve* is operating within a *slower* moving layer of air. Therefore, the relative velocity of the air passing under the surface – compared to *above* the surface – is drastically reduced. According to Bernoulli, this reduced velocity *increases* the pressure on the bottom side of the curve, thus increasing its total lift—"

"This increased lift," Charles interrupted, "allows the aero*curve* to assume a *lower* angle of attack, which in turn *reduces* the drift and improves the glide ratio!"[1]

"You've got it!"

"The question is," Charles continued, "how much does the boundary layer increase the machine's performance?"

"I think I've already measured this!" Gus said. "My *Double Decker* glider consistently demonstrated a free-air glide ratio of about 8:1... not counting the boundary layer coasting that often extends the glide. During those floating, but exhilarating flights, I've sometimes experienced a glide ratio of up to 20:1... better than double the free-air coasting!"[2]

"That seems too good to be true."

A wave of silence washed over the men as they became lost in their thoughts. Finally, Gus continued. "Now that Prandtl has me on the trail of this phenomenon, I'll be sending for a copy of his paper, 'Fluid Flow in Very Little Friction.' It's written in German... but that's no problem! I must explore the benefits and limitations of this discovery. For instance, at what altitude does the boundary layer stop being effective?

"Then there's the question of airframe configuration. The Wright machines position a prone operator atop the lower lifting surface, allowing the aero*curve* to operate very near the ground where the boundary layer is probably most effective. Back in '02, I was astonished by their long downhill glides, only two or three feet above the sand! Now I understand the physics that made that possible!"

"What about your own machines?" Manly asked.

Gus grimaced. "None of my machines were suited for efficient boundary layer skimming. With the operator dangling below the lower lifting surface, and the regulating tail trailing below that, I was fortunate to realize any boundary layer effect at all!"

Totally absorbed in their discussion, time passed quickly for the men. Before they knew it, the conductor was calling their 3 p.m. arrival in St. Louis. Their entire excursion had taken a little more than 25 hours.

*

As the train lumbered into the city, George pointed out a portside window.

"There she be, boss... the clock tower!"

Serving as a destination landmark for rail travelers, the Union Station tower soared 280 feet above the downtown streets of St. Louis. Located a mile from the western bank of the Mississippi River, the 10-year-old terminal had recently been enlarged. With an average of 85,000 people passing through its gates daily, proud city fathers boasted that their terminal was now the busiest in the world.

As the *Limited* entered the switching and turnaround area behind the station's stub end, the engineer blew the steam whistle three times prior to pulling the locomotive's *Johnson bar* controller, which initiated a reversing action that backed the elite trainset into the mammoth train shed, thus ending the 284-mile excursion at the unloading platform.

As Gus and Charles stepped from the platform onto the midway, they turned to gaze at the shed's construction. "I understand it's the largest single-span train shed ever constructed," Gus said. "They claim it's 600 feet wide and 800 feet long. That works out to an area of more than 11 acres, covering 42 sets of tracks!"

"Look at that arched roof," Charles chimed in. "It must be 150 feet tall!"

The midway consisted of a 70-foot-wide concourse that ran the full width of the shed and connected it with various parts of the headhouse, a vast open area that occupied the entire front of the terminal. The *headhouse* was made up of three sections: the centrally located grand hall, a restaurant at one end, and a hotel at the other.

Following George, who wheeled their luggage along on a low-slung cart, they soon entered the grand hall – an 8,000-square-foot edifice that contained ticketing facilities, waiting areas, concessions, and restrooms.

"Charles, look at the barrel-vaulted ceiling, the Romanesque arches and stained-glass windows! Everything has been finished in fresco and gold leaf!"

"Magnificent!" Manly crowed. "Seeing a structure like this makes me wish I had taken up civil engineering rather than mechanical—"

"Perhaps," Herring broke in, "but you would still have to work with narcissistic people. I don't have to name names, do I, Charles? Civil engineering has its own share of egotists."

Approaching the grand hall ticketing area, Herring explained their situation to an agent behind the counter. They wished to travel directly to their hotel: the Inside Inn, which was located on the southeast corner of the fairgrounds. Since Union Station was only four miles away, there were several options to choose from. After studying the overhead chalkboard for several minutes, Gus turned toward his traveling companion.

"What will it be, Charles? An electric streetcar for a nickel, a horse-drawn omnibus for a dime, a hansom cab for 25 cents a mile, a two-horse carriage for 50 cents a mile... or the *Wabash Railroad's* steam shuttle for two bits? They'll all deliver us to the Lindell Street Entrance, the fair's main point of entry. From there we'll take the trolley to our hotel."

Staring at the board, Charles massaged his chin with his thumb and forefinger. "Let's take the streetcar; we'll save some money."

Herring scowled before whispering into his friend's ear. "We just finished taking two of the most exclusive trains in the world to get here. Why would you want to waste time stopping and starting all the way across town? Let's go first class! A hansom cab, a carriage, or the steam shuttle... what do you say?"

Charles turned his head to the gold-leaf-decorated ceiling and waved his fist. "To hell with the cost! Let's take the Wabash!"

Tickets in hand, and led by their porter, the men proceeded to the west end of the train shed and clambered aboard steam shuttle #2, which was about to depart. They quickly settled in for a short ride to the stately Wabash Rail Terminal, at the Lindell Entrance. As the shuttle squealed to a halt, Herring flipped open the cover on his pocket watch. "3:45 p.m.... not a bad time, eh Charles?"

Retrieving their luggage, the men hustled across the wide entrance plaza that led to the 75-foot-tall Gate of Babylon, whose remarkably realistic-looking marble columns and yawning archway provided a sampling of the architectural splendor of the fairgrounds. Paying the 50-cent daily admission fee, Gus and Charles pushed their way through wooden turnstiles before hurrying directly to nearby Station #17 of the fair's Intramural Railway. As they waited for the next available trolley, the men could see the *St. Louis of France* statue that stood at the far end of the 1,000-foot-long Plaza of St. Louis. On either side of the Plaza's 600-foot-wide concourse loomed two enormous ornate structures, the Manufacturers' building and the Varied Industries building.

Beyond the Plaza's statue emerged the Grand Basin Waterway, a 10-acre manmade lake that projected another quarter mile, all the way to the fair's centerpiece, the magnificent Cascades – a triple waterfall that descended 60 feet from an elevated area known as the Terrace of the States. Presiding above the central Cascade was the fair's Louisiana Purchase Monument, a 100-foot-tall column, topped by a huge skeletal sculpture of the globe, on which the mammoth Bird of Peace was perched.

"With a bit of good fortune, we'll be in our hotel room before suppertime," Charles said, displaying a tired grin. *A short nap would do me wonders*, he thought.

The streetcar made short work of the 1-3/8 mile run down the eastern boundary of the fairgrounds. Along the way, they rushed past an area known as Model City, a replica U.S. War Field Hospital, and a massive community known as the Plateau of States. The sprawling, four-story Inside Inn was conveniently located adjacent to Station #13.

As the men checked in at their hotel's 100-foot-long, black marble registration desk, they marveled at the size and scope of the facility. Constructed of yellow pine, stucco and fireproof burlap, the four-story Victorian-inspired temporary building could accommodate up to 10,000 persons in its 2,257 rooms, which were serviced by a staff of 2,000.

One of the hotel's many reservations attendants, a portly middle-aged man with a Midwest accent, raised an eyebrow as Herring signed the register. Pivoting on his heel, the man waddled to the rear wall, where a great honeycomb of letter-size wooden compartments stretched the entire length of the registration area. Within seconds, he returned waving a sealed Western Union envelope above his head. "I thought there was something for you, Mr. Herring," he said, sliding the sealed letter across the counter.

Minutes later, the men were standing in front of the polished brass and glass appointments of elevator 16, a vertical transportation system that was very much appreciated at the end of a long day. Tipping the operator a penny, and the bellhop two, Gus and Charles finally entered room number 3012.

As Charles collapsed onto one of the room's two single beds, Herring sat down behind the desk. With daylight fading, he switched on the shaded electric bulb that hung conveniently from the ceiling. As he tore open the telegram, he thought, *Who besides Lillian knows that I'm here?*

TO: A.M. HERRING

ST. LOUIS WORLD'S FAIR

INSIDE INN

FROM: W. ENGBERG

BENTON HARBOR, MICH.

12 P.M. MONDAY, OCTOBER 3, 1904

RE: TRUSCOTT FIRE.

CHIEF SAUERBIER AND DEPUTY LEFT FOR FAIR VIA RAIL THIS A.M. STOP.

The news gave Herring pause. Informing Charles that he'd be back within the hour, Gus dashed from the room and down to the front desk, where he cornered the hotel's concierge. After a brief conversation, Gus used the hotel's telephone to make a long-distance call to Carl Dienstbach. He wasn't home.

Back in his room, Gus tried to explain his situation. "Charles, something unexpected has come up that requires your cooperation."

"What's going on?"

Herring perched on the edge of his bed. "I've been tipped off that the St. Joseph Chief of Police is on his way here... he may be coming to arrest me—"

"For *what*?" Manly interrupted.

"For an arson fire that happened there five years ago."

Manly shook his head in disbelief. "And... you're the prime suspect?"

Herring rubbed his forehead. "I can assure you that I had absolutely nothing to do with that fire, other than having lost all of my experimental equipment. However, at this moment, that's beside the point. Promise me that you won't tell *anyone* that I'm here."

"What are you going to do?"

"Charles, I'll fill you in later this evening. In the meantime, I'm heading back into the city."

"Now? It's almost 5 o'clock!"

"I have no choice... it's an emergency."

Donning his black canvas jacket and captain's hat, Gus hurried out of the hotel and hailed a hansom cab. Half an hour later, he entered the side door of *Sutler's*, a store that catered to people of the theater.

Four hours later, Herring returned to the lobby of his hotel, where he attempted to telephone Dienstbach again, without success. Determined to speak with his reporter friend, he provided the long-distance operator with the number for the editorial offices of the New York *Times*. Finally reaching a colleague of Dienstbach's, he learned that Carl was in St. Louis, covering the International Aeronautical Congress.

"He's *here* at the World's Fair?" Herring shouted into the mouthpiece. "Well, I'll be damned!"

*

Back at room 3012, there was a knock on the door. After waiting for what seemed a very long time, a muffled and somewhat subdued voice finally responded. "Who's there?"

"Baron Otto von Guericke!" the voice boomed, with a heavy German accent. "I am here at this regrettable hour to speak with Mr. Manly, or Mr. Herring!"

"Herring's not here. Come back tomorrow!"

The determined von Guericke shouted through the closed door. "You are speaking to a representative of the German Empire! I act on behalf of his Excellency, Kaiser Wilhelm II!"

An awkward moment of silence preceded Manly's response. "Did the Kaiser send you here?"

"Certainly not!" the German snapped. "The exposition's Chief of Transportation persuaded me to speak with you!"

Charles, needing further clarification, still refused to open the door. "Sir, state your business!"

The baron continued coolly. "The aeronautical competition rules committee needs your assistance!"

With a degree of trepidation, Manly cracked the door open a few inches. Before him stood a large middle-aged man who appeared to be dressed for an official state dinner. As von Guericke tipped his black Edwardian top hat, it revealed a full head of heavily gelled, slicked-back hair. As he straightened from an abbreviated bow and heel click, his handlebar mustache and narrow salt-and-pepper goatee flashed to prominence on his chiseled although ashen face. Except for his gray-blue eyes, the baron seemed a complete stranger. For some reason, though, the eyes seemed familiar.

"Mr. Manly, I presume?"

Opening the door wide, Manly extended his hand. "Charles Manly of Long Island, New York."

"May I enter, sir?"

"Of course, but I'm afraid things are somewhat in disarray."

"I pay no heed," von Guericke said, as he strode into the room carrying what appeared to be a gray canvas laundry bag.

As the German's wide-necked velvet cape fluttered silently behind him, Manly noticed that its hunter-green satin lining gleamed in the room's subdued light. Besides his sparkling white Victorian dress shirt with its high collar, the baron wore an Emerald Isle vest, a hunter green silk puff tie, and a black Victorian cutaway jacket. His black Callahan dress trousers were fitted with buttons that accommodated two-inch-wide black suspenders.

When he removed his cape with a flourish, his cutaway jacket revealed yet another amenity – a pair of black silk sleeve garters. Topping off his ensemble were black and white leather spats and an emerald green, glass-handled walking stick, which von Guericke carefully leaned against the oak chair where he had decided to sit.

"The hour is late, Mr. Manly, but what I have to say will not take long."

Still nervous about the stranger's presence, Charles stood tentatively near the edge of his bed. "How may I be of assistance?"

The German shifted in his seat and crossed his legs. "Mr. Smith, who I mentioned earlier, is the chairman of the Aeronautics Rules Advisory Committee. Because Santos-Dumont has departed and therefore abandoned his position on this prestigious commission, the chairman has directed me to identify a qualified individual to fill the Brazilian's little shoes! I believe a person such as you – a devoted practitioner under the splendid supervision of Professor Samuel Pierpont Langley – could serve as a suitable replacement!"

Von Guericke produced a folded single sheet of paper from his inside vest pocket, along with a handheld magnifying monocle on a delicate gold chain. After peering at the paper's contents for several moments, the baron thrust the document in Manly's direction. "Read this, if you will."

Ten seconds later, Charles lifted his eyes from the paper with confusion written across his face. "What does a restaurant menu have to do with finding a replacement for—"

"Absolutely nothing," von Guericke interrupted, as he yanked the fake mustache from his upper lip. While Gus laughed hysterically, Manly threw the menu to the floor and stomped over to where the imposter was sitting.

"Herring... you are truly a son of a bitch! You perpetrated this charade merely to test a *disguise*?"

Herring, who continued to giggle, wiped the tears from his eyes as he leaned back in his chair. "How else could I tell if the costume was any good?"

Manly retreated three steps and flopped backwards onto his bed. "I thought there was something familiar about the eyes. You should do something more to change their appearance."

Rubbing spirit gum adhesive from his upper lip, Herring had a follow-up question. "Perhaps I should use some burnt cork to add a bit of shadow... a faint smudge over the cheekbones?"

"Herring!" Manly cried, in a desperate attempt to bring his traveling companion back to his senses. "What on earth do you expect to accomplish with that getup?"

Looking glass in hand, Herring was busy applying a fresh coat of gum adhesive above his lip. "Anonymity, my good man! I must spend my time here *incognito*. As I said earlier, Chief Sauerbier is coming to arrest me on trumped-up charges! Then there's Chanute, whom I refuse to discuss—"

"Speaking of Chanute," Charles interrupted, "I have some news. After you stormed out of here, I decided to take the Intramural Railway over to the Aeronautic Concourse. Although daylight had faded, Chanute and his two assistants were still at work."

"Working outdoors... after dark?"

"Hardly. They had a large canvas circus tent... like the one you used at the Indiana dunes. I remember it from the photographs that you showed me. They were using gas lanterns for illumination and to keep the chill off."

"What were they doing?"

"The assistants were working on the launch winch. After I introduced myself, Mr. Chanute took great pains to explain how the contraption was supposed to work. Unfortunately, there was some trouble with the drive mechanism. That's what the man in the wheelchair was struggling to fix—"

"There was the old man and Bill Avery," Gus broke in. "Who was the other person?"

"Don't know. Chanute didn't bother to introduce him. He was a paraplegic with a Russian accent, who spent most of his time grumbling under his breath."

Butusov! Herring thought, with a shudder. *Another good reason to be in disguise!*

"Did the old man say when they were going to try flying the two-surface glider?"

"After the Aeronautical Congress concludes. He thought Thursday, or possibly Friday, depending on the weather. He mentioned he was having trouble getting electricity out to the winch motor, at which time Mr. Avery flew into a rage, saying 'How can an organization this large not have insulated 10-gauge copper wire on hand?' Besides being a carpenter, apparently Mr. Avery is also an accomplished electrician."

Herring stood and performed a graceful pirouette. "You do understand why this disguise is necessary, don't you, Charles?"

Manly shrugged. "I suppose so... but the outfit must have set you back a good bit."

"The whole works cost $16, and that included the hansom cab. Expensive? Yes, but I had no choice! Acquire a decent disguise... or take the next train home!"

St. Louis, Missouri

Tuesday, October 4, 1904

It was nearly 9 a.m. before the baron and his roommate arrived at the Intramural Railway Station. As they waited for the first trolley of the day, Charles broke the silence. "You couldn't ask for a more convenient location... the terminal's less than a hundred feet from the Inn."

With his cape flapping in the chilly morning breeze, von Guericke appeared annoyed. "It's less than a mile and a half as the crow flies, but almost three and a half by trolley!" he grumbled.

"To where?"

"The Hall of Congresses! We're on fairground property, but we couldn't be farther away from our destination!"

Manly shrugged off the remark. "We'll be there in less than an hour."

"That's the problem," von Guericke hissed. "The proceedings start at 10 o'clock."

"What choice do we have? The fair doesn't open for business until 9 a.m. And you know these things never start on time! Look at the positive side; if anything, you'll miss having to listen to Chanute babble about the importance of these get-togethers."

"Maybe you're right," von Guericke sighed, as they scrambled aboard the already half-filled car.

"Take those two seats on the right, Gus... I mean, *Baron*. Most places of interest are best seen from the trolley's starboard side. I learned that from yesterday afternoon's expedition."

Seated beside the window, von Guericke fidgeted with his mustache while Manly unfolded a copy of *Fair Facts*, a four-color brochure containing a map of the grounds and several pages of information.

"Where did you get that?"

"There was a stack of them on the front desk of our hotel... didn't you notice them, Otto?"

Ignoring Manly's rebuff, von Guericke had a question of his own.

"I understand the grounds are 9,500 feet long by 6,000 feet wide; that works out to be about two square miles. How are 100,000 daily visitors expected to make their way to the fair's interior?"

Manly flipped through the pages of his brochure for the answer. "Ah, here it is!... Manned rickshaws and wicker wheelchairs can be rented for 60 cents an hour. Carriages pulled by oxen are available for large groups. Open-air electric-powered cars are all over the place. Gasoline-engine-powered automobiles are also popular... although somewhat unreliable. Large steam-powered buses are highly recommended.

"Steam- and electric-powered boats are used to traverse the fair's waterways, especially the Grand Basin and various lagoons.

Then there are specialty craft such as Hawaiian surfboats, Indian canoes, dugouts, Australian catamarans, and gondolas... along with camels and elephants for the truly adventuresome!"

The baron stroked his goatee as he considered the options. "Tell you what, Charles, I'll ride atop the elephant if you promise to follow *closely* behind in a rickshaw!" As the men shared the jolly moment, the trolley shuddered into motion.

"What else does your crib sheet have to say?"

Manly turned to the front of the pamphlet and began to read.

" 'The Louisiana Purchase Exposition, commonly known as the 1904 St. Louis World's Fair, was prepared at the direction of President Theodore Roosevelt to celebrate the centennial of the purchase of the Louisiana Territory from Napoleon. Delayed by one year, the Exposition commemorates the more than doubling of America's land area. Located on a 1,300-acre tract of land that straddles the St. Louis city limits, the wilderness was transformed by 10,000 laborers into 1,500 buildings and over 75 miles of roads and walkways in less than two years!' "

The baron blinked through his artfully applied eye black. "I can't imagine how they did all this construction in only two years! Seems impossible to me – even if they did use *'staff'* to produce these temporary structures—"

"Staff?" Manly cut in. "What, pray tell, is 'staff'?"

Von Guericke gave an aristocratic nod. "All of these elaborate buildings are finished with staff. It's plaster – a mixture of lime, sand, and water with a liberal addition of hemp fibers to keep it from cracking. The glop is then troweled onto *lath* – thin wooden strips that cover the structure's framework.

"Even the statues are made from staff-filled master molds that were originally sculpted by skilled artisans. Thousands of them were required to fill the needs of all these Victorian structures."

Charles cracked a smile and plowed on. "The booklet says it takes 17 or 18 days to see everything. I hope there's time to take in a few of the major attractions—"

"Absolutely!" von Guericke exclaimed. "Let's plan on taking all of Saturday... since the fair is closed on Sundays."

Ten minutes into their journey, having traversed 200 acres of undeveloped forest that dominated the south-central portion of the fairgrounds, the trolley approached Station #12. Once there, about 30 passengers departed, heading in the general direction of the Fine Arts building complex and the Terrace of the States, a magnificent six-acre garden located 1,000 feet to the north. At the center of the Terrace, looming high above its surroundings, sat the Louisiana Purchase Monument.

"Last night, when we were on the north side waiting for a trolley," Manly said, "we could see this big monument off in the distance."

Herring gave a little laugh. "We could also see the Cascades and the Grand Basin Waterway, but I'm *really* impressed with the Plaza of St. Louis! Wouldn't it be a wonderful place to fly a glider? Not only is the area big and unobstructed, it also faces directly into the prevailing wind!"

The trolley skirted the forest by first heading west, then south, before swinging north along Skinker Street and eventually merging with University Boulevard, where it passed another 200-acre plot dedicated to Horticulture and Agriculture. Station #8 proved to be the destination for most of the car's remaining passengers.

"Look," Charles said, "the great Observation Wheel!"

Herring didn't bother to look up from Manly's map. "From what I can tell, it's the same Ferris Wheel that made its first appearance at the '93 Chicago Exposition. I've read that after the fair closes, the owners are planning to demolish it... for scrap."

Charles blinked, surprised. "Seems a pity some amusement park like Coney Island couldn't use it."

"Too expensive to disassemble and transport," von Guericke said. "That would be my guess."

Rolling past the ungainly 264-foot diameter wheel, the trolley began a slow turn to the west. An athletic field, stadium, and the Aeronautic Concourse defined the final circuitous one-mile loop around the northwest corner of the grounds to Intramural Railway Station #4, where the remaining passengers disembarked at the Convention Hall entrance. It was a five-minute stroll to the Hall of Congresses building where the International Aeronautical Congress was to be held.

After signing in, von Guericke learned that the large displays, including Thomas Baldwin's mighty gasbag and Chanute's *improved* two-surface glider, were being kept at the east end of the Aeronautic Concourse, a seven-acre dirt field situated directly west of the Hall of Congresses building. The hall was provided free of charge to the attendees. However, if they wished to enter the fairgrounds proper, the standard 50-cent daily admission fee was required.

*

Though the 10 o'clock hour had come and gone, many of the Congress's 200 attendees were still scrutinizing the aeronautical displays at the rear of the auditorium. A few, noticing the time, had taken their seats in the comfortable, 500-seat amphitheater, awaiting the commencement of activities.

Within minutes, C.M. Woodward, professor of physics at St. Louis' Washington University, called the Congress to order with three crisp whacks of his gavel. After a brief introductory statement to the assemblage of aeronauts and scientists from four countries, Woodward proceeded to outline the objectives for the two-day meeting:

"We come together today to demonstrate that progress in aerial navigation is indeed possible, to learn from the failures and experiments of others, and to compare notes and results. The lesson is this: success cannot be hoped for except through careful study and scientific investigation."

Seated by himself toward the rear of the auditorium, von Guericke thought: *Sounds like Woodward has taken a page out of an old Chanute speech. Unfortunately for both of them, the Wrights' powered flights have proven this 'crawl before you walk' method to be unnecessary.*

Determined to get things moving, Woodward called upon Professor Albert F. Zahm of American University to present his findings pertinent to aerial friction. Following a brief question and answer session, Professor Nepher described his method of measuring the pressure distribution on a curved surface. When he had finished, it was time for lunch.

Manly had a luncheon meeting scheduled with Professor Woodward in regard to his afternoon presentation, leaving von Guericke/Herring on his own to find a restaurant. As the baron exited the Hall of Congresses building, he spied Carl Dienstbach, also alone. Hurrying to his presumed countryman's side, he introduced himself and offered to buy Dienstbach lunch at the nearby Anthracite Coal Restaurant, where waiters, dressed as miners, escorted customers to underground tables situated within a replica coalmine. With the walls and floors fabricated from genuine Pennsylvania hard coal, a mine fan blew fresh air through the dining area, while battery-powered miners' lamps lighted the tables, adding an odd authenticity to the setting.

As the men consumed their predominately Dutch food, Dienstbach became suspicious when his German compatriot ordered a "large glass of cold milk" and referred to lifting surfaces as 'aero*curves*'. For a moment Carl stared in wonder at von Guericke before speaking. "Tell me, Baron... does Lillian approve of your mustache and goatee?"

Dienstbach's insight hit Herring like a bolt from the blue. Recovering quickly, he leaned across the table and whispered, "Keep your voice down, Carl. It's imperative that my true identity remain a secret."

Dienstbach laughed aloud. "This is preposterous—"

"Your articles," Herring interrupted. "Do you keep records?"

Dienstbach tilted his head. "What kind of records are you talking about?"

Not wanting to be overheard, Herring leaned in closer. "I need the dates when Chanute left France and Britain, back in '03."

Pulling a datebook from a vest pocket, Dienstbach leaned back as he casually flipped through its pages. "Aha! You're lucky that I keep a two-year planner! It seems as though Mr. Chanute left Paris for London on April 9, 1903, and he sailed from South Hampton to America aboard the *Kronprinz Wilhelm*, on April 29. These dates are accurate because I followed the Chanute party all the way back to New York."

Using a pencil to copy the dates onto a paper napkin, von Guericke became lost in thought as he contemplated their meaning. Dienstbach, expecting some sort of explanation, finally broke the silence. "Gus, why on earth are these dates important?"

Herring closed his eyes as he tried to clear his head. "Forgive me, Carl, but I can't tell you anything *yet*. I've been working on a theory, but a few pieces of the puzzle are still missing. However, I can tell you this much: there's some very fishy business going on here... and there could be serious repercussions! I'll tell you more after I interpret the facts."

*

Back at the auditorium, von Guericke paused to talk with Col. Cooper of the British Balloon Corps. Minutes later, he spoke with Major Von Tschudy of the German Balloon Section, who assumed the baron was an official observer for the German military.

Accompanied by a plethora of magic lantern projections, Chanute kicked off the afternoon session with his Soaring Birds lecture, a presentation he had been refining for the better part of a decade. At 3 o'clock, two employees of the Smithsonian carried Professor Langley's model aerodrome *Number 5* onto the stage, at which time Chairman Woodward called upon Charles Manly to give an impromptu description of the machine. After dutifully pointing out the major components of the aerodrome and describing their function, Manly suddenly changed the course of his presentation.

"Gentlemen, the remainder of my time will be dedicated to an unsung hero. The person *most* responsible for the success of Professor Langley's model aerodrome program."

An undertone of whispering rolled through the gathering.

"As you may recall," Manly continued, pointing to the model, "this particular aerodrome made history as the first combustion-engine-powered heavier-than-air machine to have successfully flown. The date was May 6, 1896... five months *after* Langley had dismissed its designer... the man who installed aero*curve* lifting surfaces, an automatic regulating tail, and a vastly improved gasoline burner and engine. Unfortunately, my predecessor did not have the professor's written *permission* to make these changes, because Professor Langley was in Europe, taking his annual three-month summer vacation! When I assumed the title of Director of Aerodromics, in January of '96, aerodromes *Number 5* and *Number 6* were ready for trial, except for minor cosmetic polishing, as directed by the professor.

"Gentlemen, this engineer has never received his due credit! Therefore, I am here today – more than eight years after the fact – to correct this neglected state of affairs! Although he is not present today, Mr. Augustus Moore Herring has my sincerest appreciation for his pioneering efforts in the emerging field of aeronautical engineering!"

As he stepped from behind the podium, the audience, taken aback by Manly's candor, began to applaud... a few at first, followed by a crescendo that persisted for the better part of a minute. Von Guericke, astonished by Manly's disclosure, thought... *perhaps there is justice in this world after all!*

Chanute continued to clap as he shambled up on stage. Extending his arms to the rafters, the grand old man of aviation nodded with an exaggerated vigor... demanding the attendees' attention. "Mr. Manly is much too modest! In that regard, I am confident that your generous ovation has been directed to his humble nature!

"Professor Langley, a longtime friend and associate, tells a somewhat different story concerning the success of aerodromes *Number 5* and *6*. According to America's premier scientist, Mr. Manly was able to produce successful flights *despite* the bungling of his headstrong predecessor! Therefore, according to the professor, it is Charles Manly who deserves the lion's share of the accolades!" With that said, Chanute hobbled off stage. Cheers, applause and renewed support reverberated throughout the hall.

Von Guericke reflected... *Justice – easy come, easy go.*

*

As the first day of the International Aeronautical Congress drew to a close, Chairman Woodward directed the congregation's attention to the Aeronautic Concourse.

"Gentlemen, I encourage you to step outside and inspect the world's most advanced lighter-than-air machines gathered here to demonstrate their speed and control. As an added attraction, Mr. Chanute's two-surface, heavier-than-air glider is also present for your examination."

Looking more like an enormous bullring than a flying field, the Aeronautic Concourse was surrounded by more than a mile of 30-foot-high wooden fencing, which the organizers claimed would "direct the wind." To the west, beyond the fence at the far end of the site, an athletic field, stadium, and gymnasium dominated the landscape. To the south, parade grounds and barracks for the Jefferson Guard were hidden from view. Also concealed were the unsightly railroad tracks that dominated the field's northern boundary.

A tall observation tower stood prominently at the concourse's northeast corner. Tethered nearby, a 12,000-cubic-foot hydrogen balloon gave paying customers the thrill of a lifetime by soaring to 800 feet in a birdcage-like gondola.

A special balloon house, also located at the concourse's east end, served to inflate the gasbags from onsite hydrogen gas generators. Fair planners had spent in excess of $36,000 to make the flight trials as welcoming as possible for the participants.[3]

Between June 1 and the beginning of the International Aeronautical Congress, no one had yet won the $100,000 prize for a controlled flight of three trips around a two-point course for a total of 15 miles, at an overall speed of at least 15 miles an hour. Wilbur and Orville Wright had declined to attend the St. Louis extravaganza because they thought the rules formulated by the advisory committee were too difficult for them to achieve in a heavier-than-air machine.[4]

*

With Chanute's encouragement and Avery's assistance, several members of the audience stepped forward to mimic the role of glider operator. "Assume the operator's position!" the old man cried.

Once situated, Avery showed the impromptu student how to lift the apparatus.

"Find the balance, my good man!" Chanute directed. "Now... assume the sprinter's stance!"

When performed as ordered, he crowed, "Perfect... this man knows how to do it!"

Standing with the 50 other spectators, von Guericke listened with increasing interest to Chanute's show-and-tell routine. Looking around the gathering, the baron thought... *If they only knew.*

"Tomorrow," Chanute began, "Mr. Avery will make several glides! Weather permitting, we hope you will be present to observe these exciting demonstrations!"

As Chanute answered questions from the crowd, von Guericke's attention drifted to his surroundings. To his surprise, he noticed at least 10 Jefferson Guards patrolling the Concourse. *An unusual number,* he thought. *I wonder if Chief Sauerbier had contacted the authorities before his arrival?*

Feeling secure in his disguise, Herring had a disquieting thought. *What if the authorities checked the registration records at our hotel?*

With growing paranoia, Herring forced himself to concentrate on more probable scenarios. *The barracks are located here,* he thought, *right next to the concourse. Curious off-duty guards – that must be what's going on.*

Convinced he had nothing to worry about, von Guericke began shuffling toward the Hall of Congresses building, when suddenly, something caused the hair on the back of his neck to stand on end.

CHAPTER 84

Rochester, New York

Wednesday, November 30, 1921

A heavy overnight snowfall delayed the start of court proceedings by half an hour. Picking up where he had left off the previous day, attorney O'Grady continued his questioning of Augustus Herring. "Mr. Herring, you testified that something at the Aeronautic Concourse caused the hair on the back of your neck to stand up. Is that correct?"

"Yes."

"As a matter of clarification, were you still disguised as the German Baron?"

"Yes."

"Tell the court... what brought about your reaction?"

Herring planted his feet and leaned forward. "The sight of Police Chief Sauerbier."

O'Grady was quiet for a moment. "You were expecting him to show up in St. Louis, weren't you?"

Herring squirmed in his chair. "Sauerbier caught me off guard. Not only was he dressed in civilian clothes, he was mingling with some of the world's most renowned aeronautical experts."

"What did you do?"

"I decided to find out what he was up to... so I edged over to where he was quizzing a member of the British balloon contingency. To my relief, they were discussing the whereabouts of the Russian – Butusov."

"What happened next?"

"I was in the process of slipping away when Sauerbier wheeled around and grabbed me by the sleeve! At first I thought my disguise had failed me... then he asked if I knew the rogue Russian!"

"What did you say?"

Herring chuckled softly before he answered using von Guericke's accent. "I said 'As German citizen and special advisor for aeronautics to Emperor Kaiser Wilhelm II, I would *never* converse with a lowly Russkie!' With that, I turned on my heel, and took my leave!"

Judge Sawyer's gavel immediately silenced the laughter spreading through the gallery. "Mr. Herring... no theatrics."

Slipping out from behind the podium, O'Grady changed his focus. "Mr. Herring, is it true that Chanute's World's Fair team consisted of himself, Bill Avery and a second assistant?"

"Yes sir."

"Who was this second assistant?"

"Mr. Manly didn't know him by name, but the man had a Russian accent and was confined to a wheelchair."

"Is it true that Mr. Chanute had complained to Charles Manly about this second assistant spending too much time on the *Pike*, a popular fair concourse?"

"Yes!"

The Pike was a mile-long, 100-foot-wide, open space that ran along the northern confines of the fair, stretching from the Plaza of St. Louis, west to Station #3 of the Intramural Railway. Primarily an after-dark destination for adventurous adults, the Pike consisted of attractions from the simple to the elaborate. Some were informational, some thrilling, while others were scary, humorous, or just plain entertaining.

Pike-goers could visit the North Pole, Cairo, or even Constantinople. They could ride a camel or a burro, or see an elephant slide down a chute. They could witness epic naval battles staged with 20-foot-long scale models, while concessionaires and barkers armed with megaphones competed with brass bands for their attention. At night the crowded Pike attracted fairgoers to belly dancing demonstrations, pubs, beer hotels, and gambling dives. Some of the more conservative visitors called the Pike an overly elaborate carnival and a shameless den of iniquity.[1]

Although the Jefferson Guards had jurisdiction within the confines of the fairgrounds, their ability to fight lawlessness and debauchery was limited; they could arrest perpetrators but couldn't charge them with a crime – a loophole that was exploited by the Pike's seedier clientele.

It was here that William Paul Butusov discovered the *Great Siberian Railway*, an attraction that featured the likeness of towns from across the Russian countryside – places such as Manchuria and Irkutsk. Boarding one of the four plush Pullman cars, paying customers were subsequently "pulled" behind a throbbing steam locomotive amid simulated hot cinders and the stench of sulfurous coal smoke. To the excitement of all, whistles blared and compressed air blew into the faces of the passengers as the cars shook and swayed, while multi-layered rolling murals raced by glazed windows.

Stopping at several typical Russian villages, the train's passengers were encouraged to disembark and mingle with actors, who performed elaborate productions of local Russian plays accompanied by musicians playing time-honored Russian melodies.

"Do you know when Butusov arrived at the fair?" O'Grady asked.

Glancing toward the courtroom windows, Gus noted that it was still snowing. "No one knows for certain, but I was told that he was seen hanging around the Aeronautic Concourse in early June, three months before Chanute and Avery arrived with their two-surface glider."

"Objection! This is all hearsay, Your Honor!" Robbins shouted, rising to his feet.

"Overruled. Continue."

"What did the Russian do after his boss arrived?"

"He spent his days working on Chanute's faulty launch winch."

"What was wrong with the winch?"

Herring flashed a knowing smile. "From what I could surmise, the winch's electric motor was inadequate. Besides voltage and overheating problems, it experienced belt slippage at its sheave. That made the motor prone to jerking the winding drum and towrope."

"What was done to correct this?"

"William Paul made himself invaluable by learning to manually *slip* the motor's drive belt... thus providing a soft startup for the launch."

"Did the Russian's skill in operating the winch solidify his position with Chanute?"

Herring donned an expression of surprise. "That had to be the case... because William Paul's after-hour tomfoolery incensed the old man."

"Describe Butusov's behavior."

After a moment, Herring carried on. "Each morning, Chanute sent Mr. Avery over to the Pike to find the drunken Russian. After sobering him up with strong coffee, Bill then wheeled Butusov back to the aeronautic courtyard."

"Why did Chanute put up with him?"

"I couldn't figure it out... then it hit me. Butusov's *work* had little to do with the *pay* he received."

"Please explain," O'Grady urged.

Herring leaned forward as if coming to attention. "Let's just say that Chanute found himself in a compromising position—"

"Objection, non-responsive!" Robbins shouted.

"Sustained! The witness will be more specific."

Herring took a deep breath. "Butusov possessed information concerning certain *illegal* activities—"

A murmur of excited whispers rolled through the visitor's section.

"Order!" Sawyer hollered, slamming his gavel down. With order restored, the judge turned his attention back to Herring. "The witness may continue."

"Butusov received a generous salary from Chanute in return for his *silence*. In short, America's renowned Father of Aviation was being *blackmailed*!"

Another wave of animated whispering washed through the gallery. Once again Judge Sawyer admonished the spectators.

Still standing, Robbins was about to object. Instead, he thrust his hands high above his head, signaling his surrender.

Sensing that he might be losing control, Sawyer called the attorneys to a sidebar meeting at the bench. "Gentlemen," he whispered, "the Court is well aware that Mr. Herring's testimony has drifted into the realm of hearsay, speculation, and conclusion. However, for the court to resolve this delicate issue, I have decided to waive the usual rules of evidence and precedent.

"Before you have a seizure, Mr. Robbins, let it be known that the Court's ruling will *only* pertain to testimony related to the Chanute-Butusov dealings."

This is the opportunity I've been waiting for, O'Grady thought.

"Presuming that Mr. Robbins will object to parts of the witness's subsequent testimony," Sawyer continued, "I will allow the defense a continuing objection—"

"Your Honor," Robbins interrupted, trying to maintain his composure, "the Court's decision is bizarre! Because the *rules* are being suspended, I request that this witness's testimony *not* be entered into the official record!"

A look of annoyance crossed Sawyer's face. "Counselor, the court will take your suggestion under advisement. In the meantime, Mr. O'Grady may continue."

"Well..." O'Grady began, and then stopped. "Mr. Herring, let us return to the beginning. What illegal activities did Octave Chanute and William Paul Butusov conspire to commit?"

Herring sat up straight. "It all began when Mr. Chanute helped to cover up Butusov's criminal record and illegal emigration to the United States—"

"Objection!" Robbins cried out. "Mr. Chanute is not on trial here! Besides, there is no evidence that he knew of Mr. Butusov's alleged crimes—"

"That's not true!" Herring interrupted. "Chanute *told* me that he hired William Paul despite some 'unpleasantness' that had occurred back in his homeland. As an aside... whenever the Russian addressed reporters, he always used his first and middle names – William Paul. By avoiding his surname, he attempted to conceal his true identity."

"Objection overruled. Continue, Counselor."

"Mr. Herring, describe William Paul Butusov's early criminal activities."

Herring tented his fingers beneath his chin. "Although I testified about this earlier, I'll try to summarize: before Butusov tried to knife me back in '96, I paid for a private investigation that revealed his repugnant past. It is now a matter of public record that Butusov, in order to avoid the charges of *murder* and *desertion* by the Russian Navy, emigrated illegally to the United States during the late 1880s."

"Why was Mr. Chanute interested in hiring this individual?"

Herring stared at O'Grady. "A friend sent Chanute a newspaper account of William Paul's alleged soaring flights with a large, bird-like contraption. Intrigued, the old man hired a translator, and the two began a correspondence. Afterward, Chanute hired the Russian to build a replica of his *Albatross* in Bill Avery's Chicago-based carpentry shop."

"What triggered the U. S. State Department's interest in Butusov?"

"In September of '99," Herring said, "Butusov repaid Chanute for his support by setting fire to the Truscott Boat Works... destroying my flying machines, engines, and tooling in the process—"

"Objection!" Robbins broke in. "Your Honor, this speculative testimony is shameful!"

Judge Sawyer slowly removed his glasses before addressing the defense attorney. "Have you forgotten our agreement, Counselor?"

"This is ridiculous, Your Honor!" Robbins snorted, before taking his seat.

"Excuse the interruption, Mr. Herring. You may continue."

Herring turned his stare from Robbins to his attorney. "As an eyewitness, I positively identified Butusov as the arsonist. This, plus the report from my investigation, persuaded Chief Sauerbier, to pursue the bastard – sorry, Your Honor. Subsequently, the hunt drew the attention of our State Department, which placed the Russian on the nation's *Most Wanted* list."

"Prior to burning down the Truscott Boat Works, what was Butusov doing?"

"He lived in Chicago with his family, while continuing to work as Chanute's assistant—"

O'Grady cut him short. "Chanute kept Butusov on the payroll?"

"Yes, even after reading the investigator's scathing report! However, when William Paul made the Most Wanted list, the old man soon helped him and his family flee back to Europe."

"Objection! Argumentative!" Robbins bellowed. "Where is the proof?"

"Overruled! Continue, Counselor."

"Mr. Herring, three years after the fire, in January of '03, you received a letter with an Italian postmark from Mr. Chanute. In it, he apologized for his tardiness in reimbursing a modest sum of money owed to you. Did Chanute happen to mention his itinerary?"

Herring nodded. "He did. Octave wrote that he and his three daughters had been vacationing in Egypt, and they were currently in Italy. Chanute added that he was about to embark on a lecture tour that would take him to Germany, France, and England, where he would enthusiastically promote the aeronautical competitions scheduled for the 1904 St. Louis World's Fair."

O'Grady paused to examine his notes. "When did you next hear about Butusov's whereabouts?"

Gus slumped in his chair. "During the winter and spring of '03, William Paul was spotted in Europe."

"Where in Europe?" O'Grady probed.

"Germany, France, and England."

"Where did that information come from?"

"Carl Dienstbach, the New York *Times* reporter who was assigned to attend the European aeronautical meetings—"

"Your Honor," Robbins, interrupted, "is the witness suggesting that Herr Dienstbach *recognized* Mr. Butusov by sight?"

Sawyer peered over his glasses. "Well, Mr. Herring?"

"Your Honor, Carl Dienstbach was present at our Indiana dunes trials back in '96... he saw and conversed with the Russian."

"Aha!" Sawyer exhaled. "Objection overruled. Continue, Counselor."

"Was Herr Dienstbach focused on anyone in particular during those aeronautical meetings?"

"Yes! Chanute!" Herring said. "He was the celebrity guest speaker in all three countries."

O'Grady feigned confusion. "How did Butusov fit into the picture?"

"Evidently, Carl had noticed that the *same* unidentified person showed up at all of Chanute's speeches!"

"In all three countries?"

"Yes."

"Why wasn't the reporter able to recognize the rascal?"

Herring shook his head. "Apparently Butusov had altered his appearance—"

"How?" O'Grady broke in.

"Later, I learned that Butusov had his missing front teeth replaced, grew a full-face beard, and began wearing civilized clothing."

"Was Dienstbach finally able to identify the *mystery* man?"

Gus winced. "Yes... but he didn't put two and two together until after he had returned home to New York for the Christmas holidays – eight months later!"

O'Grady flipped to a new section of his notes. "Mr. Herring, during the early 1900s, what was Alberto Santos-Dumont's claim to fame?"

Herring paused for a moment. "As I stated earlier, he was the world's foremost proponent of the *airship*, another name for a powered, lighter-than-air flying machine."

"Refresh the court's memory. Describe what happened to Santos-Dumont's airships on April 12th and May 3rd of 1903."

"Someone *sabotaged* the gasbags! They were both slashed to ribbons – the first in Paris, the second in London."

Judge Sawyer ignored the whispers that wafted through the courtroom.

"How did you learn about the damage?"

"From the newspapers. The mayhem was headline news in Europe, so naturally a few of the big American papers picked up on the story."

"Where was Mr. Chanute during all of this commotion?"

Herring shifted in his chair. "From my discussions with Dienstbach, it was established that Chanute left France for London shortly *before* the Paris slashing. He then left for America shortly *before* the London slashing."

"Mr. Herring... are you suggesting that Octave Chanute had a readymade *alibi* as to his whereabouts—"

"Objection," Robbins interrupted, "counsel continues to lead this witness! Once again, must I remind the Court that Mr. Chanute is *not* on trial here? For that matter, neither is Mr. Butusov!"

Judge Sawyer extended his hand toward Robbins, in a wait-a-minute gesture. "Counselor, this court is not in the habit of *justifying* what it considers to be relevant! If the defense disagrees, I suggest taking it up with the Court of Appeals! Overruled. Continue."

As Robbins sat down in a huff, O'Grady slipped from behind the podium.

"Mr. Herring, do you believe there is a *connection* between Mr. Chanute's movements and the sabotage of Santos Dumont's airships?"

"Objection! Question calls for a conclusion!" Robbins shouted from his seat.

"Overruled. The witness may answer."

Gus's reply came quickly. "Yes! There *was* a connection!"

O'Grady consulted his notes, allowing Herring's exposé to sink in. "In previous testimony you told the court of an eerily similar slashing that took place a little more than a year later. Refresh the court's memory about that incident."

Herring took a moment to gather his thoughts. "After the London incident, Santos-Dumont arrived in St. Louis with a new airship he called the *St. Louis Racer*. The next day – it was the end of June in '04 – one of his mechanics discovered that the machine's gasbag had once again been slashed!"

"Were there any suspects?" O'Grady inquired.

"Only one... but he wasn't charged."

"Why not?"

"Apparently, there was a lack of evidence."

"Was the suspect identified?"

Herring rolled his eyes. "It was Butusov... but predictably, he gave his name as William Paul."

"Why was he considered to be a suspect?"

"Long after the fair had closed for the evening, he was seen rolling around the Aeronautic Concourse in his wicker wheelchair."

"Did the authorities offer a motive for the slashing?"

"There were many theories bandied about. For instance, the Jefferson Guard suggested that Santos-Dumont, realizing that he couldn't win the grand prize, had sabotaged his own machine to save face."

"Mr. Herring," O'Grady probed, "what are your thoughts concerning a motive for the slashing?"

"At the time, the airship was threatening to become the *dominant* technology for achieving practical manned flight. I believed then, as I do now, that the motive for slashing Santos-Dumont's *St. Louis Racer*, as well as the others... was done to *discredit* the airship's significance!"

Tension could be felt in the courtroom as O'Grady pondered his next question. "Mr. Herring, what evidence was there to support your theory that Octave Chanute may have been the *mastermind* behind these crimes?"

Herring's replied quickly. "Besides all of the circumstantial evidence, there were clues in his speeches and written articles."

Leaning to his right, Gus reached into his satchel, withdrawing several old magazines. "There are a number of these publications that I have marked for the court. These are in addition to *L´Aérophile, No. 8* that has already been entered."

Over the next few minutes, three more magazine articles were placed into evidence.[2]

When testimony continued, O'Grady asked: "Mr. Herring, what incriminating evidence was found in Chanute's articles?"

"Chanute had pleaded with airship proponents to stop wasting time and resources on a 'dead-end' technology. He urged them to drop what they were doing and join forces with advocates for the heavier-than-air concept."

"Was Chanute successful in persuading the airship enthusiasts?"

Herring smiled knowingly. "To the contrary! Most of the airship experts, although polite to his face, *laughed* at the old man behind his back. Increasingly frustrated by his inability to exert influence over these experts, Chanute's response was to impede their progress, especially Alberto Santos-Dumont, and his intention to capture the fair's $100,000 Grand Prize."

St. Louis World's Fair

Saturday, October 8, 1904

Although two continuous days of rain threatened to scuttle the promised flights of his two-surface glider, Chanute was pleased to learn that many of the Aeronautical Congress' participants had remained, undeterred in their desire to observe a manned, heavier-than-air flight. Since the Aeronautic Concourse had turned into a quagmire, Chanute persuaded fair organizer Willard A. Smith to relocate the demonstration flights to the *Plaza of St. Louis'* paved concourse.

Lingering storm clouds and darkness provided a gloomy backdrop at the secluded Aeronautic Concourse, where occasional raindrops and ankle-deep mud added to the misery of the early Saturday morning move. As Bill Avery tended to the kerosene lanterns, he noted that William Paul was clearly missing. Unwilling to waste time searching for the defiant assistant, Chanute hired two roustabouts to help load the flatcar. Less than an hour later, one of the Intramural Railroad's trolleys towed the open-sided car to the Plaza of St. Louis – a trip of less than a mile. Within an hour of their arrival, Chanute's rented circus tent had been erected and the glider was moved inside. His spirits rising, the old man slapped Avery on the back. "With any luck, Bill, your first flight will be underway before the noon hour!"

*

Unaware of the glider's relocation, the two St. Joseph, Michigan law enforcement officers arrived at the Aeronautic Concourse at a leisurely 10 a.m. Learning of their blunder, they hurried over to the spectator-crowded Plaza of St. Louis, where Chanute and his boys were making final preparations for the much-anticipated flight. With his Zeiss prism binoculars in hand, Sauerbier trained his attention 400 feet upwind to where a wheelchair-bound helper sat next to the motorized launch winch, his attention focused on the distant two-surface flying machine.

Manly and Herring, who was still disguised as the Baron, stood among the spectators along the east side, watching the final minutes of preparation. "Look at all the people!" Gus said. "There must be five or six thousand... and they're lined up all the way to the winch and beyond!"

Fifty or 60 Jefferson Guards were also on hand to provide crowd control. Like the spectators, the guards were anxious to witness a manned flight that didn't require a hydrogen-filled balloon to stay aloft. At precisely 11 o'clock, Bill Avery, attired in a tight-fitting black runner's suit and rubberized white shoes, turned the stubby bill of his black sea captain's cap to the rear before stepping up and onto the pine-clad deck of the wheeled launch car. Wiggling into the operator's position amid a maze of wood and wires, he squatted, gripped the horizontal launch rails and stood up, absorbing the glider's trifling weight with ease. As he methodically jostled the machine fore and aft to find the proper balance, Chanute, cloaked in a ginger-colored, ankle-length trench coat and brown bowler, steadied the contraption from the lower portside wingtip.

Although still low in the cloudless autumn sky, the sun's morning brilliance had already warmed the Plaza to almost 60 degrees. A perfectly directed 16-mile-an-hour wind blew from the southwest. As Avery peered down the long concourse, the Louisiana Purchase Monument was plainly visible half a mile away. *Wouldn't it be grand,* he thought, *to soar across the Grand Basin waterway and circle the great statue?*

With the towrope attached to a release mechanism directly below his right hand, Avery determined that all was ready. After a final check of the wind direction, he nodded to Chanute, who signaled his winch operator by waving a bath towel-sized red flag fastened to the end of a broomstick.

The distant whine of an electric motor signaled the beginning of the much-anticipated event. An instant later, there was a gentle pull on the rope, followed by the rapid acceleration of the glider, operator, and launch cart down the 100-foot, steel-clad track. The assembly had scarcely covered 20 feet when the glider and its operator abruptly separated from the cart and soared into the air. With his legs dangling, Avery clung to the machine's arm rails as the contraption climbed steeply upward, its wings dithering from left to right and back again, like a deranged kite on a gusty day. At an altitude of almost 40 feet, he pushed all of his weight forward, causing the machine to pitch downward. As the wings approached level with the horizon, Avery pulled the towrope release lever, and the rope fell free.

William Avery prepares to be towed aloft in Chanute's modified 1897 Herring glider at the St. Louis World's Fair (1904); Wikipedia-public domain

After a free flight that carried the glider 175 feet down the concourse in about 10 seconds, the machine descended majestically to within a few feet of the unforgiving pavement. As Bill threw his weight aft, the machine reared before slowing, dropping him gently onto his feet. The crowd roared its approval and pressed closer for a better look as another of Chanute's hired hands rushed up to help carry the contraption back to the launch area.

*

Having produced their identifications to the Jefferson Guard supervisor, Chief Sauerbier and Deputy Stuckey used the lull between flights to approach the winch operator, who was tending to his equipment. Thinking the strangers were drawing near to retrieve the towrope, the man was taken aback when they flashed their badges.

"William Paul Butusov?" Sauerbier asked, as the winch man wheeled his chair out from behind the machinery.

"Who want to know?"

Bending at the waist, Sauerbier shoved his badge to within a few inches of the man's unshaven face. "The name is Sauerbier, I'm Chief of Police back in St. Joseph, Michigan... me and Deputy Stuckey have come all this way to take you into custody."

"For what?"

"For setting *fire* to the Truscott Boat Works! I have a warrant for your arrest... let's see some identification!"

"What you talk about? William Paul no set fire!"

Before leaning closer, Sauerbier broadened his stance. "Show me your identification... now!"

As the man slowly reached inside his black canvas jacket, Sauerbier momentarily shifted his gaze to Stuckey, and winked... a mistake he would soon regret.

Before Sauerbier could fully return his attention to the suspect, the Russian had delivered an arcing, underhanded slash to the Chief's protruding belly with his razor-sharp Cossack Laplander knife. As the stunned law enforcement officer involuntarily lurched backward, he narrowly missed being stabbed in the face by the lighting-fast rebound of his attacker's blade. Dumbfounded by the severity of his injury, Sauerbier slowly sank to his knees, staring in disbelief at the pulsating stream of blood that gushed onto the pavement.

The advantage of surprise gone, Butusov tossed his weapon aside and moved quickly to avoid retaliation from Sauerbier's deputy. With a powerful coordinated move, he first spun and then propelled his wheelchair down the sloped concourse that led directly to the massive Grand Basin Waterway. After a moment of confused indecision, Stuckey abandoned his wounded boss and sprinted after the speeding Russian. Horrified, the crowd looked on.

Rolling at almost 20 miles an hour, Butusov reached the water's edge in less than a minute. Although he had easily outdistanced Stuckey and the closest of his Jefferson Guard pursuers, safely stopping the out-of-control wheelchair proved to be impossible. In an attempt to avoid being hurtled into the chilly waters, the Russian yanked the chair sideways, causing it to skid and begin to decelerate. With its solid rubber tires smoking in protest, the machine skipped from the pavement onto the closely cut grass. The lead wheel dug in, flipping the contraption into the air and ejecting Butusov from his padded seat. He rolled several times before ending the ordeal on his knees.

Astonished onlookers gaped as the presumed invalid slowly rose to his feet, adjusted his cap, and hobbled toward the first of several wooden docks that extended onto the basin's choppy waters. Identifying a likely escape vessel, Butusov continued his desperate flight to freedom by forcibly pushing the rental concessionaire off the dock. During the ensuing commotion, he crawled aboard an electric motor-powered skiff and cast off amid a profusion of shouting and arm flailing by those left behind. Heading for the southwest shore at more than 10 miles an hour, Butusov had less than two minutes to plot his next move.

As the flat-bottomed boat plowed its way through the turbulent water of the Grand Basin's 600-foot semi-circular head, Butusov skillfully guided the craft and beached it beneath the three spectacular Cascades. Scrambling ashore amid the noise and chaos created by 145,000 gallons per minute of pumped water tumbling down the steep steps of the 60-foot waterfalls, he hoped to vanish amid the ensuing bedlam... and blend in with the massive crowds that occupied that portion of the fairgrounds.

During their frenzied foot pursuit of the attacker, Deputy Stuckey and members of the Jefferson Guard made good use of their service whistles to alert other guards. As a result, although Butusov was more than a hundred yards ahead of his pursuers, several more guards joined the fray, clambering down the slate steps that flanked the middle Cascade.

Panicked, the Russian retreated to the skiff he had abandoned only moments earlier. To his chagrin, the unsecured craft had been pulled offshore by one of the Grand Basin's powerful whirlpools. In a desperate attempt to reclaim his means of escape, William Paul leapt into the tornadic abyss... only to miss his mark by several feet. With arms flailing in a desperate attempt to latch onto the foundering boat, he was helpless against the powerful suction of the vortex. Also helpless, his pursuers watched the futile struggle from shore, as Butusov disappeared below the roiling waters.

Butusov's body was later found trapped in one of the Grand Basin's inlet ducts that supplied water to the Cascade's monster centrifugal pumps. Sucked onto an iron grating at the bottom of the 15-foot-deep sink, Butusov met an ironic... and some would say a well-deserved end for a criminal who had spent most of his life on the water.

*

Waiting for their waiter to arrive, Herring, Manly, and Dienstbach huddled around a small corner table at the West Cascade

Restaurant, reviewing the events of the day. Gus, still disguised as Baron Otto von Guericke, fidgeted with his glued-in-place mustache. "It's hard to believe that Butusov is gone for good. On the positive side, now I can get rid of this damn costume!"

"Service is really slow," Charles observed. "The help seems more interested in gossiping than tending to their jobs."

"Maybe," Dienstbach replied, "but you can't beat the prices! For instance, hot roast beef sandwiches are only a dime!"

Looking up from his menu, Gus stared at Charles over the top of his glasses. "Gossip is one thing, but it's not every day a criminal drowns within a stone's throw of your workplace!"

"The 35-cent baked chicken pie sounds good to me," Charles said, ignoring Gus's defense of the slow service, "as long as they don't have to chase down the chicken first."

"They've shut down the pumps," Carl said. "Probably the only way they could get the body off the inlet grate. At 45 cents, maybe I'll try the roast rib of beef... cost you double back east!"

"Wonder what they'll do with the remains," Manly mused. "Do you think Chanute will take charge?"

"At 15 cents a bowl I think I'll try the green turtle soup," Gus said, smacking his lips. "Throw in a couple of glasses of cold milk at four cents each, and the whole shebang comes to less than two bits! A pauper's burial is what the bastard deserves."

"Have all the milk you want," Carl said, "For 15 cents I'm having a big bowl of rice ice cream and a slice of Bavarian cake for dessert."

"What about the Police Chief?" Charles asked. "Last I heard, the Guards had hauled him over to the field hospital."[3]

After a minute of somber silence, Carl leaned forward and poked Gus in the chest with his forefinger. "It's time to 'fess up, my good man! What did Butusov *really* have to do with Chanute's European speaking tour?"

With a gleam in his eye, Herring reached into his vest pocket and withdrew a slip of paper. Carefully unfolding it, he took several seconds to peruse its contents. "Carl, the dates you gave me concerning Chanute's departure from France and Britain... they helped me to unlock the mystery."

Dienstbach's eyes narrowed. "Go on."

"The first slashing of Santos-Dumont's gasbag took place in Paris... three days after Chanute had left for London. The second slashing happened in London... three days after the old man had sailed for New York."

Charles pressed a finger to his temple. "Doesn't that exonerate Mr. Chanute?"

"Hardly!" Gus protested. "Octave had simply established two *alibis* for himself. It was Butusov who stayed behind to carry out his dirty work!"

A few moments of stressful silence followed.

"What are you going to do with this information?" Manly inquired.

Gus leaned back in his chair. "I *wanted* to turn over all of my findings to the authorities... but William Paul's unexpected departure from this world has left me with a dilemma!"

"Why is that?" Carl demanded.

"Allow me to explain," Gus said. "When I overheard Police Chief Sauerbier asking questions about Butusov at the aeronautical conference, I concluded that he was hot on the Russian's trail concerning the Truscott Boat Works fire back in '99—"

Charles interrupted. "If you decided that Sauerbier wasn't chasing *you*... why did you keep wearing this disguise?"

Gus hesitated a moment. "Honestly, I didn't want to muddy the waters! Besides, I didn't want to give Sauerbier ideas... like sequestering *me* as a material witness."

"Then what is the dilemma?" Dienstbach pressed.

Gus raised his hand in a "stop there" gesture. "This was my plan: I would wait until Chief Sauerbier had arrested and dispatched Butusov back to St. Joseph – then I would spring my trap—"

"Trap?" Carl broke in. "*Whom* were you trying to trap?"

"Chanute! Hear me out. Although the chief is without a doubt a glory seeker, he would still be obligated to call in the Federal authorities—"

"Why?" Edward asked.

"Because the Russian *was* still on the nation's Most Wanted list! Anyhow, my plan was to mail Sauerbier an anonymous package... a package that contained all of the incriminating evidence concerning Chanute's aiding and abetting a fugitive, and the Chanute-Butusov conspiracy to sabotage airships—"

"And..." Carl interrupted, "when Butusov was confronted with those new facts, he would probably try to save his own skin by implicating Chanute as the *brains* behind the gasbag slashings!"

"That was the plan," Gus concurred. "But now, the prime suspect and potential whistle-blower... is dead!"

Carl grimaced. "You can't go public with this, Gus. With Butusov out of the picture, where's the proof? You can speculate as to dates, times, and motives, but how can you *prove* that Chanute was behind these crimes? Absent Butusov... it's all circumstantial evidence!"

"Carl's right, my friend," Manly agreed. "Maintain your silence, lest you be ridiculed and driven from your already shaky ranks within the aeronautical community. The grand old man of aviation wields too much power!"

Rochester, New York

Wednesday, November 30, 1921

After the litigants had returned from their lunch break, Judge Sawyer wasted little time in prompting attorney O'Grady to resume questioning Herring.

"Did you speak with Chanute after Butusov's death?"

Herring smiled knowingly. "Briefly. Since Charles and I weren't scheduled to leave for home until Monday morning and the fair was closed to the public on Sunday, we hung out at the Aeronautic Concourse to watch the powered gasbags prepare for their upcoming trials. To my surprise the old man was there, acting as if nothing had happened. Overnight, he had removed the two-surface machine from the Plaza of St. Louis—"

"Why did he do that?" O'Grady interrupted.

"He told Charles that the Aeronautic Concourse offered better facilities to work on his equipment... but I knew the real reason. He needed to get away from the prying eyes of the crowd, away from the interminable questions concerning his deceased assistant."

"What did Chanute say when he saw you?"

"He seemed taken aback. He recovered by suggesting that I had missed all of the action from the previous week."

"How did you respond?"

"I told him the truth – that I had been there the entire time... in disguise."

"What did Chanute say to your disclosure?"

"His expression went blank and he simply shook his head. To his credit, he didn't pursue the matter further."

"What else did the two of you discuss?"

Herring waved his hand as if to shoo away a fly. "I limited my comments to the two-surface glider. Specifically, I condemned the addition of the severely drooped leading edge on the arc-type aero*curve*—"

"What did you say?"

"That the drooped aero*curve* was hurting the machine's performance."

"How did Chanute respond to the criticism?"

"He grumbled something to the effect that I was ill-informed."

"Did you pursue the issue?" O'Grady inquired.

"Certainly. I pointed to the replica's poor glide performance during Avery's lone flight on the previous day. It was obvious that the machine's glide ratio had suffered."

"Did that end the conversation?"

"Not quite. The old man reminded me that he would gladly sacrifice a bit of performance to ensure the operator's safety."

"Was that true?"

"Of course not! I repeated that his leading-edge droop would *not* keep the center of pressure from reversing at small angles of attack... and therefore wouldn't prevent a nosedive! I also reminded him that even his latest *students* – Wilbur and Orville – had abandoned this cockamamie idea after the failure of their 1901 glider."

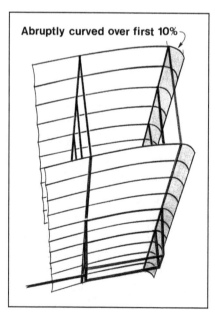

Chanute used a failed wing section from the Wright Brothers 1901 glider on the World's Fair Double-Decker (1904); Drawing by author

"Did you use the opportunity to address the uncanny coincidences of time and place among Chanute, Butusov, and the Santos-Dumont balloon slashings?"

Herring laughed at the notion. "God, no! My sarcastic remark concerning the Wrights had already caused him to turn and waddle away."

Signaling a departure from that line of questioning, O'Grady stepped away from the podium.

"When did you leave St. Louis for home?"

"Charles and I left for New York early Monday morning."

"Did the New York newspapers shed any new light on the Butusov-Sauerbier story?"

Herring glanced around the room before nodding. "There was an enlightening story in the New York *Times*, authored by Carl Dienstbach. A bit of new information came from the St. Joseph-Benton Harbor newspapers. They ran two in-depth articles about the Sauerbier attack. In one story, Deputy Stuckey was quoted as saying that the Chief had received a tip from Scotland Yard, which claimed that Butusov had again entered the United States illegally – probably as a deck hand aboard an ocean-going freighter. This information, along with his hunch that the Russian might attend the World's Fair, persuaded Sauerbier to catch a train to St. Louis. Sealing the deal was knowledge that Chanute, Butusov's longtime employer, would also be there demonstrating the slightly modified replica of my *Double Decker* glider.

"According to Stuckey, Sauerbier had been prepared to arrest Butusov for the Truscott Boat Works fire. With Butusov dead, and Sauerbier soon to be – from peritonitis – Stuckey petitioned the county's district attorney to close the case."

O'Grady, satisfied that both the Butusov and Truscott Boat Works affairs had been put to rest, addressed a related topic.

"After the Plaza of St. Louis debacle, did Bill Avery have the opportunity to fly the two-surface glider again?"

Herring relaxed a bit. "The friendly confines of the Aeronautic Concourse gave Bill the opportunity to make short glides on almost a daily basis for the next few weeks. Most were only about a hundred feet—"

"Objection, hearsay!" Robbins cut in. "The witness wasn't present! He had already departed for New York."

"Overruled," Sawyer said. "The Court will hear the witness's rendition of these events. Continue, Mr. O'Grady."

"Were there any more attempts to fly at the Plaza of St. Louis?"

Herring nodded. "After the uproar had died down, Chanute asked fair organizers to allow him to return to the Plaza. Since the fair was to close in five weeks, officials envisioned rejuvenated attendance figures from Chanute's promised flights... so they gave the old man the go-ahead."

"How did the glider perform?"

Herring scoffed. "On the afternoon of October 25, Bill made his second and *last* trial at the Plaza. He reached an altitude of 30 feet, but a defective towrope broke before he was ready to activate the release mechanism. Because of the premature release, the machine slipped sideways to the pavement. Bill fractured his ankle, ending all further demonstrations."

Shaking his head, O'Grady continued. "How did you view your weeklong experience at the Exposition?"

Gus waved his hand dismissively. "Although no one would admit it, the Aeronautical Congress had been a big waste of time – the Wrights had already solved the problem. The $100,000 Grand Prize went unclaimed, since there were no powered, heavier-than-air machines present to compete. The gasbags were fun to watch, but they were ungainly and difficult to control.

"I missed seeing Alexander Bell and his tetrahedron kite. I was told that the crowd was more interested in seeing the inventor than his kite!

"Best of all, I was exonerated from any suspicion about the Truscott fire, and I was rid of two of my biggest tormentors – Sauerbier and Butusov. In that regard, I am forever grateful!"

CHAPTER 85

Rochester, New York

Wednesday November 30, 1921

Herring, daughter Chloe and the two attorneys huddled around a table at the rear of the Empire State Restaurant, where their impromptu early evening supper would be followed by a strategy session. As a blazing hardwood fire warmed their bones, the nuisance of the late fall snowstorm slowly faded away.

"It would be a hell of a lot cozier if our government would see fit to allow me a whiskey Manhattan or two!" Maloney groused.

Catching Chloe's eye, O'Grady winked. "Then nothin' would get done; we've been down that road before, haven't we, William?"

Maloney winced, pained by the memory.

This evening, O'Grady and Maloney were buying supper, with one stipulation: everyone would order the attorneys' favorite meal – British-style onion soup, beef and potato stew, and hot Italian hard-crusted bread.

"You won't regret having *both* the soup and the stew," O'Grady said, "and the bread is splendid with gobs of melted butter! It's the house specialty!"

After everyone had ordered their drinks, Chloe retreated to the powder room, allowing Maloney the opportunity to carp about Judge Sawyer's court.

"Sawyer handed that little shit Robbins a big setback today! In all your days, have you ever heard of a more indulgent judge? The testimony he's allowing!"

O'Grady leaned back and sighed. "The St. Louis Exposition testimony went well, if for no other reason than to wrap up all of the Butusov crap... and to finger Chanute for the airship vandalisms. On the other hand, Chief of Police Sauerbier's dogged pursuit cost him dearly—"

Momentarily interrupted by the waiter, and Chloe's return, O'Grady took the opportunity to remove a small notepad from his inner vest pocket. Directing his remarks at Gus, O'Grady began. "Besides enjoying having the two of you dine with us tonight, Bill and I want to discuss the direction for your upcoming testimony. Specifically, I'm referring to the years 1905 through 1908, with the possibility of selected material from '09 and '10. However, those years are open for debate since you already testified as to the formation of the Herring-Curtiss Company in early '09—"

"However," Herring interrupted, "It's imperative that we emphasize how the formation of the Company adversely affected my ability to fulfill the conditions of the Signal Corps contract."

O'Grady nodded. "We can try to retrace our steps, but Robbins will object, and Sawyer will probably sustain. That's why we're here this evening: to form a plan of attack in dealing with this possibility, among others—"

"Preparation is the key," Maloney interrupted. "If we can tie new facts to the previous testimony, the judge might overrule the runt! I suggest we start by listing the most important events that occurred during that time frame, beginning with 1905."

All eyes turned to Herring, who squirmed at the thought of being put on the spot. "Offhand, I can't think of anything that happened in—"

"Augustus!" Chloe piped up. "What about the toy tops and model aeroplane kits that you manufactured for the department stores? Your company made beautiful toys! And what about the design work you did for Mr. Clarke's glider?"

"You remembered all that?" Maloney chuckled.

"Of course! I was only nine years old in 1905, but during summer recess I often rode the train into work with Daddy. If I promised not to make any noise, he would let me to sit on a tall wooden stool and watch him work at the drafting table. That's when Daddy made the drawings for the *Philadelphia Glider*. I remember that like it was yesterday!"

Herring grinned sheepishly and hiked his shoulders. "Chloe seems to remember '05 better than me! Now that she mentions it, I did design that towline glider back then... it almost slipped my mind."

"Should we review the glider's details?" O'Grady asked.

Herring leaned back in his chair. "We should probably wait until the '06 discussions."

"Are you ready to move ahead?"

Herring nodded. "On January 13 to 20, 1906, the Aero Club of America ran its first Exhibition of Aeronautical Apparatus, at the 69th Regiment Armory, in New York City. It was held in conjunction with the sixth Annual Automobile Show. I helped to organize the event and also participated in some of its activities. I flew a replica of my 1892 *Double Decker* model that was built from one of my commercial kits. It made dozens of flights across the Armory's huge drill floor, demonstrating perfect stability."

"What else of note happened at the show?"

Herring tilted his head. "I gave Henry Clarke the glider drawings, and he rushed off to build it. When it was ready to fly later that spring, I took a train to Philadelphia to help. There were some new power-plane ideas I wanted to try."

"Anything else of interest concerning the show?"

"Curtiss was there. In fact, that's when we first met. Someone said he was a motorcycle and engine builder who was interested in heavier-than-air flight."

O'Grady chose to ignore Curtiss for the moment. "What else happened during '06?"

"Let me think. An all-encompassing American patent for lateral control was issued to the Wrights... Professor Langley died, and I sold my one-third interest in *Gas Power* magazine—"

"What was the significance of the magazine's sale?" Maloney broke in.

"It provided financial security and allowed me to pursue other interests... or so I thought."

"When did you conceive your gyroscopic stabilization system?"

"During the summer of 1907. I remember the date, because Lawrence Sperry – Elmer Sperry's son – didn't obtain a patent for his version of *my* system until 1918."

O'Grady glanced over at Maloney, who had paused from his note taking. "We must remember to make a point of this, William. What else happened in '07?"

"Let me think," Gus said, pulling at his earlobe. "Oh yes, the stock market crashed, and Walter Brock left my employ for a better-paying job."

O'Grady shuffled his notes, a subtle signal that they should move along. "You have had some difficult years, Gus, but 1908 was especially trying. Complicating matters, William and I are at odds as to how handle the government aeroplane debacle. He's in favor of telling everything, but I'm not so sure— "

"In other words," Maloney cut in, "what are you comfortable with in terms of your testimony?"

Herring took a quick sip of his ginger ale. "Actually, the whole extravaganza began back in December of '07, when the *U.S. Army Signal Corps* called for bids on their first military aeroplane. After they announced that my proposal had been selected, along with the Wrights', we were offered contracts. At the time, there were a slew of newspaper articles written, mostly about *my* inability to produce a competitive machine. Some of the criticism was valid, but most of it wasn't. This will be my opportunity to set the record straight... to present my side of the story."

Maloney raised an eyebrow. "What were the criticisms?"

"Where should I begin?" Herring wondered aloud, reaching for his glass. "The Washington newspapers, including the Associated Press news service, were the worst. From the very beginning they ran outrageous articles, filled with errors and lies about my machine and me, for which I demanded immediate corrections. At the same time, the journalists made Orville, an Edgar Allen Poe look-alike, appear invincible. We often joked that they should have simply called him *Ares* – the Greek god of flight!"

"Interesting background material," O'Grady said, "but let's talk about the mayhem of '08. According to your deposition, besides having to design and build the government machine, you were faced with two cutting-edge technical problems: the gyroscopic stabilization system, and the two lightweight engines. Is there something we should say about their development?"

"Definitely!" Gus said. "The engines were years ahead of their time. With advice and encouragement from Charles Manly, I began a long and costly journey to produce two, five-cylinder radial engines—"

"How were your engines different from those of Langley?" Maloney interjected.

Herring perked up. "Externally, they looked somewhat alike. Internally, there were major differences. I believe that a brief discussion about my engine's unique qualities might be in order."

"Good idea," O'Grady agreed. "I've made a note to that effect."

The waiter entered the dining room with the first course of their meals.

"It's about time!" Maloney exclaimed, rubbing his hands together. "Let's eat!"

*

After the table had been cleared, coffee was served to all, with the exception of Chloe, who waited patiently for her tea to steep. O'Grady, anxious to get back to work, slapped his hands together. "Darn good meal, if I may say so myself!

"Now Augustus, let's move on to that infamous day of October 28, 1908, and the first test flight of your spanking new Signal Corps aeroplane! How should we handle this?"

Herring raised his hands as if surrendering. "By being straightforward! The year was 1908, for Christ's sake! There were only a handful of successful flying machines in the world—"

"Do you *really* want to tell this story?" Maloney broke in.

"Absolutely! I'll finally have the opportunity to address the liars and backstabbers."

"Fair enough," O'Grady said, "but we have a couple of questions. First, explain the Signal Corps' *technical delivery* policy, and what it had to do with the one-month extension you received."

Herring turned pensive. "Since my machine couldn't meet the middle-of-September deadline, I requested an extension. Signal Corps rules stipulated that if I could show significant progress in the machine's construction, extra time might be granted. Referred to as a 'technical delivery,' some aeroplane parts had to be submitted to their committee for evaluation."

"I see," O'Grady said, as he scribbled unintelligible notes. "Our second question concerns the other lengthy extension you obtained after the October 28 incident."

Herring meshed his fingers. "A month prior to my *incident*, as you call it, Orville had crashed his machine, killing Tom Selfridge and badly injuring himself. Since the Signal Corps directors voted to give Wright time to recover and rebuild his machine, they could hardly refuse my request."

"Now I get it!" O'Grady declared.

"Anything else happen during '08?" Maloney asked.

Herring took a deep breath. "I attended Curtiss's flight trials at Hammondsport, where he coaxed the *June Bug* into flying straight ahead for about a kilometer, winning the *Scientific American* trophy.

"A few weeks later, I applied for the first of my gyroscopic control patents."

"That brings us to '09," O'Grady noted.

Gus stifled a yawn... it was past 9 p.m. "When the Herring-Curtiss Company was organized in March of '09," Maloney asked, "how did this affect you in terms of meeting the Signal Corps deadline?"

Herring hesitated before continuing. "Serving *two* masters didn't allow me to do my best work. In addition, I had to meet the Patent Office's deadlines. Fortunately, my British patents were granted with only minor changes."

O'Grady set down his pencil. "How did these added responsibilities affect your ability to complete your *second* government machine?"

"By the spring of '09 I began to fall behind, so I asked for and received a *third* short extension. In an off-the-record conversation with a government bureaucrat, I was told that several of the Signal Corps evaluators weren't happy with the Wright machine's reliability... and were looking forward to me giving Orville a run for his money."

"From our previous conversations," Maloney continued, "it was apparent that you had a card up your sleeve in regard to the Signal Corps competition."

"Are you referring to my possible use of a *company* machine?"

The lawyer nodded.

"While converting one of our machines, such as the *Golden Flyer*, to carry a passenger was feasible, the suggestion never got beyond Glenn, who was using all of the Hammondsport facilities to construct the *Rheims Racer*."

O'Grady picked up where Maloney left off. "Didn't you have an important role with the *Rheims Racer*?"

"I worked on the machine's general layout and took charge of redesigning the air-cooled Curtiss cylinder heads. I changed them to liquid-cooled units with angled poppet valves that matched the curving walls of the new hemispherical combustion chamber."

O'Grady leaned back in his chair. "Speaking of engine development work, your little single-cylinder gasoline engine was still making news six years after it first flew your *Double Decker* model. What was the occasion?"

"In August of 1909, I gave the engine, along with a technical write-up, to Carl Dienstbach, who was traveling to the big technology show in Frankfurt. I wanted to show off the engine's lightweight construction and the details of its revolutionary design. In a signed and notarized letter, I gave the organizers permission to *run* the engine and measure its power. The Germans found that its power-to-weight ratio was still the best of any engine in existence at that time.

"As it turned out, specific design features of my little engine were copied by Mercedes, whose inline six-cylinder engine was

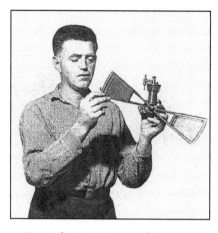

Ray Arden's 1907 copy of Herring's 1902 model aeroplane engine; Popular Science magazine

used in the dominant fighter aeroplane of World War One – Germany's *Fokker DVII*."

Removing another small yellow notepad from an inner pocket, O'Grady opened it to a page marked with one of his King George cigar wrappers. "Now for the distasteful part of 1909," the lawyer announced, pushing himself away from the table. "I'm referring to your September withdrawal from the Signal Corps competition. Should we avoid discussing this in your testimony?"

Herring shook his head vigorously. "Tell it all! The truth has never come out! I would just as soon get the gloomy facts onto the table and get it over with."

"What facts are you referring to?" Maloney probed.

"To begin with, there really *was* a second government aeroplane! Chanute, for one, insisted that it never existed—"

"There had to be witnesses!" O'Grady broke in. "Can someone corroborate the story?"

Herring pressed an index finger to his temple – he was getting a headache. "Carl and his partner MacMechen saw it. At the time, they shared space in my shop to produce their *American Aeronautics* magazine. Then there were Walter Brock and James Martin."

"Continue," Maloney urged.

"The second government aeroplane had design roots dating back to the *Philadelphia Glider*, and my design work of '05. It also looked similar to the Company's *Golden Flyer* and *Rheims Racer* – both of which incorporated my ideas."

"What happened to the second government aeroplane?" O'Grady asked."

"It became the *first* of the three Herring-Burgess aeroplanes. While it was being fabricated at my Broadway Avenue shop, Carl referred to it as the *Shop Plane*. It used my newly redesigned Herring-Curtiss four-cylinder inline engine.

"We need to stress these facts," O'Grady said.

"Third," Herring continued, "in early 1910, after successfully demonstrating the *Herring-Burgess #1* and a replica of my 1896 patent aeroplane, I again petitioned the U.S. Patent Office, demanding they reconsider my 22 claim, 14-year-old application. Inexplicably, my efforts were rejected again! Coincidently, the P.O. also delayed issuing patents for my gyroscopic control system, whose applications I had already turned over to the company.

"Claiming that my applications were worthless in the patent battle against the Wrights, Curtiss and his cohorts forced the company into bankruptcy. This we know. Soon afterward, Curtiss hired young Lawrence Sperry to pursue *my* gyroscopic stability ideas."

After several moments of silence, O'Grady moved on. "Augustus, I want you to carefully consider what I am going to say next."

After a dramatic pause, he continued. "William and I have devised a strategy we believe will disarm the defense, while casting you in a favorable light with Judge Sawyer."

"Go on," Gus said. "What's the strategy?"

"Starting tomorrow, I will begin to ask you some tough questions – questions that will probably make you angry! Some of my questions will be rather abrasive."

O'Grady lowered his voice and forged ahead. "Take no offense, Gus, but we intend to underscore your *weaknesses* – perceived or otherwise. We want to create sympathy. We want the judge to feel *sorry* for you."

Herring's ears began to feel warm. "I don't want anyone's sympathy—"

"No, Daddy," Chloe interrupted, seizing his hand, "sympathy is sorrow for another's pain... this strategy just might work!"

"Here's the key," Maloney said. "No matter how provocative the question, simply answer it to the best of your ability. Try not to show any emotion one way or the other!"

CHAPTER 86

Rochester, New York

Thursday, December 1, 1921

Winds from the southwest whipped the previous day's blizzard across Lake Ontario and into the province of Quebec. Although Rochester's streets had improved to a modest accumulation of brown slush, many lifelong residents were generally depressed at the thought of another long Western New York winter.

With all litigants accounted for, Judge Sawyer whisked through his list of preliminaries, ending with an order for the bailiff to seat Mr. Herring, the plaintiff's continuing witness.

"Mr. Herring," O'Grady opened, "what were your interests and projects in 1905?"

Herring twisted in his seat. "I was designing and manufacturing toy aeroplane kits to sell in New York department stores. I was also developing new products, such as a toy top—"

"Objection, irrelevant!" Robbins shouted, clambering to his feet. "Toys, Your Honor? How do playthings relate to the *real* world of aerial navigation... and this case?"

Judge Sawyer swiveled his chair toward Herring, an inquisitive expression on his face. "Mr. Herring? Would you care to elaborate?"

Herring nodded. "Contrary to popular belief, a model aeroplane is a *real* aeroplane. It works under the same principles as its full-size brethren. Model aeroplanes provide an outlet for budding aspirations. Model aeroplanes get kids thinking about the future—"

"Spare us the sales pitch," Robbins interrupted. "I suppose the toy top also provided an incentive to children. Really, Your Honor, does Mr. Herring take us all for fools?"

Herring continued, unperturbed. "The toy top sparked my interest in the principle of gyroscopic *precession*, which led to a method for achieving automatic stability in full-sized aeroplanes—"

"I've heard enough," Sawyer interjected. "Objection overruled. You may continue, Mr. O'Grady."

"Back in 1906, what were your activities in the field of aeronautics?"

Herring stared down at his black high-top shoes. "I designed another man-carrying glider to investigate some new ideas."

"After the Kill Devil Hills debacle of '02, you testified that you were finished with gliders. What changed your mind?"

"I didn't change my mind. I gave my drawings to Henry Clarke at the Exhibition of Aeronautical Apparatus, in New York City. He carried them back to Philadelphia and built the machine. Afterwards, except for *one* trial, he did all of the flying."

"Who organized the New York City Exhibition?" O'Grady asked.

"The Aero Club of America."

"When and where was it held?"

"It was held at the 69th Regiment Armory, on January 13 through the 20th, 1906."

"Why was this event significant?"

"For the first time in America, aeronautical paraphernalia was available for public viewing. There were bits and pieces from actual aeroplanes, there were models, there were photographic images, mechanical drawings, artistic illustrations, flying model demonstrations, lectures and even magic lantern presentations."

"Describe some of the significant hardware that was present."

Herring closed his eyes. "The Wrights sent the crankshaft and flywheel from the Kitty Hawk *Flyer's* engine. Sir Hiram Maxim sent one of the big propellers from his old steam-powered machine. Then there was Langley's steam-powered aerodrome model.

"As an aside, Carl Dienstbach and I were responsible for hoisting the aerodrome up to the rafters. When a sudden jerk caused the elastically connected tail to oscillate, Carl said 'the aeroplane is waggling its tail at you, Augustus!'

"At that point, one of the Smithsonian's security men piped up. 'Did you know that the aerodrome's stable flight was due to the right amount of springiness in its elastic tail?'

"I'm afraid that Carl and I had a good laugh at the man's expense... he didn't realize the tail 'regulator' was my idea, and that I had originally installed it in the aerodrome we were hoisting."

"Objection! Hearsay!" Robbins cried. "A likely story, Your Honor... and quite predictable coming from this witness."

Judge Sawyer raised an eyebrow. "Overruled. The court chooses to allow Mr. Herring's recollections."

"Mr. Herring," O'Grady continued, "The New York *Times* reported that one entire wall of an exposition room had been reserved for flying machine *photographs*, including those of Lilienthal, yourself, the Wrights, Langley, Maxim, Pilcher, Hargrave and others – from left to right in that order. The *Times* later reported that Wilbur Wright had written to Octave Chanute, asking him why *his* material hadn't been included. In a curtly written response to both Mr. Wright and the newspaper, Chanute replied in part that 'I am very much amused to learn how Herring arranged the exhibits. I sent the Aero Club 15 *models.*'

"It sounds as though Chanute thought you had snubbed him. Is that true?"

Herring chuckled while shaking his head. "As I recall, the wall was *only* used for photographs, which Mr. Chanute *did not* submit. On the other hand, his models – several of which I had previously constructed for him – were prominently displayed in the main exhibit room. Although he was not in attendance, Mr. Chanute felt compelled to complain, and Mr. Wright – who also didn't attend – also proved to be an accomplished agitator."

O'Grady left his place from behind the podium. "Speaking of photography, there was a photograph of your gasoline-engine-powered model hanging from the ceiling of the Exhibition. It appears to show two pusher-type propellers aft of its two lifting surfaces, and one tractor-type in front—"

"I've seen that photograph," Herring broke in. "It's an optical illusion! If you look closely, those two pusher propellers are actually part of Langley's aerodrome model that was hanging directly behind my *Flyin' Fish* model. By the way, as was my practice, the regulator-controlled cruciform tail was left off the model to save display space."

"Speaking of Langley," O'Grady proceeded, "he died less than two months after the New York show. What are your thoughts concerning his passing?"

Herring hesitated as he stared out the courtroom window. "In hindsight, I can sympathize with Professor Langley. Like myself, he was the victim of circumstance. Because of an inadequate launch system, his two attempts to fly the *Great Aerodrome* resulted in plunges into the Potomac River.

"Eleven years later, in 1914, Curtiss attempted – in a ham-handed way – to demonstrate that the machine was capable of sustained flight. Unfortunately, he butchered the original aeroplane into something unrecognizable, an error the Wright lawyers used to further invalidate Langley's claim of having the first vehicle *capable* of sustained flight.

"Truth be known," Gus continued, "the *Great Aerodrome* could easily have surpassed the performance of the Wright *Flyer* in early December of 1903."

With a look of bewilderment on his face, Judge Sawyer turned to the defense table. "Counsel for the defense does not wish to object to this witness's characterization of his client?" With more than a hint of sarcasm, the judge continued. "Quite frankly, the Court is shocked!"

Robbins slowly rose to his feet. "Your Honor, my continuing objection... it's still in effect... is it not?"

Sawyer flinched. "You never fail to disappoint, Mr. Robbins... in fact, your continuing objection *has* been carried over."

Feeling compelled to clarify, Robbins grinned sheepishly. "Certain statements do not require an *immediate* response... true in the court of law as well as life. If we choose to use Mr. Herring's meandering testimony in a petition to the Court of Appeals, it will make a wonderful statement on our behalf."

As if to distance himself from Robbins' thinly veiled threat, Sawyer pushed his chair away from the bench. "Continue, Mr. O'Grady."

"Mr. Herring, carry on with your recollections."

"That spring, as Mr. Clarke and I were concluding our testing of his glider, the newspapers reported that the U.S. Patent Office had finally granted the Wrights a pioneer patent for lateral control... elements of which were applied in their future lawsuits."

"The brothers had remained silent for three years," O'Grady said. "Was that because of their shy nature?"

Herring snorted. "Wilbur and Orville were trying to protect an invention that half the world laughed at, while the other half was scheming to steal! The Wrights, ever suspicious of their fellow man, distrusted the press, shrank from publicity, never married, and lived a protected life under the roof of their widower father."

O'Grady shuffled his notes, signaling a change in his questioning. "In 1906, Glenn Curtiss emerged as a participant in the world of heavier-than-air flight. What are your recollections of him during that time period?"

Herring stared straight ahead. "In a conversation with Lt. Selfridge at my Long Island home, he conveyed a very telling statement attributed to Mr. Curtiss. Allegedly, Glenn told him, 'I get *twice* as much money for my motors from these aeroplane cranks—' "

"Objection! Hearsay!" Robbins shouted, scrambling to his feet.

"Sustained," Sawyer concurred. "Strike the witness's last statement. Continue."

O'Grady sidled over to the witness stand. "Tell the court about *Gas Power* magazine."

"In 1906, I sold my one-third interest in *Gas Power*, so I might devote my time to other interests."

"What interests are you referring to?"

"Designing, building, and flying heavier-than-air machines that *didn't* infringe on the Wright patent. In a small way, I also dreamed of dabbling in the marine engine business. At the time of the sale, I believed the amount received would have carried my plans well into the future.[1] Unfortunately, the stock market crash of '07 changed all of that—"

"Getting back to the *Philadelphia Glider*," O'Grady cut in, "why did *you* fly the machine only once?"

"After I took an unexpected dunking in the Delaware River, Mr. Clarke performed all of the subsequent flying!"

"What influenced your decision?"

"I was feeling guilty about flying the machine! While we were still living in St. Joseph, I had promised my wife that my glider-flying days were over. On my first flight since making that vow, the powerboat's towrope broke – same as happened to Bill Avery's St. Louis Exhibition machine. Despite my best efforts, the damned thing lost all its forward momentum and slipped sideways into the river."

"Were you injured?"

"Only my pride! Fortunately, the contraption wasn't damaged too badly, and Henry was back flying it the following day. Then and there I swore never to break another promise to Lillian."

"Besides experimenting with a forward horizontal rudder, what other design features did you investigate?"

"After my slide into the river, I added jib sails to the top of the upper wing... same as my gas model from January of '03."

O'Grady flipped to another page in his notes. "During 1907, although you still worked at your engineering consulting business, aeronautical pursuits began consuming more of your time. Describe some of these activities."

Herring pricked up his ears. "When visitors came to my workshop-laboratory, we eventually talked about the gyroscope and how it could be used to automatically stabilize an aeroplane."

"What did you discuss?"

"Gyroscopic precession.[2] By harnessing the actions of two opposite rotating propellers, a system could be devised to achieve automatic stability on the aeroplane's pitch axis."

"Was your system also aimed at defeating the Wrights' patent?"

"Yes, by—"

A wave of whispers swept through the gallery, causing the judge to react.

"Order in the court!" he barked, banging his gavel. "Continue, Mr. Herring."

"As I was about to explain... by limiting the operator's control to an occasional *interruption* of the gyroscopically directed flight path, my system effectively bypassed the Wrights' patent claim that 'The machine [was] intended to be generally used with an *operator*.'"

"Was there any competition in the field of gyroscopically controlled aeroplanes back in '07?"

"Not to my knowledge. In the 1860s, the advent of the electric motor made it possible to spin a gyroscope indefinitely. This led to the first heading indicator for seafaring vessels. By 1904, the first gyrocompass was patented by a German,[3] followed by the American, Elmer Sperry, who introduced a design of his own."

Augustus Herring in Washington (1907); Library of Congress

"Were you the first to submit a patent application for a gyroscopic system designed to automatically stabilize an aeroplane?"

Herring raised an index finger. "I believe so... but here lies the irony! After dismissing my gyroscopic precession control system as 'worthless,' Curtiss immediately began collaborating with Lawrence Sperry – Elmer's son – to develop *my* idea—"

"Objection! Argumentative!" Robbins bawled. "The witness could *not* get his system to work!"

"Overruled, Counselor... you are not testifying here! Continue, Mr. O'Grady."

"When did Mr. Curtiss begin consulting with Lawrence Sperry?"

"In the fall of 1912... less than two years after the Herring-Curtiss Company had been formally declared bankrupt."

As O'Grady retreated to the sanctuary of his podium, there was a renewed rumble from the spectator's section.

"Did you employ assistants in your Broadway facility?"

"Walter Brock worked for me off and on, beginning in '07."

"Was he present for the entire year?"

Herring exhaled with a short, rueful laugh. "Walter left when the stock market failed... I couldn't pay him what he wanted. When I recouped some of my losses, he returned."

"When was that?"

"I believe it was early in '08."

O'Grady raised his hands as if surrendering. "Mr. Herring, in the interest of full disclosure... how much money did you lose in the stock market crash of 1907?"

Herring slouched in the well-worn oak witness chair. "About $10,000."

More rumbling erupted from the spectators, resulting in yet another admonishment from Judge Sawyer.

"Quite a setback," O'Grady opined, shaking his head. "How did you recover your losses?"

"Investors."

"Who was your primary investor?"

Once again Herring hesitated, noticeably uncomfortable with his attorney's doggedness. "Chloe P. Herring."

"Your daughter? Your sister? Your mother? Which one was it?"

"My m-mother." stumbled Herring, stiffening. "She almost always believed in my inventions."

"Mr. Herring... how much did your mother invest?"

Another delay in answering prompted Judge Sawyer to reprimand the witness. "Mr. Herring... Christmas is less than a month away. It would be judicious if you would answer the Counselor's questions with a degree of alacrity!"

O'Grady turned to address the judge. "The plaintiff offers its apologies, Your Honor. As you can see, this phase of Mr. Herring's testimony is difficult for him to revisit."

Turning back to Gus, O'Grady offered a reassuring smile. "Let's try this again... how much did your mother invest?"

"Seventy-five hundred dollars."

More murmuring from the spectators

"Were there any strings attached?"

Gus continued to squirm. "Only that I didn't squander any more of the family fortune."

Herring glared at O'Grady as the attorney again walked to his position behind the podium. "In the week preceding Christmas of '07, your fortunes took a turn for the better. To everyone's surprise, the U.S. Army Signal Corps called for bids to produce their first military aeroplane. Tell the court how you reacted to the news."

"After studying the requirements, I submitted a bid for $20,000... as it turned out, 20 percent less than the Wrights had offered."

"What were the requirements?"

Gus tilted his head back and gazed into the courtroom's massive crystal chandelier. "The machine had to carry the operator and a passenger for a total of 25 miles, at a speed of at least 40 miles an hour, and it had to stay aloft for a minimum of one hour before landing without serious damage."

"How many bids did the government receive?"

The hint of a smile broke across Gus's face. "More than a dozen! I don't remember the exact number, but it seemed as though every crank in the country wanted to jump on the government gravy train.

"Some newspapers reported that the Army only wanted the Wright brothers' machine, but there was this 'Herring problem.' The Corps solved the conundrum by changing the rules. They decided to acquire two machines, as long as they both met all of the requirements. *Scientific American* magazine made an analysis of the Signal Corps dilemma, and probably said it best—"

O'Grady, sensing that Herring was going to introduce the magazine, sidled over to the witness stand. "Do you have the article in question to enter into evidence, Mr. Herring?"

After several awkward moments of rummaging through his document case, Gus produced a copy of the magazine in question. Five minutes later, Judge Sawyer urged O'Grady to continue.

"Read the passage in question, Mr. Herring."

With his reading glasses in place, Gus held the magazine at eye level and cleared his throat. "This is from the editorial page:

'Another successful bidder for the government aeroplane – Mr. A.M. Herring – is a man long identified with heavier-than-air aeronautics in this country, and one who has probably done more experimenting and original research in this line than any other American.' "

Handing the magazine back to the attorney, Gus slowly removed his glasses before continuing. "Because of this reputed experience and dedication, I was also awarded a contract. To make sure that we were serious, the Wrights and I were required to make a $2,000 non-refundable deposit to the Signal Corps."

O'Grady was quick with a follow-up question. "How did you intend to avoid infringement charges from the brothers?"

"As I said earlier, by bypassing certain aspects of their patent with my gyroscopic stabilization system. In part, my system acts through the aeroplane's inter-plane control surfaces in providing automatic lateral stability – *without* the aid of an operator."[4]

*

Gus went on to specify the features of his government machine.[5] He also explained how his control system would be used to overcome the detrimental effects of wind gusts and other outside disturbances. He described how the operator could override the gyroscope when a change in heading, a takeoff, or an alighting was required.

With Herring again flush with cash, Walter Brock, who was always attracted by Gus's brilliance, left a good-paying job and returned to the Broadway Avenue shop in early 1908.[6] Added to the complexities of cobbling together a rudimentary gyroscopic control system, Herring was required to supply the aeroplane with two, five-cylinder radial engines. Relying on the advice and assistance of Charles Manly and Stephen M. Balzer, a Hungarian immigrant engine builder who had relocated to New York City, Herring was able to produce the engine's main components within six months. In the meantime, Brock and Herring fabricated the domed pistons and intricate cylinder heads, which included canted poppet valves within hemispherical combustion chambers. By August, one of the engines had been assembled, test run, problems sorted out, and tuned in preparation for the all-important torque performance test. While helping to build the government aeroplane and engines, Brock also found time to construct an intricate, strip rubber-powered flying model of the Herring design.[7]

*

"Mr. Herring," O'Grady prompted, "what were some of the problems you experienced with your radial engines?"

Leaning forward, Gus put his elbows on his knees and clasped his hands. "Where to begin?" he said, in a barely audible voice. "First, the good news! The first engine started right off the bat, but it immediately began to misfire and lose power. It also displayed a tendency to overheat, an ever-present problem with lightweight engine designs. After analyzing the situation, I found both ignition and fuel delivery difficulties.[8]

"Fortunately, I had help in resolving these problems. For instance, Mr. Manly convinced me to use a 'distributor-type' ignition system, which proved to be more reliable, easier to adjust, and simple to troubleshoot."

"How did you resolve the issue of power loss due to overheating?"

"We installed lightweight sheet-brass jackets around the individual cylinders, filled them with water and let it boil away. That didn't solve the problem, but it gave me about two minutes of full power run time."

"The canard or *tail-first* configuration was new to you in terms of design," O'Grady said. "Did you *copy* that from the Wrights?"

Herring was visibly upset by the question. "Certainly not! I was curious as to how the Wright's *unstable* horizontal rudder arrangement would act when used in cooperation with my aft-mounted cruciform tail and regulator. Specifically, I needed information about the size, location, and amount of deflection needed to produce stability and control on the pitch axis. That was one of the reasons why I designed the *Philadelphia Glider*.

"There was another reason. With two engines mounted at the rear of the lower lifting surface, the aeroplane would be very difficult to balance fore and aft. The horizontal canard rudders at the end of a long moment arm helped solve the problem... weight could be added or removed as necessary. The canard, or 'horse's head,' as we used to call it, also provided forward protection to the operator in case of a nose-first crash. Its action was similar to the springy willow hoop that Lilienthal often installed at the front of his gliders. He called it a *prellbügel*."

O'Grady nodded before continuing. "Another aspect of your government machine that was similar to the Wright *Flyer* were the side-by-side pusher propellers. Was this also a coincidence?"

Robbins and Curtiss exchanged glances of disbelief as O'Grady continued to hurl what they believed were belittling questions at their adversary.

"Totally unplanned! When the tail of my 1898 *tandem* propeller machine was tossed hither and thither by a gust of wind, the airframe reacted with strange, seemingly random gyrations. Later, after acquiring detailed knowledge as to the gyroscope's operation, I resolved the enigma. By placing the propellers side by side and rotating them in opposite directions, the unwanted motions due to gyroscopic precession could be controlled."

"Was that why the Wrights mounted their propellers where they did, because they anticipated the problems associated with gyroscopic action?"

Herring, fully aware that O'Grady was attempting to produce sympathy for his position, simply smiled. "I have no idea whether the brothers understood the ramifications of gyroscopic precession... they didn't mention the phenomenon during our Kitty Hawk discussions. If memory serves me, Wilbur only referred to the torque-reaction force-canceling properties of counter-rotating propellers. Furthermore, I believe the substandard flight qualities of *their* early powered machines speak for themselves."

Amid rumblings from the gallery, Judge Sawyer recessed for lunch.

*

Herring was back on the witness stand at 1 o'clock.

"Tell the court," O'Grady said, "about the steady stream of visitors to your fourth floor shop on Broadway and West 64th Street."

Herring leaned forward. "It was 1908, and we were busy building the government aeroplane. Most of these people were just nosy."

"Do you recall the names of the aeronautical people?"

"There were so many... let me think, there was Dienstbach, Burgess, McCurdy, Selfridge, Tom Baldwin, Dr. Bell, Manly, and of course Curtiss – and many others. Unknown to me at the time, I also entertained a secret agent of the British government! I believe his name was Alexander."

O'Grady walked over to the plaintiff's table where Maloney seemed to be dozing, his sausage-like fingers intertwined behind his substantial neck. Wheeling about, the Counselor faced the judge. "With the court's permission, a moment to confer with my associate?"

With an anemic wave of acknowledgement, Sawyer resumed scribbling his notes.

"William," O'Grady whispered in his ear, "should I introduce the quote from the *Herald*, or do you think it makes Gus sound arrogant?"

After issuing a short grunt, Maloney boomed his response. "Use the damn thing!"

Back at the podium, O'Grady held up the old newspaper. "Your Honor, the plaintiff wishes to enter the Saturday, June 6, 1908 edition of the New York *Herald* into evidence."

Within minutes, without significant objection from the defense, O'Grady began to read from an article titled, "Curtiss and Baldwin Meet with Herring":

> "'When Mr. Herring was seen yesterday, he was loath to discuss his own affairs or those of his visitors, but he admitted that Mr. Alexander had been to his shop. 'I have no doubt,' said Mr. Herring, 'that Mr. Alexander was in search of information, but I do not care to say to what extent I was able to oblige him, nor do I think it necessary to enlighten the public as to whether or not I permitted him to see the machine which I have here under construction.... I'm not looking for advertising, and have had too much of it already....' "

Looking up from the 13-year-old daily, O'Grady removed his reading glasses and addressed his client. "After reading this article, I got the distinct impression there was great interest in your aeronautical endeavors. Not only from the public, journalists, and other experimenters, but even foreign governments! Presuming my interpretation is correct, how did you and Mr. Brock get any work done?"

Gus raised his eyebrows. "Truth be known, except for the odd occasion when a visitor might provide useful input to my work, I tried to avoid them, especially those who wanted to get my photograph or to interview me."

"Back in the spring of '08," O'Grady continued, "Glenn Curtiss visited your shop. At that time, was he a person of importance in the arena of aeronautics?"

"Curtiss was an experienced motorcycle manufacturer and engine builder, and I was an experienced aeroplane designer and constructor. He was a member of Dr. Bell's Aeronautical Experiment Association and had constructed engines for Baldwin's airships. Although Glenn was an aeronautical *amateur*, we had much in—"

"Objection! Argumentative!" Robbins interrupted, leaping to his feet. "This witness's characterization of Mr. Curtiss as being an 'amateur' is strictly *his* opinion!"

"Sustained. Continue, Mr. Herring."

"As I was about to say, Your Honor, we had much in common, and therefore didn't hesitate to share information. That was the case in early June of '08, when he was about to test-fly his *June Bug*."

"Give the court an example of this information sharing."

Herring's face lit up like a lantern. "I gave Mr. Curtiss tips concerning the design and operation of his new machine. Specifically, I suggested a simple method of delivering the gasoline-air mixture to the individual cylinders of his four-cylinder engine. I also recommended lengthening the moment arms for both the front and rear horizontal rudders, thus increasing their leverage."

"Is it true," O'Grady inquired, "that less than a month later you traveled to Hammondsport to observe Curtiss's attempt at

Herring (straw hat at left) poses with Curtiss (arms folded, 3rd from left) with June Bug (Hammondsport, 7-4-1908); Glenn Curtiss Museum

winning the *Scientific American* magazine's trophy for the first officially sanctioned flight of one kilometer?"

"Yes."

"Did you observe first-hand that any of your suggested changes had been made to Curtiss' aeroplane?"

Herring leaned forward placing his elbows on his knees and his hands together. "Yes."

Robbins scrambled to his feet. "Objection! Argumentative and calls for speculation! There is nothing to indicate that the witness provided the defendant with ideas for improving the *June Bug*."

"Overruled. The witness's statements may stand. Continue, Counselor."

Muttering to himself, Robbins took his seat.

"Therefore, did your suggestions help Mr. Curtiss win the *Scientific American* trophy?"

"Objection! Calls for speculation!" Robbins shouted.

"Sustained. Move on, Counselor."

O'Grady turned to a new page in his notepad. "While in Hammondsport, did you have the opportunity to renew your acquaintance with Lieutenant Selfridge?"

"Yes."

"For the record, who was Selfridge?"

"Lieutenant Selfridge was assigned by the U.S. Army to observe the A.E.A. experiments at Hammondsport. I had been communicating with the lieutenant for the better part of a year. He had visited me at my shop and was a guest at my Freeport, Long Island home. Selfridge had written a glowing analysis of my role in the history of powered flight for Dr. Bell's group – of which he was a member.[9]

"Afterward, Selfridge wrote to take advantage of my offer to provide suggestions on how to improve the previously crashed *Red Wing*, if the A.E.A. decided to repair it."

O'Grady lowered his notes to his side. "What role did the lieutenant play in the *June Bug* saga?"

"An interesting question," Herring replied. "Selfridge, a prolific letter writer, eventually got all the A.E.A. members into hot water."

"Please explain."

"Early in '08, he wrote to the Wright brothers, asking questions about their wing-warping technique. In response, Wilbur informed the A.E.A. through Selfridge, of their patent, plus he added a warning: 'We did not intend, of course, to give permission to use the patented features of our machine for exhibitions or in a commercial way.'

"With Curtiss' success in the *Scientific American* competition, Wilbur fired off another letter to Glenn, saying: 'I understand your aeroplane had movable surfaces at the tips of the wings, adjustable to different angles on the right and left sides for maintaining lateral balance.' Wilbur went on to remind Curtiss that he and his brother had not given permission to use the patented features of their machine. This letter represented the opening volley of a protracted patent war that soon entangled the yet-to-be formed Herring-Curtiss Company."

O'Grady pressed on. "Wilbur Wright, accurately cast as the metaphorical Grim Reaper, wielded a legal scythe against all American competitors who dared produce a commercial or made-for-exhibition aeroplane. This man," O'Grady said, pointing at Herring, "was the first to challenge the brothers' attempt to monopolize the fledgling industry. Mr. Herring's contract with the Signal Corps to produce the first military aeroplane typified struggles in this arena.

"Beyond the daunting task of producing a machine that would meet the Corps requirements, Mr. Herring and his team encountered numerous extraneous problems along the way. In particular, journalists from the Washington press."

"Objection! He's bloviating again, Your Honor!" Robbins spluttered. "Does this harangue represent Mr. O'Grady's closing argument? Leading the witness is one thing, but counsel's total disregard for the rules of evidence is embarrassing!"

"Sustained!"

Pushing his glasses up the bridge of his nose, Judge Sawyer addressed the court stenographer. "Strike Mr. O'Grady's last two statements from the record!"

Sawyer turned to address the attorney. "The Court agrees with the defense. Save the rhetoric for your closing argument, sir! You may continue."

O'Grady snapped to attention, refusing to reflect his disappointment. "Mr. Herring, was your government aeroplane ready to fly by the September deadline?"

Herring wagged his head from side to side and feigned a pout. "Not quite... it had yet to be test flown."

"That brings me back to your 'technical delivery.' What parts of the government machine did you deliver to the Signal Corps committee for inspection?"

One of two 5-cylinder radial engines used for Herring's U.S. Signal Corps aeroplane (10-15-1908); New York Herald

Herring rubbed his chin. "Because it was impractical

to transport the entire airframe, Mr. Brock and I hand-delivered components of the starboard radial engine and an early iteration of my patent-pending gyroscopic stabilizer."

"What did the Corps think of your work?" O'Grady asked.

"An Army officer who saw the engine expressed great admiration. Speaking unofficially to the New York *Herald*, he said:

'It is a beautiful piece of mechanism, quite apart from what it can do as a flying machine. The machine work is like that of a watch, and the finish is most exquisite. If it flies as good as it looks, it will be far ahead of anything else of its kind in the world.'

"Subsequently," Gus continued, "the committee determined that significant progress had been made and a *one-month* extension was granted."

"Describe some of the disturbing stories that were printed about you at the time of your Fort Myer technical delivery, many of which were picked up by newspapers around the country."

There was a smile on Herring's face, but not in his voice. "I remember some of the more egregious stories. For example, the Washington *Post* wrote that in order to meet the September 13 deadline, I had planned to fly my machine all the way from New York to *Chicago*... of all places! The correction read: 'Herring to fly 2-1/2 miles to Fort Myer for delivery'.

"From that same paper, I was quoted as saying I had been making test flights *over* Manhattan Island at *night*... a tale intended to make me look like just another crank."

"Anything else?" O'Grady asked.

"Another slanderous piece stated that the House of Representatives had ordered the investigation of a mysterious $10,000 payment made to me for the technical delivery of the incomplete machine to Signal Corps officials. An exhaustive search of the *House Proceedings* by my attorney found no mention of such an order. The lie required much time, money and effort to unravel. By then, the Fort Myer trials had come and gone, but not the lasting stigma of impropriety that had been directed at me."

"Who was behind these lies?"

"Ha!" Herring exclaimed, his disgust evident. "A fellow by the name of Jerome Fanciulli, son of John Phillip Sousa's successor as head of the Marine Band. As a reporter covering the House of Representatives, first for the Washington *Post* and later for the Associated Press, he had covered Orville's demonstrations at Fort Myer. For some reason, he had a vested interest in seeing Orville succeed and me lose out. When I learned that Fanciulli was responsible for much of the published misinformation, I raised Cain with the editors and even had a meeting with the publisher of the *Post*, where I threatened to sue for defamation."

"Did you get satisfaction?"

Herring slouched a bit. "In terms of corrections, I was lucky to see one in print a week after the fact... at the bottom of the obituary page. Then again, I think I got the bastard fired – but my success was short lived. Little more than a year later, Fanciulli turned up as Curtiss' personal 'advance man' for his exhibition company. Without actually being on the payroll, he began influencing board members of the Herring-Curtiss Company. Through his lawyer friend, Ormsby McHarg, Fanciulli was the first to suggest that if our company could be shown to be bankrupt, it would be the quickest way to get rid of me—"

"Objection!" Robbins cut in. "This insidious information was detailed in the witness's prior testimony!"

"Sustained. Let's not tell the story *all over* again, Mr. Herring. Continue, Mr. O'Grady."

"After packing up your aeroplane parts, you left Virginia for New York and a few days of rest. Soon afterward, under the cover of night, your team secretly transported the government machine to Hempstead Plains for preliminary flight trials. Why did you choose that Long Island location rather than Fort Myer?"

Herring's expression was cold. "The Fort Myer site was unacceptable... it wasn't big enough for the preliminary tests. For my first attempt, I wanted to fly for long stretches a few feet above the ground, against a strong wind. This would allow me to land without danger to the machine or me. In a letter to the chief Signal Corps officer, I went on record as objecting to Fort Myer as the preliminary testing ground."

"What did he have to say?"

"The Signal Corps released me to hold preliminary tests when or where I pleased, until I was ready for my official tests the following month."

CHAPTER 87

Manhattan, New York, New York
Friday, October 23, 1908

Four stories up, employee Walter Brock sat eating his lunch in front of an open window in Gus Herring's Manhattan shop. As he enjoyed the unseasonably warm weather, the young man gazed across an area of city wilderness known as Central Park, and daydreamed of his boyhood home in rural Illinois.

Without warning, a gust of tepid air burst through the casement, flipping the Greek fisherman's cap from his head. Nudged to the present, Brock slowly stood and brushed breadcrumbs from his denim work apron before stretching his wiry six-foot frame toward the ceiling. Clean-shaven, Walter parted his wavy dark brown hair neatly down the middle, exaggerating his deep-set hazel eyes and facial frown lines that made him appear older than his 23 years. Already a master machinist and competent mechanic, Brock had followed Augustus Herring from his native Chicago to pursue a career in the emerging field of aeronautics. Under Gus's tutelage, in addition to his work on manned, powered, flying machines, Brock had turned into a talented designer for the company's expanding line of model aeroplanes and kites. Quiet and unassuming, he blended in easily with the goals and aspirations of his boss.[1]

"Back to the grindstone, Carl," Walter said, as he shuffled over to the three pine boxes sitting in the middle of the workspace.

"We'll have to hurry if we're gonna get out of here early tomorrow morning. Can't be late… have to meet Gus and Charles in *Mineola* by noon!"

Carl Dienstbach had taken a week off from the New York *Times* to help with transporting and testing the Signal Corps aeroplane.

"The *electric* delivery truck will be waiting out front at 8 a.m. sharp," Carl confirmed. "They're sending a two-ton, open-bed job with a covered top… just in case it rains—"

"How long is the bed?" Walter interrupted.

Pulling the invoice from his rear pocket, Carl skimmed its contents. "Ah, here we are… she's 12 feet long."

"What's the carrying capacity?"

"Don't know," Dienstbach said with a shrug, "but its red-oak bed is three inches thick. I measured it."

"Should be more than enough," Brock grunted.

As Walter began removing machine screws from the aeroplane's steam-bent hickory skid, part of the bicycle-wheel alighting gear, Carl resumed his task of reinforcing the shipping crates. *The wings will be easy to crate,* he thought. *Each surface breaks down into four components: two eight-foot centerpieces and two, two-foot-long wingtip extensions. They'll all fit in one of these 10-foot boxes.*

Another five minutes passed before the newspaperman broke the silence. "According to my arithmetic, the lifting surfaces have a total area of 180 square feet – not very much. What do you make of that, Walter?"

"Gus says wing area won't be a problem. He just laughs at those so-called experts who call our machine 'pocket sized.' "

"I thought the government machine was supposed to remain secret until the Fort Myer trials?"

Walter gave him a puzzled look. "Apparently, one of our *visitors* told the Wrights about our machine, and they reported to old blabbermouth; he's been spreading awful stories ever since."

"Whom are you referring to?"

"Old Man Chanute! Who else?" Brock said with a sarcastic laugh. "Anyway, Gus figures the machine will weigh 600 pounds including the operator... so the wing loading will only be 3.3 pounds per square foot. The two 22-horsepower engines will give it a nice power loading of 15 pounds per brake horsepower... better than that clumsy-lookin' orange crate from Dayton."

Carl continued driving woodscrews. "My only concern is that the machine will be too fast. Isn't that dangerous?"

"That's why Gus doesn't want to operate her in a wind of less than 25 miles an hour. He's no fool. He wants to fly the machine low and slow until he learns her tricks."

Dienstbach took a break and sauntered over to where Brock was working. "What about the gyroscopic stabilization system? Do you really believe it will work on this machine?"

Walter's eyes narrowed. "You're not gonna quote me in one of your stories, are ya, Carl?"

Feigning shock, Dienstbach took a dramatic step backward. "Absolutely not! Everything we discuss here is off the record."

"Better be," Walter said with a smirk, "otherwise I'll take pleasure in ringin' your Bavarian neck!"

A moment of strained uncertainty hung in the air like an early morning fog until Brock stood up and grinned. "Just joshin' with ya, Carl... the gyroscopic system looks great on paper, but I've learned the devil is in the details; an old saw, but true nonetheless. The system needs to be refined, but we're gonna try the *articulating engines* idea anyway, but not for the first trial. Gus wants to have full control."

"What about the gyro stabilizer on the roll axis? The one Gus thinks will invalidate the Wright patent."[2]

Walter slowly shook his head. "The electric motor gyro for lateral control will have to wait. It won't be ready."

Freeport, Long Island, New York
Friday, October 23, 1908

The sound of the fragile-looking radial engine bellowed and reverberated across the marshlands behind Herring's Long Island home. Exasperated by its performance, Gus slapped at the ignition switch with his free hand, immediately silencing the brute.

With their ears still ringing, Gus and fellow Freeport resident Charles Manly, discussed the engine's performance while inhaling the fetid fumes of scorched castor oil. Staring down at his scribbled notes, Manly relayed the numbers:

"Two hundred turns in six seconds flat, which translates into 2,000 revolutions per minute. Compared to the starboard engine, this one is down about 200... and the vibrations are *noticeably* worse."

The men had been working all day on the government aeroplane's portside engine, with little improvement to show for their efforts. Unable to elevate its performance to that of the starboard engine, Herring's voice reflected his growing frustration. "We've checked all of the systems and I've tuned her the best I can... what the hell is left? Something's amiss, and we're runnin' out of time!"

The men planned to take the westbound train out of Freeport to nearby Valley Stream the following morning. From there, they would transfer to the northern branch of the L.I.R.R. After brief stops in Hempstead and Garden City, they expected to arrive in Mineola well before noon.

"Let's try one more thing," Manly said, as he inspected the head gasket on the first of five cylinders. "There may be one or more cylinders that are running low on compression. Let's do a compression test."

Out of ideas, Gus readily agreed. While Charles began removing the engine's five sparking plugs, Herring scurried to his boathouse shop to retrieve his *Bourdon Tube* pressure gauge, the one fitted with a ball-check valve. Fifteen minutes later, with the gauge's hose-end screwed into the number-one sparking plug hole, the 5-1/2 foot diameter, two-bladed wooden propeller was pulled through manually for several revolutions.

"Fifty pounds," Herring reported, looking up from the gauge. "Not bad, let's check the next one."

Half an hour later they had their answer.

"Look here," Gus said, "cylinder number *three* is only indicating 35 pounds... it's down 30 percent! Question is, what's causing the loss? Are the piston rings not sealing, is there a leaky poppet valve, or something relatively easy to fix, like a leaking head gasket?"

"I visually checked the head gaskets," Manly said, rising to his feet, "but I think there's a better way to find out what's wrong. Get me a pump-can with some 70-weight mineral oil. I learned this trick back in '02, while working on the *Great Aerodrome's* engine."

Jogging back from the boathouse, Gus tossed the oilcan to Charles, who quickly squeezed a stream of lubricant into the suspect cylinder's sparking plug hole. After carefully screwing the pressure gauge fitting back into place, he again used the propeller to turn the engine over several times.

"Gus, I think we've got a *valve* problem!"

Herring held out his hands, palms up. "How did you come to that conclusion?"

"The oil didn't boost cylinder pressure. That tells me the rings have probably been sealing just fine. Since viscous oil *won't* seal a big gap – something you would expect to find at a leaking intake or exhaust valve – I would start by looking there."

The insight came to Herring like a bolt out of the blue. "Now I get it! If the rings were leaking, the heavy oil would have temporally sealed them, and the pressure would have spiked!"

Manly grinned like a Cheshire cat. "Hey, Augustus!" he laughed. "Not all Cornell grads are dumb!"

*

It was almost 3 o'clock before Gus grudgingly conceded the amount of work that remained to be done. If they were to leave for Mineola on time, it was going be a long night in the shop.

Lillian had lent a hand by securing the loan of a neighbor's horse and wagon, which she used to patronize Freeport's supply depot, renting a large circus-type tent and four cots. She also procured a five-gallon pail of straight-run gasoline for the engines and a gallon of kerosene for the lanterns.

"Charles," Gus said, "why don't you start pulling the head from cylinder number three, and I'll get the Mobike ready. If we need to get somewhere in a hurry, it'll come in handy."

With the cylinder head removed, Manly spotted the problem – a bent *stem* on the exhaust valve was preventing its mushroom-shaped head from sealing completely against the pressed-in-place bronze *seat*. In turn, this was causing the cylinder to leak pressure during the compression and power events. Gus surprised Charles by producing a shiny new poppet valve to replace the damaged piece. "That's Walter for you, he predicted there might be trouble with one of the valves, so he machined a few extras, just in case!"

After the time-consuming task of hand-lapping the replacement valve to its seat, the weak cylinder was returned to first-class working order. It was almost midnight before the engine had been reassembled and was ready to run again. Confirming that the replacement valve would withstand the rigors of further operation would have to wait until after they arrived at the Mineola test site.

Manhattan, New York, New York

Saturday, October 24, 1908

The *Hercules Electric* delivery truck whined its way through the 50-odd blocks between the 1931 Broadway at West 64th Street shop and the docks at the foot of East 34th Street. The trip had taken half an hour to reach the boat slips of the East River Ferry Company.

"Slip number two!" Brock hollered to the driver. "That's our boat, the *Southampton*!"

As their truck backed into the offload area, a gang of four union dockworkers scrambled to discharge the cargo. "Careful, mates!" Walter shouted. "Fragile bits inside those boxes! Treat 'em like yer mama's china!"

Impervious to taking orders from outsiders, the laborers nonetheless transferred the three crates quickly and efficiently to a large horse-drawn wagon. With Walter at the reins, the flatbed's steel-rim wheels clattered over the cobblestones leading to the loading gate while Dienstbach observed how a passenger- and freight-carrying ferryboat operated.

At the stern of the boat's main deck, a broad entrance gate led directly to the cargo gangway, a 30-foot-wide tunnel that ran the ferry's entire 180-foot length. Passenger cabins flanked both the port and starboard sides of the gangway. On busy days, horse-drawn carriages and wagons as well as the occasional automobile filled the gangway during the boat's 20-minute crossing of the East River to Hunter's Point at Long Island City.

A boat of imposing length, the *Southampton* featured a beam of 50 feet, two above-water decks, and a pilothouse on top of that. The ferry featured a compound, surface-condensing steam engine that turned twin-screw propellers. Without the old-fashioned paddle wheels, more gangway and cabin space was available. Portside cabins on the main deck were reserved for women and the occasional man who chose not to smoke cigars. The starboard cabins were set aside for men who smoked. During holidays and horseracing days, when the boat was loaded to capacity, the open-air second deck was available to both sexes, smokers or not.

Once on board, the men decided to remain with the horse and wagon.

"Not many people aboard this early," Carl said, as he climbed down from the buckboard. "Other than the half dozen employees of the Long Island Railroad's marine department, I'd wager there aren't a hundred passengers."

"I've been told that later in the day, you can't find a seat," Brock said. "Weekend races at Belmont Park, fall vacationers heading for Sheepshead Bay or North Beach will pack the ferries and rail cars... then there's the crowd headed for a day at Calvary Cemetery."

As the ferry nosed into the East River, its single tall smokestack billowed clouds of spent steam and coal smoke. Ladies, armed with fancy parasols, were particularly wary of the occasional hot cinder landing on their clothing or hair while sitting on the second deck.

"How much was the fare?" Carl asked.

"A nickel apiece for us, and half-a-buck for the horse and wagon."

A warming sun tempered the chill morning air as the *Southampton* navigated a slight chop, the minor disruption caused by a freshening southwest breeze. The stench of garbage and horse manure faded as the men inhaled the salty air. Their enjoyment of the serene passage came to a speedy conclusion when they saw a 100 foot-long sign at the entrance to the Hunter's Point terminal that announced "Long Island R.R. Ferries." Within minutes, it would be time to unload their crates.

"That was a short trip!" Dienstbach lamented. "It's no wonder people are beginning to use the Brooklyn and Williamsburg bridges. We spend all of our time loading and unloading the boat—"

"Now we have the 'tubes,'" Walter interrupted. "Tunnels dug under the river. With all the choices, it's only a matter of time before travel by ferryboat will become obsolete."[3]

By 10:00 a.m., all three crates had been loaded and lashed down aboard a flatcar that trailed a group of five passenger units on the Long Island Rail Road's main line. Headed for Mineola and points east, the train made six stops along the way, including one for the Belmont Park Raceway. The trip to Mineola would take less than an hour, but as the crow flies, it was only 20 miles from the Broadway shop. Herring had often mused, "If I could find a place to take off, my flying machine could beat the L.I.R.R. to Mineola."

"I've never been to Long Island," Brock admitted.

"Nor have I... guess there's a first time for everything. Sure hope Gus knows what he's doing by insisting on this place for the trials."

Flipping open a railroad booklet titled "Long Island Illustrated," Carl began reading aloud.

" 'In the western third of Long Island, within an hour's journey by rail from New York City, there are about fifty square miles of dry prairie called the Hempstead Plains, a treeless, grass-covered terrain that occupies the central portion of Nassau County, the first county east of New York City—' "

"Gus told me about this place," Walter broke in. "He stumbled upon it when he and Lillian were scouting the area looking for houses back in '03. He said there was plenty of room to maneuver, calling it a natural airfield."

Carl continued reading.

" 'The Hempstead Plains was created by an outwash of glacial sediment more than 10,000 years ago. The result was a vast flat open land with an altitude ranging from 60 to 200 feet above sea level. Dry prairies cover about 99 percent of the plains and there are only about 300 inhabitants per square mile.' "

As their locomotive slowed for the terminal at Jamaica, the terrain had already become flat as a tabletop, and the entire landscape had turned bluish gray.

"The booklet says that the Hempstead Plains is the only natural prairie east of the Allegheny Mountains... and it's not suited for farming." Carl continued:

" 'The soils are known as Hempstead gravelly loam with a color that might be described as washed out chocolate, full of roots and grasses near the surface. A foot or so below, the composition turns into coarse gravel of unknown depth, incapable of retaining moisture. Although there aren't many, the commonest tree is the gray birch, followed by two species of oak and pine. There are about a dozen kinds of shrubs, fifty or so herbs and a few other minor plants. The most common herb is the broom-sage, a species of grass. All of the shrubs are runts – less than knee high, when you find them at all. The trees and shrubs bloom in the spring, the herbs late in the summer. The flowers are either yellow, white or have a purple tint. None of them have any odor to speak of.'"

Freeport, Long Island, New York

Saturday, October 24, 1908

Herring and Manly's point of transfer was the Valley Stream depot, located a short three miles west of their Freeport starting point. The men, dressed in their Sunday finest, elected to carry their own suitcases from the *Westbound Limited*, lugging them 75 feet to the waiting Glen Cove branch's *Northbound Limited*. Outside their Pullman, Gus set his bag down beside Manly's and jogged toward the rear of the concrete platform. Cupping his hands over his eyes to reduce the early morning glare, he peered intently at the remaining cars in the string. A minute later, he was getting comfortable in the plush, overstuffed seat next to Charles. "Had to be sure that our boxes got switched! I'm happy to report seeing them lashed down on the deck of a flatbed carrier, three cars back."

Another 15 minutes passed before the conductor announced the train's imminent departure, followed seconds later by two short blasts from the locomotive's whistle, a sure sign they were about to continue their excursion.

"William and Chloe almost came with us," Gus said, as he gazed out of the car's stained glass window, "but Lillian put her foot down; said they would miss too much school. I thought my 15-year-old son would put up a fight, but he seemed happy to stay at home. And this comes from a youngster who hates school! On the other hand, my daughter was disappointed. She wanted to tag along."

"Chloe's interested in your work," Charles said. "I think she's your biggest admirer."

Herring had estimated they would arrive at the Mineola terminal no later than 10:30 a.m., at which time their cargo would be unloaded at the freight dock. As they passed through Garden City, Gus interrupted an otherwise somber passage across the plains. "Charles, did you know that the so-called 'Merchant Prince,' a man by the name of A.T. Stewart, founded Garden City about 40 years ago, and is thought to have been responsible for these railroads across the plains?"

"Never heard of him."

"He bought all the land around Garden City for a pittance – $55 an acre. Most of it is still held by his heirs, who lease it to wealthy individuals living in the area, mainly for equestrian sports. For years now, automobile races have been staged on the better roads, because they're level. Earlier this year, they started building the Long Island Motor Parkway, since there are no trees or farms to clutter the enterprise. Mineola is right in the middle of the Hempstead Plains and has similar qualities that will be good for flying."

Before 11 o'clock, Herring had unpacked the Mobike, got it running, and chugged off in the general direction of downtown Mineola. Half an hour later, he was back at the freight dock with a rented wagon and a team of two horses.

"Where's the motorcycle?" Manly asked.

"Left it at the livery stable. The owner said he'll keep it in the barn until I bring the wagon back later this afternoon."

*

Fifteen minutes passed before the eastbound train from Hunter's Point lumbered into the terminal. After exchanging quick greetings, the men finished loading the new boxes and luggage aboard the wagon. While Brock and Dienstbach sat astride the crates, Herring and Manly occupied the buckboard's upfront seat. The caravan of one slowly began the two-mile eastward trek over the gravel surface of Country Road to Washington Avenue. Twenty minutes later, they entered the deserted Mineola fairgrounds. Half a mile into the sprawling rural property, the wagon pulled alongside the last dilapidated carriage shed in a row of eight identical structures. Gazing to the southwestern horizon, the men were overwhelmed by the immensity of the flat, barren countryside... the perfect place for testing an unproven flying machine.

CHAPTER 88

Mineola, Long Island, New York

Saturday, October 24, 1908

Trailing a thick cloud of gray dust and blue smoke, Herring's tandem Mobike motored noisily into camp.

"That was fast!" Manly hollered as Gus pushed the motorcycle into the ramshackle shed.

"You and the boys have been busy!" Herring declared. "Any trouble raising the tent?"

"Not with Walter supervising! He claims to have worked as a roustabout for Ringling Brothers. Must have, he had her up in no time! Anyway, did you have any trouble getting the wagon and team back to the livery stable?"

Brushing road dust from his black linen shirt and canvas pants, Gus turned to address his friend. "Everything went well, but this road dust sure makes motorcycle ridin' miserable! Glad I changed out of my suit before heading back to town. The man who owns the stable said it hasn't rained around here for months. Even the wells are dryin' up."

Flipping open the canvas flap, Gus ducked his head to enter. "Reminds me of the tent we had back at the dunes in '96, but this one's bigger. It should hold the assembled aeroplane, our tools, cots, sleeping bags... the works."

While Carl, the group's volunteer cook, prepared the kerosene stove for the evening meal, Walter Brock began the time-consuming task of removing the screwed-in-place lids from the aeroplane crates. Gus and Charles, anxious to test the repaired portside engine, used long-handled coal shovels to level the wheel-rutted gravel floor of the carriage shed. The five-pound sledgehammer used for pounding home tent stakes was put into service sinking three-quarter-inch steel rods deep into the gravel to provide anchor points for the wooden test stand.

"A decent setup for testing the engine, eh Charles?"

"Beats loading her down with sandbags, that's for sure—"

At that moment a cowbell *clanged*, ending any further discussion.

"That's Carl calling us to supper!" Gus said.

Dienstbach had also been charged with purchasing food for the excursion. Along with the provisions, he arranged to have 50 pounds of ice shipped in one of the three flying machine crates. Insulated with blocks of straw and wrapped in layers of inexpensive muslin, the improvised ice chest arrived onsite intact.

"What's to eat?" Gus shouted, as he trudged around to the far side of the tent.

"Bratwurst and sauerkraut!" Carl exclaimed, rubbing his hands together. "Pork and beef sausage served with sweet German mustard, black bread and lots of butter. I even brought along a couple quarts of cow's milk!"

"Don't look at me," Walter said to Charles, "I had nothin' to do with the shopping, but I'm part Heinie anyway—"

"You won't hear me complain," Gus cut in, "but Mr. Manly might have something to say about the chow... he's a Limey, ya know."

Manly's shoulders slumped in mock disdain as he considered the consequences of having his meals prepared by a German national. "Not a problem for me," he said. "I just hope you brought along some bacon and eggs. I draw the line at sauerkraut for breakfast!"

*

By 6 o'clock the skies above the Hempstead Plains had begun to fade. In the remaining daylight, Gus and Charles retreated to a secluded spot behind the carriage shed where they used their coal shovels to dig a shallow latrine in the compacted gravel. Narrow enough to straddle, the slit trench was then supplied with a sturdy cross log and an old *Montgomery Ward* catalog. A bar of soap and a bucket of well water completed their mission of necessity... an adequate, if somewhat primitive, arrangement.

Inside the tent, the two engine men found Walter and Carl moving the aeroplane's steel tube center section onto a level portion of the water-washed gravel floor.

"Walter," Gus said, "how can you see with only one lantern burning?"

Seizing his red bandana, Brock mopped his brow before replying. "Even with the flaps open, it's too damn hot in here to work with two lanterns goin'. Are we sure it's almost November?"

"It's a perfect night for a campfire," Carl said. "Too bad there isn't any firewood around for miles!"

"Gentlemen," Gus sighed, "that's enough work for one day, let's sit outside and relax. Besides, tomorrow's Sunday. If you choose not to work for religious reasons, it's fine with me. Although I consider myself to be a Christian, I've always toiled on Sunday, and will continue to do so."

*

Sunday morning dawned cool and damp. After an early breakfast, Gus and Charles, tools in hand, retreated to the carriage shed, where the problem engine awaited their attention. Meanwhile, eager to start the flying machine's assembly, Walter and Carl first rolled up the tent's sidewalls.

By 10 o'clock the portside engine had been mounted on the test stand, fitted with its flight propeller, and its tank filled with gasoline. With a trace of anxiety etched on his face, Manly turned to Herring. "Ready?"

Typical of Gus's demeanor under stress, he was all business. "Is the revolution counter engaged? Is your stopwatch at hand?"

Charles nodded.

"Good! Are the throttle and spark advance levers in the startup position?"

As Manly leaned over the engine to check, his mind flashed back to the early days of the engine development program... so much had been accomplished in only a few months.[1]

"All right, Charles, move behind the engine, and I'll pull her through a couple of times. Switch off?"

"The switch is off."

The engine wheezed in protest as Gus slowly turned the propeller *clockwise* through four revolutions. Then he advanced one blade to the two o'clock position, where he felt a resistance... the signal that one of the engine's cylinders had reached its compression event. At that point, Herring placed his hands across the smooth wooden face of the blade, halfway between the hub and tip.

"Switch on?"

"The switch is on."

With a measured fore-to-aft swing of his right leg, Herring transferred his body weight through his shoulders and arms to the propeller, delivering a smooth and forceful downward pull. The engine coughed once before one cylinder fired, causing the propeller to whip violently through its arc of rotation. This was followed by wheezes of protest as the machinery freewheeled to a stop. Three more unsuccessful attempts at starting caused Gus to step back.

"Switch off!" he growled, wiping his brow with the back of his hand. "There must be a fouled plug or two."

True to form, cylinders number three and four, those pointing predominantly downward, had castor oil pooled inside their sparking plugs.

"In my haste," Herring lamented, "I forgot to flush the accumulated oil from the plugs on those two cylinders."

"I forgot 'em too!" Charles said, scratching his head.

*

It was almost 11 a.m. before the problem engine finally barked to life, belching a cloud of indigo smoke and atomized castor oil into the air outside the shed. Stepping back, Gus listened to the engine's firing cadence, which seemed normal. Careful to avoid the spinning propeller, he shuffled to a safe position behind the engine, where he guardedly placed an index finger on the warming crankcase. The level of vibration seemed *lower* than before.

Gus's next task was to reposition the throttle to half speed and to check the ignition timing. Using his ear as an instrument to identify speed change, he slowly advanced and retarded the spark, finally returning it to the original setting. Next, he adjusted the air-fuel mixture on the five carburetors, again judging each cylinder's degree of tune by ear.

Before long, Charles waved his arm; it was time for Gus to shut the engine down. As the machine lost its momentum and slowly hissed to a stop, Manly relayed what little information he had.

"Head temperature got up to 300 degrees in only 90 seconds. I decided that was high enough! We don't wanna caramelize the castor oil and varnish the piston rings and cylinder walls!"

Gus was feeling better. "A good decision, Charles! After she cools off, we'll run her again... this time at wide-open throttle.

I'm anxious to learn what the maximum engine speed will be on this propeller."

While the men waited for the engine to cool, they discussed *both* engines, and the rotation of their crankshafts.

"I have to admit," Manly said, "I've had difficulty getting used to the clockwise and anticlockwise crankshafts in these engines. Carving the wooden propellers must have been confusing, but pulling one the wrong way during startup... now that would be embarrassing!"

*

The wide-open-throttle run had progressed smoothly until the top cylinder began to spew scalding water and steam directly into the whirling propeller. Startled but unharmed, Herring quickly switched off the ignition. As the machinery coasted to a halt, the stricken cylinder continued to emit rivulets of boiling water and the occasional puff of saturated steam.

"That figures," Gus groused. "Now we have a water leak!"

Charles wrapped an arm around his friend's shoulders. "Look at the positive side, Gus... she's running great! The performance level is right up there with the starboard engine. We knew the cooling system was a temporary solution and would require a lot of attention. Give her a while to cool down, then we'll see what went wrong." [2]

Half an hour later, the offending cooling jacket had been removed, inspected, and the problem identified. The upper flange packing had worked its way free – an easy repair.

"It's about time something went right," Gus said, exhaling deeply. "I thought the brazing might have cracked between the jacket and the upper flange. That would have required a trip to Steve Balzer's shop!"

The engine men spent the rest of the morning installing new rope-type asbestos packing and reassembling the cylinder. After lunch, the portside engine was tested again and was found to be operating as planned. As the machinery cooled, Manly evaluated their engines.

"We need an active cooling system. This temporary arrangement will barely get us through the initial flight testing—"

Gus interrupted. "You're right, of course. Our friend Balzer is fabricating the radiators as we speak, and Walter has a good start on machining the two crankshaft-driven gear pumps."

Manly's face flashed doubt. "Let's talk about engine startup. The second engine better start right away, or the first will overheat before we can get the machine launched."

Herring took a moment to consider this. "What if we hold the first engine at idle until the second one starts? That should help... but let's give it some more thought."

*

Later, Gus and Charles removed the portside engine from the test stand and began preparing both power plants for installation. Under the tent, which had its sides rolled up for daytime assembly work, Brock and Dienstbach began the tedious process of superposing the assembled lifting surfaces above the steel tube central framework.

In the evening, with the tent sides rolled down, everyone gathered at the aeroplane to hash over the day's progress. Steadying himself with the aid of a vertical strut, Walter Brock was the first to speak. "Guess I'll start... with the central frame leveled, Carl and I assembled the lower lifting surfaces. Then we fastened them to the frame, while using sawhorses to support the wingtips. That was the easy part. Next, we used the narrow pine boards to temporarily support the assembled upper surfaces, allowing us to attach the individual vertical struts to the fore and aft wing spars.

"With all of the struts in place, we strung number-12 piano-wire through the pre-installed brass fittings, using the turnbuckles to temporarily tension the assembly, making the superposed surfaces rigid. Up until then, the entire structure was vulnerable to gusts. Normally I'd be worried... but so far we haven't had *any* wind!"

After Charles had finished reviewing the problems associated with the portside engine, Gus stepped forward to address the project's immediate future. "Two tasks must be addressed before we can install the engines. First, the trussing for the lifting surfaces must be fine-tuned, making sure everything is level, plumb, and square. Next, the takeoff and alighting gear must be bolted to the central frame to support the weight of the machine. So, the wheels and skid subassembly need to be readied; when complete, we'll add the motor mounts and engines."

Herring directed his next remark at Brock. "Walter, I want you and Carl to work on the wheels and skid."

Walter turned his head toward Carl and winked. "And what, pray tell, will you be doing, my lord?"

Gus acknowledged Walter's drollness by bowing from the waist. "After Charles and I finish fine-tuning the truss wires, we'll work on a method for temporarily disabling the articulating motor mounts. As you know, Mr. Brock, I want them *locked in place* for the first trial."

Hearing no objections, Gus continued. "When you finish with the takeoff and alighting gear, you can work on fastening the front and rear frame extensions; then the forward bi-plane horizontal rudders and the aft cruciform tail can be fastened in place. Afterward, Charles and I will need your help lifting the engines. If you run out of things to do, start rigging the cords and elastic springs for the tail regulator. Then the jib sails have to be installed..."

For the next hour the men discussed various assembly details, finally finishing with a friendly wager as to when the machine would be ready for its initial trial.

"It'll be ready by suppertime tomorrow!" Carl shouted.

"That won't happen!" Charles said, shaking his head. "I predict we'll be ready by lunchtime Tuesday!"

Walter, who had been through many assembly and disassembly cycles with the machine, had other ideas. "You're close, Mr. Manly, but I expect the machine will be ready no sooner than 2 p.m. Tuesday."

Everyone turned to Gus, who stood beaming. "Normally I'd side with Mr. Brock under these circumstances – but not this time. Remember, gentlemen, the machine will not be ready for a trial until its propeller *thrust* has been measured at wide-open throttle. We need to confirm that the apparatus is capable of an unassisted takeoff and sustained flight. I believe the procedure will take at least an hour or two. I predict she won't be ready until 5 o'clock Tuesday afternoon!"

With everyone in a jovial mood, talk eventually turned to the weather.

"What about the wind?" Carl asked. "You said the wind speed had to be at least 25 miles an hour before you would consider a trial. Any additional thoughts?"

Leaning against the machine's tubular framework, Gus scratched his head. "Herr Dienstbach, you sound just like a newspaper man!"

Blushing, Carl threw his hands up in surrender. "Old habits are hard to break... nevertheless, it's difficult for me to believe there will be any wind that strong. Look at the conditions, there's not a breath of air—"

Walter broke in. "I know a man who works for the Weather Bureau at the Belvedere Towers Observatory. Just before we left the shop Saturday morning, he told me they were tracking a tropical storm heading up the Atlantic coast. He thought it could blow into our region by the end of the month... that's only a few days from now. We're likely to get wind, and plenty of it!"

Before long, darkness had descended upon the campsite. As the men prepared to retire, Carl lit two lanterns, hung one from a hook on the tent's center pole and took the other outside. Fifteen minutes later he returned with a steaming pot of coffee, an assortment of hard rolls, and a glass of cold milk. As the boys laughed over Gus's preferred drink, he stood and called for a toast. "To success," he saluted, "and to the memory of a good friend and fellow aeronaut: Thomas E. Selfridge." [3]

CHAPTER 89

Mineola, Long Island, New York

Monday, October 26, 1908

Overnight temperatures of 70 degrees or so had produced sweltering conditions inside the big tent, causing Dienstbach to awaken from a fitful sleep drenched in sweat. Drowsy, he slipped on his trousers, flipped the enclosure's entrance flap open, and stumbled outside. As he stood inhaling the dense morning air, he detected a distinct growling-snorting noise. *Bear*, thought Carl, as he lunged for the nearby utility table and grabbed an eight-inch kitchen knife.

Cautiously avoiding tent ropes and stakes, Dienstbach crept silently to the tent's far corner... where he peeked around the stretched canvas. There, reclining atop his cot, the unconscious form of Augustus Herring was huddled inside his sleeping bag... breathing with harsh snorting noises while sleeping. Half an hour later, as the men finished their fried eggs and bacon breakfast, they laughed as Carl recounted his "growling bear" story.

"Gus... snoring? Hard to believe!" Walter said, rolling his eyes.

"Most certainly!" Dienstbach confirmed.

"You should have been with us three years ago," Manly piped up. "His snoring got our *20th Century Limited* stopped outside Cleveland— "

"Why was that?" Carl interrupted, playing along.

"The conductor thought our car had an *axle bearing* going bad!"

As the others hooted, Gus just shook his head, pretending to be upset. "Admit it, Charles... Carl paid you to say that!"

"Sir," Carl said, feigning anger, "you are speaking to a man of unsullied integrity!"

*

At the midday break, Gus proclaimed that the assembly work was ahead of schedule. "The guy wires on the lifting surfaces have been fine-tuned, and the bicycle wheel alighting gear has been assembled and installed." The alighting gear consisted of three, 20-inch bicycle wheels and a central skid. The spoke rims were fitted with white vulcanized pneumatic tires, one in front and two behind. "The forward frame extension," Herring continued, "along with its bi-surface horizontal rudder assembly, has been attached to the central frame, leveled, and guy-wired into place."

For the initial flight, substituting the Herring regulator for the intended gyroscopic action of the articulating engines would promote a degree of automatic stability. The rear frame extension, consisting of a rigid beam, was attached to the central frame by a universal joint. This "wrist-action" hinge allowed the cruciform tail, mounted to the opposite end of the beam, to swing in a cone-like pattern of limited motion. To control this action, the regulator's arrangement of cords and hardened steel tension springs are connected to the two aft vertical struts of the lifting surfaces and terminus points on the tail.

Herring's earlier application of pyroxiline varnish to the muslin-covered, open-framework components serendipitously provided the machine's distinctive amber coloring. "After rigging the throttle control and ignition switches to the operator's location," Gus declared, "everything should be in order for a thrust test—"

"Hold on, Augustus," Manly interrupted, "have you forgotten about the quart fuel tanks and shutoff valves?"

Herring's eyebrows shot up. "Charles, you're right... I'm getting ahead of myself—"

Walter Brock chimed in, "Hooking up the controls is going to take me the rest of the day, and then some. I think we should complete *everything* before running the engines."

*

The afternoon was devoted to gasoline tank fabrication and routing piano-wire to various control surfaces and levers. Walter had already installed the side-by-side wicker seats to the tubular central framework, just forward of the lower lifting surface's forward edge. The operator would sit on the left side, his passenger on the right. The controls consisted of a marine-type 12-inch mahogany steering wheel, a custom-fitted polished brass shoulder harness for the operator, and a throttle control containing the ignition shutoff switches.

The cruciform tail's vertical rudder followed the left and right turning action of the wheel, while the fore and aft horizontal rudders tracked the inward and outward motion of the pivoted base steering post, allowing the operator to control the machine's yaw and pitch attitude without taking his hands off the wheel.

Herring's method for activating the machine's roll control was simple and ingenious. By employing the snug fitting shoulder harness, he could use his side-to-side weight-shifting experience to deliver the desired results. When he leaned slightly to the left, the portside interplane control surface moved a proportional distance and assumed a negative angle of incidence. Simultaneously, the starboard side control surface moved a similar distance in the opposite direction. The result would be a subtle rolling action to port... akin to shifting his weight in the old *Double Decker* glider.[1]

The wheel, steering post, and shoulder harness were rigged with piano-wire, bellcranks, control horns, and turnbuckles to their representative control surfaces. Throttle control was more complicated. After considerable thought, Herring decided that it would be confusing for the machine's operator to have two individual throttles. Therefore, both engines, and *all 10* carburetors, were to be connected to a *single* throttle block, located to the operator's left. Earlier, back at the shop, Brock had been assigned the task of designing and machining the unit. He responded in typical fashion with a clever, all-aluminum control that incorporated positive stops at the idle, half-open, and wide-open throttle positions.

*

For his last constructive act of the day, Gus slipped a pusher propeller onto the shaft of the portside engine. "For test-stand work," he said, "I chose a propeller that would blow air *over* the cylinders to help with the cooling – a typical *tractor* setup. For flying, the propellers have to blow the air to the rear in a *pusher* configuration. Because the engines have been designed to rotate in opposite directions, the propellers end up being the mirror image of each other." Herring stepped back while using his hands to simulate the propeller's twist.

"It gets confusing!" Manly said, bending over to peer at the engine's crankcase. "Aha! Here's the arrow... the portside engine rotates clockwise when viewed from the rear. Are you sure you have the right propeller?" After some hand waving and simulated propeller turning, the two engineers finally decided that they indeed had the correct unit.

"Here's the strange part," Herring laughed. "I'll have to get used to pull-starting these propellers by holding them from what I consider to be their *back side*." Gus placed the first knuckle of his fingers around the trailing edge of the propeller, while the rear of his palms rested on its flat backside. "It feels strange to manipulate the propeller in this manner, but we'll see... tomorrow's another day!"

Mineola, Long Island, New York

Tuesday, October 27, 1908

Tuesday dawned identical to the day before – overcast, with no wind. Due to another fitful night's sleep, Gus endeavored to clear the cobwebs from his brain by sipping tepid milk from a dented tin cup. Spanner wrench in hand, a fully alert Walter Brock crawled out from under the machine's cruciform tail.

"Augustus, just the man I'm looking for!"

"Walter... you're wide awake."

Springing to his feet, Walter brushed gravel from his well-worn canvas overalls. "I'm done except for adjusting these two sets of horizontal rudders. But now I'm having second thoughts about the piano-wire; it's a long run from the front to the rear! If I had it to do over again, I would have provided more *fairleads*. I don't like these dangling wires."

Gus slowly crouched to a position that allowed him to sight along the control wires in question. "Hmmm, I see what you mean. If a wire gets snagged during takeoff... there'll be hell to pay." He thought for several seconds before rising. "Jury-rig it the best you can, Walter; before we head to Fort Myer, we'll fix it right."

Returning from the latrine, Manly sauntered over to where Gus and Walter were standing. "Whew! Our sorry substitute for an outhouse is really gettin' ripe... as I'm sure the two of you have noticed!"

"Must be the German cooking!" Herring teased, as Walter pretended to be adjusting a turnbuckle.

"If we stay here much longer, we'll have to dig a new trench," Charles persisted. "Preferably as far away from the first as possible!"

Ignoring the minor distraction, Herring tramped over to the starboard engine, where two large spring scales hung from the propeller shaft. "Charles, we've got to find good attachment points for these scales. If at all possible, I want to get the static thrust number before lunch." Manly nodded as Brock meandered over to his toolbox, leaving this decision to the engineers.

"Care to guess how much thrust we'll get?" Manly asked.

"Better be at least 125 pounds or I'll have a hell of a time gettin' off this gravel. I had to rent *two* of these hundred-pound scales; couldn't find anything bigger. They cost me *six bits* for the week! Any idea where we should attach the ropes?"

Manly walked to the machine's central frame and squatted, peering intently at the alighting gear's support structure. "Somewhere low on the frame. I wouldn't attach a rope to anything directly connected to the lifting surfaces – that's asking for trouble."

*

By 11 a.m., the flying machine had been rolled to a spot 20 feet in front of two side-by-side iron stakes that had been driven deep into the gravel. Ropes were attached to each side of the machine's central frame and extended to a stake, where the spring scales had been added. Next, the machine was pushed forward to take up the slack.

Walter and Carl volunteered to enter the tornado-like propeller blast to read the scales. After removing, cleaning, and replacing the troublesome sparking plugs from the inverted cylinders, Herring jogged back to the tent, returning seconds later with a pasteboard box. Just prior to startup, he flipped the shallow cover off the container to reveal two pairs of safety goggles.

After demonstrating how to adjust the straps, he handed each man a pair of long canvas engineer's gloves and one of Lillian's specially sewn gray canvas facemasks.

No one objected.

A minute later, after setting the throttles to the idle position, Herring signaled Manly to switch on the ignition to the starboard engine. Approaching from the machine's rear, Gus turned toward his friend, a wry smile spreading across his face. This was a major hurdle, and they both knew it.

Herring made his first counterclockwise swing at starting the starboard engine. To everyone's surprise, it coughed once, kicked forward, and continued to run at the idle setting. The portside engine didn't cooperate nearly as well. It required several pulls and a squirt of *ether* to several of the carburetors before it sputtered, backfired reluctantly, and settled into a reasonably steady idle. At this point, Herring sprinted around to the operator's station and slipped into the wicker seat and shoulder harness. Manly, standing at the starboard wingtip, signaled Herring to advance the throttle to one half.

The crescendo from 10 open exhaust stacks permeated the late morning air. As the twin engines accelerated to 1,700 revolutions per minute, their whirling propellers cast gravel and bits of debris rearward, subjecting Walter and Carl to the stinging aggression.

Almost immediately, Manly signaled Gus to advance the throttle by pointing to the heavens. At wide-open throttle, the chaos behind the machine became almost unbearable for the scale readers who, despite the protective gear, seemed fearful. Ten seconds later, their apprehension subsided when Herring cut the ignition, scrambled out of his operator's seat, and hollered to them. "Gentlemen... what are the numbers?"

"Seventy-three pounds from the portside!" Walter shouted.

"Seventy-seven from the starboard!" Carl hollered, brushing gravel from his flannel shirt. "Thank God for these goggles!"

"Let's see," Herring said, as he crept to the rear of the machine, "that's a net thrust of 150 pounds. More than enough to fly!"

As the engines cooled, Manly joined the conversation. "Gus, did you hear the whump-whump-whump noise? When two engines are not runnin' quite the same speed, the frequency of their exhausts don't match. What you hear is called the *beat frequency*. I used my stopwatch to establish the interval. I should be able to calculate the difference in engine speed—"[2]

"Too much commotion for me to hear any of that," Gus cut in, "but I'll bet we're within acceptable limits. The port-side engine was puttin' out almost as much thrust as the starboard side."

Oblivious to any further conversation, Charles worked his slide rule. A few minutes later, he had the answer. "Two beats per second equates to the port-side engine being down a bit less than 50 revolutions per minute. That shouldn't be a problem as long as they keep runnin' the way they are. You might have to turn the wheel a little to the right to keep the machine from yawing to port."

*

A lively combination of enthusiastic banter and nervous anticipation punctuated their lunch. Dienstbach asked Gus a question that had been bothering him since the thrust test. "How much more powerful is the starboard side engine?"

"Funny you should ask... I finished that calculation just before lunch. The port-side engine seems to be down 1.3 horsepower... that's 6 percent less than the starboard engine. As long as its performance doesn't drop any further, it shouldn't be a problem. Like Charles said... I might have to turn in a little right rudder."[3]

"By the way," Manly interrupted, "I guess I won the bet... we finished the machine at just about noon!"

"That's right, Charles" Carl chimed in. "There'll be no *mess duty* for you this evening!"

Herring smiled broadly as he finished a bite of his Limburger sandwich. "Gotta balance the machine fore and aft, then look everything over real good, but you won fair and square, Charles. Hope we get to fly her tomorrow!"

"Still no wind though," Dienstbach lamented.

"Carl, don't be such a wet blanket!" Walter said, throwing his hands into the air. "We'll get our wind sooner or later! Have you ever been *anywhere* where the wind didn't blow?"

"Yeah... the Hempstead Plains!"

Just then, a puff of warm air scattered dry leaves throughout the campsite.

Mineola, Long Island, New York

Wednesday, October 28, 1908

The wind – the missing commodity for the first four days – finally arrived early Wednesday morning. The rustling of the tent canvas was hardly noticeable to the groggy threesome inside. Truth be known, the breeze came as a welcome relief from the usual stagnant air. However, within 10 minutes the side curtains were rattling, and the entrance flaps began to flail.

Already outside, Herring stood with a lantern by his side and his Richard anemometer held above his head. Fumbling with the buttons on his flannel shirt, Manly joined him. "What ya got, Gus?"

Raising the lantern, Herring struggled to interpret the windspeed scale. "Steady at 18, gusting to 22."

"We're gettin' there!" Charles insisted. "We'll see what daybreak brings. In the meantime, come back inside and get some rest."

"Hold on... I need to look at the sky. First, I must turn down the wick."

As their eyes adjusted to the darkness, the first-quarter moon hanging low in the southern hemisphere came into view, followed by the stars. "Except for a few high scattered clouds, it's clear," Herring said. "Let's hope the weather stays favorable."

*

Eight o'clock came and went. Although dawn had revealed overcast, leaden skies, the winds had increased marginally to 22, with gusts to 26. "Gentlemen," Gus said, "conditions are looking good! After breakfast, we'll roll the machine out and prepare for a trial!"

The day before, Brock had driven a four-foot-long, 3/4-inch iron stake three feet into the gravel at the designated takeoff point. He had also prepared a sturdy, 3/4-inch hemp rope, 75 feet long, with loops tied at either end. The rope would tether the machine to the stake prior to the launch. When ready, Herring would lean forward, look down, and kick the loop off the short peg located between his feet.

Another hour of fretful tinkering and wind watching ensued before Herring collared Dienstbach, walking him over to the Mobike.

*

It was 10 a.m. before Herring stood; he had just finished adjusting the choke on the motorcycle's carburetor. "Carl, don't forget to click off the choke after she starts. Later, if the motor's still warm, leave the choke off altogether; otherwise, you'll flood the damn thing... then she won't start unless you remove the sparking plug."

From the starting point of the improvised runway, more than 200 feet to the south, Gus and Carl heard someone shout. It was Charles.

"Reeaaddyy!"

As Herring waved his acknowledgement, the freshening wind tried to blow his black captain's cap from his head. Grabbing it, he pulled the cap down tight – bill to the rear. Turning to face Dienstbach, he reminded the journalist of what was expected from him. "Remember, Carl, I want you to head directly upwind about a quarter mile or so. You'll have a good vantage point when I get to you. Do you have your stopwatch?"

"Yes sir, it's right here," he said, pulling the chain-tethered watch from his vest pocket.

"Good, you're the official timer," Gus said, placing his hand on Carl's shoulder. "One more thing... I chose you for this job because observing and reporting are what you do best. Later, if the occasion arises, I may need you to testify on my behalf. When the flight is over, write everything down before making your way back to camp. I'm sure there'll be some excitement one way of the other, and I don't want your candid observations watered down.

"Other than that... I'll see you soon!" With that, Gus turned and jogged toward the waiting flying machine.

"Godspeed!" Carl shouted.

CHAPTER 90

Mineola, Long Island, New York

Wednesday, October 28, 1908

Carl Dienstbach, true to his word, wrote of the flight:

Seeing Herring's Government Machine Fly

There was a cluster of men around the machine some four hundred yards to the northeast... the distance allowed me an impressive view.

Suddenly a cloud of oil smoke rose from the starting point, as though a cannon had been fired. Seconds later a second cloud burst from the spot. Less than a minute later a yellowish shape began moving directly toward me. At first it looked like a weird automobile going at ordinary speed... soon I noticed a great difference: there was a gap between the ground and that 'car'. A moment later it looked suspended 10-feet in the air, but in a most matter-of-fact way, as though floating in the air were not in the least unusual. The floating thing seemed tremendously at ease as it approached my position in the ankle-high grass. To my astonishment, the contraption was traveling at a higher rate of speed than I had originally imagined against the 24 mile-an-hour wind...Then the flying machine flew nearly overhead, and slightly to the right... the apparent slowness replaced by speed and a slight rocking from side to side that only enhanced the appearance of absolute ease and absence of difficulty.

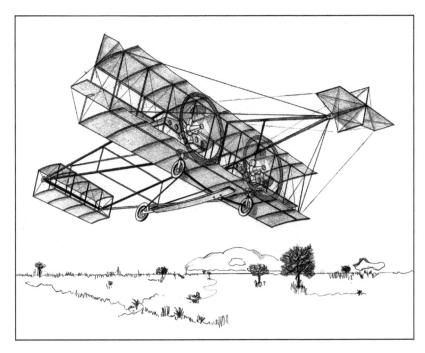

Herring's first U.S. Signal Corps aeroplane. Hempstead Plains, Long Island (October 28, 1908); drawing by author

All that changed in a few seconds of pandemonium as one of the engines began to act up. At first the difficulty was noticed by the irregular sound of the machinery, then by the pronounced yawing to port, followed by Mr. Herring's exertions to right the problem. Next, as I recall, one of the engines began backfiring, sounding similar in amplitude and frequency to a Maxim machine gun. As the machine, now several hundred-feet beyond my position, began to wander further to port, the operator apparently cut both engines.

In a valiant attempt to redirect the machine into the wind before losing the remainder of its altitude, Mr. Herring crashed into the gravel, causing a great spray of gray-blue shards to explode into the air. The bicycle wheel alighting gear, including its main frame, seemed to fold beneath the lower lifting surface as the aeroplane continued its uncontrolled slide. Both superposed lifting surfaces were the next to fail, folding upward like the wings of a great Monarch butterfly.

Throughout the turmoil of the crash, Herring fortunately remained within the confines of his wicker seat, held in place partly by the restraining qualities of his lateral control harness. Unable to start the Mobike's engine, Dienstbach sprinted to the scene, only to find Herring already examining the smoldering port-side engine. Determining that the operator wasn't seriously hurt, Carl asked what he later admitted was a frivolous question.

"What will you do now?"

With a rueful shake of his head, Herring replied. "I know what I *won't* have do, Carl... I won't have to dig that new latrine."

Rochester, New York

Friday, December 2, 1921

From his familiar seat on the witness stand, Herring had read into the record Dienstbach's moving description of his 1908 crash. Spectators and reporters alike – with the exception of Curtiss and his fellow defendants – hung on his every word as Gus described the doomed aeroplane's final plunge.

"Mr. Herring," O'Grady said, wasting little time, "why did the Signal Corps machine fail?"

Herring cleared his throat. "The port-side engine apparently overheated, lost power, and created a thrust imbalance from which I could not recover. The resultant crash pointed to weaknesses in the structure that could not have been predicted beforehand."

O'Grady left the podium and edged closer to his witness. "Were you injured?"

"A few minor scrapes and scratches – nothing serious. I wish I could say the same for the flying machine – it was a total loss. Too bad... before the untimely engine problem, it was performing admirably, proving that we were on the right path—"

"Objection, Your Honor!" Robbins broke in. "Argumentative! There is no evidence that the witness's phantom flying machine *ever* existed!"

O'Grady threw up his hands in exasperation. "Your Honor, the defense has no shame! The witness's convincing testimony aside, Mr. Robbins insists on perpetuating the debunked lies and myths attributed to Mr. Herring's Signal Corps work! However, if the court so desires, the plaintiff is prepared to subpoena additional witnesses to—"

"No need, Counselor," Sawyer broke in, "the objection is overruled. Continue."

"Mr. Herring, what did you and your team do after the crash?"

"We spent the afternoon hauling the wreckage back to camp, where we returned any salvageable parts to the crates. To make matters worse, a full-fledged wind and rainstorm rolled in from the south, making the work miserable. We used the carriage shed to store the crates and tools and struggled to save the tent, which was threatening to blow away. We spent the night in the shed.

"The following morning, I returned to Mineola on my motorcycle and rented a wagon and a team of horses. By mid-day we were standing outside the L.I.R.R. terminal, waiting for the one o'clock train back to Valley Stream and eventually Freeport, where we dumped the crates off in front of my boathouse shop. From there, everyone left for home, and I returned the tent and cots to the rental depot. The next morning, I began cleaning up the mess."

O'Grady walked back to the podium and reached for his notepad. "Depressing work... what happened next?"

Herring slouched in his chair. "Saturday was Halloween. Late that evening, someone broke into the boathouse and stole the starboard engine."

"Wasn't that the good engine?"

Herring slowly nodded his assent. "Actually, the thief did me a favor because I had yet to disassemble the port-side engine to determine exactly what went wrong."

"For the record," O'Grady asked, "why did that engine fail?"

Herring crossed his arms in front of his chest. "The replacement valve sheared off at the rocker arm. This allowed its head and most of the stem to drop into the cylinder, where it contacted the piston, punching a hole through its crown. With one of the cylinders out of commission, the power level dropped dramatically. Besides the cloud of crankcase lubricating oil that streamed from the exhaust on that cylinder, the engine began to literally shake itself off its mount. That's when I switched off the ignition to both engines and tried to alight."

"What did you learn from the catastrophe?"

Herring hesitated before tendering his response. "Back in 1908, multiple engines were an unwise idea for aeroplanes... they simply weren't reliable enough."

O'Grady skipped to a new page in his notepad. "When did you notify the Signal Corps?"

"Not until the following Monday morning. I sent a telegram informing them of the crash and requested a one-year grace period to fulfill the terms of the contract."

"Why did you suggest one year?"

"That's what they granted Orville Wright after his crash the previous month."

"What was the Corps' response?"

"My request was granted, but they only gave me until August of '09."

"What did you do next?"

"Toward the middle of November in '08, I was hard at work designing a new and improved machine."

"How did it differ from the *first* Signal Corps aeroplane?"

Herring's mood took a turn for the better. "To keep the airspeed down, the new aeroplane was designed with lifting surfaces that were 38 percent larger than the first government machine. The twin-engine scheme was scrapped for a single, 25-horsepower engine that would turn a four-bladed pusher-type propeller. Sitka spruce replaced the first machine's thin wall steel tubing, while an improved version of my gyroscopic stabilization system was applied to all three of the aeroplane's control axes."

Wilbur and Orville Wright (1909); Collier's Magazine, July 3, 1909

O'Grady paused; his questioning was about to change course. "Who helped you with the new machine?"

Herring felt a wave of fatigue wash over him. "No one. Walter Brock took a train to Montgomery, Alabama, where he signed up for flight training at the Wright Brothers Flying School. Charles Manly lost interest when I decided to use a proven Curtiss engine, and Carl Dienstbach returned to his full-time job at the New York *Times*.

"Actually, there was no need for assistants. I spent the next three months designing and working out the details for the new machine, and by the end of January of '09, Mr. Curtiss and I had entered into preliminary discussions concerning the formation of an aeroplane company."

"How did that come about?"

"When I contacted Curtiss about obtaining a reliable engine, I learned of his rather frantic search for a way around the Wrights' lateral-control patent."

O'Grady stepped around the podium and wandered in the general direction of the judge's bench. "A little more than a month later," he continued, "the Herring-Curtiss Company was incorporated—"

"Objection!" Robbins interrupted. "Does counsel for the plaintiff expect the court to hear this testimony for a *second* time?"

Sawyer's gaze was trained on O'Grady, who stood directly in front of the bench. "Well, Counselor, what do you have to say about this?"

"Your Honor, we anticipated the defense's objection and have prepared an appropriate response."

"Which is?"

O'Grady did not need to think twice about his reply. "Because Mr. Herring's work on his *second* Signal Corps aeroplane happened to coincide with the formation of the Herring-Curtiss Company, the plaintiff contends that the coincidence should not be used as grounds to *exclude* new testimony."

Judge Sawyer took the opportunity to lean far over his bench to address the attorney. "The Court recognizes the potential for duplication in Mr. Herring's testimony. In this regard, it will allow the plaintiff's questioning in terms of the government's contract with the witness... provided said testimony does not specifically *retrace* previous Herring-Curtiss Company material."

Sawyer leaned back in his chair. "Therefore, the defense objection is overruled. Continue, Counselor."

O'Grady thanked the judge and returned to the podium.

"Mr. Herring, how did the formation of the Herring-Curtiss Company affect your ability to work at completing the Signal Corps contract?"

Herring looked bewildered. "An avalanche of new work prevented me from paying proper attention to my government machine."

"Tell the court about this *new* work."

Herring glanced over to where Chloe was seated – she was smiling.

"Beginning in March of '09, I was tasked with designing the first company aeroplane... the *Golden Flyer*. Shortly afterward, the *Rheims Racer* occupied most of my time. Then there was the company's four-cylinder engine that had to be redesigned. On top of all that, my patent application *claims* for gyroscopic control had to be finalized—"

"Objection!" Robbins cut in, scrambling to his feet. "Irrelevant! Mr. Herring brought on the extra work himself!"

"Overruled!" the judge scoffed, banging his gavel for emphasis. "Mr. Robbins, sit down!" As Robbins loudly dragged his chair across the courtroom floor, the judge added, "... and be quiet about it!"

Stifling a guffaw, O'Grady continued. "Earlier, you said the key to defeating the Wright Patent was the implementation of the gyroscope. Is that correct?"

"Yes."

"You also testified that stabilizing gyroscopes would be used to control the new government machine on *all* three axes. Is that correct?"

"Yes."

"With so much at stake, how could you possibly *ignore* the Patent Office's request for claim modifications on your gyroscope application?"

"Objection! Calls for speculation," Robbins barked from his seat.

"Sustained. Restate your question, Counselor."

"Mr. Herring, due to time constraints, did you ignore the Patent Office's call for claim modifications on the gyroscopic stabilization application?"

Herring felt himself beginning to perspire. "Not at all! As I stated earlier, the original patent application was filed in August of '08. By May of '09, the Patent Office had assigned a U.S. Patent Heading to the Herring-Curtiss Company... right in the middle of my work on the company's first commercial aeroplane and 4-cylinder-engine modifications. Mr. Scherr and I worked night and day to clean up those original 51 claims. By February of 1910, all but 14 were reported to be 'possibly allowable—'"[1,2]

"Refresh the court's memory," O'Grady interrupted. "Who was Mr. Scherr?"

"He was my patent attorney."

O'Grady pressed on. "Before the Herring-Curtiss Company was formed, did Mr. Curtiss endorse the use of your gyroscopes in aeroplanes?"

Herring gave him a puzzled look. "Certainly. Curtiss believed that my patents, when backed by a strong company, could beat the Wrights in court. Among other concerns, we discussed how I would assign my North American patents for automatic stability devices to the company when they were issued."

O'Grady glanced at his notes. "Rumor has it... the reason you helped organize the Herring-Curtiss Company was to obtain a *company* aeroplane for your Signal Corps competition against Orville Wright. Was that the case?"

Stunned by the tenor of the question, Herring suppressed his urge to scream. "Not true! All of the company resources were being used to produce the *Golden Flyer*, the *Rheims Racer*, and the water-cooled V-8 engine. Even if the directors had gone along with such a proposal, Herring-Curtiss Company assets could not have met the Signal Corps' deadline. That's why I made every attempt to independently finish and submit *my* second machine for trial at Fort Myer."

"Did the company ever consider supporting your attempt to secure a Signal Corps contract?"

"Mr. Curtiss and I discussed the issue – once." There was a tinge of disappointment in Herring's voice.

"Did you come to any agreement?"

"Glenn expressed, in no uncertain terms, that we should tend to our own projects. Trouble was, *his* pet project, the *Rheims Racer*, utilized all the company's resources, while my Signal Corps project was allowed to die on the vine."

"Therefore," O'Grady continued, "after all was said and done, there was never an agreement to deliver a flight-proven *company* aeroplane to the Fort Myer competition."

"Correct. The company aeroplane *rumor* was perpetrated by Octave Chanute and the Wrights."

O'Grady strolled over to where Maloney was seated at the plaintiff's table before resuming his questioning. "Describe how your second Signal Corps aeroplane differed from the crashed Mineola machine."

Herring hesitated, deciding where to begin. "It was a single-engine, two-surface machine that featured all-wood construction, including parabolic-shaped, laminated Sitka spruce aero*curves* for the lifting surfaces."

"Mr. Herring, it was reported that your machine resembled, quote, 'a bulky skeleton sled.' Not a complimentary description, would you agree?"

Herring demurred. "Must have been some newspaperman's portrayal! Everyone who saw the finished product thought it was quite elegant! Although it resembled the Herring-Curtiss Company's *Golden Flyer* and *Rheims Racer*, its lineage reached all the way back to my design work of '05, which produced the *Philadelphia glider*. Make no mistake, I had a hand in designing *all* of those aeroplanes."

"Objection!" Robbins shouted. "Hearsay! None of this has been proven!"

"Sustained," Sawyer concurred. "Strike the witness's last sentence. Continue, Counselor."

"Mr. Herring, continue with your description of your second government aeroplane."

Weary of the constant objections, Herring took a deep breath as he attempted to stay focused. "The frame was fabricated from Sitka spruce and made rigid with piano-wire trussing. The covering material was a silk-and-rubber product originally used by Captain Baldwin for his airships. I installed 6 of my muslin-covered jib sails on the upper lifting surface to aid with lateral stability. The machine was very strong, lightweight, positively straight, and had a simple three-skid alighting gear augmented by three bicycle wheels."[3]

"What about the engine?"

"By the middle of July, the construction of my second Signal Corps aeroplane was almost complete. Concurrently, Curtiss continued to make successful flights with the *Golden Flyer*, the Aeronautic Society of New York's new aeroplane, which was the Herring-Curtiss Company's first commercial sale.

"While Curtiss' Mineola flights were reported in all the New York newspapers, including the 52-minute, 25-mile *Scientific American* Trophy flight, I sat in the Broadway Avenue shop cooling my heels, wondering when *my* engine would arrive from Hammondsport."

O'Grady next delved into the company's engine-related problems. "Because of the tribulations with your twin-engine machine, you convinced Curtiss that the company's four-cylinder engine needed upgrading. Tell the court about these changes."

"Most important, the cylinder heads were changed from air-cooled to water-cooled. The engine also needed more power. In that regard, I relocated the valves to an overhead position, and added a hemispherical combustion chamber and domed pistons to increase the compression ratio. Push-pull rods and rocker arms were incorporated into the design, along with changes to the cam profile."

"How did all these changes work out?"

Herring rolled his eyes. "The first of the redesigned four-cylinder engines was used in the *Golden Flyer*. The engine was so successful that the *remainder* of the assembled cylinder heads were used on the new V-8, which was destined for Glenn's *Rheims Racer*. All those shenanigans resulted in the delay of *my* engine build.

"Jim, take this," Herring whispered, after plucking a magazine from his document bag. "I want the judge to read what was said about the new engine…"

O'Grady walked the publication over to the bench. "The plaintiff wishes to submit the June 11, 1909 issue of the *Aeronautics* magazine."

Minutes later, plaintiff's exhibit #101 was returned to O'Grady.

"Continue, Counselor," Judge Sawyer urged.

O'Grady handed the magazine to Gus. "Mr. Herring, please read the underlined portion of the Glenn Curtiss interview on page 32."

His reading glasses in place, Gus proceeded:

> " 'We have,' Mr. Curtiss said, 'been informed by good authority that this Herring motor develops more power per displacement than has ever before been secured from a gas engine.' "

Finished reading, he flipped the periodical closed.

"Mr. Herring," O'Grady continued, "was your elusive engine the only problem keeping you from testing the second Signal Corps aeroplane?"

The reply didn't come immediately. "No, unfortunately... it was not."

"Elaborate on this."

"The gyroscopic stabilization system was giving me fits. Out of necessity, I had limited its potential use to the aeroplane's roll axis—"[4]

"Why was that?"

Herring pulled himself to an upright sitting position. "To defeat the Wright patent, the gyroscopic system was required to automatically stabilize the machine on the roll axis by controlling the interplane control surfaces."

O'Grady pushed on. "What was the significance of automatic stability?"

Herring grimaced as he launched into his explanation. "Interplane control surfaces, like wing warping and ailerons, were used to provide the flying machine with lateral control... and were erroneously covered by the Wright patent. However, patent attorney Scherr and I were convinced that an automatic stabilization system that controlled the aeroplane for the *majority* of its time in the air would effectively circumvent the legal language of the Wright patent – thus rendering it unenforceable against us."

Uncertain that Judge Sawyer had followed Herring's logic, O'Grady probed further. "Why was the automatic stabilization system giving you trouble?"

Herring stared up at the court's massive crystal chandelier as he spoke. "Many of the system's components had yet to be designed, built, and tested. There simply wasn't enough time to do all of the development work. I was stretched too thin... and then the company began to collapse."

Judge Sawyer allowed a troubled silence to engulf the court. The following day, a Rochester newspaper account noted, "if a pin were dropped, it would be clearly heard hitting the polished oak floors of the crowded courtroom."

O'Grady continued questioning his ill-at-ease witness. "What did you do about the forthcoming Fort Myer trials?"

"I delayed the inevitable as long as possible. However, in September of '09, I notified the Signal Corps that I was withdrawing from the competition, thereby forfeiting my good faith bond of $2,000."

"What became of the nearly complete Signal Corps aeroplane?"

"It became the *Herring-Burgess #1*, sometimes called the *Flying Fish* – because of its jib sails. After removing the wheels, I made the first flight of a powered aeroplane in New England on the afternoon of February 28, 1910... we used the frozen waters of Chebacco Lake, Massachusetts."

"Objection! Irrelevant!" Robbins shouted. "Let the record show that Herring's alleged 600-foot flight occurred 19 months *after* Glenn Curtiss' one-kilometer flight at Hammondsport, and six months after his 25-mile flight at Mineola, Long Island!"

Judge Sawyer turned to O'Grady with his hands upturned in a questioning manner.

The reply didn't come immediately. "Your Honor, Mr. Herring is simply stating the facts!"

"I am inclined to agree," Sawyer said. "Overruled."

"Your Honor," O'Grady said, picking up his notes, "I have no further questions for this witness."

CHAPTER 91

Rochester, New York

Friday, December 2, 1921

As the contestants filed back into the courtroom from lunch, most were surprised to find Judge Sawyer already seated. Thinking he was late in returning, Herring pulled out his pocket watch and checked the time. Immediately to his left, Bill Maloney whispered into O'Grady's ear, "Old Man Sawyer's thinkin' so hard, I can hear the gears rattlin'."

At 1 o'clock, the judge stood and scanned the courtroom, his eyes pausing at each litigant long enough to make contact. When at last he spoke, his tenor reminded many of a grizzled schoolteacher's final instructions before the big examination.

"As we progress toward a conclusion of this case, I wish to remind everyone – litigants, spectators, and reporters alike – of the court's prevailing agenda. At this point, the attorney for the defense may wish to cross-examine the plaintiff's last witness. Afterward, the plaintiff's attorney may wish to redirect questions to his witness for the purpose of clarification, call his next witness... or simply rest.

"Immediately thereafter, the defense will call its first witness. The plaintiff may or may not choose to cross-examine any subsequent witnesses, after which the defense may or may not choose to redirect. After all witnesses have been called and all cross-examination and redirect questions have been exercised, the defense will rest, and the litigants will proceed to their closing arguments!"

Inexplicably, Sawyer introduced end-of-trial issues. "The court anticipates that proceedings will end prior to the upcoming holidays! In this regard, I am keenly interested in hearing what the attorneys may have to say."

O'Grady glanced nervously around the courtroom before responding. "A very unusual request, Your Honor. Traditionally, the opposing attorneys would meet in chambers to discuss such matters—"

"Come, come, Counselor!" Sawyer interrupted. "If this were a *criminal* trial with a jury, I would agree!"

The judge swept his arm in front of his chest, causing his robe's loose-fitting sleeves to flap. "For the benefit of the spectators... a civil action suit is generally conducted in a more relaxed atmosphere. Would you not agree, Counselor?"

O'Grady stole a fleeting look at Maloney, who had buried his face in his hands. Neither lawyer had ever heard a judge speak in such a cavalier manner concerning a trial's likely conclusion date. *Sawyer is losing his grip on reality*, O'Grady thought. Aloud, he said, "Your Honor, your proposal depends upon the defense witnesses, and the magnitude of their testimony. The plaintiff does not advocate a prolonged testimony... just enough to prove—"

"The defense," Robbins interrupted, "unlike the plaintiff, will keep witnesses and their testimony to an absolute *minimum*. Unlike the plaintiff, Mr. Curtiss and the other defendants are anxious to conclude arguments *before* the upcoming holiday."

"Splendid!" the judge shouted, as he stumbled to the side of the bench, where he performed what appeared to be an abbreviated vaudevillian tap dance. "Mr. Robbins!" he chuckled, "does the defense choose to cross-examine?"

Robbins scrambled to his feet, notes in hand. "Yes sir, the defense is ready and willing, Your Hon—"

"Marvelous!" Sawyer interrupted. "Mr. Herring, you are still under oath! Return to the witness chair, sir."

Moments later, Robbins tendered the defense's first question. "Mr. Herring, in terms of your aeronautical endeavors, what were your goals?'

Herring gave him a puzzled look. "That depends – my goals have changed over the years."

Robbins shrugged and threw his hands above his shoulders. "Start from the beginning."

Herring glanced over at Chloe, who was sitting on the edge of her chair, nerves taut.

"As a young man in my twenties, my goal was to learn the science and technique of heavier-than-air flight. By my thirties, after others had laid claim to many of my achievements, my goal was to retain as much of my intellectual property as possible by obtaining patents, writing articles, and submitting to interviews. By the time I had turned forty, my goal was to be compensated for my years of work in aeronautics through the manufacture and sale of aeroplanes."

Robbins sidled out from behind the podium and approached the witness. "Mr. Herring, what is your current age?"

O'Grady scrambled to his feet. "Objection, irrelevant!"

"Overruled. Continue."

Robbins repeated the question.

"I'm fifty-four."

"Now that you are in your fifties, Mr. Herring... can it be said that your goal is to *steal* as much of Mr. Curtiss' money as possible—"

"Objection! Speculation! A sarcastic speculation, I might add!" O'Grady shouted.

"Sustained!" Sawyer shouted. "Mr. Robbins, try not to insult the court's intelligence! Strike the last question from the record. Continue."

"Sorry, Your Honor.... Mr. Herring, according to your previous testimony, the recession of '93 caused money to become tight. Is that correct?"

"Yes."

"In fact, money was so tight that your engineering consulting business *failed* and you were forced to seek work from others – is that correct?"

"Yes... I had a young family, and—"

"Sir,... do not elaborate! Is it true that Professor Langley and Mr. Chanute came to your rescue?"

Herring looked Robbins straight in the eye. "I *worked* for both of these men. It was paid employment, there was no charity—"

"Is that a *yes*, Mr. Herring?" Robbins interrupted.

Herring repeated his answer, after which Robbins turned on his heel to face the bench. "Your Honor, the witness is being non-responsive—"

"Your Honor," Herring cut in, "I *worked* for those men... I was not *rescued* by anyone!"

"Mr. Robbins," the judge implored, "... please move on."

"As you wish, Your Honor.... Mr. Herring, although those bastions of aeronautics hired you during your time of need, you claim that they took advantage of your ideas and inventions. Is that correct?"

"That is correct."

"Sir, are you aware that when an employee accepts *money* from an employer, he *forfeits* the right to retain the intellectual property he generated during his time of employment?"

Herring thought about this for a moment. "That's incorrect... not in the United—"

"Answer the question *yes* or *no*!" Robbins interrupted. "You are not to embellish—"

"Objection, Your Honor!" O'Grady barked. "Counsel is badgering the witness!"

"Overruled! The witness will answer the question."

At this point, Herring lost his patience. "It's a deceitful question, Your Honor! I shan't reply!"

Inadvertently, Herring had forced the judge into a legal corner.

Sawyer pulled out his solid-gold pocket watch and flipped open its cover. "Mr. Herring, you have exactly 10 seconds to answer the defense attorney's question... or you will be held in contempt!"

"Your Honor!" O'Grady thundered. "The plaintiff requests a sidebar."

Sawyer made a vague gesture with his hand. "The attorneys may approach the bench."

Moments later, O'Grady conveyed his concerns. "Your Honor, you speak of ending these proceedings early, yet you allow counsel to *bait* the witness into a contempt of court violation. The defense counsel's question was nothing more than the old ruse, 'Have you stopped flogging your horse yet?' Every barroom lawyer tries the tactic... why are you allowing—"

"Judge," Robbins interrupted, "that crap represents nothing more than sour grapes—"

"Mr. Robbins... watch your language and keep your voice down!" Sawyer hissed.

"Sorry, Your Honor, but this technique was only one of many I might have used. Truth be known, it wouldn't matter what method I employed, this witness would still exhibit contemptible behavior... it's his disposition. However, in the spirit of moving this suit to a conclusion, the defense proposes to end its questioning of this impertinent witness."

A long, deep silence ensued.

"The court is not in the habit of tolerating contemptible behavior, but Counselor Robbins' proposal contains a degree of merit, and will move this suit toward its natural conclusion." Sawyer turned toward O'Grady. "What are your thoughts, Counselor?"

O'Grady's lips were pursed so tightly that his words came out compressed. "The plaintiff will agree *only* if Mr. Herring is not held in contempt."

Sensing that his desire to end the trial early might slip away with a contempt ruling, Robbins ignored his sweaty palms and made a crucial, if self-deprecating, decision. "Your Honor, to satisfy both the court and the plaintiff's attorney, the defense will take partial responsibility for the witness's behavior."

Visibly pleased, Sawyer pushed his chair away from the bench. "Gentlemen, return to your stations!"

Moments later, he ordered Robbins to continue his cross-examination of Herring.

"Your Honor, the defense withdraws its previous question and requires no further testimony from this witness!"

Reluctantly, as though it cost him great effort, Judge Sawyer spoke. "Does counsel for the plaintiff wish to redirect?"

"No, Your Honor."

Sawyer blinked in surprise. "The witness is hereby dismissed! Call your next witness, Mr. O'Grady."

"Your Honor, the plaintiff had hoped to question two additional witnesses, to help *verify* important portions of Mr. Herring's testimony."

"Who are these witnesses?" Sawyer asked.

"Mr. Carl Dienstbach and Mr. Andrew Vassallo, Your Honor. Regrettably, neither individual is available to testify at this time—"

"What *verifications* are you referring to, Counselor?" the judge asked.

Checking his notes, O'Grady continued. "In regard to Herr Dienstbach's testimony: he was to provide firsthand knowledge that back in 1893, Herr Lilienthal had incorporated Mr. Herring's automatic stability *regulator* into his *Standard Number 11* glider."

"Why is that nugget of information important to the plaintiff's case?" Sawyer probed.

O'Grady was ready for the judge's query. "Because the defense had insisted that there was no *proof* that Herr Lilienthal had used Mr. Herring's regulator on his commercial machine."

Sawyer stopped scribbling notes. "What does the second verification question pertain to, Counselor?"

O'Grady again referred to his notes. "Mr. Andrew Vassallo was scheduled to confirm Mr. Herring's involvement with the New York publisher, William Randolph Hearst, and the latter's Lilienthal *Standard Number 11*—"

Sawyer interrupted. "Why is this information important to the plaintiff's case, Counselor?"

O'Grady's response was immediate. "Early in Mr. Herring's testimony, Mr. Robbins condemned the witness for his lack of proof concerning his *alleged* involvement with Mr. Hearst and his glider. The defense cited Mr. Herring's inability to produce a *signed contract* with the publisher. Making matters worse, Mr.

Hearst has refused to confirm or deny his contractual agreement with Mr. Herring. Finally, the Smithsonian Institution has *not* responded to our request to confirm the existence of Mr. Herring's *stamped initials* on Mr. Hearst's donated Lilienthal *Standard Number 11* machine."

"Counselor, what does the plaintiff suggest as a remedy?"

O'Grady leapt at the opening. "Your Honor, the plaintiff wishes to submit written *affidavits* from Herr Dienstbach and Mr. Vassallo—"

"Objection, Your Honor!" Robbins wailed. "How convenient... absentee witnesses!"

"Overruled," Sawyer added. "As you know, Counselor, written affidavits are quite legal in the eyes of the law here in New York State. I'll allow them. You may continue, Mr. O'Grady."

"The plaintiff rests, Your Honor."

Judge Sawyer slowly stood before speaking. "I wish to commend all of the litigants for their cooperation in helping to conclude the plaintiff's portion of the *Herring-Curtiss Company v. Glenn H. Curtiss, et al.*, civil action suit!" Banging his gavel, Sawyer continued. "Court is in recess until 2:45... at which time the defense will call its *first* witness."

*

With all the litigants present and accounted for, the judge wasted little time in addressing the defense attorney. "Counselor, is the defense prepared to proceed?"

Legal pad in hand, Robbins was already standing in front of the defense table. "Yes, sir!"

"Splendid! Call your first witness."

"The defense calls Glenn Hammond Curtiss."

Rising from his seat behind the defense table, Curtiss edged his way toward the aisle. As he passed behind fellow defendant

Thomas Baldwin, the portly airship entrepreneur twisted in his seat to offer his hand. Pausing briefly to acknowledge his old friend and business partner, Curtiss strode confidently to the witness box as the court clerk waited patiently.

At five feet nine inches, the 43-year-old aeroplane manufacturer and millionaire had gained weight since that windy day 12 years prior, when he first flew the Herring-Curtiss Company's *Golden Flyer*. Although he still sported a thick, dark mustache, shaggy eyebrows and steady blue eyes, a receding hairline predicted impending baldness. With all eyes fixed on the intrepid airman, the pride of Hammondsport placed his left hand on the *Bible*, elevated his right, and gazed up at the judge.

"Do you solemnly swear to tell the truth, the whole truth, and nothing but the truth... so help you God?"

"I do."

"Please be seated. Mr. Robbins, you may proceed."

Grinning broadly, Robbins sauntered up to the witness. "Mr. Curtiss! Finally! A friendly face in the witness chair!"

Curtiss, stoic as usual, barely nodded. Robbins was well aware that Curtiss's somber demeanor, accentuated by a perpetual cold frown, was neither conducive to winning over a jury... nor the judge. Although his client was not glib, especially around college graduates, rarely laughed, and often acted aloof, there was no denying he was a clever mechanic and a talented, self-taught operator of aeroplanes. In addition, Curtiss had always displayed a steadfast persistence in attaining personal goals. With these qualities in mind, Robbins wasted little time getting to the heart of their argument.

"How would you describe your partnership with Augustus Herring?"

Curtiss hesitated before becoming assertive. "He represented the *biggest* mistake of my life!"

"Mr. Curtiss, why did you enter into a business relationship with Mr. Herring?"

"So his patents, applications, and devices could be used to defeat the Wright brothers' ridiculous patent, thereby allowing us to manufacture and sell aeroplanes in North America."

"What went wrong with this partnership?"

A note of disappointment crept into Curtiss's voice. "Herring didn't have patents, there were no working devices – only worthless applications—"

"Objection! Argumentative!" O'Grady shouted, struggling to his feet. "If nothing else, the witness is misrepresenting the value of Mr. Herring's patent pending designations! Your Honor, if these applications had been worthless, why did Mr. Curtiss form a working relationship with Lawrence Sperry to develop Herring's gyroscopic stabilization system for—"

"Sustained," Sawyer broke in. "Point well taken, Counselor. Proceed."

Robbins hesitated before continuing. "Mr. Curtiss, please provide the court with an example of a failed Herring device."

Curtiss opened and closed his mouth, as if he wanted to add something. "One rather laughable device that Herring insisted we try was the string of fins along the span of our biplane's upper wing. He claimed these *sails*, as he called them, would provide lateral stability without having to resort to an *active* lateral control system—"

"Objection! Argumentative!" O'Grady broke in. "The plaintiff has established that Mr. Curtiss had steadfastly refused to try this stopgap innovation – even if the alternative was to infringe on the Wright patent—"

"That's a lie!" Robbins interrupted. "Herring's ridiculous sail idea didn't work! Mr. Curtiss had no other choice but to use his proven interplane control surfaces for lateral control."

"What say you, Mr. O'Grady?" the judge fired back.

"Curtiss had no other *choice* but to *infringe?* Your Honor, Mr. Herring's use of jib sails is well documented. Back in St. Joseph, Michigan, Mr. Herring used these sails on his successful gasoline engine-powered *Flyin' Fish* model of 1903. Seven years later, he used the innovation on the Herring-Burgess *Flying Fish*, which was the first successful powered, manned, heavier-than-air flight in New England."

Curtiss glared at O'Grady, wide-eyed with surprise, blurting out, "You must be joking! A toy aeroplane, and a puddle jumper that couldn't begin to make a coordinated turn?"

The judge, sensing that matters were getting out of hand, rattled his gavel against the bench block. "Gentlemen, enough bickering! The plaintiff's objection is sustained. Continue, Counselor."

Frustrated, Robbins shook his head. "Mr. Curtiss, from the perspective of hindsight, how would you portray your former partner?"

Curtiss thought about this for a moment. "Herring was a bungler who claimed more credit than he deserved. I also agree with those who called him an *aeronautical hyena* – an ingrate who had always sniffed around the edges of the aeronautical—"

"Objection!" O'Grady interrupted. "This is nothing more than speculation and hearsay! This portrayal borders on defamation!"

"Overruled. The Court will allow the witness's characterization. Proceed."

"Mr. Curtiss, after all these years, why did the plaintiff take legal action against you and the other defendants?"

Curtiss threw up a hand. "It's about the money! With Herring, it was *always* about the money!"

Robbins turned to face the judge. "The defense has no further questions for this witness."

Surprised at the brevity of Curtiss's testimony, O'Grady felt his body stiffen as a murmur rolled throughout the courtroom's gallery.

Cracking his gavel for silence, the judge glared down at O'Grady. "Does counsel for the plaintiff wish to cross-examine?"

"Yes, Your Honor!" O'Grady confirmed, as he hurried to his satchel to retrieve another notepad.

Annoyed by the delay, Sawyer tapped his pencil impatiently. "Any day now, Counselor!"

Lumbering to his spot behind the podium, O'Grady appeared slightly out of breath. "Sorry, Your Honor. I'm ready now—"

"Good! Let's get on with it!"

"Mr. Curtiss, I couldn't help but notice your attorney's fascination with Mr. Herring's aeronautical goals. I'm curious... what *is* your goal?"

Curtiss gave a puzzled look. "I don't understand—"

"It's a simple question... what is your goal for this lawsuit?"

"What's that got to do with—"

"Just answer the question, Mr. Curtiss!"

"Objection!" Robbins shouted. "Your Honor, counsel is being vague!"

"Sustained. Clarify your question, Counselor."

O'Grady stifled a grin. "Mr. Curtiss, in terms of your goal for this trial... is it to *keep* as much money as possible from your *looting* of the Herring-Curtiss Company?"

"Objection!" Robbins squealed. "Argumentative!"

"Overruled. The witness may answer."

Curtiss paused before answering. "My goal is to get *justice* for me and my partners!"

O'Grady paused, pressing a forefinger to his lips. "Let me see if I have this right. You say that Mr. Herring's goal is 'all about the money', but your goal is to 'get justice'... is that a correct assessment?"

"Yeah, that's it!"

O'Grady slipped from behind the podium and slowly walked up to the witness chair. "Mr. Herring's entire testimony centered on his desire to obtain *justice* for himself, and the Herring-Curtiss Company. Nowhere in his testimony did he express a desire for great wealth. Wealth that both you and the Wright Company fought over for the better part of a decade—"

"Objection! Argumentative!" Robbins blustered. "Counsel is lecturing!"

"Sustained. Get on with your cross-examination, sir."

Taking a deep breath, O'Grady continued. "Mr. Curtiss, speaking of your long legal history with the Wrights, I'm curious... are you still *paying off* the lobbyists who got your company the *Cross-Licensing Agreement of 1917?*"

"Objection!" Robbins shouted. "Scandalous speculation on the part of the plaintiff—"

"Your Honor," O'Grady cut in, "the question speaks to the witness's claim that *money* was secondary to justice."

"The Court understands your point, Counselor," the judge said, "but you're beginning to belabor the issue. Continue."

"Mr. Curtiss, this is a question you should have no trouble answering... did the Curtiss Aeroplane Company *ever* stop infringing on the Wright patent?"

Again leaping to his feet, Robbins inadvertently knocked over a folding chair. "Objection! Counsel is indulging himself by submitting unanswerable questions—"

"Your Honor," O'Grady interrupted, "I'm merely following the precedent established by my esteemed counterpart... what's good for the goose is good for the—"

"Sustained!" Sawyer broke in. "Counsel is taking advantage of the Court's good humor. Mr. O'Grady, refrain from further excesses."

Sawyer spun in his chair to face Robbins. "Speaking of excesses, Mr. Robbins... the Court is weary of the defense's continual objections! Either request a continuing objection or stifle the urge to object on every question. What is your choice, Counselor?"

Robbins threw up his hands. "If those are the choices, the defense requests a continuing objection."

"Granted! The court stenographer will duly take note. Continue, Mr. O'Grady."

O'Grady turned away from the bench, snickering to himself. "The plaintiff's next question pertains to the hastily called meeting of the Herring-Curtiss Company's board of directors. This took place in Hammondsport, New York, on Monday, October 25, 1909. All board members were in attendance, including Monroe Wheeler, Thomas Baldwin, A.W. Gilbert, Augustus Herring and yourself. Mr. Curtiss, do you remember why this impromptu meeting was called?"

"No," Curtiss said, wagging his head in a mock serious way.

"Very well," O'Grady continued, "allow me to refresh your memory. According to previous testimony, on Saturday the 23rd of October, two days prior to the formal board meeting, you and Judge Wheeler met with an outsider by the name of Jerome Fanciulli. The purpose of that get-together was to finalize the details of Mr. Fanciulli having been hired to manage your personal interests. These interests included – allow me to quote, Your Honor,

'... the sale, exhibition and exploitation of the Curtiss aeroplane.' Do you remember that, sir?"

Curtiss responded without hesitation. "Yes."

"Excellent! Directors Herring and Gilbert, having had no knowledge of the October 23rd meeting, came to Hammondsport from New York City after you had mailed them notes that said '... I think it would be a good plan for you to come up Sunday so that we can have a little director's meeting on Monday.' " O'Grady gave Curtiss a questioning look. "Would it be safe to say that Mr. Herring had no idea of what was about to happen?"

Curtiss spoke angrily. "I don't recall the details! According to you, this meeting happened more than a decade—"

"A simple *yes* or *no* will suffice, sir."

"Counselor," Sawyer interrupted, "where is this line of questioning headed?"

O'Grady glanced up at the judge. "That is about to become evident, Your Honor. In the interim, the plaintiff asks for the Court's indulgence."

"Very well, but my patience is wearing thin."

"O'Grady took a step backward. "Sorry, Your Honor, I'll withdraw the question. Mr. Curtiss, is it true that Judge Wheeler presented a *resolution* to the board that was prepared in advance and was adopted by an oral vote at this meeting?"

"I believe so."

"Did Mr. Herring vote in favor of this resolution?"

Curtiss pinched the bridge of his nose. "I don't remember."

O'Grady turned his head toward the judge. "Your Honor, from previous testimony, let the record show that Mr. Herring *did not* vote for said resolution."

Returning his gaze to the witness, O'Grady continued. "Mr. Curtiss, as the sole manager of the Herring-Curtiss Company, did you have authority to designate an assistant manager, and to fix his compensation?"

"Yes."

"Therefore, by hiring Mr. Fanciulli as your assistant manager, you could devote more of your time to the fledgling Curtiss Exhibition Company. Is that true?"

"Yes."

"Is it also true that the board of directors' resolution of October 25 stipulated that any past or future flying awards, consisting of 'trophies, medals, prizes and *compensation*, belonged solely to you, and not the Herring-Curtiss Company?"

"Yes, but there was great advertisement value for the—"

"Advertisement value?" O'Grady interrupted.

At this point Curtiss lost his patience. "If it weren't for races that I had won, along with my exhibition flights, the company wouldn't have received nearly as many orders for its aeroplanes!"

O'Grady contemplated this for a moment. "What aeroplane orders are you referring to? Are there secret books kept for aeroplane orders that the plaintiff doesn't know about?"

Curtiss gave him a befuddled look.

"Allow me to clarify this point, Your Honor," O'Grady said. "Less than a month after the passage of the October 25 resolution, Curtiss, Wheeler and Fanciulli, without any action by the board of directors, prepared and adopted for company use a *restrictive sales contract* that required a pledge, signed by the purchaser, stating that his new aeroplane would not be used for public flights intended for amusement, or advertising, or contests of any kind." Spinning about, O'Grady pointed at Curtiss. "True or false, Mr. Curtiss?"

Curtiss' face had already turned crimson. "That's true, but we were trying to avoid infringement—"

"No explanation is required, Mr. Curtiss!"

Robbins raised his hand to object but thought the better of it.

"Mr. Robbins," Sawyer interjected, "the defense may pick up on this point during redirect. Continue, Counselor."

"Other than exhibition flying, where was the market for aeroplanes back in 1909?"

Confident, Curtiss leaned forward. "Sportsmen bought our machines."

Holding up what appeared to be a letter, O'Grady marched to the bench. Within minutes, the evidence had been processed and assigned an exhibit number.

"In a letter written to Fanciulli in 1910," O'Grady said, "you expressed an opinion about the aeroplane market. If the court so desires, I will now read from this document."

The judge waved his assent.

> " 'Other than exhibition flying, there is no market for aeroplanes. A sportsman can purchase one of my Curtiss machines, as long as he flies it privately out of some backwater cow pasture. I cannot see much at present in the sales [of aeroplanes] except to exhibitors, and this would kill our exhibition business.' "

O'Grady handed Curtiss's letter to the bailiff and returned to the podium. "Mr. Curtiss," he continued, "isn't it true that the restrictive sales contract allowed you to corner the exhibition flying market by limiting the number of aeroplanes sold by the Herring-Curtiss Company, by restricting how those few machines could be used? Put another way, isn't it true that you lined your pockets with the proceeds from the Curtiss Exhibition Company, while sacrificing the mother company?"

Curtiss leapt from his seat. "That's a bunch of crap!" he exploded.

Sawyer cracked his gavel twice and briefly scolded the witness. With order restored, Curtiss was directed to answer the question. "With the Wrights breathing down my neck, how else could I make any—"

Curtiss hesitated in mid-sentence.

"*Money?*" O'Grady howled.

"Yes, money!" Curtiss growled. "That *is* why people go into business—"

"Are you finished, Counselor?" Sawyer interrupted.

O'Grady slipped from behind the podium and faced the judge. "One more question and then I'll move on, Your Honor—"

"I'll hold you to that, Counselor!"

"Mr. Curtiss, earlier you said 'With Herring, it's always been about the *money*.' In light of your present testimony, would you consider retracting that statement... or applying it to yourself?"

"Go to hell!" Curtiss shouted.

O'Grady stood speechless as the courtroom erupted. Banging his gavel repeatedly to restore order, Sawyer seemed to forget to rule on Curtiss' rude non-response.

"Move on, Counselor!"

Deciding not to object, O'Grady flipped to a new page in his notepad. "Mr. Curtiss, where did your exhibition company get its aeroplanes?"

"From the Herring-Curtiss Company – where do you think?"

"Keep your responses civil, Mr. Curtiss," Sawyer admonished.

"The Curtiss Exhibition Company owned its own aeroplanes, is that correct?"

Curtiss rolled his eyes and shook his head. "The aeroplanes in question were owned by the Herring-Curtiss Company."

O'Grady, who was standing a few feet in front of the witness, backed away with his hand up. "Are you saying that the Herring-Curtiss Company footed the bill for your exhibition aeroplanes?"

"That's right!"

"Did your exhibition company pay for the maintenance and repair of those aeroplanes?"

"No."

"Who did?"

"The Herring-Curtiss Company."

"Let me get this straight," O'Grady said. "The Herring-Curtiss Company paid the salary of an assistant plant manager, adhered to the dictates of a restrictive sales agreement, and supplied the Curtiss Exhibition Company with aeroplanes that were maintained and repaired for free... all for the opportunity to sell *fewer* aeroplanes, while allowing *you* to keep 100 percent of the exhibition proceeds! Is that an accurate representation, Mr. Curtiss?"

Curtiss seemed unperturbed. "If you say so."

"Is that a *yes* or a *no*?"

Curtiss refused to elaborate, prompting O'Grady to ask the judge to intervene.

"Entered as non-responsive, Counselor. Continue."

Annoyed, O'Grady stared at his notes. "Mr. Curtiss, early in February of 1910, the board of directors of the Herring-Curtiss Company responded to your request for $10,000 to secure a bond from the American Surety Company. The bond was needed to satisfy the federal court's temporary injunction order against the Curtiss Exhibition Company. The Wrights asked for and received the 'no fly' injunction as part of their ongoing infringement suit against you, the Herring-Curtiss Company, and The Aeronautic Society of New York. The directors then passed a resolution authorizing you to sell three of the aeroplanes in your possession, 'at the best prices obtainable and apply the proceeds, up to $10,000 to secure the bond. Any surplus money was to be turned over to the company.'

"By means of a *contrived* sale, is it true that you bought these three aeroplanes, including the famous *Rheims Racer*, for a paltry $10,000... seven to $10,000 below their true market value?"

Curtiss began to laugh, then pulled himself together. "That's false! Those three aeroplanes were worn out, and I couldn't find a buyer for them. I had to use my own money to deposit with the surety company to save the exhibition business."

O'Grady held up a finger as a signal to wait. "You *didn't agree* with the independent appraiser who valued the three machines at $20,000?"

"No."

"Mr. Curtiss, *your* $10,000 went to buy the three Herring-Curtiss Company aeroplanes. Correct?"

"Correct."

"Then the Herring-Curtiss Company purchased the $10,000 bond necessary to keep *your* exhibition company flying. Correct?"

"Correct."

"Therefore, the Herring-Curtiss Company paid $10,000 of assets it couldn't afford to lose, to keep your exhibition company flying. Is that true?"

Curtiss shook his head. "You're leaving out the advertising benefits that the exhibition company afforded the manufacturing company—"

"That wasn't the question! Answer *yes* or *no!*"

Once again, Curtiss refused to answer and was admonished by the judge – to no avail and *without* a threat of contempt. O'Grady was ordered to move on with his questioning.

"As the purchaser of these three machines, did you sign the restrictive sales agreement?"

The sarcasm, although appropriate in the eyes of the plaintiff, fell on deaf court ears. O'Grady was again ordered to continue, or to end his questioning.

"Mr. Curtiss, why has your legal team changed its tactics for this civil action suit?"

Curtiss shrugged. "I have no idea what you're talkin' about."

O'Grady slipped back behind the podium. "The plaintiff is well aware of the *delaying* tactics employed by the Curtiss Aeroplane Company in its decade-long patent infringement suit with the Wright Company. Since your side didn't have a leg to stand on, the use of that tactic was understandable. However, for this trial it seems that the defense *can't wait* to enter into deliberations! Why is that, Mr. Curtiss?"

"Objection, Your Honor!" Robbins shouted, taking his chances with a contempt charge. "Lectures aside, tactics used by the defense in this or any other case are none of the plaintiff's business!"

"Sustained. Get on with it, Counselor!"

O'Grady pondered this for a moment. "Mr. Curtiss, as a person familiar with the idiosyncrasies of the legal system, please tell me... were the Wright trials also about *justice*, or were they about the *millions* of dollars you made building aeroplanes during WWI—"

"Objection! Relevance?" Robbins screamed.

"The plaintiff has no further questions for this witness, Your Honor."

Sawyer wasted no time. "Redirect, Mr. Robbins?"

"None, Your Honor."

"Very well, the witness is excused. Call your next witness, Counselor."

Robbins clasped his fingers together in front of his chest. "The defense rests."

Judge Sawyer could hardly believe his ears, but recovered in time to recess court for 15 minutes.

*

When court reconvened and Sawyer confirmed that Robbins had no further witnesses, he spoke to the litigants. "Gentlemen, it's Friday afternoon and the hour is late. I propose we adjourn proceedings until Monday morning, at which time closing arguments will be heard. The plaintiff will present first. Are there any questions?"

None were offered.

"Wonderful! Court is hereby adjourned until 9 a.m. Monday morning."

Rochester, New York

Monday, December 5, 1921

O'Grady led off by presenting the plaintiff's closing argument. He offered a masterful rendition of Herring's years in aviation, emphasizing his goals and achievements. In summary, he reviewed the formation and demise of the Herring-Curtiss Company, along with the rise of the Curtiss aeroplane empire... and the fall of Herring's fortunes.

Less than an hour later, Attorney Robbins chronicled the detrimental effect Augustus Herring had upon the company by itemizing his purported deficiencies, especially his failure to produce the patents and devices necessary to defeat the suffocating Wright patent. A significant portion of the defense's summary was spent rebutting Herring's claim to have both designed the company aeroplanes and redesigned its engines.

By 11 a.m., it was over. In wrapping up loose ends, Judge Sawyer thanked the litigants for their professional conduct and the attorneys for effectively pursuing their clients' agendas. Finally, the judge informed the parties that his verdict would be forthcoming in a matter of months and that they should keep in touch with their attorneys, who would be notified of his decision by mail.

Cracking his gavel one last time, the judge brought the *Herring-Curtiss Company v. Glenn H. Curtiss, et al.*, hearing to its conclusion.

*

Getting together one last time for lunch, Herring and daughter Chloe, along with their attorneys, shared thoughts about the trial and their chances of winning.

"Why was Curtiss laughing and congratulating his partners?" Chloe asked. "If I didn't know better, I would have thought they had already won!"

"That's exactly what I was thinking," Gus agreed. "Strange... Glenn almost never laughs."

A troubled silence ensued.

"Let's not get ahead of ourselves," Maloney said, "The judge has yet to deliberate."

CHAPTER 92

Freeport, Long Island, New York

Wednesday, February 28, 1923

It was exactly seven months to the day after the *Herring-Curtiss Company v. Glenn H. Curtiss, et al.*, civil action hearing had concluded that Augustus M. Herring suffered a major stroke. Paralyzed on his left side, the 55-year-old was confined to bed for six months, requiring almost constant attention from his wife, along with the untiring efforts of his daughter, Chloe. Gradually, Herring regained the use of his limbs and his speech. By the end of February 1923, he began writing letters of inquiry regarding the status of his seemingly forgotten suit against Curtiss and his cronies.

"What the hell," he complained to anyone who would listen, "it's been 15 months!"

Freeport, Long Island, New York

Monday, May 7, 1923

The phone jingled three times; someone was calling Herring's house. As the cycle of rings began to repeat, Gus looked up from his morning newspaper and removed his reading glasses. Grumbling about the "damned inconvenience," he pushed his chair back from the kitchen table and struggled to his feet. With the aid of his ever-present cane, he hobbled over to the wall-mounted telephone, and snatched the receiver from its hook.

It was O'Grady. "Gus, it's finally here. The judgment arrived in the morning mail, along with Sawyer's 27-page opinion—"

"Did we win?" Herring began to sweat.

"Let me read you the *Accounting*. Are you sitting down?"

"Yes, damn it! What does it say?"

O'Grady hesitated briefly and then began to speak. "Here goes... '*Herring-Curtiss Company v. Glenn H. Curtiss, et al.* Supreme Court of New York, Monroe County.' Are you still with me?"

"I'm here! Keep going!"

"The *Headnote* summary is next. 'Charges of conspiracy resulting in bankruptcy of corporation – evidence insufficient to establish right of relief.' "

Herring felt lightheaded. He thought he might pass out. "So... we lost."

O'Grady tried to sound optimistic. "Yes, but Maloney and I believe there might be grounds for an appeal—"

Herring groaned. "Do we have to go through all that crap again?"

"No, but first we have to convince the *Appellate Division* to take the case. On average, they only hear about one in five of the requests they receive – so there's no guarantee. Look, Gus, I'm sending you a copy of the decision. Look it over... we'll talk again next week."

Freeport, Long Island, New York
Monday, May 14, 1923

As promised, O'Grady telephoned Herring a week later. With Maloney on an extension phone, the men immediately got down to business.

"Gus," O'Grady said, "I presume you've had time to read Sawyer's opinion."

"Only about 10 times!" Herring snarled.

"What do you think of the document?"

Herring hesitated for a moment. "I got the impression that Sawyer had made up his mind long before that ridiculous deliberation period started."

"Very perceptive," Maloney said. "What else?"

"On a personal note, several things caught my eye," Herring said. "Curtiss was disappointed with my *lack* of practical ability? Shit, I was designing and building aeroplanes before that asshole learned how to ride a bicycle!

"Then Sawyer claimed I was disappointed when the financial reward I expected from Curtiss's *efforts and genius* wasn't realized? Where the hell did that come from?"

There was no immediate response.

"Here's the point," O'Grady said. "Sawyer's opinion is full of biases and misinterpretations... but we have to find something *egregious* to get the Appellate Court's attention."

"What about Curtiss's acquisition of the Company's *Rheims Racer* and two other 4-cylinder machines? After making him thousands of dollars from exhibition flying, he obtained them for himself in a shady deal approved by Wheeler and Baldwin – at less than half of their market value! Sawyer's written opinion justified the deal by saying the advertising value gained by the company far outweighed any profit Curtiss may have realized. Therefore, in the eyes of the court, Curtiss was justified in making a profit, while the company was compelled to absorb the loss!"

"We agree, Gus! Although Sawyer held for Curtiss's side on all counts, we still don't believe there's enough dirt to get our side heard by the Court of Appeals! We'll keep working on it and hope to get back to you soon with better news."

Herring, utterly discouraged, hung up the phone.

Freeport, Long Island, New York

Friday, May 18, 1923

Herring waited patiently for either O'Grady or Maloney to answer the phone at their Rochester, New York office.

"James M. E. O'Grady speaking."

Herring took a deep breath. "Jim, it's Gus. I received a *copy of a letter* in the mail this morning; it was from Jerome Fanciulli... of all people!"

"What was that all about?"

"It was a letter that Curtiss wrote to Fanciulli almost a year ago... where he gloats about his victory over us in State Supreme Court—"

"Why did Fanciulli fill you in?" O'Grady interrupted. "I thought you two hated each other."

"Curtiss caught Fanciulli *skimming* money from his exhibition company almost a decade ago ... and fired him. This must be Jerome's way of getting even."

"Is that so? What did he have to say?"

"I won't recount Curtiss's rehash of the judgment report, only what he said at the end of the letter. I quote:

'Actually, Jerome, we had the verdict in the bag... Judge Wheeler and I happen to be friends with Judge Sawyer.'"

Thinking he hadn't heard correctly, O'Grady directed Herring to reread the passage. "Gus, make another copy of the letter and send it to me right away! I think this might get the Appellate Court's attention!"

Augustus Herring (1924); Meiller/Herring Collection

EPILOGUE

On May 21, 1923, attorneys for the Herring-Curtiss Company (plaintiff) filed a notice of appeal to the Supreme Court of New York, Appellate Division, seeking relief from a judgment holding for Glenn H. Curtiss, et al. (defendants), by Judge S. Nelson Sawyer of the Supreme Court.

Four and a half years later, on November 16, 1927, the court – a panel of five judges (Crouch, Hubbs, Clark, Sears and Taylor), heard the appeal arguments and reviewed briefs filed by attorneys James M. E. O'Grady and William J. Maloney for the plaintiff and Frederick A. Robbins for the defense.

Four months later, on March 14, 1928, the Appellate Division *reversed* Judge Sawyer's decision, ordering damages to be paid by Curtiss and the estates of Wheeler and Baldwin (both deceased).

Details of the Reversal

The Appellate Division unanimously made the following additional conclusions of law:

> Glenn H. Curtiss, Monroe Wheeler and Thomas S. Baldwin (defendants), are guilty of malfeasance, and misfeasance in office as directors of the Herring-Curtiss Company (plaintiff), and that:
>
> 1. Curtiss, Wheeler and Baldwin conspired to permit money, property and business profits of the Herring-Curtiss Company to be retained by Curtiss, to the loss and damage of the company, by adopting the resolution of October

25, 1909. After reviewing the Herring-Curtiss Company records, the panel of judges identified the motive for the board of directors' *resolution of October 25, 1909*:

> "It was pure selfishness on the part of defendants Curtiss, Wheeler and Baldwin,"

wrote Judge J. Crouch, summarizing the verdicts of his colleagues. The court deemed the resolution to be unwarranted, causing loss and damage to the plaintiff.

2. Next, the court condemned the *restrictive sales contract*. After the defendants had collaborated in early November of 1909, Wheeler formulated a contract to be applied to the sale of all future Herring-Curtiss Company aeroplanes. Before any sale could be finalized, the buyer was required to sign the contract, which effectively prohibited him from taking part in public flying, exhibitions or competitions. Failure to adhere to the document's conditions would void the company's guarantee against liability in patent infringement suits. Created solely to aid Curtiss' exhibition business without knowledge or consent of the board of directors, the contract limited competition from the customers of the Herring-Curtiss Co., thereby discouraging them from purchasing company machines, which resulted in further loss and damage to the plaintiff. The court concluded that the contract's intent was to prevent the sale of company aeroplanes that could have been used to compete with the Curtiss Exhibition Company.

3. By fraud and lack of consideration, the Appellate Division found that Curtiss had wrongfully obtained possession of three aeroplanes belonging to the Herring-Curtiss Company, and then used these aeroplanes for his own profit... to the company's loss and damage. It was decided that Wheeler and Baldwin, after learning of Curtiss' fraud, had failed to recover the company aeroplanes or compel

Curtiss to account for their value or the profits he realized from their use.

4. It was also decided that the court action in the name of the Herring-Curtiss Company against Herring by Curtiss, Wheeler and Baldwin had been initiated solely to prevent Herring from voting his stock, while keeping themselves in power as directors of the company. The many acts of omission, coupled with the diversion of profits from the exhibition business and the retention of the company's money by Curtiss, had brought about the Herring-Curtiss Company's needless bankruptcy. Further, through these actions Curtiss, Wheeler and Baldwin had come into possession of the company's property and assets.

5. In summary, Curtiss, Wheeler and Baldwin had fraudulently obtained the property and assets of the Herring-Curtiss Company and had illegally abused the rights of the company's creditors and stockholders. They are therefore accountable to this plaintiff.

6. Judgment is hereby ordered in favor of the plaintiff – the Herring-Curtiss Company, and against the defendants, Glenn H. Curtiss, and the estates of Monroe Wheeler and Thomas A. Baldwin.[1]

Glenn Curtiss refused to accept the decision. Determined to fight the reversal, he hired new lawyers under the direction of W. Benton Crisp – the New York attorney who had broken the Selden patent for Henry Ford. Crisp, most noted for his delaying tactics, continued the case for another four and a half years. According to James V. Martin's article, "When Will Merit Count in Aviation?" (published in the October 1924 issue of *The Libertarian* magazine), Curtiss had divided his $13 million fortune in profits from aviation with his wife.

In 1930, two years after the Appellate decision, Glenn Curtiss died at the age of 52, while awaiting surgery for appendicitis in a Buffalo, New York hospital. On the last day of December 1932, Mrs. Lena Curtiss paid $500,000 in settlement, equal currently to about $8 million.

Sadly, Augustus M. Herring didn't live to see the appellate court's final judgment. After suffering two additional debilitating strokes, he died in the summer of 1926. His daughter Chloe (Mrs. Henry H. Mason), then 30 years old, carried the suit through to its conclusion. The protracted litigation and administration expense took the majority of the settlement... only $30,000 (equal to almost $500,000 in today's dollars) filtered down to the two Herring heirs, Chloe and William.

After enduring Judge Sawyer's decision, Herring died a bitter man. Looking back at his life in aeronautics, daughter Chloe summed up her father's predicament:

> "There was nothing worse one could do in our family than lie. My dad was a gentleman. He expected others to behave like gentlemen. That's how he was [taken advantage of]."

Reflections on Herring's Work

Herring held strong opinions – opinions that caused some to view his actions and behaviors in a negative manner. Gus had once testified that he considered Chanute, at age 64, to be senile. According to longtime Herring researcher, Eugene Husting:

> "This attitude suggests impatience with (colleagues) whose actions or performance did not meet his standards. This trait showed itself in other associations before and after 1896. Herring could be difficult."

By and large, historians have not been sympathetic to Herring. In her 1949 book *Flight into History*, Elizabeth Freudenthal offered an interesting opinion about flaws in Herring's personality:

"In spite of his undoubted brilliance, and keen inventive mind, Herring's qualities and career contrasted strongly with the Wrights. [Herring's] lack of persistency... differed from the Wrights' careful tenacity of their ideas and their constant drive to reduce them to practice."

However, this lack of *persistency* argument doesn't hold up under scrutiny. When Herring's work in aeronautics is viewed from a historical perspective, it appears that his pioneering accomplishments ended late in 1899... after the devastating Truscott Boat Works fire. This is not true, although the loss of his aeroplanes, engines and tooling, and his unjustified dismissal from his engineering position at Truscott's, certainly added to his despair at the time. Herring's financial situation must be taken into consideration as well. His inheritance from his father's estate had run out, a situation compounded by the subsequent death of his partner and financial backer, Matthias Arnot, in 1901.

Although Gus did not resurface publically until 1907, when he won a Signal Corps contract to build the country's first military aeroplane... he had been active in engineering and aeronautical research during the previous eight years:

- 1900: Herring continued to develop his hemi engine (Mobike motorcycles).
- 1902: he designed and built a revolutionary miniature engine for powering his experimental *Flyin' Fish* model aeroplane.
- 1903: he successfully flew this two-surface model in free flight, using his jib-sail method of passive lateral stability. Later that year he flew a tethered version of the model inside a New York City armory.
- 1904: Herring worked out a rudimentary version of his gyroscopic automatic stability system for aeroplanes.

- 1905: he designed a two-surface towline glider to test components for a new powered machine.
- 1906: he helped to test the towrope-launched glider on the Delaware River near Philadelphia.
- 1907: Herring designed the first of his two Signal Corps flying machines.

In regard to Freudenthal's statement concerning Herring's motivation being unlike the Wrights' constant drive to move their ideas to practice, she had interviewed Herring's assistant, Walter L. Brock, who offered this point of view:

> "Herring had too many ideas, and changed them so frequently that his designs were modified each day. [His] ideas were always good, but the continual discarding of the old for a new and better plan prevented consistent, constructive work."

The Antagonists

Herring never shrank from a fight with a news organization or an overly aggressive reporter. His accounts of the emerging science of heavier-than-air flight were much sought after and widely read, especially in his home state of New York. Inevitably, his flamboyant descriptions and predictions triggered jealousy among his contemporaries, some of whom maligned him as a headstrong braggart. However, there is little doubt that leaders in the burgeoning field of aeronautics – luminaries such as Samuel Pierpont Langley and Octave Chanute – were heavy-handed in their confiscation of the younger Herring's ideas, innovations and improvements. To these legacy-driven authoritarians, granting credit to an underling was as distasteful as consuming spoiled seafood.

Adding insult to injury, these same individuals belittled Herring in their letters by expressing their unfettered opinions of his limited talents and presumptuous notions of his abilities. This correspondence has provided the raw material that historians have routinely relied on to write accounts of early aeronautical personalities, their invention and influence. Sadly, scholars have often taken such letters at face value, rather than questioning the degree of gossip, half-truths and shameless self-promotion that may have been woven into those words. Indeed, in most instances, the information shared in such letters was intended to remain confidential. In Herring's case, the opinions of him found in the letters of his contemporaries have formed a public record that has gradually evolved into accepted fact through the writing of careless authors. As many of us are painfully aware... if an *opinion* is repeated often enough, it will eventually be considered as *fact*.

Aeronautical journalist Carl Dienstbach, who knew Herring best, insisted that his friend was *not* inclined to show off. He offered additional insight concerning Herring's personality in his unpublished manuscript, *Memorial on the Birth of Flight*:

> "... Herring's outstanding peculiarity was reserve [keeping his feelings, thoughts, or affairs to himself], and in a way, being secretive. For example: it is only common sense that he flew his large gas model in that [New York] armory. It was surprising that Herring did not invite me to see the trials... he had volunteered that promise on my first visit to his Freeport home [in 1903]. At the time I did not guess how much he later would need corroborative testimony."

Those who work at the leading edge of technical innovation often experience the no-holds-barred brutality of *competition*. The invention of the aeroplane was no exception. Augustus Herring corresponded with, worked with, and fraternized with men who proved to be ruthless and unprincipled in their quest for status and fame. These individuals included:

Samuel Pierpont Langley

Always the self-serving authoritarian, Langley demanded complete subservience from his employees while implementing their ideas and taking the credit for himself. As he once told Herring, "As long as you work here, your ideas are my ideas." When Herring provided *too many* answers to the aerodrome's technical problems, he became a threat to Langley's legacy and was soon forced to resign. Claiming that Herring had nothing to do with the eventual success of aerodromes *Number 5* and *6*, Langley never missed an opportunity to malign his former employee to fellow members of the aeronautical community.

Octave Chanute

In 1960, British aviation historian Charles Gibbs-Smith, writing in the magazine *The Aeroplane*, openly discussed Chanute's meager aeronautical achievements:

> "Owing to some curious misstatements and misapprehensions, Chanute has been wrongly credited with various achievements and ideas, which he himself never claimed [or disclaimed], and which have unfortunately gained wide [belief]. His only technical contribution to aviation was his bridge-truss method of rigging a biplane."

How Chanute Used Herring

The two-surface glider, first flown by Herring at the Indiana dunes in the summer of 1896, has been acclaimed by aviation historians to be the *most successful heavier-than-air, man-carrying glider of the late 19th and early 20th centuries*. The origins of this machine have never been in doubt, but many authors have unwittingly or deceitfully rewritten its historical details.

Professor Albert Zahm, of Washington, D.C.'s Catholic University, who witnessed Herring's two-surface, rubber-band-powered model fly, said the following in 1908:

> "It is sometimes said that the best French aeroplanes are copied from the Americans... Farman's aeroplane resembles the Wright brothers'; theirs resembles Chanute's glider of 1896, and this in turn resembles Herring's rubber-powered model."

For the last 30 years of his life, Herring maintained that the man-carrying, two-surface glider was a direct descendant of the model he had designed, built and flown in 1892. However, after Herring and Avery had accumulated more than 1000 accident-free glides, Chanute gradually began to see the design's value to his *reputation* – eventually attributing much of its success to his own suggestions – especially the use of the Pratt truss, which provided the lifting surfaces with enhanced rigidity.

Over the years, primarily due to Chanute's published accounts and his 1903 European lecture tour, this two-surface glider became commonly known as the *Chanute glider*. It was rarely referred to as the *Chanute-Herring glider*, and hardly ever called the *Herring glider*.

Although patently unfair to Herring, Chanute – as historian Gibbs-Smith succinctly pointed out – *never* corrected the identification errors. When Herring challenged Chanute's version of the two-surface glider's provenance, he was met with indignation and a contorted rendition of events leading up to the second session of the 1896 Indiana dunes flight trials... to which Herring responded in his 1901 letter to Chanute:

> *"... I have never understood why you should appear to claim the whole credit for the invention of the two-surface gliding machine..."*

In a 1961 letter to Herring researcher Sherwin Murphy, historian Horace Keane wrote:

> "Herring's association with Chanute was not very pleasant. He was a qualified engineer, yet [was] treated more or less as a servant, which brought resentment."

Herring once said to Arnot, "I saw in Chanute, a sly advocate for heavier-than-air manned flight, whose goal was to charm himself into the mix... for the purpose of enhancing his legacy at the expense of any experimentalist who showed promise." Herring's comment was prescient and is clearly illustrated in Chanute's relationship with the Wright brothers.

How Chanute Used the Wright Brothers

A letter to Wilbur Wright, dated December 19, 1901, affords an excellent example of Chanute's self-aggrandizement. Disappointed that the brothers had chosen to suspend their experiments until the fall of 1902, Chanute wrote:

> "If... some rich man should give you $10,000 a year to go on, to connect his name to progress, would you do so?"

To their credit, the brothers politely declined the proposal.

In their book, *12 Seconds to the Moon*, authors Young and Fitzgerald discussed the Wrights' lack of indebtedness to Chanute.

> "Arnold Kruckman, aeronautical editor of the New York *World*, said of the Wrights in an article written December 12, 1909: 'Their persistent failure to acknowledge their monumental indebtedness to the man (Chanute) who gave them priceless assistance has been one of the puzzling mysteries of their career.'"

Wilbur answered Krukman's attack by pointing out that most of the information that leaked out about the Wrights' work had come from Chanute, a fact that led to the false impression that they were working under his direction and with his financial assistance.

In the letter to Kruckman, dated December 12, 1909, Wilbur wrote:

> "Many of the published stories have been very embarrassing because if left uncorrected, they tend to build up a legend, which takes the place of truth, while on the other hand, any attempt on our part to correct inaccuracies gives us the appearance of ungratefully attempting to hurt the fame of Mr. Chanute. Rather than subject ourselves to criticism on that score, we have preferred to remain silent, but now you find fault with our silence. We rather than Mr. Chanute have been the sufferers from this silence so far.
>
> "For many years we entrusted to him [Chanute] many of our most important secrets, and only discontinued [the practice] when we began to notice that his advancing years made it difficult for him to exercise the necessary discretion."

The Wrights became disenchanted with Chanute after his lecture tour of Europe during the early months of 1903, when he revealed their wing-warping method of lateral control. The disclosure proved detrimental to the brothers' effort to obtain patents in some European countries.

While in France on business, Wilbur had heard stories of how he and Orville had "taken up" aeronautical studies at Chanute's insistence; how the old gentleman had provided them with their first flight experience aboard one of his own gliders; how he had provided them with money and the science necessary to succeed... while they contributed only a bit of mechanical skill. When success was finally at hand, it was said that the magnanimous Chanute stepped aside, permitting the brothers to harvest the rewards of being the first to successfully fly a manned, powered aeroplane. It was not easy for the brothers to be labeled as "Chanute's pupils" or "followers of Chanute in the school of Lilienthal."

In this regard, both *Herring and the Wright brothers* had something in common... a bone to pick with the widely acclaimed Father of Aviation.

Glenn Curtiss

John Meiller, Herring's great-grandson and keeper-of-the-flame within the family, revealed his contempt for Curtiss in a 1999 correspondence.

> "Until you see an iconic figure like Curtiss shown to be a liar, a cheat, and a thief, who couldn't design his way out of a wet paper bag – especially in regard to the Herring-designed Herring-Curtiss Company's Golden Flyer and Rheims Racer – the whole situation needs airing."

Herring's business association with Curtiss unraveled after Curtiss' return from his European racing successes in the fall of 1909; the serious disagreements and overt hostility between the two men were carried into the courts, the press and throughout the aviation community, and continued well beyond their deaths. Curtiss has been portrayed as a hero, while history has cast Herring as a contentious, ungrateful figure.

The Wright Brothers

In a November 2, 1906 letter to Octave Chanute, Wilbur Wright laid down the ground rules for making a successful manned, powered flight of an aeroplane:

> "From our knowledge of the subject we estimate that it is possible to 'jump' about 250 feet, with a machine which has not made the first steps toward controllability and which is quite unable to maintain the motive force necessary for

flight....If [Santos-Dumont] has gone more than 300 ft. he has really done something; less than this is nothing."

Later, responding to critics who considered their first flight as merely a "lucky jump," Orville wrote the following in the September 1908 issue of *Century* magazine:

> "The flight lasted only twelve seconds, a flight very modest when compared with that of the birds, but it was, nevertheless, the first flight in the history of the world in which a machine carrying a man had raised itself by its own power into the air in free flight, had sailed forward on a level course without reduction of speed, and had finally landed without being wrecked."

In Joe Bullmer's 2009 book, *The WRight Story*, the author explained how Orville had "modified" Wilbur Wright's original criterion:

> "A year after Wilbur's death, Orville wrote an article for *Flying* magazine titled 'How We Made the First Flight.' In it he claimed the honor by stating that, since they had been flying into a strong headwind instead of calm air, he actually went the equivalent of 540 feet on the first flight of the 17th [1903] rather than the 120 feet actually measured."

To Orville, Wilbur's 300 feet *over the ground* criterion had suddenly morphed into the distance *flown through the air*. Using this measure, Augustus Herring's flight of 72 feet into a 26-mile-an-hour wind in 10 seconds (equivalent to about 450 feet through the air) would have qualified him as being the *first to fly*. A tribute to his character, *Herring never made this claim*. Contrary to Wilbur Wright's assertion that Herring "claimed more credit than he deserves," Herring actually downplayed his achievement, referring to it simply as having attaining "a powered airborne condition," which proved to him that the problem was *solvable*. Today, a handful of historians wish that Herring had simply said, "I flew."

Animosity

Back in 1905, worried that Herring might try to gum up the works with a priority-of-invention claim, the Wrights were dismayed to learn that the U.S. Patent Office had rejected their 1903 application. Angered, Chanute vented his displeasure in a particularly harsh letter to the federal agency:

> "Our own patent office... is raising difficulties... [and is] pursuing the very policy that drove Maxim [Sir Hiram] from the country with his inventions."

After diplomatically telling Chanute not to act on their behalf (i.e., to shut up), the Wrights finally obtained their patent in May of 1906. In the meantime, Chanute's public outburst prompted one French aeronautical expert, believed by many to be Ernest Archdeacon, the wealthy French lawyer and balloonist, to conclude:

> "The Wright machine is merely a more advantageous arrangement of other men's ideas, and a better construction of other designs – not possessing a single new or novel feature."

Many aeronautical experts, including Chanute, felt that Archdeacon had a point. Other than being responsible for introducing the practice of the operator's working from a prone position, and possibly the co-ordination of the vertical rudder with wing warping, what other innovations did they introduce?

It's enlightening to note that after Herring's application of December 11, 1896 had been rejected, Chanute had failed to express even a *whimper of protest* to the U.S. Patent Office on Herring's behalf. Because Herring lacked the funds necessary for an appeal, all of his claims were left undefended and available for others to pillage.

Orville in Europe

In August 1909, Carl Dienstbach traveled by steamship from New York City to Frankfurt, Germany to attend the International Aeronautical Exposition as a reporter for *Scientific American* magazine. Herr Dienstbach also arranged to have Herring's miniature four-stroke-cycle engine displayed, with specifications and a written description of its operating features.

When the Exposition concluded, Dienstbach took a train to Berlin to observe the flying at the new Johannisthal Field and unexpectedly bumped into Orville Wright, who had been demonstrating the brothers' Signal Corps-type machine to German military officials. Before long, their conversation turned to Herring, whom Orville roundly criticized. Dienstbach wrote:

> "He told me stories, which he probably heard from Chanute, of Herring's hesitation to fly the Chanute glider [Katydid]. In light of [my discussions] with Herring... I understood his hesitation."

The inference was clear... Orville believed that Herring was spineless. Wright added:

> "Every expert could tell that Herring's claim of making a flight in St. Joseph, Michigan, was discredited by the insufficient [resources] at his command."

Orville's opinion stood in marked contrast to Wilbur Wright's belief that "...with power enough, one could fly a *waiter's tray*." Fortunately, Herring understood the importance of a lightweight aeroplane. He designed his biplane around the lightest, most powerful engine or motor available to him at the time. Although Orville believed that 3-1/2 horsepower was insufficient to fly Herring's "waiter's tray," we are reminded that the brothers had had difficulty in grasping the geometric relationship between a flying machine's speed and the power required to fly it. When the thrust-to-weight ratio, wing loading and power loading of the two machines are compared, Herring's compressed air machine compared favorably with the much larger and heavier *Wright Flyer*. (See Appendix #2)

At the time of Dienstbach's chance meeting with Orville in Berlin, the Wrights' lawyers had already filed suit against the Herring-Curtiss Company and Glenn Curtiss. Back in America, Wilbur was infuriated by claims Herring had made in his December 1909 affidavit, writing in a December 6, 1909 letter to Chanute:

> "The affidavit filed by Mr. Herring is thoroughly characteristic of him. He has suddenly discovered that he invented in 1894, the method of controlling lateral balance by setting surfaces to different angles of incidence on the right and left sides of the machine and correcting the difference in their resistances by means of an adjustable vertical tail."

Much to Wilbur's chagrin, Herring later produced *photographs* of his 1894 Lilienthal-type glider in flight, with lever-controlled *resistance flaps* clearly seen attached to the left and right leading edges of the machine's mono-wing; piano-wires can also be seen extending from the operator's position to the aft-mounted vertical rudder.

Having read Orville's account of his epic, 120-foot flight at Kill Devil Hills on December 17, 1903, and Herring's 72-foot flight at St. Joseph, Michigan, on October 11, 1898 – it is left up to the reader to decide whether Gus's machine should be demoted to the waste heap of aeronautical waiter trays. Herring's affidavit statements in the *Wright Co. v. Herring-Curtiss Co.* lawsuit caused additional speculation about the operational effectiveness of the Wright control system – an ambiguity that lasted for decades. He stated the following regarding the machine depicted in the Wright patent:

> "... [It] is capable of gliding in a straight forward direction only... [while the aft-mounted] vertical rudder is for the purpose of

keeping the machine in a straight line ahead when the planes are warped... [and that surface] cannot be turned by the operator without warping the supporting planes."

Despite their rivalry, there is no written account of Herring having uttered a single criticism of the Wrights. In his manuscript, Carl Dienstbach quotes Herring from a 1903 conversation:

"Have you seen the Wright brothers? You must go see them. I have done so. They are wonderful. Never was there such perfection in gliding."

In a 1970s interview with Herring's daughter, Chloe, researcher Eugene Husting reported her recollections of her father's relations with the Wrights:

"Daddy was always on friendly terms with the Wrights. I never heard him say a single negative thing about them."

The Wrights' Contract with the Smithsonian Institution

After a protracted argument with Orville Wright, the Smithsonian Institution continued to refer to former Secretary Langley's *Great Aerodrome* as "The first man-carrying aeroplane in the history of the world *capable* of sustained free flight."

In 1928, after Secretary Charles D. Walcott refused to retract and amend the Institution's claims, Orville Wright sent the reconstructed 1903 *Kitty Hawk Flyer* to the London Science Museum, where it languished for 20 years.

In their book, *History by Contract*, authors O'Dwyer and Randolph state:

"Not until 1944 did the new Smithsonian Secretary, Dr. Charles G. Abbot, set the record straight by publishing a repudiation. In 1948, the year of Orville Wright's death, the Kitty Hawk machine took its place in the Smithsonian as the undisputed pioneer airplane."

In Orville Wright's eyes, repudiation required Dr. Abbot to apologize to the world for a litany of perceived wrongdoings and to recognize the Wright brothers as the *sole inventors* of the aeroplane – which he did. While the story appeared to end with the triumphant return of the *Kitty Hawk Flyer* to the friendly confines of the Smithsonian, there was one further surprise. Twenty-eight-years later, in 1976, a group of researchers forced the Smithsonian – through the *Freedom of Information Act* – to divulge the contents of a secret contract that had been executed between Orville Wright and the Institution. In part, the agreement stated:

> "... If the Smithsonian recognizes any other aircraft as having been capable of manned, powered, sustained, and controlled flight before December 17, 1903, the executors for the Wright estate have the right to take possession of the Flyer."

In the 40 years since the details of "The Contract" became public knowledge, the Smithsonian's objective has been to *protect* its most valuable aeronautical asset. As mandated under the terms and conditions of the 1944 agreement, officials of the Institution continue to deny and refute any suggestion that some other experimenter may have powered a manned heavier-than-air vehicle into the air prior to Orville's undulating 120-foot hop.

Currently, impartial aviation observers and researchers view statements discrediting pioneers such as Augustus Herring with skepticism... yet the Wright-Smithsonian contract continues to prevent the Institution from presenting the true story of American aviation history or to fully recognize the many men who contributed to the solution of the problem of heavier-than-air flight. This is best exemplified by the Smithsonian's position on *weight shifting* as a viable method of control.

Weight Shifting: Powered Hang Gliders vs. "Real" Aeroplanes

In 1981, at the height of a *new* aeronautical craze – powered hang gliders – Dr. Tom Crouch, the current (as of 2018) Senior Curator of Aeronautics for the Smithsonian Institution wrote an article for *Bungee Cord* magazine, comparing hang gliders to *real* aeroplanes. Crouch cited Herring's 1898 compressed-air-powered aeroplane as a prime example of a dead-end technology.[2]

> "Like other hang glider enthusiasts, Herring had assumed that once a stable, trustworthy glider had been constructed, the step to powered flying would be relatively simple. He discovered that such was not the case. His 1898 machine fell far short of what he had hoped, and there seemed to be no way of adding more power [and more weight] to a craft that was already overweight..."

Sadly, *historical revisionism* is at play in this commentary. As the researcher and author of *To Caress the Air*, I have found no documentation that Herring *ever* suggested that the step from gliders to powered flight would be *simple*. Readily available records show that he said something quite different... merely that his St. Joseph flights "proved that the problem of practical powered flight was solvable." As discussed earlier, when compared to the 1903 *Wright Flyer*, Herring's foot-launched, powered "hang glider" was neither overweight nor grossly underpowered.

Crouch continued:

> "Moreover, [Herring] was unable or unwilling to abandon the less than satisfactory hang glider control technique [weight-shifting]."

Clearly, Crouch chose to ignore recent history. On March 15, 1975 – 77 years after Herring's compressed-air flights at St. Joseph, Michigan – an *Easy Riser* biplane hang glider fitted with a small internal combustion engine was foot-launched from a frozen lake west of Racine, Wisconsin. It flew for 30 minutes. Four months

Modern weight-shifting, powered hang glider; public domain

later, designer John Moody demonstrated *weight-shifting* ultralight aviation to thousands of spectators at the annual EAA fly-in convention, thus starting the modern ultralight revolution in America. Soon, manufacturers began producing powered hang gliders with names such as *Doodlebug, Explorer LD, Fillo, Flyped, Raven,* and *Zenon.*

Immediately, records were set:

- In May of 1978, the first foot-launched, weight-shifting, powered hang glider crossed the English Channel.

- In August of 1978, a 168-ft^2 (wing area), weight-shifting, powered biplane (similar to Herring's layout) flew to a world record altitude of 6,000-feet.

- In May of 1979, a weight-shifting, powered hang glider flew for a record distance of 202 miles in 4 hours, tailwind assisted. This feat was accomplished with an engine producing less than 10 horsepower.

There is no doubt that the weight-shifting method of control was acceptable to the operators of modern powered hang gliders... why not for Augustus Herring's 2-surface machine of 1898? Nonetheless, Senior Curator Crouch deemed Herring's machine inadmissible as a steppingstone toward controlling a practical aeroplane. He continued:

"The seven Wright machines built between 1899 and 1905 must be seen as links in an evolutionary chain. Each of these craft was an intermediate in the sense that [they] embodied the lessons learned from experience with previous machines.

> "The Wrights progressed through a series of gliders and powered aeroplanes in a thoughtful, methodical fashion creating a firm foundation of experience and trustworthy data from which [to] take the next carefully reasoned step. Problems were not ignored or bypassed... thus with the success of the '02 glider the Wrights were prepared to tackle the... difficulties of a propulsion system. The 1903 Flyer was a natural extension of this earlier work."

Crouch's analysis begs the question: *How were the Wrights unique?* Herring employed flying models, kites, and successive full-size aeroplanes in a similar quest well before the Wrights.

Crouch's observation of the Wrights' methods being based on "... a firm foundation of experience and trustworthy data" amounts to little more than overheated rhetoric. If this observation were sound, why did the Wrights decide to use a *new and untried* aerocurve (airfoil) for their 1903 machine rather than flight-tested hardware? Does their decision serve as an example of taking the "... next carefully reasoned step"?

In terms of the '03 *Flyer* being a natural extension of their '02 glider, an inconvenient question presents itself: *why hasn't anyone been able to coax the Wright Flyer replicas to... fly?* A plethora of Wright machines were built and tried out for the centennial celebration of Orville's 120-foot benchmark hop. Oddly, not one of these "knock-offs" could match Orville's jump... and most crashed while trying.

The Curator refers to "other experimenters" ignoring or bypassing problems before they applied power to their gliders – another thinly veiled attempt to discredit Otto Lilienthal and Augustus Herring's weight-shifting method of control. In yet one more attempt to justify his contention that powered hang gliders were inferior to *real* aeroplanes, Crouch wrote:

> "... all three men [Pilcher, Lilienthal and Herring] believed that they could 'leap-frog' the aerodynamic problems that the

Wrights had struggled so far to overcome, proceeding immediately to the construction of a powered hang glider, a sort of proto-airplane that would enable them to extend their time in the air.

"Once this had been solved, they could return to solve the difficulties that they had bypassed.

"Thus, while the 1903 Flyer was indeed an intermediate step toward a practical flying machine, the powered hang gliders were not."

Crouch avoids *specifying* what other difficulties Pilcher, Lilienthal and Herring had bypassed. Straining to justify his rejection of the weight-shifting method of control, Crouch puts himself in the unenviable position of appearing to be *biased*. Because the Wrights followed one path of control for powered flight does not preclude the existence of another. If not for his bad luck, especially the devastating fire of 1899 and the death of his financial partner, Herring's achievements would have *required* modern-day museum keepers to sing an entirely different tune.

In my mind's eye, a scene from about 1900 looms clear: as the graceful, manned hang glider cruises high above their heads in sustained powered flight, a person in the crowd of onlookers first points and then shouts, "That doesn't count! The fool is shifting his weight for control. *Real* aeroplanes use wing warping!"

The Pursuit of Justice

Although no evidence exists that anyone in the United States had achieved a powered, airborne condition before Herring's, on October 10, 1898, Gus always insisted that, "The aeroplane has a good many daddies." But in the final years of his life, Augustus Herring believed that his role in the invention of the aeroplane had been deliberately ignored. Former Herring assistant James V. Martin wrote in support:

> "Why is Herring's name so seldom heard and why isn't he a multi-millionaire?... since all of the aeroplanes [already] built have infringed on [his] patents – had he obtained any patents."

Martin had hit the mark: patent application #615,353. In 1898, the U.S. Patent Office, after more than a year of arguing back and forth with Herring's lawyer, rejected his December 1896 application for a manned, powered, heavier-than-air flying machine. The agency claimed there was no evidence that Herring or anyone else had successfully flown such a machine.

However, eight years later, in 1906, the Patent Office reluctantly issued the Wrights their pioneer patent for lateral control. Immediately, Herring's lawyer petitioned the office to revive his previously rejected application. To Herring's dismay, the agency continued to demand a full-size working model of his machine. With none available, the petition to reopen was again denied.

Four tumultuous years passed before Herring finally arranged to have a full-size replica of the patent application aeroplane constructed. Then, in April of 1910, he successfully flew the 3-surface machine from the sand dunes of Plum Island, Massachusetts. Armed with written eyewitness accounts, Herring's attorney, Scheer, petitioned the Patent Office to revive his almost 14-year-old application, only to have the bureaucracy claim that a *fatal delay* in responding had occurred (i.e., that he was less than a year too late), thus rendering the most recent request dead on arrival.

With the demise of #615,353, almost two dozen pioneer-status claims were surrendered to the public domain. If Herring had been granted his patent, it may well have forced the Wright Brothers to negotiate a compromise regarding the use of their lateral control patent... thus providing the Herring-Curtiss Company with much-needed leverage to compete in the fledgling North American aeroplane market.

Historian Eugene Husting wrote about Herring's final days:

"Practically unknown, in poverty and bad health from multiple strokes, letters he attempted to type... are pitiable, the last with many mistakes and his apology for them. An early 1926 stroke was a severe one, and he died July 27, 1926 at the age of 59, in a Brooklyn flat he had rented to be near his doctor."

He deserved better.

C. David Gierke May 2018

FOOTNOTES

Chapter 56

1. Walter Brock was a longtime employee of Augustus Herring's. He worked at the Broadway Ave. shop.

2. Herring's new government machine was designed to utilize his gyroscopic control system to actuate *interplane surfaces* that he hoped would skirt the Wright patent. Realizing there wouldn't be time to finish the second aeroplane, Herring planned to deliver one of the proven Herring-Curtiss aeroplanes with his control system installed. Curtiss turned down Herring's request because all of Hammondsport's resources were needed to build the *Rheims Racer* along with its V-8 engine. Curtiss and the racer had to be in France by August 1, 1909. Frustrated, Herring withdrew from the Signal Corps competition.

   Ten months after Orville's crash of September 1908, the younger Wright exceeded the Army's stated performance criteria... and the government had its first military aeroplane, sometimes referred to as *Signal Corps No.1*.

3. Kirk W. House, author and former curator of the *Curtiss Museum* in Hammondsport, N.Y., wrote the following in his article, "Into the Air", on p.51 of the July 2009 issue of *Aviation History* magazine:

> "Glenn Curtiss began building aircraft that were dramatically different from the AEA [Aerial Experiment Association] designs and the Wright machines. There is some evidence that his highly successful ideas originated from the AEA's forgotten stepchild, the [Chanute-Herring] hang glider. Comparisons of the dimensions of contemporary aircraft suggest that he essentially added an engine and control surfaces to the hang glider when he developed his famed Curtiss Pusher."

AEA members (including Glenn Curtiss) copied the Augustus Herring-designed "Chanute-type two-surface glider" in Curtiss's shops to gain experience in gliding down the wintry slopes near Hammondsport, before attempting to fly their first crude powered machines (*Red Wing* and *White Wing*).

4. The bowl-shaped combustion chamber (hemispherical) allowed diagonally positioned (canted) poppet valves to seat effectively within its confines, while avoiding the piston as it passed through the top of its stroke.

5. After watching Herring's flight, a circus performer (C.W. Parker) from Abilene, Kansas purchased the aeroplane for $6,000. Back in Kansas, after re-assembling the machine, Parker thoughtlessly started its Herring-Curtiss engine without a propeller or flywheel. Racing out of control, the engine shook the aeroplane to pieces.

6. One of the Herring-Curtiss Company's allegations in this civil action suit against Curtiss was that Curtiss wrote the following statement regarding the Herring injunction to a confidant:

> *"If it had not been for our intimacy with the judge, we would not have gotten the injunction."*

7. Herring's four suppressed patent applications, serial numbers 615,353, 499,611, 500,262 and 567,563, if

granted, could have hobbled the Wrights' control over the American aeronautics industry.

Chapter 57

1. The Patent Office allowed Herring's attorney to prosecute the 1896 application for over a year, based on his declaration and evidence that he had successfully flown a rubber-strip-powered *model*, and a *man-carrying glider*. However, the agency denied that Herring's *powered man-carrying machine* "as a whole" was operative.

2. Herring wrote of his inability to obtain funding for his manned, powered aeroplane:

   "I cast about in many directions for aid to continue my experiments… and I met with many disappointments."

3. At the end of the 19th century, the Arnots were the leading family in Elmira, N.Y. Today, the family name is Faulk, and they are still a prominent part of Elmira's business community. Their bank has weathered all economic storms since its inception in 1832, when Matthias C. Arnot's grandfather first opened the doors to accept customers. Matthias, much to his father's distress, turned out to be more of an engineer than he was a banker. Known in town as an innovator and master dabbler, he had both the ambition and the money required to carry out his schemes.

4. The *Herring-Arnot glider* had a wingspan of 16 feet, with a 5-foot chord, for a total of 160 square feet of wing area. Each wing contained 11 ribs, with the outer struts positioned at the wingtips, unlike the '96 machine where the wing continued for several inches beyond the struts. The '97 machine also contained inward-slanted supports for the pilot rails.

Chapter 58

1. The following represent the general steps in constructing the two-surface machine of 1897:

   - Seal all wooden surfaces with two coats of varnish.
   - Assemble each wing panel flat with ribs and main spars (2).
   - Attach full-span cotton cord to the leading and trailing edges of ribs. This provides the fore and aft attachment points for nainsook covering.
   - Bow individual ribs to their intended curvature using lightweight cotton cord (like stringing a bow).
   - Clamp the lower wing to sawhorses that have been leveled and checked to be warp-free.
   - Position nainsook covering (wrinkle- and crease-free) to the topside of wing panel; lock covering into place with a penetrating coat of pyroxiline dope – wherever the material *contacts* the wood and cotton cord.
   - The similarly completed *upper wing* is then superposed above its counterpart by fastening struts to the anchor ribs with individual bolts, nuts, and flat washers. Temporary wooden strips, clamped diagonally between individual struts, insure that the upper wing is both square and plumb with the lower wing.
   - The time-consuming task of cutting and attaching diagonal piano-wire guys in familiar *Pratt Truss* configuration begins. The completed truss will strengthen and stiffen the entire wing structure, later allowing the tail assembly and pilot's framework to be securely attached.
   - Using a three-inch camelhair brush, apply three uniform coats of pyroxiline varnish (dope) to the nainsook covering of both wings, allowing 12 hours between

coats. The wing panels remain clamped to the sawhorses to ensure that they don't warp as the covering shrinks to a tight surface.

- The vertical and horizontal tail rudders are treated in a similar manner to that of the wings' assembly, where warped surfaces are considered intolerable. While guyed and clamped to level and plumb sawhorses, the dope coating is applied as before.

- The pilot's parallel bars are attached to the inner wing structure by nuts, bolts, and flat washers.

- The cables and springs for the automatic regulator are attached between the wings and rudders during the final assembly.

- After measuring and checking for warp-free straightness, the entire machine is then broken down into its component parts before being placed into a custom wooden box for shipping to the dunes via steamboat.

2. Back in St. Joseph, Herring had anticipated the aeroplane's on-site assembly problem. After a few sketches and a trip to the lumberyard, he fabricated four foldaway sawhorses that could be easily transported to the dunes. Three of these "horses" were used to support the lower wing off the irregular surface of the sand. A four-foot-long bubble level was then used to jostle and shim the sawhorses into positions that produced a wing that was level in both span and chord. The only other special instrument required was a plumb bob. It was used at each strut position along the span of the wing to confirm that the upper surface was perfectly superposed over the lower while installing the tensioned guy wires. While being attached to the machine's mainframe by booms, the cruciform tail assembly was supported by the fourth sawhorse.

3. Herring explained the changes in the 1897 glider to the reporters as:

> "Superior strut alignment, coupled with a refined piano-wire truss system that has provided a much more rigid superposed wing arrangement. The pilot's support framework is now angled and reinforced to provide superior weight shifting, without the flexure and response time deficiencies of the previous machine. Subtle changes such as the new tailskid and simplified elastic spring system for the automatic tail regulator make the machine more responsive to sudden changes in the wind direction, plus the inevitable gusts that the operator must contend with."

4. Chanute supplied blueprints of the 1897 *Herring-Arnot glider* to the French magazine *L'Aérophile* in 1903, calling it the "Two-Surface Machine of Chanute and Herring, '96/'97." However, in analyzing almost two-dozen of Chanute's magazine articles that were published in the United States, France, Great Britain and Germany, between 1897 and 1910, it was found that as time passed, he gradually took more credit for the inspiration and design of the *Double Decker* glider. He eventually eliminated Herring from his writing altogether, and the apparatus eventually became known as the *Chanute Glider*, which he said was "...much to my personal gratification." ("La navigation aérienne aux États-Unis." *L'Aérophile*, No. 8, April 1903, Paris, pp. 171-183.)

5. In 1904, Chanute commissioned William Avery to build a version of *his* two-surface machine – with parabolic aerocurves. The apparatus was exhibited and flown (as a towline glider) at the 1904 St. Louis World's Fair.

6. Herring's wastebook notes indicate that he continued to experiment with full-size '97-type machines as late

as 1906. He towed them behind a motorboat to measure the drag of different configurations. From these measurements, he was able to calculate the machine's power requirements – an essential factor needed for designing his Government Aeroplane of late 1907.

7. A gelatin silver print is produced on paper coated with a gelatin emulsion containing light-sensitive silver salts. These prints are "developed out" instead of "printed out." The paper registers a latent image that only becomes visible when developed in a chemical bath. The "developing out" procedure allows for a much shorter exposure time, and its image is less susceptible to fading. Perfected toward the end of the 19th century, gelatin silver printing was becoming the dominant black-and-white photographic process of the 20th century.

Chapter 59

1. These were new foods in the 1890s:

   The National Biscuit Company, of Niagara Falls, N.Y, introduced the first commercial dry cereal: Shredded Wheat. The Campbell Soup Company, whose red and white label was inspired by Cornell University's football uniforms, introduced condensed, canned soup, in 1897. Tomato was its first.

2. Herring had fallen into another of Chanute's traps. The old man had successfully *bought* his way into the Herring-Arnot gliding trials, which he would write about until his death 13 years later. From these writings and lectures around the world, the trials would be forever misnamed as the "Chanute glides of 1897."

3. During the summer of 1899, the Herring-Arnot two-surface glider of 1897 was shipped to Arnot's home in Elmira, New York. His well-to-do family preserved the

machine before eventually allowing it to be uncrated and reassembled for display at the National Soaring Museum's administration building, in their hometown. Tragically, the most *rare* of American aeronautical relics was later destroyed by yet another fire.

4. In a letter to Means (November 1896), Chanute spoke of his former assistant, Augustus Herring:

   > "He left me... I think the real reason is that he knew I would never consent to apply a motor to a full size machine."

   This begs the question: did James Means ignore the content of Chanute's letters, or perhaps did he not believe the old man was serious in his rejection of dynamic flight?

5. In an article in *McClure's* magazine, titled "Experiments in Flying" (June 1900, Vol. 15, pp. 127-133), Chanute all but disregarded the accomplishments of Herring, while publishing photographs of the '97 *Herring-Arnot glider* and referring to it as the "Chanute machine."

   When Herring informed his patron, Matthias Arnot, that Chanute was preparing an article on "Flying Machines" for the London *Times* Supplement of the *Encyclopedia Britannica*, Arnot wrote a scathing letter, dated November 29, 1900, to the editors of the *Britannica*, suggesting that they carefully review Chanute's material. Arnot pointed to Chanute's slighting of Herring's work in the *McClure's* article as evidence of his deceitful tactics. Ensuing letters between the *Britannica* and Chanute caused him great embarrassment, requiring a revision of his account of the '96-'97 trials, whereby Herring was reluctantly given more credit than he would have otherwise received.

6. For the rest of his life, Chanute wrote articles and lectured around the world – in part about the 1896 two-surface glider that was illustrated from photographs taken of

the *Herring-Arnot glider of 1897*. Chanute never admitted to the substitution, which was a constant source of aggravation for Herring. Although Chanute reluctantly turned over the first set of glass plate negatives to Arnot – who had paid for all of the photographic supplies – he retained the second set for his exclusive use.

Chapter 60

1. Chanute wrote to James Means on the day of Herring's visit (September 29, 1897):

    "I heartily agree to your proposal [to build a power machine]...but you are mistaken in assuming that I control the whole of the machine, which you saw in gliding flight. I originally made the general design, it is true, but Mr. Herring supplied the automatic regulator, and I should not feel justified in using it without his consent.... He is so sensitive, and has so many illusions as to the value of the device, that the praise that you unduly bestow upon me, and the hint that he might be left out, might cause him to boil over; more particularly as you propose to use the Langley engine. He was here this morning; told me that the Patent Office had again rejected his application."

    This letter demonstrates Chanute's method of handling a ticklish conflict: A- Appear to agree with another person's contention. (Although vehemently against powered flight, Chanute *agrees* with Means' proposal to build a manned, powered, flying machine.) B- Blame others for not taking action (Herring is "... so sensitive, and has so many illusions...") C- Take no action. (Chanute uses his perceived inability to "control the whole of the machine"

2. In 1910, the new Director of Patents stated that there was *never* any valid reason why Herring's patent application of 1896 could *not* have been granted.

3. Herring's hemispherical combustion chamber would become the cylinder head of choice for engineers interested in producing the maximum power for their designs. The hemi-head began showing up in WWI aeroplane engines and remains popular to this day. Like many of his other innovations, Herring's contribution to the hemi engine has also been ignored over time.

4. "Herring appeared odd and mysterious to most people," declared a longtime St. Joseph resident in a 1935 interview, "but that was understood. Like most inventors, his mind was centered on his work to the exclusion of everything else. When he talked about building a machine that would fly through the air, the people just couldn't believe him – it sounded too much like a fairy story. He was so many years ahead of us folks [back then] that we just couldn't understand him." ("World's First Airplane Flight Made In St. Joseph," *St. Joseph Herald Press*, December 31, 1935.)

5. Today, *solenoids* are referred to as cooling fins.

Chapter 61

1. The engine that Herring described during his testimony in 1921 was acknowledged to be at the leading edge of engine development for 1898. His use of the single overhead valve in conjunction with exhaust ports was a decade ahead of its time. Herring intended to use a so-called automatic magneto ignition system for the engine's anticipated high-speed operation. Due to availability problems, Herring substituted a battery ignition

system that used the "jumping spark" from a Rühmkorff induction coil and a set of breaker points in the secondary (high-voltage) circuit, actuated by the engine's half-crankshaft-speed timing gear. Although heavier than desired, the system proved to be reliable during the time that it was required to run the engine.

The poppet valve assembly was kept relatively cool by the vaporized gasoline at the inlet runner, thereby utilizing an effect known as *latent heat of vaporization*. In addition, Herring's use of gasoline continued a trend that he established while acting as an assistant to Professor Langley.

The size of the engine was calculated from its bore (three inches) and its stroke (three inches), giving a total displacement of 42 cubic inches. The poppet valve was 1.5 inches in diameter; the thickness of the cylinder wall was 0.035-inch; and the thickness of the brass sheet oil jacket was a mere 0.006-inch.

2. Herring's review of the four events in a *standard* two-poppet-valve Otto Cycle engine included the following:

    Inlet operation: With the inlet poppet valve open and the exhaust poppet valve closed, air and gasoline vapors are pushed into the cylinder by atmospheric pressure as the piston moves away from the cylinder head on the inlet stroke.

    Compression operation: With the inlet and exhaust poppet valves closed, the air and gasoline vapors are compressed as the piston moves toward the cylinder head on the compression stroke. Squeezed to about one-quarter of its original volume, the flammable mixture is ignited near the end of the stroke by a timed spark, initiating the combustion process and the onset of the expansion operation.

    Expansion operation: With the inlet and exhaust poppet valves still closed, the piston is forced away from the cylinder head by the expanding gasses of combustion on the power stroke. The expansion operation is the only operation that delivers power to the engine's crankshaft.

Exhaust operation: With the inlet poppet valve closed and the exhaust poppet valve open, the piston moves toward the cylinder head on the exhaust stroke, forcing the spent gasses of combustion from the engine.

3. Herring's argument covered the following points:

   A single poppet valve can be positioned in the best possible location within the cylinder head's hemispherical combustion chamber; this promotes the best cylinder filling during the inlet operation.

   Toward the end of the power stroke, the crown of the descending piston uncovers a row of radially drilled holes, or ports, that are located around the circumference of the cylinder. The resulting pressure "blow-down" and rapid release of exhaust gasses to the surrounding atmosphere reduces the work required by the engine to complete the following exhaust stroke. The small amount of exhaust gasses that remain are purged through the open poppet valve in the cylinder head and offer little resistance to the fresh mixture charge that awaits induction during the following inlet stroke.

4. In operation, the overhead camshaft *oscillates* back and forth, while an elliptical cam-lobe acts against the end of the valve stem. The valve has to open at the beginning of the exhaust operation and remains open through the end of the inlet operation, representing one complete revolution of the crankshaft.

   The compression spring that surrounds, and is attached to the poppet valve's stem ensures that the valve closes at the end of the inlet stroke. For the next revolution of the crankshaft, during the compression and expansion operations, cylinder pressures force the head of the valve tightly against its seat inside the cylinder head.

The poppet valve is made of tool steel and resembles a mushroom with a flattened, umbrella-shaped cap borne on a long stem. The three-inch-long stem protrudes through a close-tolerance guide-way hole in the cylinder head to the outside of the engine.

Inside the cylinder head's hemispherical combustion chamber, the head of the poppet valve acts like a silver-dollar-size metallic cork that opens and closes a drilled passageway called a runner. The half-inch-diameter runner ends at Herring's rudimentary *spoon type* carburetor.

At the end of the overhead camshaft, a first-class lever known as a "horn" is connected to a pushrod that communicates with a half-crankshaft-speed cam gear and cam. This dedicated cam provides the correct reciprocating motion needed to oscillate the overhead camshaft. (See drawing).

Chapter 62

1. The concept of dynamic balancing had yet to be addressed by engine designers of the late 19th century. In order to dampen vibrations, early experimenters in the field of internal combustion engines relied upon heavy flywheels.

2. To apply the load, the *Prony brake* dynamometer uses a wide, flat leather belt that is partially wrapped around the engine's flywheel. One end of the strap is anchored, while the other end is movable. The harder the movable end of the strap is pulled, the more load the engine is required to overcome. By controlling the amount of pull on the strap, the operator can force the engine to run at any speed within its operational range. The engine's torque can then be determined by finding the product of

the pull – the applied force expressed in *pounds* – and the radius of the flywheel in *feet* (t = F x r).

3. Torque is measured in *pound-feet*. Shaft speed is measured in revolutions per minute. Therefore, the units of measure for power are pound-feet (or, foot-pounds) per minute. Horsepower, the theoretical power exhibited by a common field animal, was determined to be 33,000 ft-lb of work per minute (hp = t x rpm ÷ 33,000). Example: If an engine produces 55 pound-foot of torque at 1400 revolutions per minute, its horsepower would be: 55.0 x 1400 ÷ 33,000 = 2.33 bhp (*brake* or *measured* horsepower).

4. Advantages of the compressed air motor:
    - The compressed air motor develops maximum torque to its crankshaft from the instant the air valve is opened – unlike the gasoline engine, which requires an elevated speed to obtain its maximum torque and horsepower. This characteristic is similar to that of the steam engine but requires fewer mechanical changes to obtain the desired results.
    - Each cylinder is pressurized on every rotation of the crankshaft, effectively providing twice the power impulses of the gasoline engine.
    - There are no heavy flywheels to contend with. The compressed-air engine produces smooth, almost vibration-free performance without the problems associated with fuel, ignition, and cooling systems.

5. During the mid-1890s, Matthias C. Arnot's first wife, the former Alice Updegrapf, died unexpectedly at the age of 21.

Chapter 63

1. A *surface carburetor* consisted of a box compartmentalized into a labyrinth of half-inch-square passages. Air, pushed through the device by atmospheric pressure, coursed

through gasoline vapor that rose from the warmed fuel in the carburetor's base. Turbulence created by the passageway design served to combine the two components into a heterogeneous mixture, which was then piped to the engine's induction system. If the mixture accidentally ignited outside the engine, the flame would invariably surge back to the carburetor body, where the internal burning of the air-gasoline mixture caused an explosion.

2. Pesch and Herring spoke of: Morey's *surface carburetor*, Maybach's *spray carburetor*, battery and magneto ignition systems, and 4-stroke and 2-stroke cycle designs, among other technical topics of interest.

3. Little did Maybach realize how the Father of Aviation, Octave Chanute, would continue to conspire against Herring, as Langley had earlier, preventing his former assistant from realizing the recognition he deserved.

Chapter 64

1. Goldbeaters skin: treated animal membrane (ox gut), first used to separate sheets of gold being hammered into gold leaf. Later used by some early aeroplane constructors to cover the open areas of their frame structures.

2. The Beaufort Scale of wind speeds:

MPH	CALLED
3. 1-3	Light air
4. 4-7	Light breeze
5. 8-12	Gentle breeze
6. 13-18	Moderate breeze
7. 19-24	Fresh breeze
8. 25-31	Strong breeze
9. 32-38	Near gale

10. 39-46   Gale

11. 47-54   Strong gale

12. 55-63   Storm

13. 64-72   Violent storm

    73+   Hurricane

3. Shroud: One of a set of ropes or wire cables stretched from the masthead, smokestack, or comparable structure to a vessel's sides for support.

4. Horsepower calculations:

Drift (drag) and Thrust = cosine 45 deg. (56.6 lb)

$$= 0.7071 \times 56.6 \text{ lb}$$

$$= 40 \text{ lb}$$

Cruise horsepower required for 4 mph ground speed in 20 mph wind.

20 mph + 4 mph = 24 mph (35.2 ft/sec)

35.2 ft/sec × 40 lb = 1407.84 ft-lb/sec ÷ 550 ft-lb/sec = 2.56 bhp

Propeller efficiency = 94%

2.56 hp ÷ 0.94 = 2.72 hp

Chapter 65

1. A skilled machinist, George Housam was able to fabricate the needed compressor parts, and refurbish many of the others. He fabricated new cylinder heads, restored the inlet and exhaust valves and repacked the piston seals. He also cast new bearing inserts and finished machining a new cast-iron flywheel-fan for the engine.

2. By slowly turning the distributor – first in one direction, and then in the other – the best ignition advance was determined by listening for the highest exhaust fre-

quency – the compressor engine's greatest speed. After locking down the distributor, Herring next turned to the spray carburetor, where its main metering jet was adjusted by a solitary needle valve. The mixture strength was varied from weak to strong and back again, until the maximum engine speed was again attained – as determined by the engine's exhaust note.

3. While working as an instructor of physics and mathematics at Notre Dame University, Zahm suggested calling a meeting of scientists and engineers who were interested in solving the problem of flight as part of the 1893 Columbian Exposition in Chicago. Failing to convince a reluctant Octave Chanute, he contacted prospective participants, called for papers to be written, and scheduled them to be read at the subsequent Conference on Aerial Navigation. In a magnanimous gesture, Zahm named Chanute the conference's chairman and master of ceremonies.

4. Zahm's Ph.D. dissertation, a study of the air resistance encountered by a projectile moving at high speed, established his reputation in American scientific circles.

5. Between the years 1895-1900, Zahm, with the generous financial contributions of Hugo Mattulath (a wealthy, eccentric aeroplane experimenter), helped construct America's first well-equipped aeronautical laboratory at Catholic University. During those years, although the wind tunnel was the largest research tool of its kind, its capabilities were wasted on preliminary studies devoted to Mattulath's huge flying machine design.

Chapter 66

1. In a letter to Octave Chanute on September 30, 1898, Herring described the nature of his current problems with the air motor:

> "I have met with an unexpected disappointment in the [motor's] balanced valves which blow open and exhaust the tank when the air is turned on to the motor at higher pressure than about 80 psi. These valves must be changed before a trial of the flying machine can be made. Just how easy or how hard this change will be to make I cannot say – it will take nearly a week anyhow."

In a second letter to Chanute on October 5, 1898, Herring reports on his progress to reconfigure the motor's inlet valves:

> "...On first testing it was found that the inlet valves blew open and exhausted the tank. This necessitated reconstructing [the] valves so the tank pressure should [now] keep them shut. This work has been done by myself, having hired shop privileges. Today I finished the valve on one side – the tank no longer exhausts but the propellers are not driven as fast as they should, presumably [because] the passageway through [the] new valves is smaller and... that the openings and pressure back of the reducing valve mechanism has yet to be changed for the new conditions.... The pressure back of [the] reducing valve and also its opening can easily [be] increased... it will take from two to three days to make a new inlet valve for the second cylinder and much more time to do the other necessary puttering."

2. Known as the *pendulum*, the device was first used by Professor Langley to improve the chances for a successful flight with his large, steam-powered aerodrome models. In 1895, he established a rule that no aerodrome could be launched until it had demonstrated an ability to

lift 50% of its *total dead weight* (TDW) for at least two minutes. We now recognize that this arbitrary figure was much greater than necessary.

Herring used the pendulum test to confirm or refute what his kite experiment had already revealed: that 40 pounds of propeller thrust would be needed to produce a successful flight. The question was: could the compressed air motor spin the propellers fast enough to provide the necessary thrust, and if so, for how long?

The pendulum device consisted of an eight-foot-long arm whose pivot-end rested on two low-friction, knife-edge supports atop a stationary wooden frame. The flying machine was rigidly secured to the swinging end of the arm at its center of gravity; its propeller axis was then carefully aligned at right angles to the arm.

In operation, the propeller thrust (T) pushes the flying machine on an arc to a new equilibrium point. The angle formed between the arm and the vertical is displayed on a stationary scale, or in Herring's case – a recording device. The sine of this angle multiplied by the flying machine's total dead weight (TDW), equals the *dead lift force* (DLF) produced (DLF = sine θ x TDW). Because all of the forces are in balance, the propeller thrust (T) is equal to the dead lift force (DLF).

By dividing the DLF by the TDW, a percentage of the total dead weight that is capable of being lifted is determined (% = DLF ÷ TDW x 100).

Since Herring's compressed-air motor was fitted with a pressure-reducing valve, the tank pressure could be regulated to a relatively steady 450 psi from its initial 600 psi and still maintain 40 pounds of propeller thrust for 12 seconds. By continuously recording the angle formed by the pendulum arm, versus the runtime in seconds, the flying machine's powered flight time could be determined. Herring's recorder consisted of a two-rpm rotating drum, to which a spring-loaded pen was engaged to a sheet of grid paper.

Rather than ballast the flying machine to its total flying weight of 251 pounds, Herring used its pilotless weight of 88 pounds to conduct the test. The difference was simply the amount of deflection produced by the pendulum arm when the propellers were producing their thrust – about 27 degrees compared to about 9 degrees at 251 pounds.

Compared to Langley's overkill requirement of 50% dead lift force, Herring's 16% realized that his thrust requirements, and therefore the machine's DLF, only needed to overcome its aerodynamic drag, while producing a modest flying speed over the ground. The required wind speed of 20 miles an hour *provided most of the lift* necessary to overcome the machine's total dead weight.

3. The measurements that Herring referred to were the distance between the flying machine's center of gravity and pendulum's pivot point, along with the distance between the propeller thrust line and the pivot. These were taken into account to balance the *moments* around the pivot and provide an accurate thrust (vertical lift) reading for a given sine of the angle's rotation.

4. Broad Street ran east and west through the city of St. Joseph, ending at Silver Beach. The pavilion – a natural extension of Broad Street – extended far out onto the sand. Herring's takeoff point was 50 feet west of the pavilion and 75 feet from the shore of Lake Michigan. Placed onto the sand, the machine was pointed into the prevailing northeast wind that ran up the shoreline.

Chapter 67

1. Chanute, "Some American Experiments," *Aeronautical Journal*, January 1898, n.p.
2. William Murphy, a correspondent for the Chicago *Tribune* and Chicago *News*, later became postmaster in St. Joseph, Michigan.
3. The first newspaper report of Herring's successful powered flight of October 22, 1898, at St. Joseph, Michigan's Silver Beach:

   > "Professor A.M. Herring... today made a second successful test of his power driven flying machine on which he succeeded in making a flight upwards of 70-feet against a strong northerly wind... a small engine furnishes power to two large screw propellers that propel the machine at a high rate of speed. In the flight of today it was able to advance against a wind estimated at 26 miles per hour. The speed of the machine in the one trial attempted this afternoon was remarkably slow. During the flight which lasted eight to ten seconds, professor Herring's feet seemed to almost graze the ground while the machine skimmed along on a level path over the beach.
   >
   > "The landing was characterized by a slight turning to the left and slowing of the engine, when the machine and operator came as gently to rest on the sand as a bird instinct with life. Professor Herring expresses himself as "Well pleased with the result. The present machine," he explains, "was constructed merely for experimental purposes, being designed only for short flights in order to work out some of the difficulties of the problem.

"He expects to continue his work through the winter with the idea of constructing a machine capable of extended flight. "Today's test is still of greater importance," he explains, "that it has removed any question of the possibility of a man operating and controlling a self-propelled machine in high winds." ("Professor Herring Made a Successful Flight This Morning," The Benton Harbor (MI) *Evening News*, October 24, 1898, n.p.)

Chapter 68

1. Early success with the twin-cylinder gasoline engine prompted Herring to use one of its cylinders for powering the popular two-wheel "safety bicycle," which he referred to as a *motor bicycle*. Later in 1899, he also built a prototype *motorcycle* with an integrated single-cylinder engine and transmission.

2. The 1900 "Mobike" was the *first commercially available motorcycle designed, manufactured, and sold in the United States*. To the best of Herring's recollection, there were 226 manufactured; eight of these were destroyed in testing.

3. A description of the Mobike:

    "The engine for the motor bicycle is placed on one side of the frame, whereas the flywheel is on the other – between the two are located the change gears and the small igniting dynamo, especially designed for the purpose of furnishing [spark] ignition. Near the front forks are two levers with pedals or foot rests on their lower ends. The motor is controlled entirely from these rests, pressure on the right pedal throwing the motor in and out of friction mesh, allowing any proportion of slip desired and consequently any speed from zero to the maximum.

"The left pedal is used for changing the ratio of the gearing from the ordinary to the slower one used for hill climbing or sandy stretches of road. The average country road permitted speeds of only about twelve miles an hour, so that the gearing and the governor are proportional to this velocity." (*Horseless Age* magazine, June 7, 1899, n.p.)

4. Herring's steam motor had a cylinder bore of two inches and a stroke of three inches. Most of its components were made of tempered tool steel and were ground and lapped to fit; all of the motor's rotating shafts were supplied with expensive, anti-friction ball bearings. Working at steam pressures of up to 250 psi, the double-acting slide-valve motor produced 5-1/2 brake horsepower at 2500 rpm.

The feed water supply for the "flash-type" boiler, along with the gasoline burner's flame size, was fully automatic. Should the steam pressure rise above 200 psi, the fire was slowed. Should the pressure be less than that, the fire was increased. If too much water was being carried over to the steam, shortening the force pump's stroke during each revolution of the motor diminished its output. Should steam come to the engine too dry (super heated) the stroke was lengthened until the steam was just right (saturated). The operator could manually turn the burner's fire up or down, put it out instantly, or he could easily light it again.

The boiler's steam space was composed of half-inch-diameter tubing which was evaporated every 12 seconds. Although there was a possibility that the tubing might rupture at elevated pressures, the operator was in little danger... the quantity of water was so small that it would convert instantly into a cloud of cold steam.

Chapter 69

None

Chapter 70

1. A description of the single cylinder engine and transmission:

    "The motors will be of the Otto-type (single cylinder), with variable speed. All working parts including a 2-speed transmission or change gears, being ball bearing and enclosed in oil tight cases. The gearing is novel in that there are no cogwheels, and though friction alone will pull up to 32-pounds at the rim of the wheel – which ought to enable a rider to climb a 10% grade on a poor road. The motor fits on the crank hanger of the bicycle. $250.00 for the whole outfit." ("Herring's Bicycle Engine," *Horseless Age*, May 17, 1899, n.p.)

2. Parts such as Herring's engine crankcase are made by a method called sand-casting, in which molten iron is poured into a cavity (mold) made of damp sand. The metal is allowed to cool until it solidifies, then the sand is broken away, freeing the part. The mold is made by compacting sand around a pattern (model) of the part, after which it is carefully removed. Because shrinkage occurs when metals cool, patterns are made slightly oversize. To compensate for this phenomenon, the pattern maker uses a measuring tool called a *shrink rule*, which has slightly oversize units. Different shrink rules are used for different metals.

3. Matthias Arnot's letter of November 29, 1900:

To: Editor – *Encyclopedia Britannica* (London *Times*)

Re: Chanute's disservice to Herring.

"Dear Sir,

Information has reached me to the effect that the article on ['Flying Machines'] in the new edition has been given to Mr. Chanute of Chicago, Ill. to prepare;... as one of your subscribers to the new edition I desire to say a few words to you upon the matter, that the article may be complete in all respects.

"Three years ago Mr. A.M. Herring of Saint Joseph, Michigan, U.S.A. whose contributions to the Aeronautical Journal published in Boston, gave him a high position as one of the foremost inventors and experimenters in the art of aerial navigation, built for me a flying machine upon his own plan – a gliding machine with which we conducted experiments at Dune Park, Indiana, in September 1897...

"Next year we continued our work, and built engines to propel the machines and in October 1898, Mr. Herring made the first flight from the ground with a powered machine, self-propelled.

"The next year the factory, where the new engines suitable for a long flight were completed and awaiting the rest of the apparatus, was destroyed by fire, with the engines and machinery, a catastrophe that will require time to recover from.

"The above facts Mr. Chanute disregarded in his recent article in McClure's magazine, although he did publish photographs of our gliding machine, and merely gave Herring passing mention.

> "In justice therefore to Mr. Herring, I ask that when you look over Mr. Chanute's article for publication in so important a work, in case his (Herring's) work is slighted, you can write to him for some of the articles published in the Horseless Age describing his experiments, and would offer in the humble spirit of a man of science the suggestion that Mr. Herring be allowed a short appendix to describe the work he has devoted much of his time to, and risked his life to attain: the first [powered] flight known in history.
>
> Matthias C. Arnot
>
> Chemung County Bank, Elmira, N.Y."

Chapter 71

1. The following is the Hensleys' interpretation of Huffaker's original explanation for how lift is generated by a curved surface:

    > "Imagining a wing as stationary and the air as moving, one can envision a column of air, stratified horizontally, encountering the leading edge of a wing and being separated by it into two streams, each affected differently by the wing. Sandwiched by the undisturbed airstream above it and the solid wing below, the air moving over the maximum camber of the wing's surface must (in order to make way for the mass of air coming behind it) accelerate to clear the confined channel. In doing so, its pressure drops until it clears the trailing edge of the wing, reduces speed, expands to its original volume, and rejoins the stratum of air that passed below the wing.
    >
    > "That lower layer, in contrast to the one above, maintains its original pressure as it passes below the wing's

leading edge. Responding to the reduced pressure above it and the steady pressure below, the wing will rise." (Steven and Julia Hensley, *The Unwelcome Assistant: Edward C. Huffaker and the Birth of Aviation*, The Overmountain Press, ©2003, p. 77.)

2. Unknown to Huffaker, Chanute had written Langley of his skepticism concerning Huffaker's hypothesis about soaring birds, saying in effect that more evidence was needed to support his conclusions. This letter spurred Langley to design new *curved surface* lift experiments that led to Huffaker's unhappy resignation.

Edward Huffaker returned to Chuckey City, Tennessee, where he resumed work as a surveyor. Less than a month after his departure Chanute hired him part time to design and build model gliders that emphasized automatic stability concepts that were near and dear to the old man's heart. Nine months later, in October of '99, Chanute began sending Edward additional money to continue the experiments and to begin the planning phase of a man-carrying glider.

3. In March of 1900, an unknown named Wilbur Wright wrote to Chanute requesting ideas for the construction of a glider, a list of appropriate materials, and the best place to experiment. This letter represented the beginning of a long and twisted relationship where Chanute, then 68 years old, served as the Wrights' most trusted confidante. The association would prove to be one of the brothers' most regrettable blunders.

In May, two months after his first contact with Wilbur, Chanute visited Huffaker in Chucky City to discuss the most promising methods for achieving automatic stability. While they conferred, Chanute warned his assistant to keep all information about their new glider strictly confidential. This bit of paranoia was peculiar since

Chanute's forthcoming article in *McClure's* magazine emphasized that:

> "Experimenters who wish to advance the final solution of the quest... must work without the expectation of rewards... other than being remembered hereafter."

This snippet of uncharacteristically candid insight offered a rare glimpse into the psyche of a man who valued his legacy above all else.

Following the return of the Wright brothers from their first excursion to Kitty Hawk, North Carolina in the fall of 1900, Chanute and Wilbur exchanged several more letters before the end of the year. Besides providing stimulating technical dialogue for Chanute, the dispatches also inspired him to pursue a long-delayed project. In mid-January of 1901, the old man wrote to Huffaker and emphasized that:

> *"We will proceed to the building of a full-sized machine upon the following plans."*

Within the week, Huffaker had received $50 for materials.

In a later letter to Chanute, Wilbur indicated that he and his brother were planning to visit Kill Devil Hills (on the outskirts of Kitty Hawk) in September 1901, where they would stay for a total of six weeks. With those confidential dates in hand, Chanute was confident that Huffaker would have enough time to complete *their* glider and have it shipped to the Outer Banks. It was there that Chanute hoped to demonstrate his expertise to the aeronautical upstarts from Ohio. Then the Wrights changed their plans! Citing upheavals within their bicycle business, they decided to visit Kill Devil Hills two months *early*, in July and August.

Back in Chuckey City, Huffaker questioned his ability to finish the machine two months earlier than originally planned. Worse yet, Chanute didn't think that it was necessary to explain *why* the completion deadline had been moved up.

On Wednesday, June 26, 1901, Chanute visited the Wrights for the first time at their Dayton, Ohio home. He left the following afternoon for Chuckey City, where he would also view his new gliding machine for the first time. On Saturday, he wrote to Wilbur saying that the new apparatus was a disappointment, and that Huffaker didn't appear to be up to the task of correcting the situation.

In a Letter to Wilbur Wright, dated June 29, 1901, from Chuckey City, Tennessee, Chanute wrote:

> "The mechanical details and connections of the gliding machine which Mr. Huffaker has been building for me are so weak, that I fear they will not stand long enough to test the efficacy of the ideas in its design.
>
> "If you were not about to experiment I should abandon the machine without testing, but perhaps it will stand long enough to try it as a kite, and to make a few glides from a height of 15 to 20 feet. If you think you can extract instruction from its failure... I will send Mr. Huffaker and his machine to your testing grounds at my expense, and pay his share of camp expenses.
>
> "... Huffaker is a trained experimenter, but he lacks mechanical ability."

Although the men had collaborated on the glider design for months, Chanute decided that the apparatus was too weak to fly, blaming Huffaker's use of non-traditional construction methods as the most likely reason. Chanute had his reputation to protect and wanted nothing to do with a probable failure. In the meantime, the Wrights agreed that both Huffaker and another of Chanute's protégés, George Spratt, would be welcome at their Kill Devil Hills campsite anytime after July 10. Huffaker arrived on Thursday, July 18, with his crated gliding machine in tow.

4. The most legitimate reason that the Wright brothers continued to use their front-mounted horizontal rudder must be credited to *serendipity*. At first they argued that the canard (tail first) location allowed them to achieve near-neutral longitudinal stability and avoid pitch oscillations. This proved not only to be *incorrect*, but earlier experimenters such as Lilienthal, Pilcher, and Herring had used the Pénaud and cruciform tails (at the rear) to achieve this very objective. However, as fate would have it, while the brothers were gliding their 1901 machine, Wilbur was saved from a nosedive plunge (caused by what we now refer to as a 'stall') by the action of the canard rudder.

5. In September of 1901, at the suggestion of Chanute, Wilbur Wright gave a talk to the *Western Society of Engineers* in Chicago. During his speech he gave thanks to both Spratt and Huffaker for their accurate diagnosis of the glider's pitching problem. In all of the Wrights' future writings and speeches, there was *never* another mention of the two men's Kill Devil Hills contribution.

6. "Throughout 1900 and 1901, the Wrights believed that the unexpected movement of the center of pressure on their wings [that caused the bucking] was due to a peculiar property of the shape of their wing cross sections, i.e., incorrect camber shapes. So they struggled

to find a shape that would not exhibit the anomaly. Unfortunately, in doing so, they were greatly hindered by their misconception concerning how wings generate lift....they thought only in terms of air impinging on surfaces. So they believed that the travel of the center of pressure with changes in angle of attack on a gradually curved wing section was due to its presenting substantially more or less of its upper surface for the wind to impact upon. Consequently, their solution to limiting the adverse movement of the center of pressure was to put all of the rising camber at the very front of the airfoil. Then, they reasoned, moderate changes in the angle of attack would not change how much of the upper surface the airflow would impact. And this would pretty much cause the center of pressure to stay put. All that would then be necessary to achieve stability would be to properly locate the vehicle's center of gravity.

"Unfortunately, as seems to happen so often, their solution had just the opposite effect of what they desired. By curving the wing section over only the first 10 percent or so, making the remaining 90 percent of the wing profile nearly flat, they had created... a very sudden airflow separation characteristic... Instead of the flow separating over the last 10 percent or so of the upper surface and the separation point gradually creeping forward on a smoothly curved surface as the angle of attack is increased, it would suddenly separate, or reattach, over a substantial portion of the large flat section of their wings [causing pitch instability]." (Joe Bullmer, The WRight Story, Create Space, 2009, p. 168.)

The following are Mr. Bullmer's reasons for the rearward movement of the center of pressure (lift) at lower angles of attack:

> "…At lower angles, the airflow is accelerated – and thus the pressure is reduced over a longer stretch of the wing's upper surface, and also any flow separation and resulting lift loss from the upper surface is diminished." (Ibid. p. 131.)

7. The day before (July 31, 1901), Matthias Arnot died in Elmira, New York.

8. "Katharine, the younger sister of Orville and Wilbur, was very close to both brothers. When [their mother died], she assumed the task of running the household and its family. The Wrights' home on Hawthorn Street became her responsibility, and in the years following her mother's death, her devotion to Orville and Wilbur grew. Bishop Wright was more and more on the move with his [church] duties, and Katharine came to assume all responsibilities for the family and hearth, acting as mother, sister, cook, seamstress, friend, confidante and scold, and later as an unsuccessful matchmaker. She held this role for eight years, until 1897, when she left for two years to attend Oberlin College to get her credentials as a teacher. Except for this interruption, the family remained a tightly knit group, and even brief intervals of separation were filled with almost constant letters between them." (Harry Combs. *Kill Devil Hill*, Houghton Mifflin Company, 1979, pp. 33-34.)

Chapter 72

1. Chanute's function with the Wright brothers:

> "In March 1900, at age 68, Chanute assumed the role of most trusted confidant with the Wright brothers... Chanute, although he didn't contribute anything technical to their work...quickly realized their true merit and spared no pains through voluminous correspondence and several meetings in advising, assisting and encouraging them in everything they did, even to offering financial assistance... Octave Chanute was the most productive 'middle man' in the history of flying." (Charles H. Gibbs-Smith. *The Aeroplane, An Historical Survey*, Her Majesty's Stationary Office, ©1960, P. 33.)

Chanute cast a wide net. He never gave up on his very successful ploy of providing cash to the most promising aeronautical experimenters from around the world, in exchange for partial credit in solving the problem of the ages.

2. By 1909, George A. Spratt developed the "controlwing" idea for achieving automatic stability:

   > "Controlwing theory: Aircraft with movable wings so hinged that they would yield to the forces of turbulent air without transmitting these loads to the structure. At the same time these wings would be so controlled that it would not be necessary to overcome the stability of the aircraft in changing the flight path. Nor would it be possible to fall into an uncontrolled flight condition such as a stall or spin. In other words, the object was an aircraft inherently stable under any and all conditions." (George G. Spratt [son]. "The Controlwing Aircraft," *Sport Aviation*, June 1974, n.p.)

3. Also in his sworn deposition of January 13, 1920, in the case of *Montgomery v. the United States*, Orville Wright

stated that their *drift/lift balance* utilized an idea suggested by Dr. Spratt during his visit to Kill Devil Hills in the summer of 1901.

Chapter 73

1. According to aeronautical researcher Donald C. Paulson:

    "The Wrights' promise in 1909 (letter) to give Spratt credit for his contributions was never fulfilled. Quite the opposite, while admitting to Spratt in private that they owed him a debt, in public they made every effort to conceal his contributions....Orville died in 1948 insisting to the end that he and his brother invented the aeroplane without the help of anyone." (truth in aviationhistory.blogspot.com)

2. Dr. Spratt had suggested using a *lift balance* instrument, which allowed the brothers to mathematically find the *lift coefficient* from the angle of deflection produced by wind passing over the test surface. A second balance (and the key to Spratt's idea), measured the test surface's *drift-to-lift ratio* (D/L). Knowing the lift coefficient and the drift-to-lift ratio allowed the *coefficient of drift* to be calculated with great accuracy.

3. Following Orville Wright's death, researchers were unable to find Spratt's letter of November 21, 1901, in the brothers' voluminous files. Fortunately, Spratt had sent an *identical letter* to Octave Chanute, a fastidious keeper of correspondences.

4. Wilbur Wright's letter to Dr. Spratt on December 15, 1901, continued:

    "...If I understand you properly, the machine is intended for locating the center of pressure at *any angle (or rather locating the angle for any center of pressure), and for finding the direction of the resultant pressure as measured in degrees from the wind direction, so that the ratio of lift to drift is easily obtained, the lift being the cotangent and the drift being the tangent of the angle at which the arm stands. Does the machine also measure the lift, in terms of percent of the pressure at 90 degrees? So that you can make tables like that of Lilienthal?"*

5. Dr. Spratt's letter of September 29, 1909, was also missing from the Wright files. As a result, Spratt's caustic criticism of the Wright brothers was never recorded in McFarland's *The Papers of Wilbur and Orville Wright, Volume Two, 1906-1948.*

6. In a later letter, Spratt called Wilbur and Orville Wright:

   "Secretive, obstructive and lacking in vision and generosity."

7. Wilbur's explanation for what now is called *adverse yaw* and *adverse roll*:

   "A disruptive gust from the left-front rolls the machine to its right. To raise the right wing and restore equilibrium, we employ the wing warping control – a downward-twisted right wingtip. At first the wing begins to rise, but then it reverses direction. The reason? Because of the twist, the right-side wingtip is now flying at a higher angle of attack than its left wing counterpart. Therefore, the right wing's drift [drag] increases, slowing it down. Slowing causes a loss of lift on that side as the machine rotates or 'yaws' to the right.

> "Simultaneously, the left wing has speeded-up, increasing its lift. The result causes the machine to roll right... the opposite direction from what was intended."

8. On October 3rd, in the middle of the night, Orville envisioned changing the vertical tail from a fixed vane to a rudder that could be moved to restore lateral balance. By moving the rudder away from the down-turned wingtip, a restorative force would help to keep that wing from slowing down.

9. In 1899, the Smithsonian sent literature to the Wrights on the subject of human flight, including reprints of E.C. Huffaker's "On Soaring Flight," Louis Pierre Mouillard's "Empire of the Air," Otto Lilienthal's "The Problem of Flying and Practical Experiments in Soaring," and Samuel P. Langley's "The Story of Experiments in Mechanical Flight."

    In addition, the Smithsonian suggested further reading: Octave Chanute, *Progress in Flying Machines*; Samuel P. Langley, *Experiments in Aerodynamics*; James H. Means, *The Aeronautical Annual* for the years 1895, '96, and '97.

Chapter 74

1. On May 25, 1902, Chanute responded to Maxim's letter concerning Herring:

   > "[Herring] has had much experience in experiments in aerial navigation, he possesses considerable ability, knowledge and mechanical instincts – but he cannot easily be managed. I think that he can be useful to you provided that you make him follow your instructions strictly

and establish clearly which are your ideas and which are his."

2. Interplane surfaces are aileron-like devices that were typically placed between the upper and lower lifting surfaces of what we today call a biplane. They are located at the extreme ends of both the right and left wings and function in a manner similar to that of conventional ailerons, *without* the accompanying drag restrictions that produce adverse yaw and the accompanying adverse roll.

Chapter 75

1. In a statement concerning the fourth Lilienthal-type glider that he constructed and flew in 1894, Herring said:

   > "...The device was provided with a pivoted horizontal tail, and also a vertical pivoted rudder; cords led from it to underneath my elbows, so by moving one elbow or the other I could turn the rudder to either side so as to steer the machine to the right or left.
   >
   > "There were two bars between the wings, which came under my armpits and supported me when the machine was in flight. Midway between my body and the forward end of these bars were two pivoted, movable 'auxiliary surfaces'. They were pivoted at one end to these bars, and at their outside points to the forward beams [of the main lifting surfaces]. They were so pivoted that they could be moved independently to present a greater or lesser angle of incidence to the air. They were worked by two short, upstanding levers, which I grasped – one in each hand. From such levers, wires went to the pivoted framework of the two auxiliary surfaces, so as to incline them at will. Similar wires led from the bottom ends of the levers.

"In the first two machines which were built, these auxiliary surfaces were not provided, and for example, if a gust of wind from the side tilted one side of the machine up when it was in flight, it was necessary for me to throw my legs toward the side which was tilted up in order to bring it back to the horizontal. In this machine... I provided these auxiliary surfaces so as to obviate or reduce this movement of the body. If the machine tilted out of the normal horizontal, it was only necessary to move one lever in one direction and the other in the other, so as to rock the auxiliary surfaces in such a way as to present the one on the down side at a greater angle of incidence to the air than the other one, which would give an increased lifting effect on the down side, and if the other surface were lifted in the opposite direction, it would give a decreased lifting effect on the higher side to bring the machine back to normal horizontal. This is what I meant when I said in my 'Soaring Experiments' article, in the January issue of the 1895 American Engineer & Railroad Journal: 'I have been able to preserve my balance without the extreme muscular effort which was necessary with the first machine.

"I made many flights in this machine of over 100-feet in length, and it was a successful man-carrying glider, and the auxiliary movable surfaces operated to accomplish the same result as do the side-balancing rudders of the Herring-Curtiss Co. machine. In fact the efficiency of this construction was where we got the idea of using such balancing rudders." (From: Augustus M. Herring's affidavit, given on October 29, 1909, to the United States Circuit Court, at the Western

District of New York, regarding: the *Wright v. The Herring-Curtiss Co., and Glenn H. Curtiss* lawsuit.)

2. "It [has been] asserted that Augustus M. Herring was the inventor of the adjustable forward rudder as a means of regulating the flying angle of an aeroplane, and [implies] that the Wrights had borrowed this device from him and failed to acknowledge their debt. As evidence, a photograph copyrighted by Herring in the United States in 1894, shows a glider of more or less the Lilienthal-type, with small horizontal surfaces to the right and left of the longitudinal axis of the machine and in front of the main supporting surfaces." (From the German publication, *Illustriete Aeronautische Mittelungen*, March 1904, pp. 101-102.)

Besides their ability to control a flying machine's side-to-side roll, Herring's auxiliary surfaces could simultaneously control the flight angle, or pitch. In modern airplanes, control surfaces that combine these functions are called *elevons* – a word derived from elevator (pitch) and aileron (roll).

3. Using smoke to determine the nature of airflow around aero*curve* sections:

    "Holding large wing sections at the proper angle from one end, Herring would slowly rotate them around his body in a room filled with motionless, stratified smoke. Charging to memory the motion of smoke "streamlines" that flowed around the model, he would immediately illustrate them as a series of sweeping pencil lines onto a previously prepared cross-sectional drawing. These experiments would then be repeated multiple times to confirm his observations." (From: Herring's wastebook, Broadway Avenue shop experiments, March 1895, unpaged.)

Chapter 76

1. At the Kill Devil Hills trials of 1902, Chanute told Wilbur:

    "I suggest that you take out a patent on those principles of your machine that are important, not that money is to be made by it, but to save unpleasant disputes as to priority."

    Before leaving for Dayton, the Wrights began drawing up their patent application.

2. Charles Lamson of California built the three-surface glider according to Chanute's specifications. Herring flew it with difficulty, minus the "horse's head" front horizontal rudder. Historians (e.g., Crouch) have misidentified this glider as being a Herring design, and therefore have used its failure as "proof" that Gus's machines couldn't compete with those of the Wrights.

3. Shortly after the Wrights' 1902 trials, Langley (through Chanute) invited the brothers to visit him in Washington – at his expense. They refused, allegedly because of time constraints. Shortly thereafter, the Wrights refused the old professor's offer to lend them money "to assist with your continuing experiments." In 1903, knowing a good thing when he saw it, Samuel Cabot, a wealthy aeronautics advocate from Boston, offered the Wrights a large sum of cash to have his name associated with what he perceived to be a winner; the brothers also refused his offer as well.

To their credit, the Wrights saw through the offers of monetary support by Chanute, Langley and Cabot as thinly veiled attempts to *buy* into their future accomplishments.

4. In a letter to Langley dated October 21, 1902, Chanute wrote:

> "I have lately gotten out of conceit with Mr. Herring, and I fear that he is a bungler... when the [rebuilt] machine was tried by him in N. Carolina it proved a failure, and he said that he did not know what was the matter.... Since the work was done in St. Joseph, Michigan, I could give it no attention."

Sometime between July 7 and 17, 1902, Chanute had visited St. Joseph and inspected the reconstructed *Katydid*. Prior to his visit (in a letter dated May 30, 1902), Chanute had expressed approval of Herring's plans to alter the wing size and decrease the overall weight. In a letter to Wilbur Wright dated July 17, 1902, Chanute wrote:

> "I have been over to St. Joseph to see Mr. Herring. He is putting excellent workmanship on the machine."

Either Chanute had forgotten about his visit, or he had lied.

Chapter 77

1. World's first gasoline engine for model aeroplanes.

    Herring's miniature 4-stroke-cycle engine incorporated several original features that became state of the art for internal combustion engine design in the decades following its 1902 inception.

Features:

- Cam actuated push rod for overhead exhaust poppet valve in head.
- Atmospherically operated overhead inlet poppet valve in head with light coil spring closure.
- Hemispherical combustion chamber.
- Canted (angled) poppet valves that protrude into the cylinder at a normal angle to the hemispherical surface.
- Make and break spark ignition utilizing a miniature ruby mica insulator sparking plug.

2. Herring's miniature engine had a cylinder bore of 1-1/4-inches, a stroke of 1-5/8-inches, and a displacement of 2 cubic inches. Operating at a peak speed of 3,000 revolutions per minute, it produced a precedent-setting 0.47 brake horsepower. With a weight of 2 pounds minus fuel, propeller, and ignition accessories, his goal of 4 pounds per brake horsepower had been achieved. Later, with all of its systems optimized, the little engine purportedly produced 0.76 bhp on a prony brake dynamometer (2.63 lbs/hp).

3. Herring's checklist:

- Check tightness of propeller and engine to test stand.
- Set up revolution counter onto rear crankshaft.
- Ready and check the operation of stopwatch.
- Secure gas tank with rubber strips at correct height to carburetor.
- Turn gas shut-off valve to 'off' position.
- Mix gas to oil (3:1) and fill tank.
- Open carburetor needle valve to 5-turns from fully closed position.

- Lock throttle valve in completely open position.
- Check rubber gas line for leaks.
- Ignition switch to the 'off' position.
- Connect two #6 doorbell cells (1-1/2 v) in series with DPDT switch, primary circuit of ignition coil, and moving half of breaker points. Stationary point is grounded to aluminum crankcase of engine. Negative side of battery is grounded to engine High voltage conductor connects from secondary circuit of ignition coil to the center electrode of sparking coil. Body of sparking plug is grounded to engine by its installation.
- Check cotton insulated copper wire for bare spots.
- Check and adjust mechanical advance lever (breaker point assembly) for proper frictional resistance to its movement.
- Set mechanical advance lever to 'full retard' for starting.
- Check for sparking plug leakage. Place drop of oil at intersection of fused quartz insulator and steel body; slowly turn crankshaft through compression stroke and watch for telltale bubbles.
- Verify that copper flat washer has been installed under sparking plug and that it is not leaking compression.

Chapter 78

1. Known medically as tetanus, lockjaw is a non-contagious, insidious spore infection that attacks the nervous system and ultimately the body's skeletal muscles. Spores from horse manure – an abundant waste material in most neighborhoods of the early 1900s – or the intestines of many other non-human animals such as sheep,

cattle, dogs, cats, rats, chickens, etc., probably entered Housam's puncture wound at the time of his accident. In any case, when doctors cleansed the wound, they inevitably missed some of the deadly invaders.

2. Today, Herring's dynamometer is referred to as a *fan brake*. In his case, it used the flight propeller to place a load on the engine's output shaft. In actual practice, the engine is mounted at the end of a semi-stationary shaft that rides in a set of low-friction ball bearings that are anchored to a sturdy bench or frame. This shaft is allowed to partially rotate in either a clockwise or counterclockwise direction, but is limited by a heavy weight attached to the end of a pendulum – an arm that is rigidly affixed to the end of the semi-stationary shaft.

In operation, the engine's speed is allowed to stabilize at a point that corresponds with the load (propeller) attached to its output shaft. For example, a clockwise-turning propeller produces torque in that direction. Simultaneously, an equal and opposite *torque reaction force* acts upon the engine, its motor mount, etc., in accordance with *Newton's third law*. The torque reaction force is therefore transmitted to the semi-stationary shaft and pendulum in that direction. In other words, as the shaft and pendulum arm slowly rotate under the influence of the torque-reaction force, a weight is raised against gravity until the equilibrium point is reached... and the operator notes its position.

After the engine is shut down, a calibration wheel with a known radius is attached to the pendulum-arm end of the semi-stationary shaft. Weights of known mass are then hung from the wheel's circumference until the test's arc of rotation is exactly duplicated. Since torque

(T) is defined as force (F) acting through a radius (r), it is an uncomplicated matter to calculate the torque-reaction (T_r = F x r), which is equal to the engine's torque at a specific speed (rpm), and therefore the engine's power may be calculated.

3. Theoretically, an engine would have to produce only 0.05 brake (tested) horsepower to deliver 0.86-pound of propeller thrust. When Herring's dynamometer showed that the engine required 0.065 bhp to produce the required thrust, he could then calculate the propeller's efficiency in converting power into thrust: 0.05 bhp (theoretical) ÷ 0.065 bhp (actual) = 0.769 x 100 = 77%

Chapter 79

1. Known in modern times as a *lake-effect* snowstorm.
2. The model's minimum lifting capacity in a 22-mile-per-hour wind is equal to the model's weight divided by the throttled horsepower needed to produce 0.86-pound of propeller thrust (7.0 lb ÷ 0.065 bhp = 108 lb/bhp).

    Knowing the *minimum* horsepower required to fly the model and the engine's *maximum* horsepower, allows for the calculation of the *horsepower fraction* – expressed as a percentage (0.065 bhp ÷ 0.47 = 0.138 x 100 = 14%).
3. The percentage of the model's total dead weight required to fly the machine is determined by dividing the minimum pounds of thrust produced by the propeller, by the total dead weight (0.86 lb ÷ 7.0 lb = 0.122 x 100 = 12%).
4. The use of chord-wise, vertical fin-like surfaces (side-force generators), located near the fore-aft center of gravity of modern aerobatic airplanes, takes advantage of a phenomenon that Herring identified more than a hundred years earlier: that undesired slippage due to the

forces of the wind and gravity could be counteracted in a fashion that suggests automatic stability.

Chapter 80

1. When Herring arrived at the snowy flying site in January of 1903, he decided to play it safe and increase the engine's horsepower by opening the throttle a bit. He wanted to double the horsepower from 0.06 at 1410 rpm – the minimum required to fly the model in a 20-mile-an-hour wind (as determined from his earlier pendulum testing), to a slightly higher 0.12 hp. The question then became: while using the same propeller, what rpm would represent 0.12 horsepower?

    Herring knew that additional dynamometer testing would not be necessary; he had all of the required information to determine the unknown rpm by utilizing mathematics. Here is his *power factor* method:

    - Horsepower increases as the cube of rpm. $Hp = rpm^3$
    - In its basic form, if two shaft speeds are known, along with a single horsepower, the unknown horsepower can be determined. $Hp_2 = hp_1 (rpm_2 \div rpm_1)^3$
    - To determine the rpm required to generate 0.12 hp, a variation of the *power factor* formula is used. $Rpm_2 = rpm_1 (\text{cube root of } bhp_2 \div bhp_1)$.
    - $Rpm_2 = 1410 \text{ (cube root of } 0.12 \div 0.06) = 1780$

2. On January 3, 1903, Augustus Herring *successfully* flew the world's first gasoline-engine-powered, heavier-than-air, free-flight model aeroplane. Herring's flight preceded the first successful trial of Langley's gas-engine-powered, quarter-size aerodrome model by seven months. Aviation historians often credit Langley as the first to accomplish the feat. However, entries from the profes-

sor's own *Langley Memoir on Mechanical Flight* contradict these claims, to wit:

> "A test of [the quarter-size model] in free flight was made on June 18, 1901... After it had gone about 100-feet, it began to descend slowly... and finally touched the water about 150-feet from the houseboat, having been in the air between 4 and 5 seconds.
>
> "[On a second trial, the] aerodrome flew straight ahead about 300-feet, when it again began to descend slowly and finally touched the water about 350-feet from boat, in about 10-seconds."

A year later, in a letter dated June 17, 1902, Langley's chief engineer, Charles Manly, wrote to his employer,

> "The quarter-size model was given a trial free flight last summer (1901), but proved to be insufficiently supplied with power... the power of the [model's engine] has been greatly increased since last summer, and it now probably has sufficient [power] to fly."

In another letter dated August 8, 1903 [after Herring's success], Manly wrote to a colleague (Mr. Rathbun) at the Smithsonian Institution,

> "I made a trial of the model at 9:31 this morning. The experiment was entirely successful... the machine was in the air for only 18-seconds."

It is important to note that Herring's *Flyin' Fish* would sustain flight *until its fuel supply ran out*.

Chapter 81

1. Herring's description of the world's first successful gasoline-powered free-flight model aeroplane was preserved

in the Herring/Chanute correspondences that were later donated to the Smithsonian.

2. Herring applies for a position at the Smithsonian.

   A) Herring letter to Chanute (Oct. 17, 1902):

   > "Yesterday, I went over to see Mr. Watkins at the National Museum; he asked me to put some of our conversation in writing, as he wanted to talk to Prof. Langley about me, as he thought my knowledge could aid the secretary in his work.
   >
   > "The secretary was interested in the letter and will doubtless refer to you as to your opinion of what value my information – acquired by experiment since leaving the institution – would be to him.
   >
   > "I trust you will do as well by me in your reference as you conscientiously can – in fact I feel certain of that and beg to thank you beforehand."

   Note: Following the Kill Devil Hills excursion, Chanute belittled Herring during a brief meeting with Langley. After returning to Chicago, he reiterated his feelings in a letter (see below).

   B) Chanute's "recommendation" letter to Langley (Oct. 21, 1902):

   > "I have lately gotten out of conceit with Mr. Herring, and I fear that he is a bungler..." (From: *The Papers of Wilbur and Orville Wright, Vol. 1*, © 1953, Footnote, p. 274.)

3. The office for *Gas Power* magazine was located at 42nd Street and 5th Avenue in Manhattan.

4. From: "The Wright Brothers Aeroplane," *Century* magazine, Vol. LXXVI, no. 5, September 1908, n.p.

5. Herring's letter to the Wrights on December 28, 1903:

> "Mess. Wright Bros.
>
> Dear Sirs:
>
> I want to congratulate you on your success at Kitty Hawk, and want to write a frank, straightforward letter.
>
> "To begin with, I do not know the construction of your power machine any more than you know the details of my last, but since it is operative too, and represents the gasoline-driven two-surface machine reduced to near its simplest form, it seems more than probable that our work is going to result in interference suits in the patent office, and a loss of value of the work owing to there being competition. I do not think litigation would benefit either of us, if we can come to agreements otherwise; because there will be enough money to be made out of it to satisfy us all. My position is about this: I have worked long and faithfully at the problem... spent my fortune and all my earnings on it, even almost to my very last dollar on this last machine.
>
> "You have perhaps adopted Mr. Chanute's version, but I know that when you come to sift the evidence down, you will see that I was the originator of the Dune Park two-surface machine... among the evidence on this score is the old rubber-powered model of this machine, which antedates the glider. This model is still in existence. There is also enough of the 1897-99 machine, on which I flew 72-feet, for it to be reconstructed and flown.

"Now I recognize your invention and ingenuity in developing your power machine, and I think you will recognize too that my work has contributed a tangible amount to your success, because my work with the two-surface machine gave you your interest in the problem and provided you with a starting point. Now the point is this: If you turn your invention into a company, I want to be represented in it; not solely because it is to my interests, but because it would probably be equally much to yours.

1. Because we both have similar machines.

2. Because my machine carries nearly 100-pounds per horsepower.

3. I have an efficient means for keeping the equilibrium automatically. A new method for quickly righting the machine from side tilting.

4. My machine is operative and can be carried to large sizes.

5. I have reduced the engine problem to the lightest weight and simultaneously to the most reliable running for long flights.

6. Though we have worked independently, it is not improbable that we will be found to interfere on several points, and perhaps a contest would invalidate some good claims of both.

"Well, so much for the preamble. The further news is that I am negotiating for the sale of the information and plans of my machine with two foreign governments, with considerable prospect for success. If we can get together, I think all of us could realize much more per individual than either party could alone – it eliminates competition, and enables a stronger covering by patents.

"Would you consider our joining forces and acting as one party in order to get best terms, broadest patent claims, and to avoid future litigation? The basis being a 2/3 interest for you two and a 1/3 interest for me. Please think the matter over and let me hear from you.

Yours very truly,

A.M. Herring

P.O. Box 216

Freeport, Long Island, N.Y."

6. The Wrights did not respond to Herring's overture to form a mutually beneficial company. Instead, Wilbur forwarded a copy of the letter to Chanute who replied with indignation on January 14, 1904:

"I am amazed at the imprudence of Mr. Herring in asking for one-third of your invention... I think that your interests are quite safe. The fact that Herring visited your camp... which I subsequently regretted will certainly upset any claims which he may bring forth. I suppose that you can do nothing until an interference is declared. If it is, please call me, and in the meantime try to find out who is his patent attorney."

Herring, having 20-plus claims of his own with the 1896 patent application, had no intention of stealing the Wrights' lateral control invention – the basis for their 1903 application. Instead, contrary to Chanute's slanderous letter, Herring's "joining forces" correspondence suggested the formation of a strong company that would potentially dominate the domestic aeroplane market. The telling point in Chanute's letter to Wilbur was his suggestion to *learn the name of Herring's patent attorney...* a thinly veiled insinuation that political and/or economic pressure could be applied against his cause.

7. Letter: Chanute to Moedebeck, May 6, 1904. In response to Dienstbach's well-researched article in *Illustrierte Aeronautische Mitteilungen* (Moedebeck's aeronautics magazine) concerning Herring's claim to have started the practice of gliding independently of and simultaneously with Lilienthal, and that he was the pioneer of many other aeronautical inventions, Chanute felt it necessary to "correct" the record with the publisher... *plus add the vindictive charge of blackmail to Herring's résumé.*

> "... Your New York correspondent (Carl Dienstbach) has been misled into giving first class notices to Mr. Herring in the February and March [issues of your magazine for 1903]. I may say to you <u>confidentially</u> [emphasis mine] that Mr. H., having seen the Wright machine through my own fault, has been engaged in an attempt at 'Chantage' [blackmail], by threatening to interfere with the Wrights in the patent office."

Besides being deliberately misleading, the duplicitous Chanute tried to avoid retribution by requesting confidentiality.

8. The record confirms that Chanute wrote at least two letters in which he accused Herring of attempting to interfere with the Wrights at the U.S. Patent Office, and that he also attempted to blackmail them.

Chapter 82

1. #13 piano-wire is 0.031-inch in diameter.
2. $900 in today's money.
3. The Cleveland, Cincinnati, Chicago and St. Louis railway.
4. On December 21, 1904, Chanute wrote a testy letter to Balsan:

   > *"The glider you ordered on November 6, 1903 was finished except for the fabric in March of 1904, the time at which you were coming to America. Receiving no letter from you, I had to take the glider at my own expense."*

   Chanute continues, saying that he covered the glider, and that Avery flew it at the St. Louis World's Fair in September and October of that year,

   > *"The glider is back here [Chicago]. It is in good shape, but it will cost you 500 Francs if you want it."*

   Balsan reconsidered, and Chanute was successful in selling the machine – the oldest flying machine to be purchased outside of French soil – as indicated by a second letter (Feb. 28, 1905) that provided detailed information about shipping the machine to Paris. In 1925, Balsan donated the Herring-designed glider to the Paris Air Museum, where it remains on display in Le Bourget today.

5. Track pans were mounted on the wooden ties between the rails – they were almost a mile long and were filled with soft water from lakes and streams. Locomotives reduced their speed to about 40 miles an hour, and the fireman dropped the scoop below the tender to take on water.

6. Only two trainsets were required to provide the *20th Century Limited's* promised daily schedule.

Chapter 83

1. Contrary to Herring's initial analysis, Prandtl's *boundary layer* theory did not completely explain why an aerocurve's performance improved when flown in close proximity to the ground. Twenty years later, the *ground effect* theory explained the phenomenon better: the *cushion effect* occurred when wingtip vortices and downwash behind the lifting surface were interrupted. This lowered the induced drag (drag due to lift), which in turn increased the aircraft's speed and/or lift.

   In 1904, Prandtl's discovery put Herring on the right track. He thought that an aero*curve* flying close to the ground experienced an increased pressure on the bottom of its surface – and he was correct. He also was correct in saying, "When an aero*curve's* lift to drift [drag] ratio improves near the ground, the surface may be flown at a lower, more efficient angle of attack." Finally, he observed, "The closer to the ground an aerocurve flies, the better its performance."

2. An aeroplane's glide ratio (e.g., 20/1) compares the horizontal distance traveled through the air (20 feet) for each foot of altitude lost (1).

3. $1,500,000 in today's dollars.

4. Wilbur Wright thought that landing within 50 yards of the starting point was the most difficult, along with starting their machine within the confines of the Aeronautic Concourse. He also objected to the layout of the course, feeling it was too close to obstructions.

Chapter 84

1. The term "Coming down the Pike" was coined from the 1904 Louisiana Purchase Exposition.

2. Other examples of Chanute's writing that criticized airships:

   "L'aviation en Amerique." *Rev. Gen. Sci.*, No. 22 (Nov. 1903) Paris, p. 1133-1142.

   "Octave Chanute in England". *Aero World*. Vol. 1, No. 11, 1903, Glenville, Ohio, p. 256.

   "Octave Chanute in Wien" Wien. Luftsch. Zeit, II Jahrg., No. 4 (April 1903) Wien, pp. 73-74.

3. Police Chief Sauerbier, gravely wounded and suffering from excessive blood loss, was taken to the fairground's Army Field Hospital, where he seemed to be making a splendid recovery. A week after the attack, however, having been moved to the St. Louis General Hospital, he took a turn for the worse and died of peritonitis. The knife wound didn't kill him, but the bacterial infection did.

Chapter 85

None

Chapter 86

1. *Gas Power* magazine eventually became *Mechanix Illustrated* magazine in 1928.

2. An aeroplane's spinning propeller makes an excellent gyroscope and displays all the same physical properties as those of a spinning top. When a force is applied to deflect the propeller from its plane of rotation, the effects of gyroscopic precession will be apparent: as the

plane of rotation axis is pushed upward by the applied force – as a gust of wind pushes the tail of an aeroplane upward – an effective force pushes the top of the plane-of-rotation disc forward. The resultant force pushes on the rear of the plane of rotation at a point 90° in advance of the direction of rotation... causing a yawing moment to the left (with a counterclockwise rotation of the propeller, as viewed from the front). This motion is known as *precession*. As a result of gyroscopic precession, any pitching of the aeroplane around its longitudinal axis will result in a yawing moment, and any yawing around the aeroplane's vertical axis will result in a pitching moment.

3. Hermann Anschütz Kaempfe
4. The interplane surfaces are located between the port and starboard wingtips of a two-surface machine.
5. Herring's first government aeroplane consisted of two stacked lifting surfaces that straddled both a canard (forward) and an aft-mounted horizontal rudder. Mechanically linked, these rudders were fixed to opposite ends of a delicate-appearing lattice framework made of tempered-steel tubing that extended far out in front of and behind the wings. The canard rudders, arranged biplane-style, initiated a climb when moved to a positive angle in relation to the wings, while the single aft horizontal rudder (part of the standard Cayley-type cruciform tail) helped continue the climb when simultaneously moved to a negative angle; opposite actions resulted in a dive.

The pusher-propeller engines (2) were mounted above the aeroplane's lower lifting surface, on both the port and starboard sides, near the trailing edges. Designed to rotate in opposite directions, the engines were flexibly mounted on the yaw (vertical) axis and were connected by rods and wires to the machine's horizontal rudders.

In addition to the gyroscopically controlled pitch axis, two electric motor-driven, rotor-type gyroscopes were designed to stabilize the aeroplane laterally (longitudinal axis) by automatically actuating the interplane control surfaces at each wingtip.

The machine was fitted with a sturdy wooden (ash) takeoff and landing skid as part of its central frame structure. Three wire-spoke, bicycle-type wheels were outfitted with pneumatic tires; one in front, and two toward the rearward extremities of the frame, thus providing limited protection to the canard rudders and aft tail.

The operator and his passenger were seated side by side in wicker-style seats, above the forward portion of the lower wing.

6. In an interview conducted many years later, Brock confided:

> "... although Herring's ideas were always good, his continual discarding of the old for the new [aeroplane], prevented any consistent, consecutive work."

7. The *Signal Corps* government-model aeroplane still exists today, the only known artifact to have survived the tumultuous pre-Herring-Curtiss Company years. It is currently part of the late John Meiller's estate – Herring's great-great grandson.

8. The following list represents some of the problems and solutions associated with Herring's 5-cylinder radial engines:

Problem: In 1908, it was nearly impossible to find properly insulated wires that could withstand the high voltages found between the spark coil and individual sparking plugs (soon to be known as spark plugs), whose wires often short-circuited, causing the engine to misfire.

Solution: Using a low-viscosity lubricant, two or more layers of telescoping rubber tubing were slid over the high-voltage conductors.

Problem: Sparking plugs, although more readily available, tended to form a carbon bridge between the porcelain insulator and the ground electrode. This was especially noticeable within the lower cylinders of radial engines, where lubricating oil would inevitably run down the cylinder walls and accumulate within the sparking-plug cavity, causing them to foul (not fire) during engine startup, and later to form carbon bridging.

Solution: The sparking plug's center electrode, accompanied by its porcelain insulator, was further extended into the combustion chamber, allowing the heat to effectively burn away carbon deposits. Sparking plugs in inverted cylinders were routinely removed, cleaned, and reinstalled prior to startup.

Problem: The crude carburetors exacerbated the fouling of sparking plugs – especially those that delivered an overly rich mixture to the individual cylinders.

Solution: Substitute five simple "bicycle-type" carburetors for the single large unit usually used on those early engines. As a result, at the expense of time-consuming tuning, the air-fuel mixture problems were kept under control.

Problem: Ignition system reliability. The original ignition system used five individual induction coils, one for each cylinder, and a single set of camshaft-driven breaker-contact points that distributed low-voltage current to the primary winding of each coil, where it was converted to the high-voltage required to fire each sparking plug.

Solution: Use a single spark coil, with its secondary winding delivering high voltage current to individual sparking plugs through a camshaft-driven rotary distributor and insulated conductors.

Problem: Overheating, leading to power loss and engine damage.

Solution: A temporary fix involved brazing thin sheets of brass into water jackets, and installing them onto the individual tool steel power cylinders. For the initial test flights, Herring didn't plan to use a circulating pump and tube-type radiator – primarily to save time and weight. By allowing the water simply to boil away, the latent heat of vaporization would keep things cool enough to enable a two-minute run without power loss.

9. "Selfridge was not convinced that the Wrights were the first to fly [a powered machine], and passed this on to his A.E.A. associates... he referred to Herring as 'the first man so far as we know, to have made a flight in a power-driven machine.' He gave credence to Herring's claim of having left the ground twice with a compressed air motor at St. Joseph, Michigan, back in October of 1898... he qualified the Wrights as the second to get into the air." (C.R. Roseberry. "Red Wing," *Glenn Curtiss Pioneer of Flight*. Syracuse University Press, 1991, p. 87).

Chapter 87

1. Walter Lawrence Brock, who considered Herring a genius, became a member of the "Wright Exhibition Team" two years after leaving Herring's employ. He later

suffered a serious and painful injury in an accident while flying at an Asbury, N.J. air meet. After recovering, he journeyed to England, where he distinguished himself as an instructor and champion air racer before the onset of WW I. After winning approximately $500,000 (today's value), Brock returned to his Chicago roots and purchased suburban land on which he established an airport, where he worked until his death in 1964 at the age of 78.

2. From Carl Dienstbach's unpublished manuscript:

> "The gyroscope was Herring's old 'regulator' all over again: the regulator helped to control his *Double Decker* gliders, while the precessing propellers obviated the potentially fatal delay of the [operator], who often applied his remedies too late."

3. Tunnels and bridges finally killed off the famous route in 1925.

Chapter 88

1. Herring's radial five-cylinder engine had individual carburetors – one for each cylinder. Each carburetor had a rotating throttle plate that filled the air induction passageway, controlling the intake of air and, ultimately, the cylinder's power.

   A mechanical throttle-linkage system that would allow the individual throttle plates to move in unison – from idle to wide-open (full power) – had to be designed and fabricated. The engine's spark-advance system was synchronized to the throttle's position: retarded (firing at or near top-dead-center) for startup, and advanced (firing much earlier, before top-dead-center) for full power. As might be imagined, each incremental throttle position required an experimental determination of the

2. The engine's elementary cooling system was Herring's compromise to getting the long-delayed government machine into the air. It consisted of individual sheet-metal bronze jackets that were wrapped around each of the five cylinders. Each jacket contained three pints of water that would boil away at atmospheric pressure, thus allowing it to remove great amounts of *latent heat*. The method provided for only about two minutes of operation before the cylinder head's temperature would surge out of control, thus causing the pistons to either to "stick" or catastrophically seize in their cylinders due to excessive expansion and friction.

3. Thomas Etholen Selfridge, a 1st Lieutenant in the U.S. Army, was the first person to die in a crash of a powered flying machine (Sept. 17, 1908). He was the passenger on an aeroplane operated by Orville Wright. Previously, Selfridge had designed the *Red Wing* – the first powered aeroplane of the Aerial Experiment Association, which flew briefly on March 17, 1908. On May 19 of that year, Selfridge became the first U.S. military officer to pilot a modern machine when he took to the air alone in the AEA's newest craft, the *White Wing*.

Chapter 89

1. From: Affidavit of Augustus M. Herring (October 29, 1909), to the *U.S. Circuit Court at Western District of New York*. Re: *Wright v. Herring-Curtiss*, lawsuit:

   > "...The two balancing rudders [interplane surfaces] in normal flight of the defendants machine, present

to the atmosphere equal but opposite angles of incidence, no matter to what extent they are moved, and therefore each side of the machine is held back exactly the same, and there is no tendency caused thereby to turn around the vertical axis. Consequently, there is no necessity or reason for simultaneously turning the rear vertical rudder, as the case [cited] in the patent suit, in order to keep a straight course."

2. Beat frequency: Manly used his stopwatch to determine the number of beats the engines produced in 10 seconds as 20, or 2.0 beats per second (20/10 = 2). He had previously measured the rpm of the starboard-side engine (using his stopwatch and rotation counter) as being 2200.

    Because it requires two complete revolutions of the crankshaft to complete one cycle of a 4-stroke-cycle engine, each of the 5 cylinder engines emitted 2-1/2 *exhaust discharges* per revolution. 2200 x 2-1/2 = 5500 exhaust discharges per minute = 91.7 per second.

    For a beat frequency of 2.0 per second to be present during the operation of the engines, the portside engine had to be producing 89.7 exhaust discharges per second (91.7 – 89.7 = 2.0). Therefore, 89.7 exhaust discharges per second equals 5382 discharges per minute, or 2153 rpm... 47 rpm less than the starboard-side engine.

    Although beat-frequency physics wouldn't be fully understood for years to come, mechanical engineers such as Manly and Herring knew how to utilize the phenomenon to their advantage... in this case, to determine the speed of an otherwise inaccessible engine.

3. Power Factor: $P_1 = hp_2 \, (rpm_1 / rpm_2)^3$

When two engines operating at wide-open throttle with the identical load produce different shaft speeds, the horsepower of one engine can be determined if the horsepower of the other is known.

Stated another way: the unknown horsepower is equal to the product of the known horsepower, and the *cube* of the *rpm ratio* of the unknown hp engine to the known hp engine.

$P_1 = hp_2 (rpm_1 / rpm_2)^3$

$P_1 = 22 (2153 / 2200)^3$

$P_1 = 22 (0.98)^3$

$P_1 = 22 (0.94)$

$P_1 = 20.7$ hp

$20.7 / 22 = 0.94$

$100 - 0.94 = 0.06 = 6\%$ *power loss*

Chapter 90

1. During the course of the lawsuit, Herring's patent attorney (Scherr) testified that the Herring gyroscope control "patents pending", if granted had a potential value of $500,000 (in 1921), but the expense of litigating interferences could be sizable – especially against Elmer Sperry Sr., who held more than 400 patents.
2. According to Herring researcher Eugene Husting:
   > "A better result came from England. Herring later assigned the *Herring-Curtiss Company* his patents that were pending in the United States, reserving for himself any and all patents applied for in foreign countries."

3. Herring's second *Signal Corps* aeroplane:

Two seats: operator & passenger.

Three skids: takeoff and alighting.

A larger, single-engine version of the first *Signal Corps* design.

Two main lifting surfaces:

Span .... 26ft., 9 in.

Chord .... 5ft., 6 in.

Gap .... 4ft. 4 in.

Total area (main surfaces) .... 292 ft^2.

Jib fins (6) (top wing) 2ft.x2ft .... 12 ft^2 total.

Canard (tail-first) bi-surface horizontal rudder:

Span .... 6ft., 2 in.

Chord .... 2ft., 3 in.

Total area .... 28 ft^2.

Cruciform tail (horizontal rudder):

Span .... 6ft., 2 in.

Chord .... 2ft., 3 in.

Area .... 14 ft^2.

Cruciform tail (vertical rudder):

Height .... 2ft., 6 in.

Chord .... 2ft., 0 in.

Area .... 5 ft^2.

- Overall length .... 33ft., 0 in.
- Distance from the centerline pivot point of the canard horizontal rudder to the leading edge of the main lifting surfaces .... 12ft., 0 in.

- Distance from the trailing edges of the main lifting surfaces to the forward-located pivot point of the cruciform tail's horizontal rudder ... 12ft., 0 in.
- 4-blade wooden propeller:

Diameter .... 6ft., 0 in.

Pitch .... 5ft., 0 in.

  Operating speed (designed) .... 1200 rpm.
- Weight (empty), approximate with 200-lb engine .... 500 lb.
- Herring-Curtiss engine: 4-cylinder, in-line, water-cooled, 25 horsepower, featuring Herring's overhead valve arrangement, hemispherical combustion chamber, and domed pistons.
- Controls:

  Gyro-stabilized front and rear horizontal rudders on lateral (pitch) axis.

  Gyro-stabilized interplane control surfaces on longitudinal (roll) axis.

  Gyro-stabilized vertical rudder on yaw axis.

Note: For heading changes, takeoffs or alighting, the operator could override all gyroscopically stabilized control surfaces. In this regard, the operator was provided a steering wheel for yaw, a fore and aft swinging steering wheel post for pitch, and a shoulder harness connected to the interplane surfaces for roll. A throttle-control lever was located to the left of the operator and was actuated by his left hand. The passenger was seated to the right of the operator. They were both provided with wicker seats.

4. Herring's first gyroscopic stabilization system consisted of a *gyroscopic rotor* driven by a *direct-current electric motor*, powered by a *lead-acid battery*. It performed reasonably well, but was inherently heavy and was prone to damaging acid spills.

As an alternative, he developed a *propeller-driven electric generator* to power the gyroscope's direct-current motor. When placed in the slipstream of the aeroplane engine's propeller, it developed enough power to spin the gyroscope at close to 10,000 revolutions per minute... but the speed fluctuated.

He was well underway in developing an *electric controller* that sensed when the aeroplane had changed its position due to the gyroscopic rotor's precession. *Electric contacts*, installed within close proximity of the rotor, opened or closed electric circuits that operated electromechanical devices called *servos*, which provided the force necessary to move the control surface in question... thus restoring equilibrium to the aeroplane. Problems involving the servomotors, the lead-acid battery power supply, and other difficulties, delayed progress beyond the Signal Corps' August, 1909 deadline.

Chapter 91

None

Chapter 92

None

Epilogue

1. Herring-Curtiss Company, Appellant, v. Glen H. Curtiss and Others, Respondents*

*Modifying 120 Misc. 733. [No number in original] Supreme Court of New York, Appellate Division, Fourth Department 223 A.D. 101; 1928 N.Y. App. Div. LEXIS 6144.

2. Tom Crouch, "The Origins of a Classic: Octave Chanute, Augustus Moore Herring and Their 'Two-Surface' Glider of 1896-1897." *Bungee Cord*, Volume 7, Number 1, Spring 1981, pp. 7-11.

SIGNIFICANT CHARACTERS: BOOK 2

Chronologically as they Appeared

Matthias E. Arnot (1867-1901) commissioned the construction of the 1897 *Herring-Arnot Double-Decker glider*; he partnered with Herring to produce the manned, compressed-air-powered aeroplane of 1898.

Henry Clarke (n.d.) was Herring's Philadelphia-based assistant.

George Housam was Herring's fictitious assistant in St. Joseph, Michigan.

William Engberg (n.d.) was a confidant of Herring; he owned a machine shop in St. Joseph, Michigan.

Edward Truscott (n.d.) was one of Thomas Truscott's (founder of Truscott Boat Works) three sons; he ran his father's experimental engine shop where Herring was employed between the summer of 1897 and the fall of '99.

Arthur Pesch was a fictitious character.

Dr. Albert F. Zahm (1862-1954) was an early aeronautical experimenter and professor of physics; he was a confidant of Herring.

Charles Saurerbier was St. Joseph's chief of police.

Dr. George A. Spratt (1870-1934) was the inventor-confidant who gave the Wright brothers a method for accurately determining a test surfaces drift to lift ratio (D/L); the data retrieved from Spratt's unique wind tunnel balance allowed the coefficient of drift (drag) to be calculated with great accuracy.

Wilbur Wright (1867-1912) was credited with co-inventing, building and flying the first successful heavier-than-air, manned aeroplane.

Orville Wright (1871-1948) was credited with co-inventing, building and flying the first successful heavier-than-air, manned aeroplane.

Charles M. Manly (1876-1927) was an American engineer who succeeded Herring at the Smithsonian Institution, where he helped build and pilot Secretary Langley's *Great Aerodrome*.

Walter L. Brock (1885-1964) was Herring's assistant prior to the formation of the Herring-Curtiss Company; he learned to fly at the Wright school, before becoming a celebrated racing pilot in England prior to the Great War. He returned to his hometown of Chicago, where he established and operated his own airport. Brock was also celebrated for promoting the emerging hobby of model airplane building and flying in the State of Illinois.

APPENDIX #1

The Accomplishments of Augustus Moore Herring

First demonstration of his classic two-surface, rubber powered model (1892).

- Invented the "Herring regulator". The first active control device for achieving partial automatic stability of a heavier-than-air flying machine (1892-8).

- First gilder equipped with leading edge "resistance flaps" used in combination with an aft-mounted vertical rudder for lateral balance (1894).

- Engineered and executed the changes necessary to enable Langley's steam engine-powered model aerodrome *(No. 5)* to make its first successful flight (1895).

- Designed, built and flew the revolutionary 2-surface glider used at the Indiana dunes; considered to be the "beginning of the modern aeroplane" (1896-7).

- First patent application (23 claims) for a practical, powered, heavier-than-air flying machine in the United States (1896).

- First to utilize the combination of an aerocurve for lift and propellers for propulsion in a heavier-than-air flying machine patent application (1896).

- First to use towed, manned, heavier-than-air gliders to measure their performance characteristics (1898, 1906).
- The most experienced glider pilot in the world – more than 2000 glides (1898).
- First to make a manned, powered, controlled flight of any significant duration – 72-feet across the ground/ 450-feet through the air (1898).
- First overhead valve, hemispherical (hemi) combustion chamber, domed piston, 4-stroke cycle internal combustion engine (1898).
- First commercially available motorcycle in the United States – Mobike (1899).
- First internal combustion, gasoline-fueled model aeroplane engine (1902).
- First internal combustion engine-powered model aeroplane flights – tethered and free flight (1903).
- Was awarded one of two contracts issued by the government to develop the U.S. Army's first military aeroplane (1907).
- First twin-engine, heavier-than-air flying machine to attain sustained flight (1908).
- Formed the *Herring-Curtiss Company*, the first commercial aeroplane company in the United States (1909).
- Responsible for the design work on the *Herring-Curtiss Company's* first aeroplane, the *Golden Flyer*, and the competition *Rheims Racer* (1909).
- Redesigned the Curtiss 4-cylinder inline air-cooled aeroplane engine into a water-cooled assembly that utilized his Hemi combustion chamber, overhead valve configuration. Herring was also instrumental in the development the subsequent Herring-Curtiss v-8 engine (1909).

- Herring's Hemi-engine and valve configuration was licensed to Mercedes for use in the German *Fokker D-7* fighter plane. Rumors persisted that Allied forces acquired the engine from a crashed *D-7*, whose key features were then used in America's Liberty engine program (1909-18).
- First gyroscopic control system patent issued for a heavier-than-air flying machine (1910).
- Invented and developed the "wing sail" method of passive automatic lateral stability in powered, heavier-than-air aeroplanes, thus avoiding the Wright patent (1903, 1909).
- First successful manned, powered, heavier-than-air flight in New England (1910).

APPENDIX #2

Comparing the 1898 Herring aeroplane to the 1903 Wright Flyer

Similarities:

- Both machines used two aerocurve lifting surfaces.
- Both machines used piano wire bridge trusses for enhanced rigidity between the lifting surfaces.
- Both machines used propellers to provide thrust.

  A) The Herring aeroplane used two counter-rotating propellers in tandem, providing a tractor-pusher action.

  B) The Wright aeroplane used two adjacent, counter-rotating propellers, providing a pusher action.

- Both machines used aft-mounted vertical rudders.

Differences:

- Herring machine used an aft-mounted cruciform tail with a limited motion "regulator" for partial automatic stability on the pitch and yaw axis.
- Wright machine used a forward-mounted horizontal rudder (pitch), with the before-mentioned aft vertical rudder (yaw).
- Herring machine used weight shifting for lateral and pitch control.

- Wright machine used active control for the horizontal rudder, and a coupled wing warping/vertical rudder for lateral and yaw control.
- Herring machine used a 2-cylinder compressed air motor.
- Wright machine used a 4-cylinder internal combustion engine.
- Herring machine required the operator to be located at the fore-aft balance point, in an upright position within the forward framework.
- Wright Flyer required the operator to recline (belly-down, head forward) at the central-forward portion of the lower lifting surface – also at the fore-aft balance point.
- Herring machine used the operator's feet for take-off and skids for alighting. Wright machine used skids for support and alighting.
- Launching the Herring machine was achieved by running into the prevailing wind.
- Launching the Wright Flyer was achieved by its propellers, which thrust it down a level 60-foot length of iron-capped 2 x 4 wooden rail that faced into the prevailing wind. The machine rested atop a small-wheeled truck that ran on the rail. An assistant ran alongside to support a wingtip, thus preventing the machine from tipping side-to-side.

Specifications:

- Herring machine's wingspan: 20 ft.
- Wright machine's wingspan: 40 ft.
- Herring machine's wing area: 162 ft^2.
- Wright machine's wing area: 520 ft^2.
- Herring machine's weight w/operator: 251 lb.
- The Wright machine's, w/operator: 710 lb.

- Herring's 12-lb. motor: 3.5 bhp at 450 psi (average air pressure).
- Wright's 150-lb. engine: 12 bhp @ 1400 rpm.
- Herring machine's propeller thrust: 40 lb.
- Wright machine's propeller thrust: 132 lb.
- Herring machine's thrust to weight ratio: 0.16 lb./lb.
- Wright machine's thrust to weight ratio: 0.19 lb./lb.
- Herring machine's wing loading: 1.55 lb/ft^2.
- Wright machine's wing loading: 1.37 lb/ft^2.
- Herring machine's power loading: 0.014 bhp/lb.
- Wright machine's power loading: 0.017 bhp/lb.
- Herring machine's speed (wind + ground): 30 mph (26 + 4).
- Wright machine's speed (wind + ground): 33.8 mph (27 + 6.8).

ACKNOWLEDGEMENTS

At an early age, thanks to aeronautics authors, I became fascinated with the pioneers of flight and their stories of invention and heroism. Names such as Montgomery, Hargrave, Maxim, Ader, Langley, Whitehead and Santos-Dumont were eagerly pursued as part of my high school reading list. Of course, the Wright brothers dominated all of the other published aviation personalities by a ratio of at least ten to one! As the years passed and new aeronautics writers added their in-depth perspectives, it soon became obvious that many of these wordsmiths – like others from the distant past – were purposefully snubbing the work of one particular principal – Augustus Moore Herring. It seemed as though the aviation community was piling-on in an almost universal condemnation of Herring, whom I would later consider to be a genuine pioneer. I first became curious and then somewhat annoyed at what I had learned... Gus had been called an aeronautical hyena – someone who sniffs around the edges of the accepted aviation community, an egotistical loner and everything in-between; this was the 1990s and the beginning of my in-depth personal research, which eventually led me to write *To Caress the Air*.

Before long, I discovered that I was not alone in my quest for knowledge about Herring. Long Island's Eugene E. Husting, a man I had never met, turned out to be an unfailing supporter of Herring and his exploits. A retired Boston banker, Gene proved to be a relentless researcher and disseminator of aeronautical information related to Gus. Everywhere that my wife,

Carolyn, and I traveled in pursuit of the Herring story... Gene had preceded us. Libraries and history museums from Newburyport, Massachusetts, to St. Joseph, and Kalamazoo, Michigan, the Glenn Curtiss Museum (of all places), and even the Smithsonian Institution, Library of Congress, and the District of Columbia's Air and Space Museum had research materials previously embedded by Husting. Gene had also written a number of meticulously researched magazine articles about Herring and his exploits in prestigious journals such as *W.W. I Aero*, which I ravenously consumed.

In 1997, not knowing that he was grievously ill, I was finally able to contact Gene's wife (Betty) who directed me to Herring's great grandson, John Meiller of San Diego, California. All of Gene's research materials including irreplaceable glass-plate negatives had been summarily shipped to Mr. Meiller. The Holy Grail of Herring research was now in my sights as I dialed Meiller's telephone number.

It didn't take long for me to realize that John and Scott Meiller were the proverbial "keepers-of-the-flame" for everything aeronautical related to their great-grandfather. I worked with John for about five years; during that time he provided me with powerful insights, copies of Gus's letters, scores of documents, and dozens of illustrations, including prints from original glass plate negatives. With a tremendous amount of research material on hand, I began an intensive two-year program to chronologically catalogue and reference more than 100-pounds of Herring resources onto index cards; the effort produced a packed shoebox... an ideal starting point for planning scenes, verifying content, and to begin writing a first draft of Herring's story.

Research: Thanks goes out to the following institutions and individuals for their help in fulfilling this project:

Buffalo and Erie County Library

Manatee County Library System

The History Center at Courthouse Square, Berrien Springs, MI (Robert C. Myers), where I reviewed the Tom Millar (Herring researcher) files.

Benton Harbor Michigan Public Library

Plum Island Airport (Aerodrome) at Newburyport, MA (Charles Eaton)

Museum of Old Newbury, MA

Newburyport Public Library

Kalamazoo Aviation History Museum (Air Zoo) at Kalamazoo, Mich. (Gerard Pahl – Education Assistant)

The Heritage Museum and Cultural Center at St. Joseph, Mich. (Caitlin Dial)

National Soaring Museum at Harris Hill, NY (Peter Smith)

Cornell University at Ithaca, N.Y. Rare Books and Manuscript Collection (A.M. Herring. *Scrapbook No. 1*)

American Aviation Historical Society at Santa Ana, CA (Burgess research)

Simine Short, Lamont, IL (Octave Chanute question)

Kirk W. House, Curator of the Glenn H. Curtiss Museum (photos)

Nancy Mendola – genealogy research

Readers:

Richard Brox

Art Pesch

Dee Serrio

Carolyn Gierke

Cover Art:

Richard Thompson

Editors:

Jeanette Willert

Molly O'Byrne

TRUTH OR SPECULATION

Although *To Caress the Air* is based on fact, readers may be curious as to how much of the book is *truth* and how much is *speculation?*

Truth: dates, places, major characters, and scientific/technological principles have been steadfastly adhered to.

Speculation: fictionally dramatized to maintain the story's continuity, most conversations between the characters are imagined; other than the identified correspondences and publications, no record of such interactions exist.

Request to Reader
(From the Author)

Dear reader: If you enjoyed **Book Two** of *To Caress the Air*, consider *leaving a review on the site where you purchased your copy.* It helps to share your experience with other readers. Potential readers depend on comments from people like you to help guide their purchasing decisions... ***and by all means, tell your friends!***

For more information:

About Me

My Other Books

Giveaways

My Mailing List

Contacting Me

Go to my website at: **www.davegierkebooks.com**

CPSIA information can be obtained
at www.ICGtesting.com
Printed in the USA
LVHW031244171218
600738LV00003B/570